RNA

RNA

Editors

Professor Dieter Söll,
Yale University, USA

Professor Susumu Nishimura
Banyu Research Institute, Japan

Professor Peter Moore
Yale University, USA

Pergamon
An imprint of Elsevier Science

Amsterdam - London - New York - Oxford - Paris - Shannon - Tokyo

ELSEVIER SCIENCE Ltd
The Boulevard, Langford Lane
Kidlington, Oxford OX5 1GB, UK

First edition 2001

Library of Congress Cataloging in Publication Data
A catalog record from the Library of Congress has been applied for.

British Library Cataloguing in Publication Data
A catalogue record from the British Library has been applied for.

Several Chapters of this book are incorporated from Volume 6 of Comprehensive Natural Products Chemistry

ISBN: 0 08 043408 8

⊗ The paper used in this publication meets the requirements of ANSI/NISO Z39.48-1992 (Permanence of Paper).
Printed in The Netherlands.

Contents

Preface

The importance of the role RNA plays in all aspects of gene expression has been understood by molecular biologists and biochemists since the late 1950s. Nevertheless, relative to what was going on in the DNA and protein fields, RNA biochemistry remained a backwater for many years primarily because RNA is hard to work with. For example, unless handled carefully, RNAs are rather prone to hydrolytic degradation, and most of the RNAs abundant in nature have molecular weights so large that for a long time it seemed unlikely that anything useful could be learned about them using the physical and chemical techniques of the day. In addition, many biologically important RNAs are so rare that it is difficult to prepare enough of any one of them from natural sources to do all the experiments one would like. Finally, for many years, by comparison with the protein world, the RNA universe appeared to he very small, consisting only of transfer RNAs, ribosomal RNAs, messenger RNAs, and a few viral RNAs. Why spend one's career struggling to understand the properties of a class of macromolecules so difficult and so limited?

The mind set of those in the RNA field has slowly been transformed from a somewhat pessimistic resignation to near manic optimism by the events of the last twenty years. Powerful methods have been developed for sequencing RNA, and a rich variety of chemical and genetic methods is now available for determining the functional significance of single residues in large RNAs, and even that of individual groups within single residues. On top of that, the supply problem has been solved. Chemical and enzymatic methods now exist that make it possible to synthesize RNAs of any sequence in amounts adequate for even the most material-hungry experimental techniques. In many other respects, RNA is easier to work with today than protein. In addition, the RNA universe has expanded. Scores of new RNAs have been discovered, most of them in eukaryotic organisms, that perform functions of which the biochemical community was entirely ignorant in the 1960s, when the first blossoming of the RNA field occurred. Additional stimulus was provided in the 1980s by the discovery that two different RNAs possess catalytic activity, and several additional catalytic RNAs have since been identified. Their existence has led to renewed interest in the possibility that the first organisms might have used RNA both as genetic material and as catalysts for the reactions required for their survival. Francis Crick's reflection (in 1966) on an RNA molecule's versatility ("It almost appears as if tRNA were Nature's attempt to make an RNA molecule play the role of a protein") can now be extended to many RNA species. One interesting offshoot of these developments has been the invention of a new field of chemistry that is devoted to the production of synthetic RNAs that have novel ligand binding and catalytic activities. Finally, belatedly NMR spectroscopists and X-ray crystallographers have begun solving RNA structures.

This volume covers the full range of problems being addressed by workers in the RNA field today. Fach chapter has been contributed by a scientist expert in the area it covers, and is thus a reliable guide for those interested in entering the field. The Editors hope that those patient enough to read the entire book will come away with an appreciation of the rapid progress now being made in the RNA field, and will sense the excitement that now pervades it. RNA biochemistry is destined to catch up with DNA and protein biochemistry in the next 10 or 15 years, and it is certain that important new biological insights will emerge in the process.

DIETER SÖLL
Editor

Contributors

Dr. S. Altman
Department of Molecular, Cellular and Developmental Biology, Yale University, New Haven, CT 06511, USA

Dr. B.L. Bass
Howard Hughes Medical Institute, University of Utah, 6110a Eccles Institute of Human Genetics, Building 533, Salt Lake City, UT 84112, USA

Dr. J.H. Cate
Departments of Chemistry and Molecular Cell Biology, Sinsheimer Laboratories, University of California, Berkeley, CA 95064, USA

Dr. D.M. Crothers
Department of Chemistry, Yale University, New Haven, CT 06520, USA

Dr. A.E. Dahlberg
Department of Molecular Biology, Cell Biology and Biochemistry, Brown University, Box G-J4, Providence, RI 02912, USA

Dr. J.A. Doudna
Department of Molecular Biophysics and Biochemistry, Yale University, 266 Whitney Avenue, New Haven, CT 06520-8114, USA

Dr. F. Eckstein
Abteilung Chemie, Max-Planck-Institut für Experimentelle Medizin, Hermann-Rein-Strasse 3, D-37075 Göttingen, Germany

Dr. A.D. Ellington
Department of Chemistry, Institute for Cellular and Molecular Biology, University of Texas at Austin, 26th and Speedway, Austin, TX 78712, USA

Dr. C. Florentz
UPR 9002 du CNRS, Institut de Biologie Moléculaire et Cellulaire, 15, rue René Descartes, F-67084 Strasbourg-Céex, France

Dr. M.A. Garcia-Blanco
Departments of Genetics, Microbiology and Medicine, Duke University Medical Center, Durham, NC 27710, USA

Dr. S. Ghosh
Department of Genetics, Duke University Medical Center, Durham, NC 27710, USA

Dr. R. Giegé
UPR 9002 du CNRS, Institut de Biologie Moléculaire et Cellulaire, 15, rue René Descartes, F-67084 Strasbourg-Cédex, France

Dr. S.T. Gregory
Department of Molecular Biology, Cell Biology and Biochemistry, Brown University, Box G-J4, Providence, RI 02912, USA

Dr. M. Helm
UPR 9002 du CNRS, Institut de Biologie Moléculaire et Cellulaire, 15, rue René Descartes, F-67084 Strasbourg-Cédex, France

Dr. H. Kawasaki
National Institute for Advanced Interdisciplinary Research, AIST, MITI, Tsukuba Science City 305-8562, Japan

Dr. Y. Komatsu
Graduate School of Pharmaceutical Sciences, Hokkaido University, Sapporo 060-0812, Japan

Dr. T. Kuwabara
Institute of Applied Biochemistry, University of Tsukuba, Tennoudai 1-1-1, Tsukuba Science City 305-8572, Japan

Dr. L.A. Lindsey-Boltz
Program in Molecular Cancer Biology, Duke University Medical Center, Durham, NC 27710, USA

Dr. D.H. Mathews
Department of Chemistry, University of Rochester, Rochester, NY 14627-0216, USA

Dr. J.A. McCloskey
Department of Medicinal Chemistry, University of Utah, 30 S. 2000 East, Salt Lake City, UT 84112-5820, USA

Dr. P.B. Moore
Departments of Chemistry and Molecular Biophysics and Biochemistry, Yale University, New Haven, CT 06520-8017, USA

Dr. M. O'Connor
Department of Molecular Biology, Cell Biology and Biochemistry, Brown University, Box G-J4, Providence, RI 02912, USA

Dr. J. Ohkawa
National Institute for Advanced Interdisciplinary Research, AIST, MITI, Tsukuba Science City 305-8562, Japan

Dr. M. Öhman
Department of Molecular Genome Research, Stockholm University, SE-106 91 Stockholm, Sweden

Professor E. Ohtsuka
Graduate School of Pharmaceutical Sciences, Hokkaido University, Sapporo 060-0812, Japan

Dr. R. Parker
Departments of Molecular and Cellular Biology & Biochemistry and Howard Hughes Medical Institute, University of Arizona, Tucson, AZ 85721, USA

Dr. M.P. Robertson
Department of Chemistry, Institute for Cellular and Molecular Biology, University of Texas at Austin, 26th and Speedway, Austin, TX 78712, USA

Dr. J.K. Strauss-Soukup
Department of Molecular Biophysics and Biochemistry, Yale University, 260 Whitney Avenue, New Haven, CT 06520, USA
Current address: *Chemistry Department, Creighton University, 2500 California Plaza, Omaha, NE 68178, USA*

Dr. S.A. Strobel
Department of Molecular Biophysics and Biochemistry, Yale University, 260 Whitney Avenue, New Haven, CT 06520, USA

Professor R.H. Symons
Department of Plant Science, Waite Campus, The University of Adelaide, Glen Osmond, SA 5064, Australia

Dr. K. Taira
Department of Chemistry and Biotechnology, Graduate School of Engineering, The University of Tokyo, Hongo, Tokyo 113-8656, Japan

Dr. S. Tharun
Departments of Molecular and Cellular Biology & Biochemistry and Howard Hughes Medical Institute, University of Arizona, Tucson, AZ 85721, USA

Dr. D.H. Turner
Department of Chemistry, University of Rochester, Hutchinson B08, Rochester, NY 14627, USA

Dr. N.K. Vaish
Ribozyme Pharmaceuticals, Inc., Boulder, CO 80301, USA

Dr. S. Verma
Department of Chemistry, Indian Institute of Technology, Kanpur 208016 (U.P.), India

Dr. A. Vioque
Instituto de Bioquímica Vegetal y Fotosíntesis, Universidad de Sevilla – CSIC, Americo Vespucio s/n 41092 Sevilla, Spain

Dr. M. Warashina
Institute of Applied Biochemistry, University of Tsukuba, Tennoudai 1-1-1, Tsukuba Science City 305-8572, Japan

Dr. T. Xia
Department of Chemistry, University of Rochester, Rochester, NY 14627-0216, USA

1

A Spectroscopist's View of RNA Conformation: RNA Structural Motifs

PETER B. MOORE
Yale University, New Haven, CT, USA

1.1 INTRODUCTION

Biologically, RNA mediates between DNA and protein — DNA makes RNA makes protein — and RNA is also intermediate between DNA and protein chemically. Some RNAs are carriers of genetic information, like DNA, and others, e.g. transfer RNAs and ribosomal RNAs, are protein-like. Their functions depend on their conformations as much as their sequences, and some even have enzymatic activity.

Even though RNA biochemists have recognized their need for structures almost as long as protein biochemists, far more is known about proteins than RNAs. Coordinates for over 8000 proteins have been deposited in the Protein Data Bank, but the number of RNA entries is of the order of 100, and many of them describe RNA fragments, not whole molecules.

All the RNA structures available before 1985 were crystal structures, and X-ray crystallography remains the dominant method for determining RNA conformation. By the late 1980s, nuclear magnetic resonance (NMR) had emerged as a viable alternative, but for many years, only a few structures a year were being solved spectroscopically. In the last two years, the production rate has risen to roughly a structure a month, and because the field is taking off, it is time RNA biochemists understand what the structures spectroscopists provide are all about.

This chapter describes how RNA conformations are determined by NMR. The description provided is intended to help biochemists understand what NMR structures are, not to teach them how to do it. The chapter also summarizes what NMR has taught us about RNA motifs. For these purposes, a motif is any assembly of nucleotides bigger than a base triple that has a distinctive conformation and is common in RNAs.

1.2 THE DETERMINATION OF RNA STRUCTURES BY NMR

The behavior of all atoms that have non-zero nuclear spins can be studied by NMR, and the predominant isotopes of two of the five elements abundant in RNA qualify in this regard: ^{1}H and ^{31}P. Both have spins of $^{1}/_{2}$. Those not content with the information ^{1}H and ^{31}P spectra provide, can easily prepare RNAs labeled with ^{13}C and/or ^{15}N, which are also spin-$^{1}/_{2}$ nuclei (see below). Thus NMR spectra can be obtained from all the atoms in a nucleic acid except its oxygens, for which no suitable isotope exists. What can be learned from them?

The answer to this question, of course, can be mined out of the primary NMR literature, but it is vast and much of it too technical for non-specialists. For that reason, rather than fill this chapter with references its intended readers will find useless, I direct them here to a few secondary sources. For NMR fundamentals, Slichter's book is excellent.[1] It is complete, and its verbal descriptions are good enough so that readers need not wade through its (many) derivations. Those interested in multidimensional NMR, about which almost nothing is said below, can consult Goldman's short monograph,[2] or the treatise of Ernst and coworkers.[3] Although a bit dated at this point, Wüthrich's book on the NMR of proteins and nucleic acids is so useful the cover has fallen off the local copy.[4] A more technically oriented text on protein NMR appeared recently, which is also useful.[5]

1.2.1 NMR Fundamentals

Nuclei that have spin (and not all do) have intrinsic magnetic moments, and thus orient like compass needles when placed in magnetic fields. Because nuclei are very small, their response is quantized. Spin-$^{1}/_{2}$ nuclei orient themselves in magnetic fields in only two ways: parallel to it or antiparallel to it. Because the energy associated with the parallel orientation is only slightly lower than that of the antiparallel orientation, the number of nuclei in the parallel orientation is only slightly larger than the number in the antiparallel orientation in any population of magnetically active atoms that has come to equilibrium in a magnetic field. In the strongest available magnets, the excess is only a few per million. The tiny bulk magnetization their collective alignment produces is what NMR spectroscopists study. Sensitivity is not one of NMR's selling points!

An NMR spectrometer consists of a magnet to orient the nuclei in samples, a radio frequency transmitter to perturb nuclear orientations in controlled ways, and a receiver to detect the electromagnetic signals generated when the orientations of the magnetic moments of aligned populations of nuclei are perturbed. NMR spectrometers produce spectra, which are displays of the magnitude of these electromagnetic signals as a function of perturbing frequency. A peak in such a display is a *resonance*.

1.2.2 Chemical Shift

Electromagnetic radiation causes the reorientation of spin-½ nuclei that have become aligned in external magnetic fields most efficiently when the product of Planck's constant and the frequency of the reorienting radiation equals the difference in energy between their two possible orientations, i.e. $h\nu = \Delta E$. That frequency, the *resonant frequency*, is the one at which the intensity of a resonance in a spectrum is maximum. The energy difference that determines a resonant frequency is the product of the strength of the orienting magnetic field a nucleus experiences and its intrinsic magnetic moment.

The magnetic moments of all the nuclides relevant to biochemists were measured long ago, and they differ so much that there is no possibility of confusing the resonances of one species with those of another, a hydrogen resonance with a phosphorus resonance, for example. (N.B.: The resonant frequency of protons in a 500 MHz NMR spectrometer is 500 MHz.)

The reason NMR interests chemists is that the magnetic field a nucleus experiences, and hence the frequency at which it resonates, depends on its chemical context. Nuclei in molecules are surrounded by electrons, which for these purposes are best thought of as particles in continual motion. When a charged particle moves through a magnetic field, a circular component is added to its trajectory, and charged particles moving in circles generate magnetic fields. Thus when a molecule is placed in a magnetic field, it becomes a tiny solenoidal magnet the field of which (usually) opposes the external field. As you would expect, both the direction and the strength of the field induced in a molecule this way depend on its structure and on its orientation with respect to the inducing field. In solution, where rapid molecular tumbling leads to averaging, orientation effects disappear, and the atom-to-atom variations in the strength of the induced magnetic field within a molecule are reduced to a few millionths the magnitude of the inducing field. Small though these induced field differences are, the contribution they make to the total field experienced by each nucleus is easily detected because the receivers in modern NMR spectrometers have frequency resolutions of about 1 part in 10^8. Thus the proton spectrum of a biological macromolecule is a set of resonances differing modestly in frequency, not a single, massive resonance. Incidentally, all else being equal, the magnitude of each resonance produced by a sample is proportional to the number of nuclei contributing to it.

The frequency differences that distinguish resonances in a spectrum are called *chemical shifts*, and their importance cannot be overstated. Chemical shift differences are the primary way the resonances of atoms at different positions in a population of identical molecules are distinguished from each other, and if a spectrometer cannot *resolve* a large fraction of the resonances in an RNA's proton spectrum, little progress can be made. (A *resolved* spectrum is one in which each resonance represents an atom in a single position in the molecule of interest.)

Chemical shifts are usually reported using a relative scale the unit of which is the *part per million* (ppm). The chemical shift of a resonances is 10^6 times the difference between its resonant frequency and the resonant frequency of an atom of the same type in some agreed-upon standard substance, divided by the frequency of the standard resonance. A virtue of this scale is its independence of spectrometer field strength. If a resonance has a chemical shift of 8 ppm in a 250 MHz spectrometer, its chemical shift will be 8 ppm in a 800 MHz spectrometer also. By convention, if the frequency of a resonance is less than that of the standard, its chemical shift is positive, and it is described as a *down field* resonance. *Up field* resonances have negative chemical shifts. The proton spectrum of an RNA spans about 12 ppm, and spectrometers can measure chemical shifts to about 0.01 ppm.

1.2.3 Couplings and Torsion Angles

The resonant behavior of atoms is also affected by its interactions with the magnetic fields generated by all of the atomic nuclei in its neighborhood that have non-zero spins. In solution, most of these interactions are averaged to zero by molecular tumbling, leaving the solution spectroscopist only a single kind of internuclear interaction to worry about: *J-coupling*. If in a solution of identical molecules, a spin-$^1/_2$ atom at one position is J-coupled to a single spin-$^1/_2$ atom at another, that atom will contribute two closely spaced resonances to the molecule's spectrum, not the single resonance otherwise expected. More complex splitting patterns arise when an atom is J-coupled to several neighbors.

J-coupling results from the magnetic interactions that occur when electrons contact nuclei, which they do when they occupy molecular orbitals that have non-zero values at nuclear positions. For example, electrons in σ molecular orbitals contact both of the nuclei they help bond, but electrons in π molecular orbitals do not. Electrons are spin-$^1/_2$ particles, and have intrinsic magnetic moments, just like spin-$^1/_2$ nuclei. If the spins of an electron and the nucleus it contacts have the same orientation, the orbital energy of the electron will be slightly lower than it would be if their spins were antiparallel because of favorable magnetic interactions. If electrons having both spin orientations contact a nucleus equally, the sum of their interaction energies is zero.

Why does contact lead to splitting? Suppose two spin-$^1/_2$ nuclei, A and B, are bonded by a molecular orbital that contacts them both and contains two electrons, one spin up and the other spin down. If the spin of A is parallel to the external magnetic field, electronic configurations that put the spin-up electron close to A will be favored because they have lower energies. Since the electrons are paired, if the spin-up electron is close to A, its spin-down partner must be close to B, and B will experience a small net magnetic field because it is not "seeing" both electrons equally. If the orientation of the spin of nucleus A were reversed, the bias in the spin orientation of the electron contacting nucleus B would also be reversed, as would the magnetic field experienced by B. Thus within a population of identical molecules, nuclei of type B will resonate at two slightly different frequencies, one for each of the two possible orientations of the spin of A. The difference in resonant frequency between the two resonances is called a *splitting* or a *coupling constant*, and splittings are mutual. The splitting of the resonance of B due to A is the same as the splitting of the resonance of A due to B.

Four facts about J-coupling are relevant here. First, J-coupling effects are transmitted *exclusively* through covalent bonds. Second, J-couplings between atoms separated by more than 3 or 4 bonds are usually too small to detect. Third, splittings are independent of external magnetic field strength, and vary in magnitude from a few Hz to 100 Hz in biological macromolecules, depending on the identities of the atoms that are coupled, and the way they are bonded together. Fourth, macromolecular torsion angles can be deduced from coupling constants because the magnitudes of three- and four-bond couplings vary sinusoidally with torsion angle. Thus experiments that explore the couplings in a molecule's spectrum can identify resonances arising from atoms that are near neighbors in its covalent structure, and determine the magnitude of torsion angles.

1.2.4 Spin–Lattice Relaxation: Nuclear Overhauser Effects and Distances

Every resonance in an NMR spectrum has two times associated with it: a *spin–lattice relaxation time*, or T_1, and a *spin–spin relaxation time*, or T_2. Spin–lattice relaxation is important because a phenomenon that contributes to it is an important source of information about interatomic distances. Spin–spin relaxation is important in a negative way because it limits the sizes of the RNAs that can be studied by NMR.

It takes time for the magnetic moments of nuclei to become oriented when a sample is placed in a magnetic field, or to return to equilibrium, if their equilibrium orientations have been disturbed. Both processes proceed with first-order kinetics, and their rate constants are the same. The inverse of a first-order rate constant is a time, of course, and in this case, that time is called the spin–lattice relaxation time, or T_1.

Spin–lattice relaxation is caused by magnetic interactions that make pairs of neighboring nuclei in a sample change their spin orientations in a correlated way. The rates at which such events occur depend on a host of factors, among them the magnitudes of the magnetic moments of the atoms involved, the external magnetic field strength, the distances between atoms, and the speed of their relative motions. Everything else being equal, the slower a macromolecule rotates diffusionally, the longer the T_1-values of its atoms. For RNAs the size of those being characterized by NMR today, proton spin–lattice relaxation times range from 1 to 10 s.

Transmitters in modern spectrometers can be programmed to irradiate samples with pulses of electromagnetic radiation that under favorable circumstances can instantaneously upset the spin orientation of all the atoms in a molecular population that contribute to a single resonance, without disturbing the orientations of any others. Suppose this is done to the H1′ resonance of nucleotide n in some

RNA. What happens next? As the disequilibrated H1′ population returns to equilibrium, exchanges of magnetization that occur between its members and protons adjacent to them, cause the latter to "share" in their disequilibrium. The H2′ protons of nucleotide *n* are certain to be affected, as are nearby protons belonging to nucleotide $(n + 1)$. In molecules the size of an RNA, a reduction in the magnitude of the resonances of adjacent protons results that becomes more pronounced with the passage of time out to hundreds of milliseconds after the initial disequilibration, and then fades away. These changes in resonance intensity are called *nuclear Overhauser effects*, or NOEs, for short.

NOEs are transmitted through space, and everything else being equal, their magnitude is inversely proportional to the distance between interacting nuclei raised to the sixth power. In modern spectrometers, proton–proton NOEs, which are the ones usually studied, are large enough to measure if the distance between nuclei is less than 5 Å. Thus by studying NOEs, you can determine which protons are within 5 Å of any other proton in an RNA, and even estimate their separation.

1.2.5 Spin–Spin Relaxation: Molecular Weight Limitations

Spin–spin relaxation exists because NMR spectrometers detect signals only when the magnetic moments of entire populations of nuclei are aligned, and moving in synchrony. This condition is met at the outset of the typical NMR experiment, but as time goes on, the motions of the magnetic moments of individual nuclei vary from the mean due to random, molecule-to-molecule differences in environments. As the variation in the population grows, the vector sum of their magnetic moments decays to zero. Since nuclear magnetic signals also lose intensity when individual nuclei return to their equilibrium orientations, all processes that contribute to spin–lattice relaxation contribute to spin–spin relaxation also. T_2 is always shorter than T_1.

NMR signals decay with first-order kinetics, and their characteristic times, T_2s, can be estimated by measuring the widths of resonances in spectra. If T_2 is short, resonances will be broad. If T_2 is long, resonances will be narrow. For RNAs in the molecular weight range of interest here, T_2s are of the order of 20 ms, and the more slowly a macromolecule tumbles, the shorter its T_2. Thus big molecules have broader resonances than small molecules.

The broadening of resonances that accompanies increased molecular weight contributes to the difficulty of resolving the spectra of large RNAs. The chemical shift range over which RNA atoms resonate is independent of molecular weight. Since large RNAs contain more atoms in chemically distinct environments than small RNAs, the larger an RNA, the more resonances per unit chemical shift there are in its spectra, on average, and the more difficult its spectra are to resolve. T_2 broadening adds insult to injury. The bigger the RNA, the broader its resonances, and broad resonances are harder to resolve than narrow resonances. Since resolution of spectra is a *sine qua non* for spectroscopic analysis, spectral crowding and resonance broadening combine to set an upper bound to the molecular weights of the RNAs that can be studied effectively by NMR. The molecular weight frontier stands today (1999) at about 45 nucleotides.

There is nothing permanent about this frontier. For example, the higher the field strength of a spectrometer, the better resolved the spectra it produces. Thus as long as the field strengths of the spectrometer magnets available continue to increase, as they have in the past, the frontier will continue to move forward. The sensitivity improvement that accompanies increases in field strength is an important added benefit of this very expensive approach to improving spectral resolution.

Isotopic labeling can also contribute. When multidimensional experiments are done on samples labeled with ^{13}C and ^{15}N, spectra can be obtained in which proton resonances that have identical chemical shifts are distinguished on the basis of differences in the chemical shifts of the ^{13}C or ^{15}N atoms to which the protons in question are bonded, and hence J-coupled. Surprisingly, these techniques have had a much bigger impact on NMR size limits for protein than they have for RNA. Proton T_2s in macromolecules labeled with ^{13}C and ^{15}N are always shorter than those in unlabeled macromolecules because of ^{1}H–(^{13}C, ^{15}N) interactions, and the sensitivity of all experiments degrades as T_2s decrease. For reasons that have yet to be fully articulated, this isotope-T_2 effect is more important in RNAs than it is in proteins, and so in contrast to what protein spectroscopists have experienced, only modest increases in the molecular weights of the RNAs that can be studied have resulted from the application of heteronuclear strategies. What they have done is increase the reliability and completeness of the assignments that are obtained for the spectra of RNAs of "ordinary" size.

RNA T_2s can be reduced by selective deuteration because the relaxation rates of protons are determined mainly by their interactions with neighboring protons. Thus when some protons in a molecule are replaced with deuterons (^2H), which have much lower magnetic moments, the relaxation rates of the remaining protons decrease. Note that because deuterium resonates at frequencies well outside the proton range, site-specific deuterium labeling can also be used to remove specific resonances from the proton spectra of macromolecules, which can also help solve assignment problems (see below).

The molecular weight frontier is also being pushed forward by advances in experimental techniques that do not depend on costly expedients like the construction of new instruments or complex isotopic labeling schemes. The physics of relaxation in molecules containing several different kinds of magnetically active nuclei is a good deal more complicated than the description given above might lead one to believe. By taking appropriate advantage of the opportunities this complexity affords, experiments can be devised that produce macromolecular spectra similar to those less sophisticated experiments would supply if T_2s in samples were significantly longer than they really are (e.g. Pervushin *et al.*[6] and Marino *et al.*[7]). Novel experimental approaches like these, applied to isotopically labeled samples in ultra-high field spectrometers, may make the analysis of 100-nucleotide RNAs possible in the next 5 years.

1.2.6 Samples

A single sample consisting of 0.2 ml of a 2 mM solution of an RNA can suffice for its structural analysis. However, contrary to what is sometimes said, not all RNAs can be investigated under all possible solvent conditions by NMR. A structure will not emerge from a spectroscopic investigation unless the RNA of interest is monomeric under the conditions chosen, and has a single conformation. As already suggested, it is sometimes convenient to study RNA samples that are labeled with ^{13}C, ^{15}N and ^2H, either generally or site-specifically. Samples like this are not hard to make. The technology required is constantly improving, and the cost continues to fall.[8–11]

1.2.7 Multidimensional NMR

The modern era of macromolecular NMR began in the late 1970s, when the two-dimensional spectra first began being obtained from proteins.[4] Among the first experiments done were the COSY (or COrrelation SpectroscopY) and NOESY (or Nuclear Overhauser SpectroscopY) experiments. The former generates a two-dimensional spectrum in which resonances that are J-coupled are displayed, and the latter does the same for resonances that cross-relax and hence give NOEs. The more complicated multi-dimensional experiments introduced subsequently accomplish similar ends by different means. Happily, there is not the slightest reason for the consumer of NMR structures to worry about the details.

1.2.8 Assignments

Ribonucleotides contain 8–10 protons of which 7–8 are bonded directly to carbon atoms, and hence do not exchange rapidly with water protons. The remainder are bonded to nitrogens and oxygens, and exchange rapidly. The resonances of an RNA's non-exchangeable and slowly exchanging protons can be observed in spectra taken from samples dissolved in H_2O, and the resonances of its non-exchangeable protons can be studied selectively using samples dissolved in D_2O. As Figure 1 shows, RNA resonances cluster in four groups, depending on chemical type. Note, however, that the chemical shift separations between groups of resonances are about the same size as environmental chemical shift effects that disperse resonances within groups, and hence resonances can appear between clusters or even in the "wrong" cluster. The ^{13}C and ^{15}N spectra of RNAs are similarly complex, but since the chemical shift separation between groups is significantly larger (see Varani and Tinoco[12]), "misplacement" of resonances is less likely (but still not impossible).[13] An RNA's ^{31}P spectrum is always its worst dispersed because all its phosphorus atoms appear in a single chemical context. Fortunately, there is only one phosphorus resonance per residue.

The first order of business for the RNA spectroscopist is assignment of spectra, and this is invariably the most time-consuming phase of any NMR project. A resonance is assigned when the atom (or

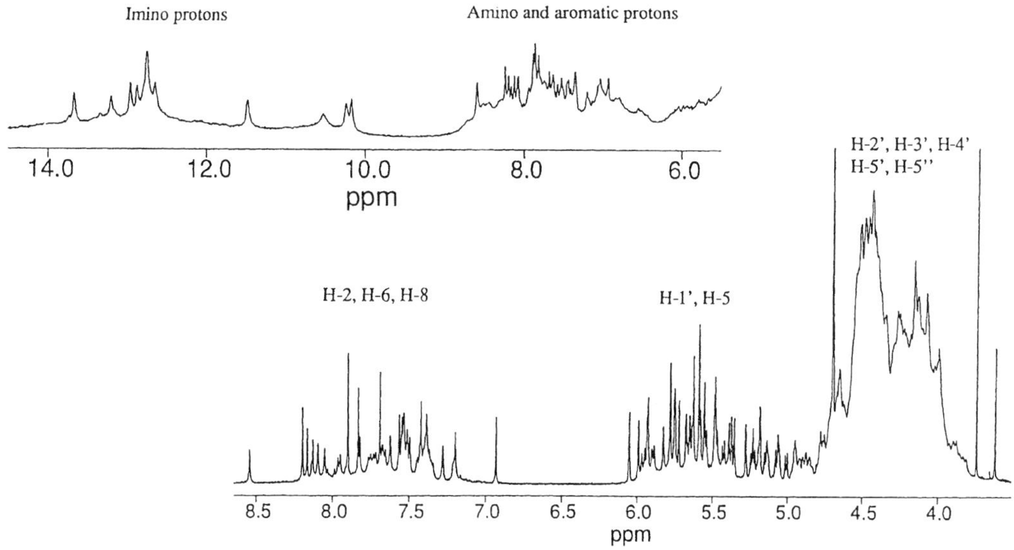

Figure 1. The proton spectrum of a typical RNA. The lower spectrum shows resonances that can be observed in D_2O, and the upper spectrum shows the additional resonances observed when an RNA is dissolved in H_2O. The types of protons that contribute to each region of the spectrum are indicated. This figure is copied, with permission, from the Ph.D. thesis of A. Szewczak, Yale University.

atoms) responsible for it have been identified. Assignments are vital because until they are obtained, nothing can be inferred about molecular conformation from NOESY and COSY crosspeaks. A number of strategies for assigning RNA spectra are available, all derived from techniques pioneered by protein spectroscopists (see Varani and co-workers,[11,12] Nikonowicz and Pardi,[14], and Moore[15]). As is the case with multidimensional spectroscopy, there is no need for the non-specialists to worry about the details.

1.2.9 Helices and Torsion Angles

By the time NMR spectroscopists get involved, the A-form helices of an RNA have usually been identified by other means, and their existence is easy to confirm spectroscopically. The imino proton resonances of AU, GC, and GU base pairs are easily distinguished on the basis of their chemical shifts and the NOEs they give to other kinds of protons. Furthermore, imino–imino NOEs, which are characteristic of double helices, can be used to determine the order of base pairs in helices, and a distinctive pattern of NOEs involving non-exchangeable proton resonances is observed in double-helical RNAs.[4] Note that an experiment now exists that makes it possible to identify directly groups that are the hydrogen-bonding partners of base imino protons.[16]

In principle, the conformation of the non-helical parts of an RNA could be determined by measuring the glycosidic and backbone torsion angles of each nucleotide (see Figure 2).[17] As a practical matter, it is hard to measure coupling constants that speak to many of these torsion angles, and difficult to measure any of them with sufficient accuracy. Nevertheless, data that define the rotamer ranges of torsion angles are relatively easy to obtain, and that information is immensely helpful. The two torsion angles that are easiest to access spectroscopically are δ and χ.

The glycosidic torsion angles of nucleotides (χ) fall into two, non-overlapping ranges, *syn* and *anti*, which are easily distinguished. The intranucleotide distance between pyrimidine H6 or purine H8 protons and H1′ ribose protons is short if nucleotides are *syn*, and long if they are *anti*, and since NOE intensities are proportional to r^{-6}, the difference in ((H6 or 8) to H1′) NOE intensity is huge. The only way to get χ wrong is by misassigning resonances.

Sugar pucker, which corresponds to δ, is also easy to determine. The riboses of most nucleotides in RNA have a C3′-endo pucker, but some are found in the DNA-like C2′-endo configuration. Sugar puckers can be deduced from H1′–H2′ coupling constants, which are large for C2′-endo riboses, small for C3′-endo riboses, and intermediate if a ribose is exchanging rapidly between the two alternatives. H1′–H2′ crosspeaks in COSY-like spectra fall in a distinctive chemical shift range, and because their

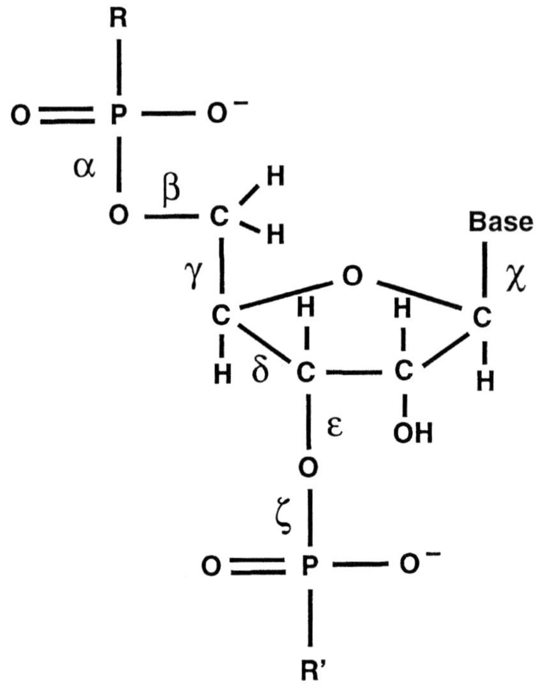

Figure 2. Definitions of the torsion angles in RNAs.

appearances are determined by the magnitude of couplings they represent, coupling constants can be estimated by measuring their substructures. H1′–H2′ coupling constants also have a bearing on ε. For steric reasons, ε is never found in the +gauche range, and if a ribose is C3′-endo, the *trans* rotamer is impossible also.[18] (See Saenger[17] for rotamer definitions.)

Soft information about α and ζ can be gleaned from an RNA's ^{31}P spectrum because ^{31}P chemical shifts are sensitive to both [19]. ^{31}P chemical shifts fall in a narrow range in A-form RNA, and thus α and ζ are likely to have A-form values when ^{31}P shifts are within that range.[11,20] If an unusual ^{31}P chemical shift is observed, neither angle can be constrained.

1.2.10 Distance Estimation

In simple situations, the initial rates at which crosspeaks increase in intensity in proton–proton NOESY spectra are proportional to the distances between the protons they relate, raised to negative sixth power. RNA NOESY spectra are internally calibrated because every pyrimidine contributes an intranucleotide H5/H6 crosspeak to it, and the separation between those protons is fixed covalently. Thus if the intensity of each crosspeak in an RNA's NOESY spectrum is evaluated relative to the intensities of its H5–H6 crosspeaks, estimates of a proton–proton distances can be obtained:

$$\mathrm{NOE}_{i,j} \cdot d_{i,j}{}^{6} = \mathrm{NOE}_{\mathrm{H5,H6}} \cdot d_{\mathrm{H5,H6}}{}^{6}$$

where $\mathrm{NOE}_{i,j}$ is the intensity of the crosspeak assigned to protons i and j, $d_{i,j}$ is the distance between them, and $\mathrm{NOE}_{\mathrm{H5,H6}}$ and $d_{\mathrm{H5,H6}}$ are the corresponding quantities for pyrimidine H5 and H6 protons.

Unfortunately, distance estimates obtained this way are quite crude. First, it is usually impractical to collect RNA NOESY spectra under conditions that prevent the alteration of crosspeak intensities by transfers of magnetization to protons other than the two each crosspeak represents. When "third party" protons are involved, the conversion of crosspeak intensities into interatomic distances outlined above is invalid. Techniques exist for taking these effects into account during the computation of NMR structures (for an application, see White *et al.*[21]), but they are imperfect. The relative motions of the protons in a molecule have to be understood in detail if the rates at which magnetization transfer between them are to be estimated accurately. Because the detailed information required is invariably lacking, *faut de mieux*, it is assumed that the dynamics of all the protons in an RNA can be characterized by a single correlation time, which is a gross oversimplification.

Second, the distances between non-bonded protons in a molecule fluctuate all the time due to thermal motions. If the fluctuations are fast, a molecule will look as though it has a unique conformation spectroscopically, and average NMR data will be measured. Unfortunately, because NOE intensities depend on distance raised to the negative sixth power, the average NOE intensity observed for a pair of protons whose separation fluctuates will always be *greater* than the intensity that would be observed if their separation was fixed at the average value. (Averaging is also a problem when torsion angles are estimated quantitatively using coupling constants (see Varani *et al.*[11]). The reason is fundamentally the same as for NOEs. The conformational parameter sought is not linearly related to the data used to estimate it.)

Third, NOE intensities relate to distances in a simple way only if NOESY spectra are acquired under conditions that allow sample magnetization to equilibrate completely between each iteration of the experiment that is averaged to produce them. It takes about 5 times T_1 for this to occur, and for many RNAs, this implies the need for (at least) a 50 s(!) wait between iterations. Since multidimensional experiments commonly consume days of spectrometer time when cycle times as short as 5 s are used, fully relaxed spectra are seldom accumulated. If all the protons in a molecule have the same T_1, the effect of hyper-fast data collection is the same for all NOEs, and hence is not a problem, but this is not the case for RNA. For all of these reasons, most RNA spectroscopists are content to classify their NOE crosspeaks as being "weak", "medium", and "strong", and to assign broad, overlapping distance ranges to them on that basis.

1.2.11 Structure Calculations

Once distances and torsion angles have been estimated, structure computations can begin. There are several algorithms for extracting conformations from NMR information. Debate about their relative merits remains lively, but there is no need for the non-specialist to worry about the details. However, it is important for the non-specialist to realize that even though the objective is to find the single structure that best accounts for the data, unique structures never emerge. What is produced instead are families of structures that are consistent with all, or almost all the information available, within error. If the data are sufficiently constraining, the members of the family will be closely similar, and the spectroscopist responsible will claim that the structure is solved.

1.3 SOLUTION STRUCTURES AND CRYSTAL STRUCTURES COMPARED

Both crystallographers and spectroscopists deposit lists of atomic coordinates in data banks, and publish molecular images that look exactly alike. Thus biochemists can be forgiven for acting as though the information in an NMR structure is equivalent to that in a crystal structure. It is not, and it is important to understand why.

1.3.1 On the Properties of Crystallographic Structures

Crystallographic analyses produce molecular images equivalent to what an X-ray microscope of large numerical aperture would produce, if such a thing existed. Almost no assumptions are made in generating these images, which are called electron density maps, and the ones that get published are seldom wrong. One reason is that atomic resolution electron density maps are easy to verify. A map of a nucleic acid, for example, had better contain density that looks like nucleotides, and if the number of nucleotides present in a map is not the same as the number of nucleotides in the sequence crystallized, something is wrong.

Macromolecular electron density maps are interpreted by fitting into them representations of the biopolymer of interest that have appropriate bond lengths and bond angles. (Note that the bond lengths and angles used derive primarily from small molecule crystallography!) The lower the resolution of an electron density map, i.e. the longer the wavelength of the shortest-wavelength Fourier components included in its computation, the less detail it contains, and the harder it is to build models into it unambiguously. Once the initial fitting process is finished, the conformation of the model is adjusted to optimize the correspondence between the diffraction pattern it implies and that actually observed. The refined product is the structure that gets published. When published structures are wrong, which they sometimes are, model building errors are invariably to blame.

Not surprisingly, the quality of the product depends on the resolution of the electron density map on which it is based. A 4 Å map of an RNA may be difficult to interpret unambiguously, but can be useful. A 3 Å RNA map should lead to a structural model that accurately depicts the overall shape of the molecule, and reliably reports the placement of its bases and the trajectory of the backbone. Some bound waters and metals may be evident. A map in the low 2 Å range will provide additional information about waters and metals ions, and will accurately define all torsion angles. An RNA map that has a resolution in the low 1 Å range should be totally unambiguous, and specify atomic positions with an accuracy of a few tenths of an ångstrom.

1.3.2 Solution Structures

Like the crystallographer, the spectroscopist interprets his data assuming the molecule of interest is a biopolymer of the appropriate type and sequence, and that it has standard bond lengths and bond angles. Nevertheless, as the preceding discussion has made clear, spectroscopic structures are *not* interpretations of molecular images.

Because spectroscopic structures are not based on images, they do not have resolutions. The experimental constraints that determine a solution structure are all local: estimates of individual torsion angles, and interatomic distances less than 5 Å. Thus NMR structures tend to depict small-scale detail more accurately than they depict overall molecular shape, which is exactly the reverse of crystallography. Spectroscopic models do have precisions, which can be estimated by computing pairwise root- mean-squared deviations (rmsds) between atomic positions within families of structures. Well-determined families of RNA structures have average pairwise rmsds around 1.0 Å, and computational studies suggest that structures determined using large numbers of crudely specified constraints are likely to be more precise and more accurate than structures determined using small numbers of precisely specified constraints; quantity is more important than quality [22].

Spectroscopic constraints are seldom distributed evenly throughout a molecule's volume, and hence some parts of a spectroscopically derived model will be more precisely determined than others. In regions where the data are highly constraining, the members of a structure family will be closely superimposable. Where the data are sparse, the scatter between independently computed structures will be large. The reader is warned that there is an alarming tendency of authors to describe the poorly determined regions of their solution structures as "flexible". Information about molecular dynamics can be obtained by NMR, but it is extracted from measurements of relaxation times, not from COSY and NOESY spectra. Regions of structures where rmsds are large may be flexible, but then again, they may not. (Crystal structures often suffer from a similar problem. Because of local, static, crystal disorder or dynamic disorder, the data obtained from a crystal may determine the conformation of one part of a structure less well than it determines others. Here too there is no simple way to determine the degree to which the local lack of structural definition is due to dynamics or not.)

1.3.3 Constraints and Computations

By protein NMR standards, the number of constraints per unit molecular weight that can be extracted from RNA spectra is small because the number of protons per unit molecular weight of RNA is (relatively) small. Furthermore, they are not evenly distributed. Many of the easiest intranucleotide NOEs to observe are determined by χ, and a large fraction of the easily observed internucleotide NOEs are determined primarily by the distance between the H2′ of nucleotide n and the H6 or H8 proton of nucleotide $(n + 1)$. For both reasons, RNA solution structures tend to be less accurate than protein solution structures. In fact, most NMR-derived RNA structures would be of poor quality indeed if the only information used in their computation was their covalent structures and the spectroscopic data.

Reasonably precise RNA solution structures emerge nevertheless because lots of additional information is fed into the computations that produce them. The lengths of hydrogen bonds in standard base pairs are often specified exactly, for example, and most structure-producing programs attempt to minimize the conformational energies of the structures they produce. In fact, some of them can fold nucleic acid sequences into compact conformations in the total absence of experimental information! The contributions made by these programs to published structures would not be objectionable if one could be sure that they are capable of evaluating conformational energies accurately, but they are not, and it would take an entire treatise to explain why. Thus in addition to helping these programs select the right conformation from the set of "low energy" alternatives, the experimental data also have to keep them honest.

The interpretation of NMR structures is further vexed by the fact that no two laboratories compute structures the same way, and each structure produced by a single laboratory is likely to have been computed differently from its predecessors. At this point in the field's development, it is perfectly possible that were two laboratories to produce models for the same RNA starting from the same data, the models that resulted would differ by much more than the precisions ascribed to them. Thus when two laboratories publish solution structures for the same RNA (e.g. Huang *et al.*[23] and Fountain *et al.*[24]), the only reliable way to decide whether differences between their models are real is to compare the spectra they publish. If the spectra differ, the differences may be real. Otherwise, differences in data treatment must be looked to.

Finally, the unfavorable ratio of experimental observations to coordinates characteristic of RNA spectroscopy makes NMR-derived RNA models hypersensitive to assignments. A single, misassigned NOE crosspeak can have a devastating impact on the conformation proposed for an RNA because important qualitative features of structures are often supported by single NOE crosspeaks! (For a modest example of this effect, compare Cheong *et al.*[25] with Allain and Varani.[26])

Most of the shortcomings of RNA spectroscopy are characteristic of a physical technique still in its infancy. There is reason to hope that many of them will be ironed out in time, and that standards of practice will develop that reduce the impact of the rest. Until that day arrives, however, *caveat emptor*.

1.3.4 Experimental Comparisons of Solution and Crystal Structures

Until recently, there were no RNAs whose structures had been determined by both NMR and X-ray crystallography, and hence no way to assess the accuracy of NMR structures. It was not for lack of trying. Several oligonucleotides that had been characterized in solution were crystallized so that comparisons could be made, but, frustratingly, their conformations changed radically during crystallization (e.g. Cheong and Varani[25], Holbrook *et al.*[27], Baeyens *et al.*[28], and Heus and Pardi[29]). Fortunately, there are now four systems where comparisons can be made: (1) the anticodon stem-loops of tRNAs;[30–34] (2) fragment 1 from *Escherichia coli* 5S rRNA;[35,36] (3) a cobalt hexamine-binding stem-loop from the group I intron;[37,38] and (4) the sarcin/ricin loop from 28S rRNA.[13,39] The news they convey is that spectroscopists have been doing quite well.

The tRNA study cited was motivated by the absence of unambiguous information about anticodon loop conformation in the two initiator tRNA crystal structures published previously,[40,41] and concern that initiator anticodons might differ conformationally from elongator anticodons. The structure of the anticodon loop of yeast initiator methionyl tRNA was compared spectroscopically with that of *E. coli* elongator methionyl tRNA, which has the same sequence. In solution, both anticodon loops have conformations resembling that seen crystallographically in the anticodon loop of yeast phenylalanyl tRNA, which is an elongator tRNA; the rmsd between the anticodon backbone atoms of yeast phenylalanyl tRNA and the yeast initiator tRNA NMR model was 1.2 Å. The bases on the 3′-side of the initiator loop do not stack as neatly as those in phenylalanyl tRNA, but the difference is real. All the anticodon riboses in the yeast phenylalanyl tRNA crystal structure are C3′-endo, but several in the solution structure are C2′-endo.

In 1996–1997, both crystal and solution structures were obtained for several molecules containing the helix IV–helix V-loop E region from *E. coli* 5S rRNA. The 18-nucleotide, loop E regions of both structures superimpose with an all-atom rmsd of about 1.0 Å, and the irregular, non-Watson–Crick pairing in the middle of loop E seen in the crystal structure is faithfully represented in the NMR structure. When longer segments of the two models are compared, the superposition degrades because the relative orientations of distant segments of the 42-base RNA studied by NMR are not well-determined.[20] This is bound to be a problem in any elongated structure that is determined using a method that measures only short distances.

The third comparison is provided by a small stem-loop from the P4–P6 domain of the group I intron from *Tetrahymena*. Crystallographic studies have shown that this loop binds cobalt hexamine when it is part of the larger RNA, and it binds cobalt hexamine in isolation also. The conformation of the loop in solution closely resembles that seen in the P4–P6 crystal structure, and cobalt hexamine binds to both molecules in the same position.

The sarcin/ricin loop (SRL) from rat 28S rRNA provides the last comparison. It is the only example so far of an RNA where the oligonucleotide crystallized is identical to the one characterized spectroscopically. The molecule is organized the same way in both structures. The same base pairs are seen in both, but the relationship between its loop and its stem is not well-determined spectroscopically [20]. Even though the rmsd difference between the loops of the two models is only about 1.5 Å, the solution structure of SRL was not close enough to its crystal structure so that the structure of the crystal to be solved by molecular replacement using the solution structure as the starting model (C. Correll, personal communication).

These comparisons demonstrate that solution structures describe an RNA's topology correctly, i.e. accurately specify its base pairs, and the approximate trajectory of its backbone. At least locally, solution structures are likely to superimpose on corresponding X-ray structure with rmsds less than 2 Å. For many purposes, this level of accuracy is good enough for biochemists and molecular biologists, and it is not clear that the differences between solution structures and crystal structures should all be attributed to error in solution structures.

1.3.5 New Approaches

No one familiar with the history of NMR spectroscopy would dare suggest that the NOESY/COSY approach just described will turn out to be the only way to determine RNA solution structures, or even the best way. NMR spectroscopy has shown an amazing capacity for growth and renewal over the years, and recent developments in the protein NMR field suggest that improved methods for RNA structure determination will soon be available.

NMR has been used to study solids for decades, and in recent years several solids-related methods have emerged that have important applications to RNA. As the reader will recall, the magnetic field of each magnetically active nucleus in a molecule propagates through space like any other magnetic field and contributes to the total magnetic field experienced by all of its neighbors. Solution spectroscopists ignore these interactions because they are averaged to zero by the rotational diffusion of the macromolecules they study. Solid-state spectroscopists cannot ignore them because their molecules do not rotate. Techniques exist for detecting these through-space dipolar interactions in solids, and it is clear that their effects are measurable over a much wider range of distances than the NOEs on which solution spectroscopists dote. Solids-derived methods are already available determining interatomic distances that exceed 10 Å in proteins with 0.1 Å accuracy (see Griffin[42]). In addition, it has been discovered that macromolecules orient slightly in magnetic fields when they are dissolved in liquid crystal solvents. When oriented this way, through-space dipolar nucleus–nucleus interactions can be observed that cannot be detected in regular solutions, and information can be obtained about the relative orientations of the interatomic vectors within molecules.[43] Clearly, if one were to add some accurate, long-range interatomic distances and information on relative bond orientations to the traditional mix of COSY and NOESY data, RNA solution structures would emerge that are significantly more accurate than those available today.

1.4 LESSONS LEARNED ABOUT MOTIFS BY NMR

For reasons already elucidated, RNA spectroscopists cannot determine the conformations of entire, naturally occurring RNAs. Consequently, RNA spectroscopists have concentrated on three classes of RNAs: (1) small, synthetic oligonucleotides that contain interesting base-pairing irregularities; (2) RNA aptamers; and (3) domains excised from large, natural RNAs.

The work done on synthetic oligonucleotides has been motivated by the belief that RNA structures are modular, which is to say that the conformations of motifs in small oligonucleotides of otherwise arbitrary sequence are identical to the conformations of the same motifs in all other RNAs. Aptamers are RNA sequences selected from random populations *in vitro* on the basis of their capacity to bind specific ligands or to perform other selectable functions (see Gold *et al.*[44]). In order that sequence space be sampled thoroughly, the lengths of oligonucleotides in the RNA populations from which aptamers are selected must be quite small, and consequently, most aptamers are small enough for spectroscopists to study intact (see Cech and Szewczak[45] and Marshall *et al.*[46].) Those who concentrate on domains do not need to invoke modularity to justify their activities. By definition, a domain is a portion of a macromolecule that is conformationally autonomous; the conformation determined for a domain in isolation has to be the same as that in the larger RNA from which it derives. The only problem students of domain structure confront, therefore, is proving their oligonucleotides are domains in the first place.

1.4.1 RNA Organization in General

Qualitatively, the way single-stranded RNAs organize themselves was understood almost 40 years ago.[47,48] They fold so that the short sequences they contain that are "accidentally" complementary form short double helices to (approximately) the maximum extent possible. The dominant structural element that results is the *hairpin loop*, or *stem-loop*, which is produced when an RNA chain folds back on itself so that complementary sequences close to each other in its sequence can pair. Thus most RNAs have secondary structures that consist of a series of stem-loops separated by sequences of less certain conformation that are usually represented as single-stranded.

Inevitably, in RNA stems where strands of "random" sequence are aligned to maximize Watson–Crick pairing, bases are juxtaposed that cannot form canonical pairs, and because stems are stabilized if hydrogen bonds form and bases stack, they pair anyway. GU pairs within otherwise regular helical stems are a case in point. They are so common that wobble GUs, which fit easily into helices, are considered "honorary" Watson–Crick pairs. In addition to occasional non-canonical base pairs, helical stems are often interrupted by *bulged bases*, which is to say bases on one strand that have no partner to pair with on the other, and by *internal loops*, in which longer sequences on both strands are juxtaposed that cannot obviously be paired. Some internal loops have sequences long enough to include stem-loops of their own; they are called *junctions*. Whether the stem of a stem-loop contains irregularities or not, it must have a *terminal loop*, i.e. a sequence that links the 5′- to the 3′-strand of its stem, and their conformations cannot be predicted *a priori* either. The terminal loops of some stem-loops are big enough to contain stem-loops of their own.

The evidence available suggests that most stem-loops are domains, and since many of them contain less than 45 nucleotides, and those that do not can often be "trimmed", stem-loops derived from natural RNAs are favorite targets for spectroscopic investigation. By characterizing them one is investigating the conformations of important elements of RNA secondary structure.

Large RNAs have tertiary structures, of course; some of them are as compactly folded as globular proteins. The interactions that stabilize RNA tertiary structures involve both stem-loops and the "unstructured" sequences that link them together, but they are unusual in RNAs of the sizes RNA spectroscopists can study. For that reason, NMR has provided little insight into this aspect of RNA conformation.

1.4.2 Terminal Loops

A great deal has been learned about terminal loop structure by NMR, particularly about the conformations of terminal loops that have short sequences. Short terminal loop sequences play the same role in RNA as β-turns in proteins. They are concise structures that stabilize 180° changes in backbone direction.

1.4.2.1 U-turns

The U-turn is a four-base, terminal loop motif, the consensus sequence of which is UNRN. (N.B.: N stands for any nucleotides, and R means any purine.) They were first characterized in the mid-1970s

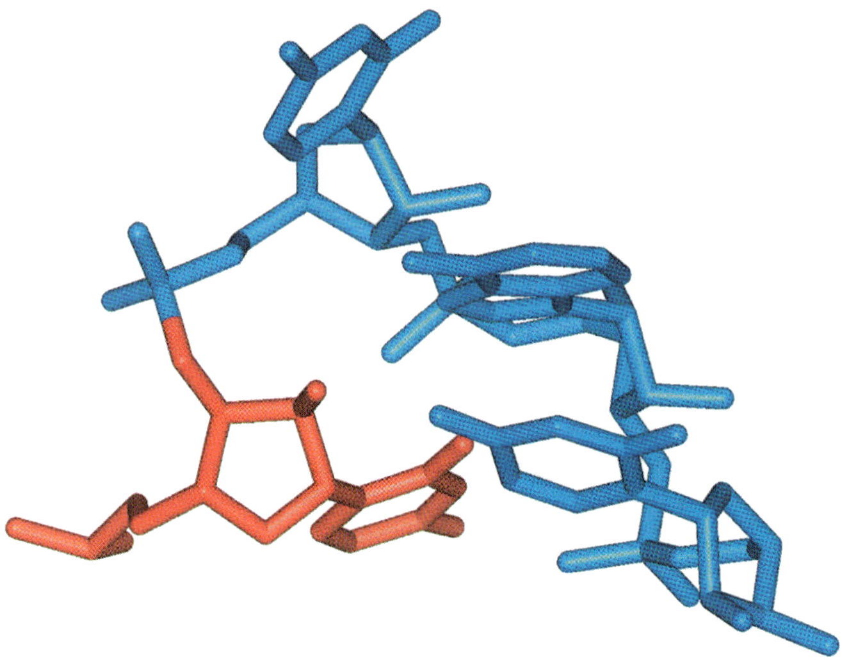

Figure 3. The conformation of a typical U-turn.[24] The U at the 5′-end of the motif is shown in red. It points away from the viewer. The three bases that follow (blue) form a stack the bases of which point out towards the viewer.

by crystallographers working on transfer RNAs,[31–34] and their existence in tRNAs in solution has been confirmed.[30] Recent spectroscopic studies have demonstrated that they occur in other contexts. The L11 binding region of 23S rRNA includes a U-turn[23,24] as does loop IIa in yeast U2snRNA.[49]

Figure 3 shows a typical U-turn. Like all other U-turns, it is stabilized by a hydrogen bond between the imino proton of U1 and an oxygen belonging to the phosphate group of R3, and the 2′OH of U1 and N7 of R3.[50] All of the U-turns characterized so far are components of larger terminal loops.

1.4.2.2 Tetraloops

In the late 1980s, it was noticed that helical stems terminated by 4-nucleotide loops, or *tetraloops*, having the sequence UNCG are unusually abundant in rRNAs, and it was demonstrated that they are unusually stable.[51] Further analysis revealed the existence of two other "special" tetraloops sequences: GNRA and CUNG.[52] Spectroscopic studies done subsequently have demonstrated that each of these tetraloops has a distinctive conformation, as expected, and those who work with short RNA oligonucleotides now routinely include them in sequences intended to form stem-loops.

The conformation of the UNCG motif was analyzed initially in Tinoco's laboratory in 1990,[25] and five years later, their structural proposal was revised using a larger set of NMR-derived restraints.[26] The most striking feature of the UNCG turn is the unusual syn–anti pair that forms between U1 and G4, which has a phosphate–phosphate distance so small it can be spanned by the middle two residues, N2 and C3.

GNRA tetraloops have also received a great deal of attention, and, as expected, they all have similar conformations.[29,53] As is the case with UNCG tetraloops, the "secret" of these structures is the slipped, or side-by-side pair that forms between G1 and A4, which greatly reduces the distance between the backbones of the two strands of the loop being capped. Interestingly, the trajectory of the backbone in GNRA tetraloops is so similar to that in U-turns that some now refer to GNRA tetraloops as U-turns. It would be wiser to apply that phrase only to turns whose sequence is UNRN.

A GNRA tetraloop has recently been observed in an entirely unexpected context: that provided by an aptamer which binds AMP.[54,55] In the presence of AMP, an otherwise unstructured internal loop in this RNA folds so that the AMP can interact with the RNA as though it were A4 in a GNRA tetraloop. The similarity between the conformation of the resulting loop and that of a normal GNRA tetraloop is striking.

The structure of the last member of the set, CUNG, is quite different from that of the other standard tetraloops.[56] C1 and G4 form a Watson–Crick base pair, and U2 reaches down into the minor groove of the helical stem being capped, and interacts with its last base pair. This interaction appears to require that the last base pair be a GC, an inference strongly supported by phylogenetic data. Thus conformationally, CUNG tetraloops are really UN diloops, but they have a consensus sequence that is 6 bases long: G(CUNG)C.

1.4.2.3 Other terminal loops

Many terminal loops are not, or do not appear to be, motifs. It would be a mistake to assume they lack structure, however. Conformations have just been obtained for two such loops: the conserved UGAA loop found at the 3′-end of all 18S rRNAs,[57] and the UGGGGCG loop that is a universal component of the peptidyl transferase region of 23S-like rRNAs.[58] We will not discuss their conformations here because they are not motifs, but the reader should examine them anyway. Both are highly structured, and contemplation of them should induce a sense of humility. No one could possibly have predicted their conformations in advance.

1.4.3 Internal Loops

The dominant motif in RNA stem-loops is the A-form helix, the conformation of which was well-understood long before NMR spectroscopy was mature enough to contribute in any way. It is a two-stranded, antiparallel, double helix of indefinite length having geometry so well-known it need not be described here.[17] There is no restriction on the nucleotide sequence in either of the two strands of an A-form helix, provided the sequence of the other strand is its Watson–Crick complement. If GU wobble pairs are accepted as equivalent to Watson–Crick GCs and AUs, roughly two-thirds of the bases in an RNA like a ribosomal RNAs are involved in A-form helix.

As pointed out earlier, the helical continuity of many stem-loops is interrupted by internal loops, only a small number of which have been characterized spectroscopically (or crystallographically, for that matter). They come in two varieties: symmetric and asymmetric. In a symmetric internal loop, the number of loop nucleotides is the same in the two strands, and in an asymmetric loop, it is not. Only a small number on internal loop motifs have been identified so far; there are bound to be more.

1.4.3.1 Symmetric internal loop motifs

Recent NMR and crystal structures provide numerous examples of internal loop motifs called "cross-strand purine stacks". In A-form helix, the bases in each strand form a continuous stack that runs the length of the helix. In cross-strand purine stacks, a purine in one strand stacks on a purine from an adjacent base pair that belongs to the other strand. This alters the relative sizes of the major and minor grooves.

The first cross-strand purine stacks observed spectroscopically were the cross-strand A stacks found in loop E from eukaryotic 5S rRNA[59] and the sarcin/ricin loop from rat 28S rRNA.[60] The consensus sequence for this kind of stack is 5′(G or C)GA paired with 5′UA(G or C), and the pairing is a Watson–Crick GC, in either orientation, followed by a slipped GA and a reverse Hoogsteen AU. The six-membered ring of the A in the GA stacks on the six-membered ring of the A in the AU (Figure 4). Two more examples have been found in loop E from prokaryotic 5S rRNA.[36]

Loop E also contains a cross-strand G stack that is composed of two wobble GU pairs sandwiched between two Watson–Crick GCs. In this motif, 5′UG is paired with 5′UG, and the six-membered rings of its Gs are stacked (Figure 5). Note that since GUs embedded in helices are thought of as equivalent to GCs and AUs, it may be somewhat surprising that this motif has a distinctive conformation. It is clear from crystallographic studies that the sequences other than those mentioned here also cause cross-strand purine stacks (e.g. Cate *et al.*[38]).

As it happens, loop E from *E. coli* 5S rRNA is one of the only symmetric internal loops whose

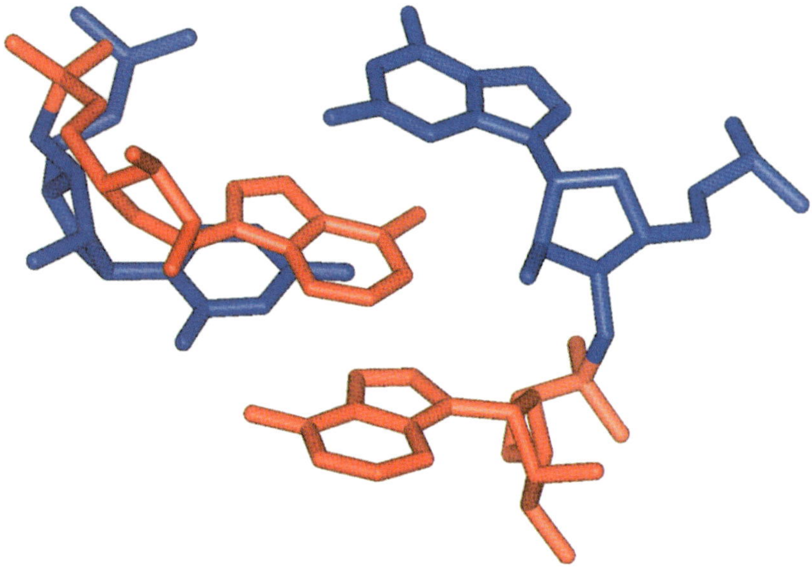

Figure 4. A cross-strand A stack.[30] The reverse-Hoogsteen AU belonging to this stack lies below its side-by-side AG in this diagram. The two A's are red and the G and U with which they pair are blue. The stacking of the six-membered rings of the As is obvious.

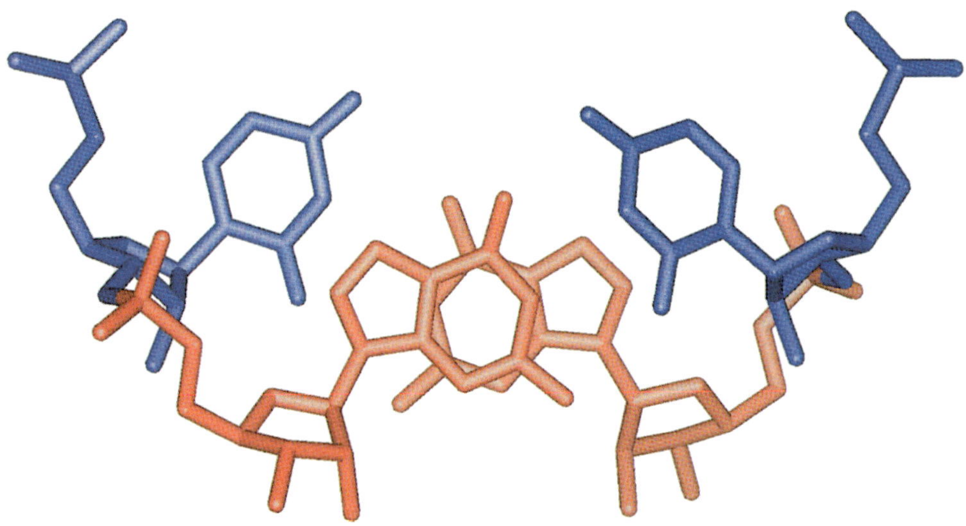

Figure 5. A cross-strand G stack.[30] The two successive GU wobble pairs that constitute this motif are viewed down the axis of the double helix to which they belong. The six-membered rings of the two G's (red) stack almost perfectly. There is an approximate two-fold axis in this motif running between the planes of the two G's, perpendicular to the helix axis.

conformation is known.[36] Thus even though the conformation adopted of the six bases in the middle of this loop are not a motif, its conformation is worth examining.

1.4.3.2 Asymmetric internal loop motifs

Both prokaryotic loop E and the sarcin/ricin loop include a three-base structure called a "bulged G motif".[13,59,60] The sequence is 5′(G or C)GAA paired with 5′AGUG(G or C). G2 of the second strand reaches across the minor groove of the motif so that its imino proton can hydrogen bond to the phosphate group that links G2 and A3 in the first strand. The remaining bases (5′(G or C)GA, and 5′...UG(G or C)) form a cross-strand A stack, and A4 from the first strand forms a symmetric, parallel, *anti–anti*-pair with A1 of the second strand. It is not clear what nucleotides can follow the AA pair, but so far, only antiparallel, *anti-*, *anti-*, all pyrimidine pairs have been found at that position. Because the AA

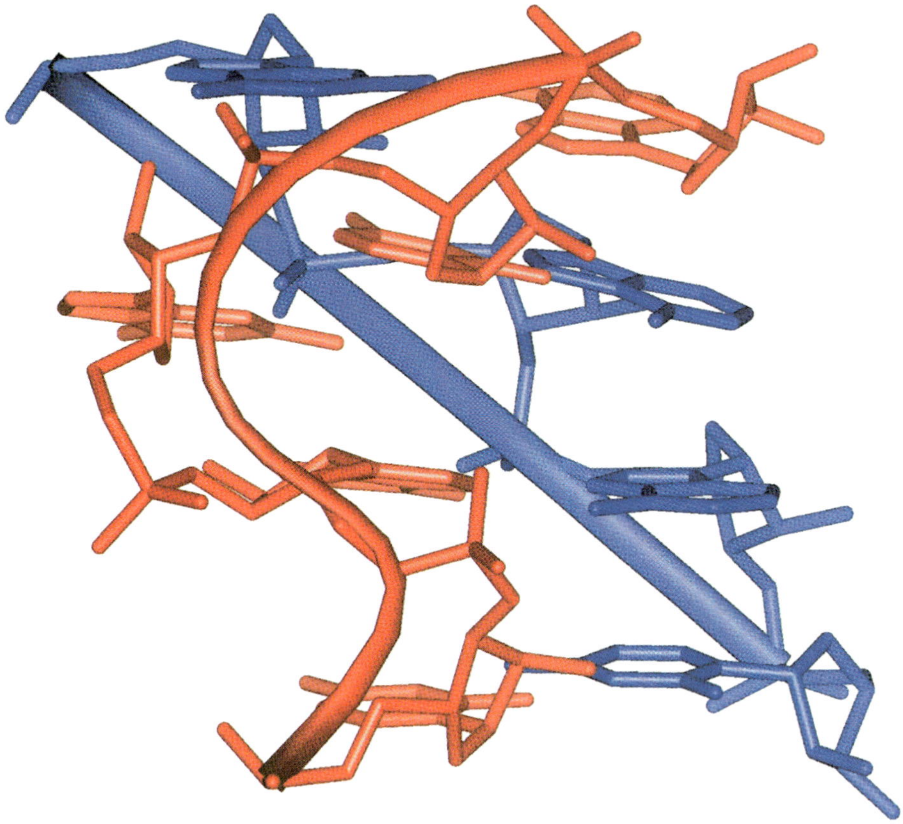

Figure 6. The S-turn in the backbone of bulged-G motifs. The bulged-G motif ion the sarcin/ricin loop is shown.[53] The 5′-strand of the motif, which contains the bulged G, is shown in red, and the 3′-strand is blue. The backbone trajectories of both strands are indicated by continuous oval lines.

pair is symmetric, the backbone of this motif has a distinctive, S-shaped trajectory on its bulged G side (Figure 6).

This motif is just one example of how "extra" bases in asymmetric internal loops get "taken care of". A rich variety of alternatives is on display in the many structures of aptamers and ligand-binding natural RNAs that have been published recently, none of which are motifs.[61–72] Examination of these structures leaves one with a single strong impression. A remarkable fraction of these loops are distorted double helices that interrupt the regular helices they separate without breaking their continuities.

1.4.4 Pseudoknots

Many RNAs contain pseudoknots, which are structures in which the loop of some stem-loop forms a double helix by pairing with other nucleotides from some other part of the same molecule. When the sequence that base-pairs with the loop starts immediately after the stem of the stem-loop, the object that results is two stem-loops joined side by side, like Siamese twins, because the loop bases of both stem-loops are one strand of the stem of their partners (for details see Wyatt and Tinoco[73]). The structures that result are motifs topologically, even though their sequences vary a lot. In the late 1980s, a series of synthetic pseudoknots were studied by NMR,[74] and recently a natural pseudoknot has been characterized.[75]

1.5 REFERENCES

1. C.P. Slichter, "Principles of Magnetic Resonance", 3rd ed., Springer, New York, 1989.
2. M. Goldman, "Quantum Description of High-Resolution NMR in Liquids", Oxford University Press, Oxford, 1988.

3. R.R. Ernst, G. Bodenhausen and A. Wokaun, "Principles of Nuclear Magnetic Resonance in One and Two Dimensions", Oxford University Press, Oxford, 1987.
4. K. Wuthrich, "NMR of Proteins and Nucleic Acids", Wiley, New York, 1986.
5. J. Cavanagh, W.J. Fairbrother, A.G. Palmer, III and N.J. Skelton, "Protein NMR Spectroscpy. Principles and Practice", Academic Press, San Diego, 1996.
6. K. Pervushin, R. Riek, G. Wider and K. Wuthrich, *Proc. Natl. Acad. Sci. USA*, 1997, **94**, 12366.
7. J.P. Marino, J.L. Diener, P.B. Moore and C. Griesinger, *J. Am. Chem. Soc.*, 1997, **119**, 7361.
8. E.P. Nikonowicz, A. Sirr, P. Legault, F.M. Jucker, L.M. Baer and A. Pardi, *Nucl. Acids Res.*, 1992, **20**, 4507–4513.
9. R.T. Batey, M. Inada, E. Kujawinski, J.D. Puglisi and J.R. Williamson, *Nucl. Acids Res.*, 1992, **20**, 4515.
10. R.T. Batey, J.L. Battiste and J.R. Williamson, *Methods Enzymol.*, 1995, **261**, 300.
11. G. Varani, F. Aboul-ela and F.H.-T. Allain, *Progr. Nucl. Magn. Reson. Spectrom.*, 1996, **29**, 51.
12. G. Varani and I. Tinoco Jr., *Q. Rev. Biophys.*, 1991, **24**, 479–532.
13. A.A. Szewczak and P.B. Moore, *J. Mol. Biol.*, 1995, **247**, 81.
14. E.P. Nikonowicz and A. Pardi, *J. Mol. Biol.*, 1993, **232**, 1141–1156.
15. P.B. Moore, *Acc. Chem. Res.*, 1995, **28**, 251.
16. K. Pervushin, A. Ono, C. Fernandez, T. Szyperski, M. Kainosho and K. Wuthrich, *Proc. Natl. Acad. Sci. USA*, 1998, **95**, 14147.
17. W. Saenger, "Principles of Nucleic Acid Structure", Springer, New York, 1984.
18. C. Altona, *Recueil. J. R. Neth. Chem. Soc.*, 1982, **101**, 413.
19. D.G. Gorenstein, "Phosphorous-31 NMR. Principles and Applications", Academic Press, Orlando, FL, 1984.
20. J.P. Rife, S.C. Stallings, C.C. Corell, A. Dallas, T.A. Steitz and P. B. Moore, *Biophys. J.*, 1999, **76**, 65–75.
21. S.A. White, M. Nilges, A. Huang, A.T. Brunger and P.B. Moore, *Biochemistry*, 1992, **31**, 1610.
22. F.H.-T. Allain and G. Varani, *J. Mol. Biol.*, 1997, **267**, 338.
23. S.G. Huang, Y.X. Wang and D.E. Draper, *J. Mol. Biol.*, 1996, **258**, 308.
24. M.A. Fountain, M.J. Serra, T.R. Krugh and D. Turner, *Biochemistry*, 1996, **35**, 6539.
25. C. Cheong, G. Varani and I. Tinoco, *Nature*, 1990, **346**, 680.
26. F.H.-T. Allain and G. Varani, *J. Mol. Biol.*, 1995, **250**, 333.
27. S.R. Holbrook, C. Cheong, I. Tinoco and S.-H. Kim, *Nature*, 1991, **353**, 579.
28. K.J. Baeyens, H.L. De Bondt, A. Pardi and S.R. Holbrook, *Proc. Natl. Acad. Sci. USA*, 1996, **93**, 12851.
29. H.A. Heus and A. Pardi, *Science*, 1991, **253**, 191.
30. D.C. Schweisguth and P.B. Moore, *J. Mol. Biol.*, 1997, **267**, 505.
31. B. Hingerty, R.S. Brown and A. Jack, *J. Mol. Biol.*, 1978, **124**, 523.
32. S.R. Holbrook, J.L. Sussman, R.W. Warrant and S.-H. Kim, *J. Mol. Biol.*, 1978, **123**, 631.
33. E. Westhof and M. Sundaralingam, *Biochemistry*, 1986, **25**, 4868.
34. E. Westhof, P. Dumas and D. Moras, *Acta Crystallogr. Sect. A*, 1988, **44**, 112.
35. C.C. Correll, B. Freeborn, P.B. Moore and T.A. Steitz, *Cell*, 1997, **91**, 705.
36. A. Dallas and P.B. Moore, *Structure*, 1997, **5**, 1639.
37. J.S. Kieft and I. Tinoco Jr., *Structure*, 1997, **5**, 713.
38. J. Cate, A.R. Gooding, E. Podell, K. Zhou, B.L. Golden, C.E. Kundrot, T.R. Cech and J.A. Doudna, *Science*, 1996, **273**, 1678.
39. C.C. Correll, A. Munishkin, Y. Chan, Z. Ren, I.G. Wool and T.A. Steitz, *Proc. Natl. Acad. Sci. USA*, 1998, **95**, 13436.
40. N.H. Woo, B. Roe and A. Rich, *Nature*, 1980, **286**, 346.
41. R. Basavappa and P.B. Sigler, *EMBO J.*, 1991, **10**, 3105.
42. R.G. Griffin, *Nat. Struct. Biol.*, 1998, **5**, 508.
43. N. Tjandra and A. Bax, *Science*, 1997, **278**, 1111.
44. L. Gold, B. Polisky, O. Uhllenbeck and M. Yarus, *Annu. Rev. Biochem.*, 1995, **64**, 763.
45. T.R. Cech and A.A. Szewczak, *RNA*, 1996, **2**, 625.
46. K.A. Marshall, M.P. Robertson and A.D. Ellington, *Structure*, 1997, **5**, 729.
47. J.R. Fresco and B.M. Alberts, *Proc. Natl. Acad. Sci. USA*, 1960, **46**, 311.
48. J.R. Fresco, B.M. Alberts and P. Doty, *Nature*, 1960, **188**, 98.
49. S.C. Stallings and P.B. Moore, *Structure*, 1997, **5**, 1173.
50. G.J. Quigley and A. Rich, *Science*, 1976, **194**, 796.
51. C. Tuerk, P. Gauss, C. Thermes, D.R. Groebe, M. Gayle, N. Guild, G. Stormo, Y. d'Aubenton-Carafa, O.C. Uhlenbeck, I. Tinoco, E.N. Brody and L. Gold, *Proc. Natl. Acad. Sci. USA*, 1988, **85**, 1364.
52. C.R. Woese, S. Winker and R.R. Gutell, *Proc. Natl. Acad. Sci. USA*, 1990, **87**, 8467.
53. F.M. Jucker, H.A. Heus, P.F. Yip, E.H.M. Moors and A. Pardi, *J. Mol. Biol.*, 1996, **264**, 968.
54. F. Jiang, R.A. Kumar, R.A. Jones and D. Patel, *Nature*, 1996, **382**, 183.
55. T. Dieckmann, E. Suzuki, G.K. Nakamura and J. Feigon, *RNA*, 1996, **2**, 628.
56. F.M. Jucker and A. Pardi, *Biochemistry*, 1995, **34**, 14416.
57. S.E. Butcher, T. Dieckmann and J. Feigon, *J. Mol. Biol.*, 1997, **268**, 348.
58. E.V. Puglisi, R. Green, H.F. Noller and J.D. Puglisi, *Nat. Struct. Biol.*, 1997, **4**, 775.
59. B. Wimberly, G. Varani and I. Tinoco Jr., *Biochemistry*, 1993, **32**, 1078.
60. A.A. Szewczak, P.B. Moore, Y.-L. Chan and I.G. Wool, *Proc. Natl. Acad. Sci. USA*, 1993, **90**, 9581.
61. J.D. Puglisi, R. Tan, B.J. Calnan, A.D. Frankel and J.R. Williamson, *Science*, 1992, **257**, 76–80.
62. F. Aboul-ela, J. Karn and G. Varani, *J. Mol. Biol.*, 1995, **253**, 313.
63. J.D. Puglisi, L. Chen, S. Blanchard and A.D. Frankel, *Science*, 1995, **270**, 1200.

64. X. Ye, R.A. Kumar and D.J. patel, *Chem. Biol.*, 1995, **2**, 827–840.
65. J.L. Battiste, R. Tan, A. Fraenkel and J.R. Williamson, *Biochemistry*, 1994, **33**, 2741.
66. J.L. Battiste, M. Hongyuan, N.S. Rao, R. Tan, D.R. Muhandiram, L.E. Kay, A.D. Frankel and J.R. Williamson, *Science*, 1996, **273**, 1547.
67. K. Kalurachchi, K. Uma, R.A. Zimmermann and E.P. Nikonowicz, *Proc. Nat. Acad. Sci. USA*, 1997, **94**, 2139.
68. D. Fourmy, M.I. Recht, S.C. Blanchard and J.D. Puglisi, *Science*, 1996, **274**, 1367.
69. Y. Yang, M. Kochpyan, P. Burgstaller, E. Westhof and M. Famulok, *Science*, 1996, **272**, 1343.
70. P. Fan, A.K. Suri, R. Fiala, D. Live and D. Patel, *J. Mol. Biol.*, 1996, **258**, 480.
71. G.R. Zimmerman, R.D. Jenison, C.L. Wick, J.-P. Simorre and A. Pardi, *Nat. Struct. Biol.*, 1997, **4**, 644.
72. L. Jiang, A.K. Suri, R. Fiala and D.J. Patel, *Chem. Biol.*, 1997, **4**, 35.
73. J.R. Wyatt and I. Tinoco, Jr. in "The RNA World", eds. R.F. Gesteland and J.F. Atkins, Cold Spring Harbor Laboratory, Cold Spring Harbor, NY, 1993, p. 465.
74. J.D. Puglisi, J.R. Wyatt and I. Tinoco Jr., *Nature*, 1988, **331**, 283.
75. Z. Du, D.P. Giedroc and D.W. Hoffman, *Biochemistry*, 1996, **35**, 4187.

2

Thermodynamics of RNA Secondary Structure Formation

TIANBING XIA, DAVID H. MATHEWS and DOUGLAS H. TURNER

University of Rochester, NY, USA

2.1 INTRODUCTION

RNA is an active component in many cellular processes.[1] For example, RNA alone can act as an enzyme to catalyze RNA transformations.[2–4] It is also possible that the RNA in ribosomes[5,6] and signal recognition particles[7] is actively involved in protein synthesis and protein translocation across membranes, respectively. Retroviruses, including HIV, are RNA–protein complexes.

Nucleic acids are now being sequenced at a rate of more than one million nucleotides per day,[8,9] and the entire three billion bases in the human genome are now known.[10] This is providing sequences for many important RNA molecules. While such sequence information facilitates investigations of RNA, in-depth understanding of structure–function relationships requires knowledge of three-dimensional structure, energetics, and dynamics.

Due to their complexity and dynamic behavior, it is difficult and time-consuming to determine three-dimensional structures for natural RNA molecules. Thus from 1973 to 1996 the only three-dimensional structures determined by X-ray crystallography (see Chapter 3) for natural RNAs longer than 30 nucleotides were tRNAs,[11–13] hammerhead ribozymes,[14–17] and one domain of a group I intron.[18] Structures of some natural fragments of RNA have also been determined by NMR.[19–32] These methods cannot keep pace with the rate of discovery and sequencing of interesting new RNA molecules. Thus there is a need for other reliable methods of determining RNA structure. If the energetics of RNA were completely understood, it would be possible to predict their folding, reactivity, and functional properties directly from their sequences.

The first stage in predicting RNA structure is determination of secondary structure, essentially a listing of base pairs contained in the folded structure. Determination of secondary structure also defines the various loops present in a given RNA. Figure 1 shows a secondary structure illustrating most of the loop motifs. Often, these non-Watson–Crick regions of an RNA are particularly important for function since unusual arrays of functional groups are available there for tertiary interactions[33] or recognition of other cellular components.[34] Thus determination of secondary structure helps identify nucleotides that may be important for function.

Phylogenetic sequence comparison is one way to determine RNA secondary structure, provided large numbers of homologous sequences from different organisms are available.[35,36] When not enough related sequences are known, however, alternative methods must be used, and the most popular is free energy minimization. It is based on two assumptions: (i) the dominant interactions responsible for RNA structures are local,[37–41] presumably hydrogen bonding between bases and stacking between adjacent base pairs;[42–44] (ii) the conformations RNA adopts are equilibrium, lowest free energy conformations.[42,45]

At least two factors limit the success of secondary structure prediction by free energy minimization. First, algorithms do not exist that include all possible folding motifs and deal efficiently with the enormous numbers of possible secondary structures for a long sequence.[42,46–48] Second, our knowledge of the contributions of various RNA motifs to the total free energy of RNA structures is still incomplete.

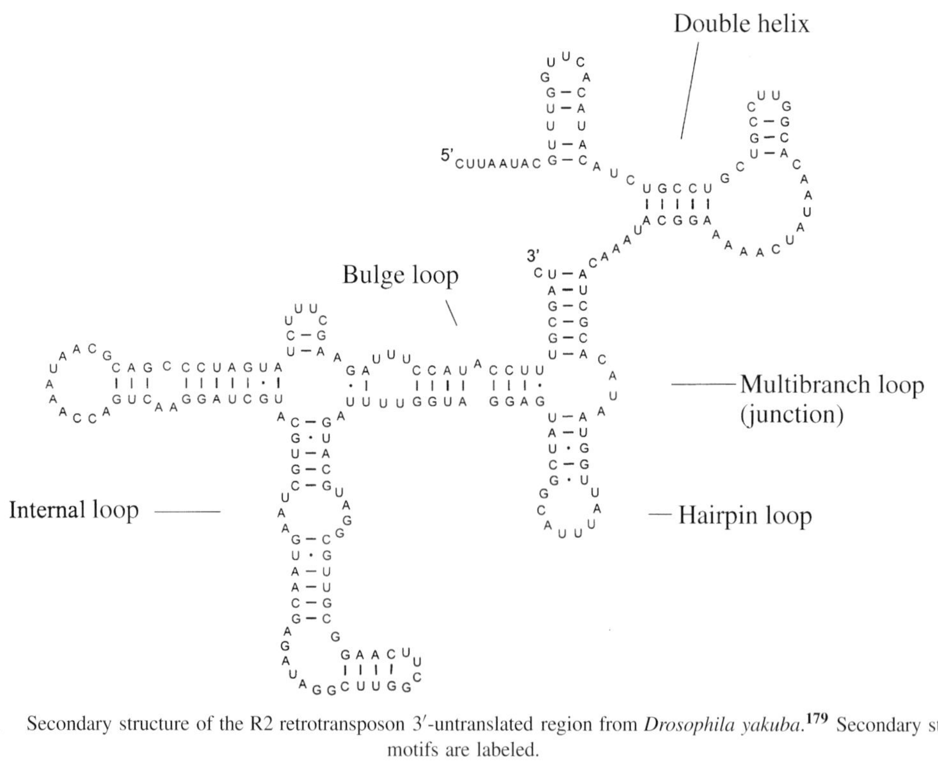

Figure 1 Secondary structure of the R2 retrotransposon 3′-untranslated region from *Drosophila yakuba*.[179] Secondary structure motifs are labeled.

Rapid methods for the synthesis of oligonucleotides[49,50] (see Chapter 6) make it possible to study the sequence dependence of RNA secondary structure thermodynamics in a systematic way. Accumulation of this knowledge has steadily improved predictions[51,52] and incorrect predictions often occur at motifs for which little experimental data are available. Thus, understanding of the thermodynamics of RNA secondary structure is crucial for successful structure prediction. This chapter reviews the methods available for measuring the thermodynamics of RNA motifs, the known sequence dependence of these thermodynamics, and applications to predicting RNA secondary structure, modeling tertiary structure, and designing therapeutics.

2.2 THERMODYNAMIC ANALYSIS OF RNA STRUCTURAL TRANSITIONS

2.2.1 Hypochromism: Basis of Transition Analysis

Many techniques for investigating order–disorder structural transitions follow changes that occur in a spectroscopic property when the transition is induced thermally. A convenient property to follow for nucleic acids is UV absorption, which results from complex $n \rightarrow \pi^*$ and $\pi \rightarrow \pi^*$ transitions of the bases.[53–55] A decrease in UV absorption is observed in nucleic acids upon duplex formation. This decrease is called "hypochromism".[55] For short oligonucleotides, 30–40% hypochromicity at 260 or 280 nm is typical.

Hypochromism is largely due to interactions between electrons in different bases.[56–61] In particular, the transition dipole moment of the absorbing base interacts with the light-induced dipoles of neighboring bases. For a polymeric array of chromophore residues, such as bases in a nucleic acid, this interaction depends on the relative orientation and separation of bases. If the bases are stacked parallel so that the transition dipole moments of adjacent bases are oriented more or less head-to-head (helical form), the probability of photon absorbance by a base is reduced due to light-induced dipoles in neighboring bases. Because the shape of the UV absorption of nucleotide bases is not significantly affected by these interactions,[56] the order–disorder transition of RNA can be followed by monitoring the UV absorption at a single wavelength, typically at 260 nm for AU-rich or 280 nm for GC-rich sequences.[62–64]

2.2.2 Equilibrium Transition: Two-state Model

A UV absorption vs. temperature profile is called a "UV melting curve" by analogy with true phase changes. A typical experimental curve for a duplex to random coil transition for a short oligonucleotide is shown in Figure 2. Often, short duplexes melt in a two-state, all-or-none, manner, i.e., an RNA strand is either in a completely double helical or in a completely random conformation state; no partially ordered states are significantly populated. This is because the initiation step in helix formation is unfavorable compared to helix growth steps.[54]

General formulas have been presented for analyzing melting curves.[65] The majority of equilibria of interest to molecular biologists are bimolecular or unimolecular in nature:

$$\mathrm{A} \rightleftharpoons \mathrm{B} \quad \text{(unimolecular)} \tag{1}$$

$$2\mathrm{A} \rightleftharpoons \mathrm{A}_2 \quad \text{(bimolecular, self-complementary)} \tag{2}$$

$$\mathrm{C} + \mathrm{D} \rightleftharpoons \mathrm{E} \quad \text{(bimolecular, non-self-complementary)} \tag{3}$$

The equilibrium constant for duplex formation is

$$K = \frac{\alpha/2}{(C_\mathrm{T}/a)(1-\alpha)^2} \quad \text{(bimolecular)} \tag{4}$$

where C_T is the total single strand concentration, a is 1 for self-complementary duplexes and 4 for non-self-complementary duplexes; α is the mole fraction of single strand in duplex form. The equilibrium constant is related to the free energy change at temperature T, $\Delta G°(T)$, of the transition by

$$K = \mathrm{e}^{-\Delta G°(T)/RT} \tag{5}$$

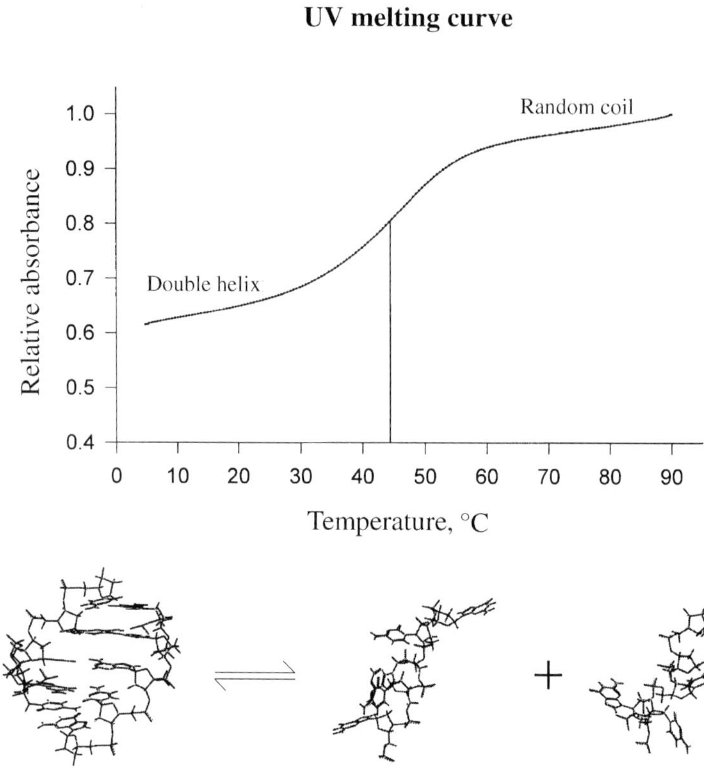

Figure 2 Typical melting curve for a double helix to random coil transition. The rate of heating must be much slower than the rate of conformational relaxation of the RNA, i.e., equilibrium is established at each temperature during measurement of the melting curve. The vertical line indicates the melting temperature, T_M, where half the strands are in double helix and half in random coil conformations.

Here, T is temperature in kelvins, $T = 273.15 + t$, where t is temperature in °C. The free energy change is related to the enthalpy and entropy changes, ΔH° and ΔS°, by:

$$\Delta G^\circ(T) = \Delta H^\circ - T\Delta S^\circ \tag{6}$$

The melting temperature, T_M, is defined as the temperature at which $\alpha = 1/2$. At the T_M, equilibrium constants are given by:

$$K_{T_M} = \frac{a}{C_T} \text{ (bimolecular)} \tag{7}$$

Equations (5)–(7) predict a concentration dependence of T_M for duplexes:

$$\frac{1}{T_M} = \frac{R}{\Delta H^\circ} \ln(C_T/a) + \frac{\Delta S^\circ}{\Delta H^\circ} \text{ (bimolecular)} \tag{8}$$

For unimolecular transitions, the corresponding equations are

$$K = \frac{\alpha}{1-\alpha},\ K_{T_M} = 1, \text{ and } T_M = \frac{\Delta H^\circ}{\Delta S^\circ} \text{ (unimolecular)} \tag{9}$$

Note that the T_M is concentration independent for unimolecular transitions.

2.2.3 Data Analysis

Experimental techniques for measuring melting curves have been described in detail.[64,66] Given an absorbance vs. temperature melting curve of a duplex to single strand transition like the one in Figure 2, thermodynamic information can be derived by analyzing α as a function of temperature. Under the two-state transition assumption, the measured extinction coefficient, $\varepsilon(T)$, at any temperature can be expressed as a mole-fraction weighted linear combination of two components, ε_{ss} and ε_{ds}:[66]

$$\varepsilon(T) = \varepsilon_{ss}(1-\alpha) + \varepsilon_{ds}\alpha \tag{10}$$

where ε_{ss} is the average extinction coefficient for the single-stranded states and ε_{ds} is the extinction coefficient per strand for the double-stranded state. Since lower and upper baselines for a typical melting curve are relatively straight, ε_{ss} and ε_{ds} are usually approximated as linear functions of temperature,[66]

$$\varepsilon_{ss} = m_{ss}T + b_{ss}, \quad \text{and} \quad \varepsilon_{ds} = m_{ds}T + b_{ds} \tag{11}$$

When melting temperatures are low, base stacking in the single strands can sometimes produce nonlinear upper baselines. For non-self-complementary duplexes, it is sometimes possible to separately measure the temperature dependence of ε_{ss} instead of using the linear approximation.[67,68] The fraction of strands in duplex, α, can be expressed as follows:

$$\alpha = \frac{\varepsilon_{ss}(T) - \varepsilon(T)}{\varepsilon_{ss}(T) - \varepsilon_{ds}(T)} \tag{12}$$

The parameter α as a function of temperature is related to ΔH° and ΔS° through the equilibrium constant K by Equation (4).

At high temperature, each strand can exist only in the random coil single-stranded state. Thus, the total single-strand concentration can be estimated from the absorbance at high temperature (normally >80 °C), and extinction coefficients of single strands calculated from published dimer and monomer extinction coefficients[69] using the nearest-neighbor approximation.[70] Thermodynamic parameters can be derived by using a nonlinear least-squares routine[71] to fit experimental curves to the two-state model (Equation (10)) with m_{ss}, m_{ds}, b_{ss}, b_{ds}, ΔH°, and ΔS° being the adjustable parameters.[29,66]

Thermodynamic parameters of duplex formation can be averaged over melting curves measured at different concentrations or obtained from plots of the reciprocal of the melting temperatures, T_M^{-1} vs. $\ln(C_T/a)$, (Equation (8)).[72] The data are normally taken as being consistent with a two-state transition if the ΔH° values calculated by the two analysis methods agree within 15%.[42,66,73,74] A 100-fold range in strand concentration is normally explored. Typical discrepancies in ΔH°, ΔS°, and ΔG°_{37} obtained from the two analysis methods are 5.8%, 6.5%, and 1.8%, respectively.[75] Note that the 15% criterion is a necessary but not sufficient condition for proving two-state behavior. The derived parameters are indirect and model dependent.

Methods for estimating errors in thermodynamic parameters have been described in detail.[75,76] Instrumental fluctuations contribute negligibly to the uncertainties. Standard deviations of parameters for single measurements are typically about 6.5%, 7.3%, and 2.4% for ΔH°, ΔS°, and ΔG°_{37}, respectively. Standard deviations for parameters calculated from Equation (8) based on 7–10 measurements are typically 2.9%, 3.3%, and 1.0% for ΔH°, ΔS°, and ΔG°_{37}, respectively. The relative uncertainty in ΔS° is usually about 13% larger than that in ΔH°, because ΔS° depends on more experimental parameters than ΔH°. Uncertainty in T_M is normally about 1.6 °C.[75] The errors in ΔG° and T_M are less than those in ΔH° and ΔS° because errors in ΔH° and ΔS° are highly correlated, with average observed correlation coefficients being greater than 0.999.[75] Thus, ΔG° and T_M are more accurate parameters than either ΔH° or ΔS° individually.[66,75–78]

2.2.4 Complications and Caveats

The above treatment assumes that ΔH° and ΔS° are temperature independent, which need not be the case. ΔH° and ΔS° will be temperature dependent if ΔC°_P, the difference in heat capacity (where $C^\circ_P = (\partial H^\circ/\partial T)_P$) between single- and double-stranded states, is not zero. A simulation of how a temperature-independent, nonzero ΔC°_P affects van't Hoff analyses[79] showed that a small ΔC°_P can make a hidden contribution to data analysis that biases the slope of van't Hoff plots. Since the curvature that should result from the small ΔC°_P is likely to be lost within the noise, this may lead to systematic errors in ΔH°_{vH}. In principle, one could explicitly include a nonzero ΔC°_P in the fitting function.[66,79] The error associated with the ΔC°_P, however, is likely to be as large as the parameter itself.[79]

While many assumptions and simplifications have been used in the analysis of RNA optical melting data, the results obtained have proven useful to predict the stabilities of many new sequences. Evidently, totally accurate values for the thermodynamic parameters are not required.

2.2.5 Calorimetry

Calorimetry is another technique for investigating the energetics of biomolecules. Experimental techniques for differential scanning calorimetry (DSC) and isothermal titration calorimetry (ITC) have been described in great detail.[80–83] Compared to the model-dependent $\Delta H^{\circ}_{\mathrm{vH}}$ values indirectly derived from the measurement of temperature-dependent spectroscopic properties, transition enthalpies determined calorimetrically do not depend on the nature of the transition. DSC measures the excess heat capacity, $\Delta C_{\mathrm{P}}^{\mathrm{ex}}$. From the $\Delta C_{\mathrm{P}}^{\mathrm{ex}}$ vs. temperature profile, one can obtain ΔH° and ΔS° directly, after subtracting baselines appropriately:

$$\Delta H^{\circ} = \int \Delta C_{\mathrm{P}}^{\mathrm{ex}} \, \mathrm{d}T \tag{13}$$

$$\Delta S^{\circ} = \int \frac{\Delta C_{\mathrm{P}}^{\mathrm{ex}}}{T} \, \mathrm{d}T \tag{14}$$

The shapes of such curves depend on the nature of the transitions they represent, but it is the area underneath them ($\Delta C_{\mathrm{P}}^{\mathrm{ex}}$ vs. T or $\Delta C_{\mathrm{P}}^{\mathrm{ex}}/T$ vs. T) that gives ΔH° and ΔS°.

If a transition actually proceeds in a two-state manner, ΔH° values determined from optical melting curves and calorimetry will agree. If intermediate states are significantly populated, a transition will be broadened and this will make the apparent $\Delta H^{\circ}_{\mathrm{vH}}$ smaller than the true ΔH° as determined calorimetrically. The ratio $\Delta H^{\circ}_{\mathrm{vH}}/\Delta H^{\circ}_{\mathrm{DSC}}$ provides a measure of the size of the cooperative unit, i.e., the fraction of the structure that melts cooperatively.[65] If the ratio is 1, then the transition is two-state; if the ratio is less than 1, then the transition involves intermediate states.There are, however, exceptions.[84] Calorimetry has not been as widely used as optical methods in studies of RNA because it requires more material. In DSC, moreover, errors in ΔH° and ΔS° appear to be uncorrelated. Thus, errors in ΔG° are larger for DSC than for optical experiments. For example, in one study, ΔG° values reported from calorimetric and optical melting of duplexes differ by 5 kcal mol^{-1}, which corresponds to more than a 1000-fold difference in equilibrium constant.[85] Methods have been developed, however, for using calorimetric and optical data simultaneously to determine thermodynamic parameters.[86,87]

2.2.6 Statistical Treatment of Transitions

The two-state assumption is normally only applicable to relatively short oligonucleotides (less than 20 base pairs). In long oligomers or polymers, the helix growth steps dominate the initiation step; therefore intermediate states are significantly populated, and the two-state model is no longer valid. Since the helix growth steps are unimolecular, the concentration dependence of T_{M}, which is characteristic of the multimolecular initiation step, is not observed with polymers. Even some short oligomer sequences do not have two-state transitions. For example, base pairs at the end of a double helix may open before central base pairs.[88] Statistical models must be used to analyze transitions that are not two-state.[53,54,88–95]

The general procedure for a statistical treatment is first to write the partition function q for the molecule's conformations, which by definition, contains a complete description of the thermodynamics of its transitions. From the partition function, the expected average properties of the system can be expressed as a function of relevant parameters like equilibrium constants. These parameters can then be extracted by fitting predictions derived from the partition function to the experimentally accessible data.

Assuming that an RNA sequence can adopt random coil and n different duplex conformations, the molecular partition function is

$$q = \sum_{i=0}^{n} \exp(-G_i^{\circ}/RT) \tag{15}$$

where G_i is the free energy of the ith conformation and the summation is over all possible conformations including all the different duplex conformations and the random coil conformation. If we set the free energy for random coil, G_0, at 0, and remove the contribution to the partition function from the random coil state, we are left with the conformational partition function, q_c,

$$q_c = q - 1 = \sum_{i=1}^{n} \exp\left(-\frac{\Delta G_i^\circ}{RT}\right) = \sum_{i=1}^{n} K_i \tag{16}$$

where ΔG_i is the free energy difference between the ith duplex conformation and random coil and K_i is the corresponding equilibrium constant.

A simple statistical model, the zipper model, is probably adequate for most transitions of small RNAs.[53,54,94,96] This model assumes that each residue exists in either a double helical or coil state, that initiation of a base-paired region can occur at any residue in the sequence, and that all of the base-paired residues occur contiguously in a single region, i.e., only one double helical region is allowed. To calculate K_i, we further assume that only perfectly aligned duplexes make significant contributions to q_c (perfectly matching zipper model); the equilibrium constant for initiating the duplex is $\kappa = \sigma \cdot s$, where s is the equilibrium constant for adding one base pair to an existing double helical region. To simplify the presentation, we assume that s is independent of sequence, although this is not generally true.

With these assumptions, the equilibrium constant for forming a duplex with j base pairs is κs^{j-1}. If we ignore the symmetry number for simplicity, the degeneracy of a duplex with j base pairs is $g_j(L) = L - j + 1$, where L is the length of the polymer. If the summation is taken over the energy levels instead of over the individual conformational states, then the conformational partition function becomes:

$$q_c = \sum_{j=1}^{L} g_j(L)\kappa s^{j-1} = \kappa\left[(L+1)\sum_{j=1}^{L} s^{j-1} - \sum_{j=1}^{L} j s^{j-1}\right] \tag{17}$$

The first term is just a simple geometric series,

$$\sum_{j=1}^{L} s^{j-1} = \frac{s^L - 1}{s - 1},$$

and the second term can be converted into one since it is equal to $\partial\left(\sum s^j\right)/\partial s$.[54] Thus the equation reduces to:

$$q_c = \kappa \frac{s^{L+1} - (L+1)s + L}{(s-1)^2} \tag{18}$$

The two-state assumption, where the only duplex conformation that contributes to q_c has L base pairs, corresponds to the condition of large s and finite L, i.e., short oligomers with favorable helix growth steps. In this case, the conformational partition function is just the equilibrium constant, $K_{eq} = q_c = \kappa s^{L-1}$.[54]

When analyzing UV melting data, the absorbance A is assumed to be linear in the fraction of total bases paired, X_b:

$$X_b = \frac{A - A_s}{A_d - A_s} \tag{19}$$

where A_s and A_d are the absorbances of the mixture of single strands and the pure fully-formed duplex, respectively. Since q_c is a summation over equilibrium constants, it can be written as

$$q_c = \frac{\sum_i [H_i]}{[SS]^2} = \frac{X}{2(1-X)^2 C_T} \quad \text{(self-complementary)} \tag{20}$$

$$q_c = \frac{\sum_i [H_i]}{[SS_1][SS_2]} = \frac{2X}{(1-X)^2 C_T} \quad \text{(non-self-complementary)} \tag{21}$$

where H_i are the possible helical conformations, SS are single stranded states, C_T is the total strand concentration, and X is the fraction of strands in all possible duplex states. The average number of base pairs per duplex, $\langle j \rangle$, can be obtained by calculating the expectation value of j,

$$\langle j \rangle = \frac{1}{q_c} \sum_{j=1}^{L} j \kappa g_j(L) s^{j-1} = \frac{1}{q_c} \frac{d}{ds} \left(\sum_{j=1}^{L} \kappa g_j(L) s^j \right) = \frac{1}{q_c} \frac{d(sq_c)}{ds} \tag{22}$$

This allows X_b to be expressed in terms of X and q_c:

$$X_b = \frac{X \langle j \rangle}{L} = \frac{X}{Lq_c} \frac{d(sq_c)}{ds} \tag{23}$$

We can substitute X in terms of q_c and C_T from Equations (20) and (21) and obtain

$$X_b = \frac{1 + 4q_cC_T - (1 + 8q_cC_T)^{1/2}}{4Lq_c^2C_T} \frac{d(sq_c)}{ds} \quad \text{(self-complementary)} \tag{24}$$

$$X_b = \frac{1 + q_cC_T - (1 + 2q_cC_T)^{1/2}}{Lq_c^2C_T} \frac{d(sq_c)}{ds} \quad \text{(non-self-complementary)} \tag{25}$$

Thus, κ and s can be determined by fitting to X_b, which is experimentally accessible by Equation (19).

Real transitions in RNA are usually more complicated than the zipper model proposes, e.g., more than one helical region may be present at the same time, and s is sequence dependent.[54] In that case, additional κ and s values have to be added. The number of parameters increases rapidly for realistic models. On the other hand, it is sometimes possible to reasonably approximate a transition with a single or a few intermediate states.[41,97]

2.3 THERMODYNAMICS OF RNA SECONDARY STRUCTURE MOTIFS

As discussed above, thermodynamic parameters for structural transitions can be determined by analyzing the melting behavior of RNA molecules. This information can be used to predict the thermodynamic properties of other RNA molecules of known sequence, if the measured energetics can be partitioned to separate contributions for various structural motifs. For example, the free energy change and therefore equilibrium constant (Equation (5)) for forming an ordered structure from a random coil can be predicted if it is assumed that the total free energy change is simply the sum of the free energy changes for forming the separate motifs.

2.3.1 Watson–Crick Helical Regions

Watson–Crick base pairs are the most abundant motif in RNA structures, typically accounting for about 50% of nucleotides in a sequence. They form the helical framework for RNA and are also central to RNA–RNA recognition processes.[1,98,99] Therefore, they are the most extensively studied motif. Measurements on the sequence dependence of energetics of Watson–Crick duplex formation by RNA oligonucleotides and polynucleotides show that base pair composition (mole fraction of AU and GC base pairs) cannot alone explain the thermodynamics.[72,100,101] Sequences with identical composition but different permutations of base pairs may have different free energy changes for duplex formation. For example, the duplexes $(5'\text{GUUCGAAC}3')_2$ and$(5'\text{UUGGCCAA}3')_2$ both have four AU and four GC base pairs, but have free energy changes of duplex formation at 37 °C of −8.8 and −11.0 kcal mol^{-1}, respectively.[49,75] This difference of 2.2 kcal mol^{-1}translates into a 35-fold difference in equilibrium constants for duplex formation at 37 °C (Equation(5)). Thus, stability depends on more than the number of hydrogen bonds formed. Presumably, vertical stacking interactions between neighboring base pairs are the sequence-dependent variable.

The sequence dependence of duplex formation involving only Watson–Crick base pairs is approximated well by a nearest-neighbor model.[42,72,74,75,100,101] Several such models have been proposed.[72,75,102]

The INN–HB (Individual Nearest Neighbor-Hydrogen Bonding) model of Xia *et al.*[75] assumes that the stability of a duplex depends on an initiation term for formation of the first pair, 10 helix propagation terms for the 10 possible nearest-neighbor combinations of adjacent Watson–Crick pairs, a base pair composition term that accounts for the number of hydrogen bonds in the duplex, and a symmetry term that is zero except when the duplex is self-complementary. The base composition term in this model is simply a term associated with each terminal AU pair, since duplexes with identical nearest neighbors but different base pair composition must have different numbers of terminal AU pairs.[75]

For example, the duplexes $\frac{5'\text{ACGCA}3'}{3'\text{UGCGU}3'}$ and $\frac{5'\text{GCACG}3'}{3'\text{CGUGC}5'}$ have the same nearest neighbors, but different base compositions and therefore different terminal base pairs. The measured free energies of duplex formation at 37 °C are −4.97 and −6.17 kcal mol^{-1}, respectively. Duplexes with terminal GC pairs are consistently more stable than duplexes with the same nearest neighbors but with terminal AU pairs. Thus, the free energy term for terminal AU pairs is unfavorable.[75] This INN–HB model is statistically identical to the ISS (Independent Short Sequence) nearest-neighbor model of Gray,[102] although the physical models differ.

In the INN–HB model, the free energy change for formation of a duplex with only Watson–Crick pairs is given by:

$$\Delta G^{\circ}\,(\text{duplex}) = \Delta G^{\circ}_{\text{init}} + \sum_{j} n_j \Delta G^{\circ}_j(\text{NN}) + m_{\text{term-AU}} \Delta G^{\circ}_{\text{term-AU}} + \Delta G_{\text{sym}} \quad (26)$$

Each $\Delta G^{\circ}_j(\text{NN})$ term is the free energy contribution of the jth nearest neighbor with n_j occurrences in the sequence. The $\Delta G^{\circ}_{\text{init}}$ term is the free energy of initiation. Among other factors, initiation includes translational and rotational entropy loss for converting two particles into one. This free energy of initiation is assumed to be independent of length, although this assumption is not rigorously correct. The parameters $m_{\text{term-AU}}$ and $\Delta G^{\circ}_{\text{term-AU}}$ are number of terminal AU pairs and the free energy parameter per terminal AU pair, respectively. The ΔG_{sym} term is required because there is two fold rotational symmetry in self-complementary duplexes, but not in single-stranded states or non-self-complementary duplexes. The maintenance of this symmetry makes self-complementary duplexes less favorable by a factor of two in conformational space,[54,103] i.e., there is an extra entropy loss of $R \ln 2 = 1.4$ eu upon duplex formation. At 37 °C, $\Delta G_{\text{sym}} = -T\Delta S = -(310.15)(-1.4) = 0.43$ kcal mol^{-1} for self-complementary duplexes and zero for non-self-complementary duplexes.[74,75] Equations similar to Equation (26) can be written for ΔH° and ΔS°, except there is no symmetry term for ΔH°.

Parameters for the INN–HB model are obtained by applying multiple linear regression analysis of experimental data to the regression function of Equation (26).[75] Parameters based on a set of 90 RNA duplexes with only Watson–Crick base pairs are listed in Table 1. On average, these parameters predict ΔG°_{37}, ΔH°, ΔS°, and T_{M} of duplex formation within 3.2%, 6.0%, 6.8%, and 1.3 °C. This accuracy is compatible with the limits of a nearest-neighbor model as determined from measurements on different sequences predicted to have identical thermodynamic parameters.[49,75,104,105] Although these parameters are derived from duplexes with two-state melting behavior, they can also provide useful approximations for RNA duplexes that are not two-state.[75]

From Table 1, the ΔG°_{37} values of AU-only, one-AU/one-GC, and GC-only nearest-neighbors have average values of 1.1, 2.2, and 3.0 kcal mol^{-1}. Evidently, base composition is an important factor. The ΔG°_{37} for $\frac{5'\text{GC}3'}{3'\text{CG}5'}$ propagation (−3.42 kcal mol^{-1}) is more favorable than for $\frac{5'\text{CG}3'}{3'\text{GC}5'}$ propagation (−2.36 kcal mol^{-1}) by more than 1 kcal mol^{-1}, suggesting that orientations of base pairs and therefore stacking patterns are also important. In the INN–HB model, each nearest-neighbor parameter includes contributions from stacking between the two base pairs of the doublet, and half of the hydrogen bonding interaction from each base pair.

2.3.2 GU Pairs

GU base pairs are the most common motif besides Watson–Crick base pairs. GU pairs are often conserved in RNA secondary structures, suggesting that they may play either a structural or functional role[106,107] rather than simply being replacements of Watson–Crick base pairs. For example, GU pairs are conserved in the P1 helix of group I introns[33] because they allow the formation of tertiary interactions with the intron's catalytic core.[108]

Table 1 Thermodynamic parameters for Watson–Crick[75] and GU[48] base pairs in 1 M NaCl, pH 7.

Parameters	ΔG°_{37} (kcal mol^{-1})	ΔH° (kcal mol^{-1})	$\Delta S^{\circ\,a}$ (eu)
Propagation of Watson–Crick nearest-neighbors			
5′AA3′ / 3′UU5′	−0.93	−6.82	−19.0
5′AU3′ / 3′UA5′	−1.10	−9.38	−26.7
5′UA3′ / 3′AU5′	−1.33	−7.69	−20.5
5′CU3′ / 3′GA5′	−2.08	−10.48	−27.1
5′CA3′ / 3′GU5′	−2.11	−10.44	−26.9
5′GU3′ / 3′CA5′	−2.24	−11.40	−29.5
5′GA3′ / 3′CU5′	−2.35	−12.44	−32.5
5′CG3′ / 3′GC5′	−2.36	−10.64	−26.7
5′GG3′ / 3′CC5′	−3.26	−13.39	−32.7
5′GC3′ / 3′CG5′	−3.42	−14.88	−36.9
Propagation of GU-containing nearest-neighbors			
5′GU3′ / 3′UG5′	1.29 (−1.06)[b]	−14.59 (−14.14)[b]	−51.2 (−42.2)[b]
5′GG3′ / 3′UU5′	0.47	−13.47	−44.9
5′UG3′ / 3′GU5′	0.30	−9.26	−30.8
5′AG3′ / 3′UU5′	−0.55	−3.21	−8.6
5′UG3′ / 3′AU5′	−1.00	−6.99	−19.3
5′GA3′ / 3′UU5′	−1.27	−12.83	−37.3
5′GU3′ / 3′UA5′	−1.36	−8.81	−24.0
5′CG3′ / 3′GU5′	−1.41	−5.61	−13.5
5′GG3′ / 3′CU5′	−1.53	−8.33	−21.9
5′GG3′ / 3′UC5′	−2.11	−12.11	−32.2
5′GC3′ / 3′UG5′	−2.51	−12.59	−32.5
Duplex parameters			
Initiation[c]	4.09	3.61	−1.5
Each terminal-AU or GU[d]	0.45	3.72	10.5
Symmetry correction[e]	0.43	0	−1.4

[a] Calculated from parameters for ΔG°_{37} and ΔH°. The units eu are cal K^{-1} mol^{-1}.

[b] Parameters in parentheses are for propagation of $\frac{5'\text{GU}3'}{3'\text{UG}5'}$ in context of $\frac{5'\text{GGUC}3'}{3'\text{CUGG}5'}$ only.

[c] Includes potential GC end effects.

[d] Parameter per terminal AU or GU pair.

[e] Only for self-complementary duplexes.

The thermodynamics of GU pairs have been systematically studied.[29,109–112] The results show that GU pairs are often thermodynamically similar to AU pairs. The data can be reasonably fit to a nearest-neighbor model except that the tandem $\frac{5'\text{GU}3'}{3'\text{UG}5'}$ motif has to be treated differently depending on context.[112] The $\frac{5'\text{GGUC}3'}{3'\text{CUGG}5'}$ motif is found to have remarkable stability compared to $\frac{5'\text{GU}3'}{3'\text{UG}5'}$ in other contexts. Parameters of GU-containing nearest-neighbors have been evaluated from a database of duplexes containing single or tandem GU pairs,[48] and are listed in Table 1. Note that it is assumed terminal GU pairs should be penalized the same as terminal AU pairs (0.45 kcal mol^{-1} per terminal GU pair) because both have only two hydrogen bonds.

The parameters in Table 1 allow reasonable predictions of the thermodynamic properties of RNA duplexes with Watson–Crick and GU pairs. Examples of applications of these parameters in predicting ΔG°, ΔH°, ΔS°, T_M, and dissociation constants, K_D, are shown in Figure 3 for a self-complementary and a non-self-complementary duplex.

2.3.3 Dangling Ends and Terminal Mismatches

The average length of an uninterrupted helix of Watson–Crick and GU base pairs in RNA secondary structures is about seven base pairs.[42] Therefore there are many helix ends in RNA secondary structures. Thus it is important to understand the effects of terminal unpaired nucleotides on the stability of a helix. A 5′ or 3′ single unpaired nucleotide is called a dangling end, and a pair of 5′ and 3′ unpaired nucleotides at the same helix end is called a terminal mismatch.

Effects of dangling ends and terminal mismatches can be directly determined by comparing stabilities of helices with and without them in model systems, i.e., they can be treated as additional nearest neighbors. For example, the effect of a 3′ dangling A is taken as half of the difference in free energy change of duplex formation between a duplex with two 3′ dangling As and the core duplex without a 3′ dangling A:[66]

$$\Delta G^\circ\,(3'\,\text{dangling A}) = \frac{1}{2}\left[\Delta G^\circ\begin{pmatrix}5'\text{GCGCA}\\ \text{ACGCG}5'\end{pmatrix} - \Delta G^\circ\begin{pmatrix}5'\text{GCGC}3'\\ 3'\text{CGCG}5'\end{pmatrix}\right] \tag{27}$$

The effect of 5′ dangling ends and terminal mismatches can be obtained similarly. Many of these parameters have been measured[48,66,110,113–119] and the values are summarized in Tables 2 and 3.

The parameters in Table 2 show that a 5′ dangling end contributes little to the stability of a duplex, averaging $\Delta G^\circ_{37} = -0.2$ kcal mol^{-1}. This is similar to the effect of a 5′ phosphate alone, probably

Table 2 Thermodynamic parameters for unpaired dangling nucleotides in 1 MNaCl, pH 7.[a,b,c]

	$X = A$			$X = C$			$X = G$			$X = U$		
	ΔH°	ΔS°	ΔG°_{37}	ΔH°	ΔS°	ΔG°_{37}	ΔH°	ΔS°	ΔG°_{37}	ΔH°	ΔS°	ΔG°_{37}
3′-Dangling nucleotides												
$\overrightarrow{\text{CX}}$ / G	−9.0	−23.4	−1.7	−4.1	−10.7	−0.8	−8.6	−22.2	−1.7	−7.5	−20.4	−1.2
$\overrightarrow{\text{GX}}$ / C	−7.4	−20.0	−1.1	−2.8	−7.9	−0.4	−6.4	−16.6	−1.3	−3.6	−9.7	−0.6
$\overrightarrow{\text{RX}}$ / U	−4.9	−13.2	−0.8	−0.9	−1.2	−0.5	−5.5	−15.0	−0.8	−2.3	−5.4	−0.6
$\overrightarrow{\text{UX}}$ / R	−5.7	−16.4	−0.7	−0.7	−1.8	−0.1	−5.8	−16.4	−0.7	−2.2	−6.8	−0.1
5′-Dangling nucleotides												
$\overrightarrow{\text{XC}}$ / G	−2.4	−6.0	−0.5	3.3	11.8	−0.3	0.8	3.4	−0.2	−1.4	−4.3	−0.1
$\overrightarrow{\text{XG}}$ / C	−1.6	−4.5	−0.2	0.7	3.1	−0.3	−4.6	−14.8	0.0	−0.4	−1.2	0.0
$\overrightarrow{\text{XR}}$ / U	−1.6	6.1	−0.3	2.2	7.9	−0.3	0.7	3.4	−0.4	3.1	10.6	−0.2
$\overrightarrow{\text{XU}}$ / R	−0.5	−0.7	−0.3	6.9	22.8	−0.1	(0.6)[c]	(2.7)[c]	(−0.2)[c]	(0.6)[c]	(2.7)[c]	(−0.2)[c]

[a] Parameters are in units of kcal mol^{-1} for ΔH° and ΔG°_{37} and eu for ΔS°.
[b] R is A or G. Dangling ends on AU have been measured and those on GU are estimated as the same.
[c] Parameters in parentheses are estimated.[203]

(a) Non-self-complementary duplex

$$\Delta G^\circ_{37}\begin{pmatrix}\text{5'AUGACU3'}\\\text{3'UACUGG5'}\end{pmatrix}=\Delta G^\circ_{\text{init}}+\Delta G^\circ_{37}\begin{pmatrix}\text{5'AU3'}\\\text{3'UA5'}\end{pmatrix}+\Delta G^\circ_{37}\begin{pmatrix}\text{5'UG3'}\\\text{3'AC5'}\end{pmatrix}+\Delta G^\circ_{37}\begin{pmatrix}\text{5'GA3'}\\\text{3'CU5'}\end{pmatrix}$$

$$+\Delta G^\circ_{37}\begin{pmatrix}\text{5'AC3'}\\\text{3'UG5'}\end{pmatrix}+\Delta G^\circ_{37}\begin{pmatrix}\text{5'CU3'}\\\text{3'GG5'}\end{pmatrix}+2\times\Delta G^\circ_{37}\begin{pmatrix}\text{5'A}/\text{G}\\\text{3'U}/\text{U}\end{pmatrix}+\Delta G_{\text{sym}}$$

$$=4.09+(-1.10)+(-2.11)+(-2.35)+(-2.24)+(-2.11)+2\times 0.45+0$$

$$=-4.92\ \text{kcal}\ \text{mol}^{-1}$$

$$\Delta H^\circ\begin{pmatrix}\text{5'AUGACU3'}\\\text{3'UACUGG5'}\end{pmatrix}=\Delta H^\circ_{\text{init}}+\Delta H^\circ\begin{pmatrix}\text{5'AU3'}\\\text{3'UA5'}\end{pmatrix}+\Delta H^\circ\begin{pmatrix}\text{5'UG3'}\\\text{3'AC5'}\end{pmatrix}+\Delta H^\circ\begin{pmatrix}\text{5'GA3'}\\\text{3'CU5'}\end{pmatrix}$$

$$+\Delta H^\circ\begin{pmatrix}\text{5'AC3'}\\\text{3'UG5'}\end{pmatrix}+\Delta H^\circ\begin{pmatrix}\text{5'CU3'}\\\text{3'GG5'}\end{pmatrix}+2\times\Delta H^\circ\begin{pmatrix}\text{5'A}/\text{G}\\\text{3'U}/\text{U}\end{pmatrix}$$

$$=3.61+(-9.38)+(-10.44)+(-12.44)+(-11.40)+(-12.11)+2\times 3.72$$

$$=-44.72\ \text{kcal}\ \text{mol}^{-1}$$

$$\Delta S^\circ\begin{pmatrix}\text{5'AUGACU3'}\\\text{3'UACUGG5'}\end{pmatrix}=\Delta S^\circ_{\text{init}}+\Delta S^\circ\begin{pmatrix}\text{5'AU3'}\\\text{3'UA5'}\end{pmatrix}+\Delta S^\circ\begin{pmatrix}\text{5'UG3'}\\\text{3'AC5'}\end{pmatrix}+\Delta S^\circ\begin{pmatrix}\text{5'GA3'}\\\text{3'CU5'}\end{pmatrix}$$

$$+\Delta S^\circ\begin{pmatrix}\text{5'AC3'}\\\text{3'UG5'}\end{pmatrix}+\Delta S^\circ\begin{pmatrix}\text{5'CU3'}\\\text{3'GG5'}\end{pmatrix}+2\times\Delta S^\circ\begin{pmatrix}\text{5'A}/\text{G}\\\text{3'U}/\text{U}\end{pmatrix}+\Delta S_{\text{sym}}$$

$$=-1.5+(-26.7)+(-26.9)+(-32.5)+(-29.5)+(-32.2)+2\times 10.5+0$$

$$=-128.3\ \text{eu}$$

$$T_M=\frac{\Delta H^\circ}{\Delta S^\circ+R\ln(C_T/4)}=\frac{-44.72\times 1000}{-128.3+1.987\times\ln\left(2\times 10^{-4}/4\right)}=302.2\ \text{K}=29.1\ ^\circ\text{C}$$

$$K_D=\exp(\Delta G^\circ_{37}/RT)=0.34\ \text{mM}$$

Figure 3 Calculation of predicted thermodynamic properties for duplexes with Watson–Crick and GU base pairs. (a) Non-self-complementary duplex from binding of a short oligonucleotide (5′AUGACU3′) to the internal guide sequence (5′GGUCAU3′) from a group I intron of *Pneumocystis carinii*,[177] T_M is calculated for total strand concentration of 2×10^{-4} M.

because a 5′ phosphate restricts the conformation available to the following sugar. Evidently, 5′ dangling nucleotides do not interact with the adjacent helix.[116] In contrast, 3′ dangling end effects are very sequence dependent, ranging from -0.1 to -1.7 kcal mol^{-1}, with dangling purines more favorable than pyrimidines. Certain 3′ dangling ends can stabilize a helix almost as much as a base pair. For example, a 3′ dangling A adjacent to a terminal CG pair, $\begin{smallmatrix}\text{5'CA3'}\\\text{3'G}\end{smallmatrix}$, stabilizes a duplex by 1.7 kcal mol^{-1}, compared to the nearest-neighbor parameter of -2.1 kcal mol^{-1} for $\begin{smallmatrix}\text{5'CA3'}\\\text{3'GU5'}\end{smallmatrix}$.

The difference between stabilizing increments of 5′ and 3′ dangling ends can be rationalized by structural considerations.[44,116] In A-form geometry, a 5′ dangling base is not close to a base on the opposite strand, while a 3′ dangling base can stack directly on the base of the opposite strand in the terminal pair, thus helping to hold the duplex together. Dangling end effects can help rationalize and predict local three-dimensional structure.[42,113,117,120] Comparisons of free energy increments measured

(b) Self-complementary duplex

$$\Delta G^\circ_{37}\begin{pmatrix}\text{5'UGGCCA3'}\\ \text{3'ACCGGU5'}\end{pmatrix} = \Delta G^\circ_{\text{init}} + 2\times\Delta G^\circ_{37}\begin{pmatrix}\text{5'CA3'}\\ \text{3'GU5'}\end{pmatrix} + 2\times\Delta G^\circ_{37}\begin{pmatrix}\text{5'GG3'}\\ \text{3'CC5'}\end{pmatrix}$$

$$+\Delta G^\circ_{37}\begin{pmatrix}\text{5'GC3'}\\ \text{3'CG5'}\end{pmatrix} + 2\times\Delta G^\circ_{37}\begin{pmatrix}\text{A3'}\\ \text{U5'}\end{pmatrix} + \Delta G_{\text{sym}}$$

$$= 4.09 + 2\times(-2.11) + 2\times(-3.26) + (-3.42) + 2\times 0.45 + 0.43$$

$$= -8.74 \text{ kcal mol}^{-1}$$

$$\Delta H^\circ\begin{pmatrix}\text{5'UGGCCA3'}\\ \text{3'ACCGGU5'}\end{pmatrix} = \Delta H^\circ_{\text{init}} + 2\times\Delta H^\circ\begin{pmatrix}\text{5'CA3'}\\ \text{3'GU5'}\end{pmatrix} + 2\times\Delta H^\circ\begin{pmatrix}\text{5'GG3'}\\ \text{3'CC5'}\end{pmatrix}$$

$$+\Delta H^\circ\begin{pmatrix}\text{5'GC3'}\\ \text{3'CG5'}\end{pmatrix} + 2\times\Delta H^\circ\begin{pmatrix}\text{A3'}\\ \text{U5'}\end{pmatrix}$$

$$= 3.61 + 2\times(-10.44) + 2\times(-13.39) + (-14.88) + 2\times 3.72$$

$$= -51.49 \text{ kcal mol}^{-1}$$

$$\Delta S^\circ\begin{pmatrix}\text{5'UGGCCA3'}\\ \text{3'ACCGGU5'}\end{pmatrix} = \Delta S^\circ_{\text{init}} + 2\times\Delta S^\circ\begin{pmatrix}\text{5'CA3'}\\ \text{3'GU5'}\end{pmatrix} + 2\times\Delta S^\circ\begin{pmatrix}\text{5'GG3'}\\ \text{3'CC5'}\end{pmatrix}$$

$$+\Delta S^\circ\begin{pmatrix}\text{5'GC3'}\\ \text{3'CG5'}\end{pmatrix} + 2\times\Delta S^\circ\begin{pmatrix}\text{A3'}\\ \text{U5'}\end{pmatrix} + \Delta S_{\text{sym}}$$

$$= -1.5 + 2\times(-26.9) + 2\times(-32.7) + (-36.9) + 2\times 10.5 + (-1.4)$$

$$= -138.0 \text{ eu}$$

$$T_M = \frac{\Delta H^\circ}{\Delta S^\circ + R\ln(C_T)} = \frac{-51.49\times 1000}{-138.0 + 1.987\times\ln(1\times 10^{-4})} = 329.4\text{ K} = 56.3\,^\circ\text{C}$$

$$K_D = \exp(\Delta G^\circ_{37}/RT) = 0.69\ \mu\text{M}$$

Figure 3 (continued) (b) Self-complementary duplex, T_M is calculated for total strand concentration of 1×10^{-4} M. Note that thermodynamic parameters are given for association, but the dissociation constant $K_D = K_A^{-1}$, the inverse of the association constant.

for dangling nucleotides in oligonucleotides with X-ray structures of large RNAs show that dangling ends that favor oligonucleotide duplex formation by more than 0.8 kcal mol^{-1} in short RNA duplexes are usually stacked in structures of large RNAs. Dangling ends that favor duplex formation by less than 0.4 kcal mol^{-1} in short RNA duplexes are usually unstacked in large RNAs unless they are in GA pairs. Unstacked dangling ends can also allow the RNA backbone to make a turn.

Stability increments for terminal mismatches are essentially the sum of increments for the constituent dangling ends (Table 3) when they are pyrimidine–pyrimidine and CA or AC mismatches. For purine–purine mismatches, the increment is less than the sum of constituent dangling ends, indicating that they do not interact synergistically to increase helix stability.

To use the parameters in Tables 2 and 3 to calculate the thermodynamic properties of duplexes with dangling ends or terminal mismatches, simply add the terms for dangling ends or terminal mismatches to the duplex parameters predicted by the INN–HB model using Table 1.[75] Note that when dangling ends or terminal mismatches follow terminal AU or GU pairs, the penalty for a terminal AU is still applied.

Table 3 Thermodynamic parameters for terminal mismatches in 1 M NaCl, pH 7.[a,b]

Base pair	$X\downarrow$	$Y\rightarrow$	*A*	*C*	*G*	*U*
5′AX3′		ΔH°	−3.9	(2.0)	(−3.5)	
3′UY5′	A	ΔS°	−10.2	(9.6)	(−8.7)	–
		ΔG°_{37}	−0.8	(−1.0)	(−0.8)	
		ΔH°	−2.3	(6.0)		(−0.3)
	C	ΔS°	−5.3	(21.6)	–	(1.5)
		ΔG°_{37}	−0.6	(−0.7)		(−0.7)
		ΔH°	−3.1		(−3.5)	
	G	ΔS°	−7.3	–	(−8.7)	–
		ΔG°_{37}	−0.8		(−0.8)	
		ΔH°		(4.6)		(−1.7)
	U	ΔS°	–	(17.4)	–	(−2.7)
		ΔG°_{37}		(−0.8)		(−0.8)
5′CX3′		ΔH°	−9.1	−5.6	−5.6	
3′GY5′	A	ΔS°	−24.5	−13.5	−13.4	–
		ΔG°_{37}	−1.5	−1.5	−1.4	
		ΔH°	(−5.7)	(−3.4)		−2.7
	C	ΔS°	(−15.2)	(−7.6)	–	−6.3
		ΔG°_{37}	(−1.0)	(−1.1)		−0.8
		ΔH°	−8.2		−9.2	
	G	ΔS°	−21.8	–	−24.6	–
		ΔG°_{37}	−1.4		−1.6	
		ΔH°		−5.3		−8.6
	U	ΔS°	–	−12.6	–	−23.9
		ΔG°_{37}		−1.4		−1.2
5′GX3′		ΔH°	−5.2	(−4.0)	−5.6	
3′CY5′	A	ΔS°	−13.2	(−8.2)	−13.9	
		ΔG°_{37}	−1.1	(−1.5)	−1.3	
		ΔH°	−7.2	(0.5)		(−4.2)
	C	ΔS°	−19.6	(3.9)	–	(−12.2)
		ΔG°_{37}	−1.1	(−0.7)		(−0.5)
		ΔH°	−7.1		−6.2	
	G	ΔS°	−17.8	–	−15.1	–
		ΔG°_{37}	−1.6		−1.4	
		ΔH°		(−0.3)		(−5.0)
	U	ΔS°	–	(−2.1)		(−14.0)
		ΔG°_{37}		(−1.0)		(−0.7)
5′GX3′		ΔH°	3.4	(2.0)	(−3.5)	
3′UY5′	A	ΔS°	10	(9.6)	(−8.7)	–
		ΔG°_{37}	0.3	(−1.0)	(−0.8)	
		ΔH°	(−2.3)	(6.0)		(−0.3)
	C	ΔS°	(−5.3)	(21.6)	–	(1.5)
		ΔG°_{37}	(−0.6)	(−0.7)		(−0.7)
		ΔH°	0.6		(−3.5)	
	G	ΔS°	0.0	–	(−8.7)	–
		ΔG°_{37}	0.6		(−0.8)	
		ΔH°		(4.6)		(−1.7)
	U	ΔS°	–	(17.4)	–	(−2.7)
		ΔG°_{37}		(−0.8)		(−0.8)
5′UX3′		ΔH°	−4.0	−6.3	−8.9	
3′AY5′	A	ΔS°	−9.7	−17.7	−25.2	–
		ΔG°_{37}	−1.0	−0.8	−1.1	
		ΔH°	−4.3	−5.1		−1.8
	C	ΔS°	−11.6	−14.6	–	−4.2
		ΔG°_{37}	−0.7	−0.6		−0.5
		ΔH°	−3.8		−8.9	
	G	ΔS°	−8.5	–	−25.0	–
		ΔG°_{37}	−1.1		−1.2	
		ΔH°		−1.4		1.4
	U	ΔS°	–	−2.5	–	6.0
		ΔG°_{37}		−0.6		−0.5

continued

Table 3 (continued)

Base pair	*X* ↓	*Y* →	*A*	*C*	*G*	*U*
5′UX3′		ΔH^o	−4.8	(−6.3)	(−8.9)	
3′GY5′	A	ΔS^o	−12.1	(−17.7)	(−25.2)	–
		ΔG^o_{37}	−1.0	(−0.8)	(−1.1)	
		ΔH^o	(−4.3)	(−5.1)		(−1.8)
	C	ΔS^o	(−11.6)	(−14.6)	–	(−4.2)
		ΔG^o_{37}	(−0.7)	(−0.6)		(−0.5)
		ΔH^o	−3.1		1.5	
	G	ΔS^o	−11.2	–	2.1	–
		ΔG^o_{37}	0.5		0.8	
		ΔH^o		(−1.4)		(1.4)
	U	ΔS^o	–	(−2.5)	–	(6.0)
		ΔG^o_{37}		(−0.6)		(−0.5)

[a] Parameters are listed in units of kcal mol^{-1}, eu, and kcal mol^{-1} for ΔH^o, ΔS^o, and ΔG^o_{37}, respectively. Blanks are Watson–Crick or GU nearest-neighbor parameters.
[b] Parameters in parentheses are estimated from measured numbers.[48,203]

2.3.4 Loops

RNA loops are regions of sequence not involved in canonical pairs but are flanked by one or more canonical paired regions. Here a canonical pair is defined as a Watson–Crick or wobble GU pair. The types of loops common in RNA are illustrated in Figure 1. The thermodynamics of RNA loops have not been fully investigated due to their enormous sequence diversity. With more and more experimental data available, however, our understanding of their contributions increases and models for approximating their stabilities are becoming more and more realistic.

2.3.4.1 Hairpin loops

Hairpin loops occur when nucleic acid strands fold back on themselves to make base pairs. Hairpin loops are a large part of RNA secondary structure. For example, nearly 70% of the small subunit rRNA of *Escherichia coli* is found in small stem-hairpin loop structures. Hairpins can provide nucleation sites for overall three-dimensional folding, and be involved in tertiary interactions.[18,30,33,121,122]

Hairpin loops in RNA can be very large, or, in at least one case, as small as two nucleotides.[122] Tetraloops (hairpins with four nucleotides) are the predominant hairpin loops in ribosomal RNAs, with GNRA and UNCG being the most common sequences,[123] where N is any nucleotide and R is purine. In general, stabilities of hairpins depend on the stem, the first mismatch on the closing base pair, the size of the loop, and occasionally intraloop interactions dependent on sequence. These factors have been extensively studied.[21,119,124–128] The free energy of loop formation is obtained by measuring the free energy of forming the entire hairpin with the stem, and subtracting out the contribution from the stem, as approximated by the parameters of the INN–HB model, including GU parameters (Table 1), but without the duplex initiation or symmetry term. Note that hairpin stems with an AU or GU base pair at either end of the stem are penalized by the terminal AU term. The free energy calculated for the hairpin loop ΔG^o_{HL} is assumed to be the sum of various interactions:

$$\Delta G^o_{\mathrm{HL}} = \Delta G^o_{\mathrm{hairpin}} - \Delta G^o_{\mathrm{stem}} = \Delta G^o_{\mathrm{init}}(n) + \Delta G^o\,(\text{first mismatch}) + \Delta G^o\,(\text{bonus/penalty}) \quad (28)$$

The contributions of first mismatches are approximated by the values for terminal mismatches (Table 3), except for loops smaller than four nucleotides which are too constrained to allow the same stacking possible for terminal mismatches at the end of a duplex. The relative sequence independence of stability for loops of three is consistent with this model.[125] First mismatches of GA or UU stabilize more than terminal mismatches, so a bonus of −0.8 kcal mol^{-1} is added.[48] The free energy for hairpin loop initiation is unfavorable, due primarily to the unfavorable entropy associated with constraining the nucleotides in the loop. The initiation values depend on loop length n and are listed in Table 4.[48]

Several effects can be included in the ΔG^o(bonus/penalty) term. For example, Giese *et al.*[119] found that hairpins closed by GU have an enhanced stability of 2.1 kcal mol^{-1} if the G is directly

Table 4 Free energy changes at 37 °C (kcal mol^{-1}) for initiation of various loops in 1 M NaCl, pH 7.

Types of loop	*Number of nucleotides in loop*								
	1	*2*	*3*	*4*	*5*	*6*	*7*	*8*	*9*
Hairpin loops [a]	–	–	5.7	5.6	5.6	5.4	5.9	5.6	6.4
Bulge loops [b]	3.8 [b]	2.8	3.2	(3.6) [b]	(4.0) [b]	(4.4) [b]	d	d	d
Internal loops	–	–	–	1.7 [c]	1.8	2.0	d	d	d

[a] Hairpin loops with less than three nucleotides are prohibited. For hairpin loops larger than nine nucleotides, the initiation can be approximated by $\Delta G°(n > 9) = \Delta G^{\circ}_{\text{init}H}(9) + 1.75RT\ \ln(n/9)$.[204]

[b] For a bulge loop of 1, stacking of adjacent Watson–Crick or GU pairs has to be included. Based on limited experimental data, the free energies are made 0.4 kcal mol^{-1} less favorable for the next larger size from 4 to 6. Larger bulge and internal loops can be approximated by $\Delta G°(n > 6) = \Delta G^{\circ}_{\text{init}B}(6) + 1.75RT\ \ln(n/6)$.[204]

[c] Value for loops of 4 are for 3 × 1 internal loops only, 2 × 2 internal loops are treated differently (Tables 5 and 6).

[d] Larger internal and bulge loops can be approximated by $\Delta G°(n > 6) = \Delta G^{\circ}_{\text{init}I}(6) + 1.75RT\ \ln(n/6)$.[204]

preceded by two Gs.[48] Interestingly, hairpins closed by GU are often preceded by two Gs in known secondary structures. Some tetraloops also have enhanced stability that can be added as part of the $\Delta G°$(bonus/penalty) term. The UNCG tetraloops are the most stable, but GNRA loops are also somewhat more stable than random sequence tetraloops.[126–129] The enhanced stability can be attributed to hydrogen bonding in the loop.[19–21,25,26,128,130–133] Tetraloops are sometimes involved in tertiary interactions.[14,16,18,33,134–136] Bonuses for specific tetraloops based on their phylogenetic occurrences are sometimes applied in RNA secondary structure prediction to mimic the effects of such tertiary interactions.[48] Poly-C hairpins are less stable than other hairpin loops of the same size.[137] When $n > 3$, the penalty can be fit to a linear equation: $\Delta G°(n) = An + B$, where A and B are found to be 0.3 and 1.6 kcal mol^{-1}, respectively.[48] When $n = 3$, this penalty is 1.4 kcal mol^{-1}.

With the above parameters, the thermodynamic properties of hairpins can be approximated. Calculations of ΔG°_{37}, $\Delta H°$, $\Delta S°$, K, and T_M for a hairpin are illustrated in Figure 4. The hairpin transition is unimolecular, therefore the T_M is independent of concentration.

2.3.4.2 Bulge loops

Bulge loops have unpaired nucleotides on only one strand of a double helix (Figure 1). Bulge loops can be either extrahelical or intrahelical. In natural RNA, one-nucleotide bulges are the most common bulge loops, and single purine bulges tend to stack in the helix and bend the helix, whereas single pyrimidine bulges are extra helical.[138–140] Single bulged As are known to be important for protein binding and perhaps for tertiary folding.[141–143] It is thought that single nucleotide bulges do not interrupt nearest-neighbor stacking, while larger bulges do interrupt stacking.[51,144] Bulges become more destabilizing as the number of nucleotides in the bulge increases.[68,144] Due to lack of data, the destabilizing effects of bulge loops are taken as sequence independent.[68,145,146] Free energy parameters for bulge loops of different size are given in Table 4.[48]

2.3.4.3 Internal loops

Internal loops are flanked by two helices with canonical pairs, and contain nucleotides on both strands that are not in canonical pairs (Figure 1). RNA internal loops may play important roles in tertiary interactions[18,33,134] and in protein recognition.[34] Stabilities of internal loops are very dependent on the identity and orientation of closing base pairs, on the sequence in the loop, and on the size and symmetry of the loop. The contributions of internal loops can be estimated by comparing stabilities of duplexes with and without the loop, corrected for the Watson–Crick nearest-neighbor interaction absent in the duplex with the loop and present in the duplex without the loop:[88]

$$\Delta G^{\circ}_{37}\ (\text{internal loop}) = \Delta G^{\circ}_{37}\ (\text{duplex w/ loop}) - \Delta G^{\circ}_{37}\ (\text{duplex w/o loop}) + \Delta G^{\circ}_{37}(\text{NN}) \quad (29)$$

where ΔG°_{37}(NN) is the relevant nearest-neighbor parameter of the INN–HB model (Table 1). This assumes that the loop does not affect the regions beyond the closing base pairs. Compared to the

Hairpin

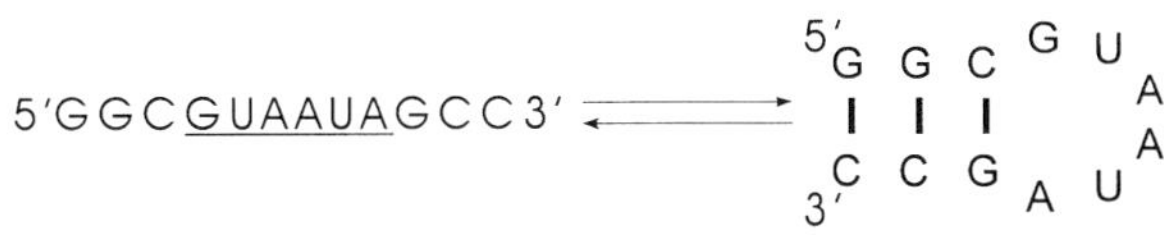

$$\Delta G^\circ_{37}(\text{Hairpin}) = \Delta G^\circ_{init.}(\text{n}=6) + \Delta G^\circ_{37}(\text{First Mismatch}) + \Delta G^\circ_{37}(\text{Stem})$$

$$= 5.4 + (-1.4) + (-0.8) + (-3.26) + (-3.42) = -3.48 \ \text{kcal/mol}$$

$$\Delta H^\circ(\text{Hairpin}) = \Delta H^\circ_{init.}(\text{n}=6) + \Delta H^\circ(\text{First Mismatch}) + \Delta H^\circ(\text{Stem})$$

$$= 0.0 + (-8.2) + (-0.8) + (-13.39) + (-14.88) = -37.27 \ \text{kcal/mol}$$

$$\Delta S^\circ(\text{Hairpin}) = \Delta S^\circ_{init.}(\text{n}=6) + \Delta S^\circ(\text{First Mismatch}) + \Delta S^\circ(\text{Stem})$$

$$= -17.4 + (-21.8) + (-32.7) + (-36.9) = -108.8 \ \text{eu}$$

$$T_M = \frac{\Delta H^\circ}{\Delta S^\circ} = \frac{(-37.27)\times 1000}{(-108.8)} = 342.6 \ K = 69.4 \ ^\circ\text{C}$$

$$K = \exp(-\Delta G^\circ_{37}/RT) = 283.4$$

Figure 4 Calculation of predicted thermodynamic properties for a hairpin that is highly conserved in the L11 protein binding region of the large subunit ribosomal RNA.[27] Nucleotides in the loop are underlined. Note that the bonus for the GA mismatch is approximated as completely due to a favorable ΔH° and that ΔH° for hairpin initiation is assumed to be zero. The measured values of ΔG°_{37}, ΔH°, ΔS°, and T_M are -3.42 kcal mol^{-1}, -36.9 kcal mol^{-1}, 107.8 eu, and 68.7 °C, respectively.[124]

enormous possible sequence dependence of internal loops, available experimental data are only sufficient for making rough approximations, and future improvements are therefore expected.

(i) Single mismatches (1 × 1 internal loops)

Single mismatches in a helix are the smallest internal loops. A few single mismatches have been studied.[67,88,147] Most single mismatches that have been studied destabilize duplexes, but some are stabilizing. A reasonable approximation for single mismatches with two adjacent GC pairs is that $\Delta G^\circ_{37} = 0.4$ kcal mol^{-1} when the mismatch is not GG and $\Delta G^\circ_{37} = -1.7$ kcal mol^{-1} when the mismatch is GG.[148] The stability is less favorable by roughly 0.65 kcal mol^{-1} per adjacent AU or GU pair, largely due to the 0.45 kcal mol^{-1} term assigned to such "terminal" pairs of a helix.

(ii) Tandem mismatches (2 × 2 internal loops)

Tandem mismatches are formed when there are two opposing nucleotides on each strand that are not involved in Watson–Crick or GU pairs. These 2 × 2 internal loops can be symmetric or nonsymmetric in terms of loop sequence and closing base pairs. Tandem mismatches are the most extensively studied internal loops,[76,77,149–151] and the results show that their stabilities are very sequence dependent.

Symmetric tandem mismatches contain adjacent identical mismatches closed by the same base pair,

Table 5 Free energy changes for symmetric tandem mismatches $\frac{\text{5}'\text{PXYQ3}'}{\text{3}'\text{QYXP5}'}$ in 1 M NaCl, pH 7.[a,b]

Mismatches *XY* → *YX*; *Closing base pairs* *P* ↓ *Q*	*UG GU*	*GU UG*	*GA AG*	*AG GA*	*UU UU*	*GG GG*	*CA AC*	*CU UC*	*UC CU*	*CC CC*	*AC CA*	*AA AA*
G/C	−4.9	−4.1	−2.6	−1.3	−0.5	(0.8)	1.0	1.1	(1.0)	(1.0)	0.9	1.5
C/G	−4.2	−1.1	−0.7	−0.7	−0.4[c]	0.8[c]	1.1[c]	1.4[c]	1.4[c]	1.7[c]	2.0[c]	1.3
U/A	−2.6	−0.3	0.7	0.9[e]	1.1	2.3[d]	1.9	2.2	2.8	(2.8)	(2.8)	2.8
A/U	−1.9	0.2	0.3	1.7[e]	0.6	1.9[d]	2.3	(2.2)	(2.2)	(2.2)	2.5	2.8

[a] Parameters are based on results of Wu *et al.*[151] and references cited therein except for $\frac{\text{GG}}{\text{GG}}$. Free energy changes are calculated by Equation (29).
[b] Parameters in parentheses are estimated.
[c] Ref.[149].
[d] Burkard *et al.*.[206]
[e] S.J. Schroeder and D.H. Turner, submitted.

e.g., $\frac{\text{5}'\text{GGAC3}'}{\text{3}'\text{CAGG5}'}$, $\frac{\text{5}'\text{CUUG3}'}{\text{3}'\text{GUUC5}'}$, etc. The dependence of stabilities of symmetric tandem mismatches and tandem GU pairs on closing base pairs and loop sequences fits a pattern,[151] as shown in Table 5, which lists values recalculated[48] with parameters of the INN–HB model. Loops closed by GC base pairs are more stable than those closed by AU base pairs. The decreasing order of stabilities of mismatches is

$$\frac{\text{UG}}{\text{GU}} > \frac{\text{GU}}{\text{UG}} > \frac{\text{GA}}{\text{AG}} \geq \frac{\text{AG}}{\text{GA}} > \frac{\text{UU}}{\text{UU}} > \frac{\text{GG}}{\text{GG}} \approx \frac{\text{CA}}{\text{AC}} \approx \frac{\text{CU}}{\text{UC}} \approx \frac{\text{UC}}{\text{CU}} \approx \frac{\text{CC}}{\text{CC}} \approx \frac{\text{AC}}{\text{CA}} \approx \frac{\text{AA}}{\text{AA}}.$$

Including tandem UG sequences, the range in free energy changes is −4.9 to 2.8 kcal mol^{-1}, corresponding to about a 270 000-fold range in equilibrium constants for folding at 37 °C. GC closing base pairs give enhanced stability over CG closing pairs, especially with GA mismatches. For example, $\frac{\text{5}'\text{GGAC3}'}{\text{3}'\text{CAGG5}'}$ is almost 2 kcal mol^{-1} more stable than $\frac{\text{5}'\text{CGAG3}'}{\text{3}'\text{GAGC5}'}$. Tandem UG, GA, and UU mismatches can be stabilizing, partly due to two hydrogen bonds between mismatched bases as evidenced by NMR spectra.[23,29,32,149–153] Other mismatches, CA, CU, CC, and AA are destabilizing and there is no NMR evidence for two hydrogen bonds between mismatched bases, suggesting they are more flexible. Formation of these destabilizing tandem mismatches is also associated with a more favorable entropy change than for GU, GA, or UU, consistent with greater flexibility.[151]

There are over 2000 possible tandem mismatch sequences, most of which are nonsymmetric, where two different mismatches are adjacent or the two closing base pairs are different.[77] At present, it is impossible to study exhaustively all the sequence dependence of tandem mismatch stability. Thus it is necessary to identify the most important factors that determine the stability of nonsymmetric tandem mismatches and develop a model that can reasonably approximate their stabilities. Xia *et al.*[77] studied a series of nonsymmetric tandem mismatches in the context of $\frac{\text{5}'\text{GXYG3}'}{\text{3}'\text{CWZC5}'}$, where XW and YZ can be either the same or different mismatches. The results are more complicated than for symmetric tandem mismatches.

The stabilities of nonsymmetric tandem mismatches cannot be predicted by simply averaging values for symmetric tandem mismatches.[77] The contribution of one mismatch to the free energy increment for nonsymmetric tandem mismatch formation depends on the identity of the other mismatch. This is partly because the structure of a mismatch is dependent on the structure of the adjacent mismatch and partly because mismatches can have different sizes, e.g., purine–purine, purine–pyrimidine, pyrimidine–pyrimidine. Comparison of stabilities of nonsymmetric tandem mismatches to the average values for symmetric tandem mismatches reveals a pattern. In general, it is more favorable to have two stabilizing mismatches of the same size adjacent to each other than to have two stabilizing mismatches of different sizes. For example, two adjacent GA mismatches stabilize the helix regardless of their orientations. In contrast, GA and UU are two stabilizing mismatches of very different sizes, and any combination of

Table 6 Penalty terms for nonsymmetric tandem mismatches.[a]

Category of combinations[b]	*Penalty* (Δ_P) (kcal mol^{-1})
Two stabilizing mismatches of different size	1.8
One stabilizing and one destabilizing mismatch (not AC) of different size	1.0
Other combinations	0.0

[a] To be used in Equation (30). Penalties are the same for GC and AU closing pairs.
[b] See text.

them destabilizes the helix.[77]

From these results, a model has been developed that can approximate the stabilities of any tandem mismatch. The general equation is:[77]

$$\Delta G^{\circ}\begin{pmatrix}5'\text{PXYS}3'\\3'\text{QWZT}5'\end{pmatrix} = \frac{1}{2}\left[\Delta G^{\circ}\begin{pmatrix}5'\text{PXWQ}3'\\3'\text{QWXP}5'\end{pmatrix} + \Delta G^{\circ}\begin{pmatrix}5'\text{TZYS}3'\\3'\text{SYZT}5'\end{pmatrix}\right] + \Delta_P \quad (30)$$

where PQ and ST are the closing base pairs, and XW and YZ are any mismatch combinations. The free energy changes in the bracket are values for symmetric tandem mismatches (Table 5). Δ_P is a penalty term that depends on the identity of the mismatches. In general, combinations of two different stabilizing mismatches have larger penalties, and combinations of destabilizing mismatches have smaller penalties. These penalties are collected in Table 6.[48]

(iii) 2 × 1 internal loops

Internal loops of three contain two unpaired nucleotides opposing one nucleotide. Schroeder *et al.*[154] investigated internal loops of three closed by GC base pairs, and found a considerable sequence dependence to the stabilities. Loops with potential for forming GA and UU mismatches generally have more favorable free energies of formation than loops without such potential. For example, $\begin{smallmatrix}\text{C A C}\\\text{GAAG}\end{smallmatrix}$ is 1.2 kcal mol^{-1} less stable at 37 °C than $\begin{smallmatrix}\text{C A C}\\\text{GGAG}\end{smallmatrix}$. This enhancement, however, depends on the orientation of the closing base pairs and of the potential GA mismatch in the loop. Loops without potential GA or UU mismatches fall in a relatively narrow range with an average free energy of loop formation of 2.2 kcal mol^{-1}. This trend is similar to the results for symmetric tandem mismatches. The results allow approximations for loops that have not been measured.[48,154] These approximations are listed along with measured values in Table 7.

(iv) Other asymmetric internal loops

Asymmetric internal loops containing only adenines[67] or cytosines[144] have been studied. They are found to be more destabilizing than symmetric loops with the same total number of nucleotides. This trend was predicted by Papanicolaou *et al.*[155] based on comparisons of predicted and known RNA secondary structures. From the experimental results, a loop asymmetry penalty can be defined as:

$$\Delta G^{\circ}\text{ (asymmetry penalty)} = \Delta G^{\circ}\text{ (asymmetric loop, } n) - \Delta G^{\circ}\text{ (symmetric loop, } n) \quad (31)$$

Experimental data[67] for internal loops with only adenines is fit well by a model where ΔG°_{37} (asymmetry penalty) $= 0.48|n_1 - n_2|$ kcal mol^{-1},[48] where n_1 and n_2 are the number of unpaired nucleotides on each strand. It is possible, however, that the asymmetry penalty depends on the sequence as well as the size.

(v) Larger internal loops

Measurements cannot currently be made on all possible internal loop sequences. A simple model has been developed to approximate the free energy changes of internal loops with more than four nucleotides.[48] This model includes the loop size, closing base pairs, first mismatches, and loop asymmetry. The free energy change is approximated as:

$$\Delta G^{\circ}_{\text{loop}} = \Delta G^{\circ}_{\text{init}}(n_1 + n_2) + \Delta G^{\circ}_{\text{asymm}}(|n_1 - n_2|) + \Delta G^{\circ}_{\text{AU/GU closure}} + m\Delta G^{\circ}_{\text{GA/UU}} \quad (32)$$

Table 7 Free energy increments (kcal mol^{-1}) at 37 °C for 2 × 1 internal loops, $\begin{smallmatrix}5'\mathrm{P\ X\ P}3'\\3'\mathrm{QZYQ}5'\end{smallmatrix}$ in 1 M NaCl, pH 7.[a,b]

3′ZY5′	*X = A*	*X = C*	*X = G*	*X = U*
AA	2.3 2.5	2.3	1.7	
AC	2.1	(2.2)		
AG	0.8		0.8	
AU		2.5		
CA	(2.2)	(2.2)		
CC	1.7	2.5		2.2
CG	(0.6)			
CU		1.9		1.7
GA	1.1 2.1		0.8	
GC	(1.6)			
GG			(2.2)	
UA		(2.2)		
UC		(2.2)		1.5
UU		(2.2)		1.4

[a] Parameters are for two GC closing pairs. Each AU closing pair is penalized by 0.45 + 0.2 = 0.65 kcal mol^{-1}, where the 0.45 is the INN–HB model penalty for terminal AU pairs in a helix (see text). When there is only one number, it is for $\begin{smallmatrix}5'\mathrm{C\ X\ C}3'\\3'\mathrm{GZYG}5'\end{smallmatrix}$; when there are two numbers, the top one is for $\begin{smallmatrix}5'\mathrm{C\ X\ C}3'\\3'\mathrm{GZYG}5'\end{smallmatrix}$, and the bottom one is for $\begin{smallmatrix}5'\mathrm{G\ X\ G}3'\\3'\mathrm{CZYC}5'\end{smallmatrix}$.

[b] Parameters in parentheses are estimated.[48,154,206]

where n_1 and n_2 are the numbers of nucleotides on each side of the loop, $\Delta G^{\circ}_{\mathrm{init}}(n_1 + n_2)$ is a penalty term for closing the loop (Table 4), $\Delta G^{\circ}_{\mathrm{asymm}}(|n_1 - n_2)|$ is the asymmetry penalty term discussed above, $\Delta G^{\circ}_{\mathrm{AU/GU\ closure}}$ is a penalty of 0.2 kcal mol^{-1} for loops closed by either AU or GU base pairs. This penalty is applied in addition to the terminal AU/GU term of the INN–HB model. $\Delta G^{\circ}_{\mathrm{GA/UU}}$ is a favorable bonus for each GA or AG (−1.1 kcal mol^{-1}), or UU (−0.7 kcal mol^{-1}) first mismatch, and m is the number of GA, AG, or UU first mismatches. For nucleotides in the loop other than in the two first mismatches, the sequence dependence is neglected. This is likely an incomplete model for some sequences. An example of using Equation (32) to estimate the free energy increment for an internal loop is shown in Figure 5.

2.3.5 Coaxial Stacks and Multibranch Loops (or Junctions)

Coaxial stacking can occur when two helices are directly adjacent or separated by a mismatch. Coaxial stacks have been seen in crystal structures of tRNAs[11–13,156,157] and are essential for the three-dimensional shape of tRNA. Stability increments from coaxial stacking of Watson–Crick pairs and of GA or CC mismatches have been investigated in model systems where a short oligomer binds to a four- or five-nucleotide overhang at the base of a hairpin stem.[52,158,159] This binding creates an interface

Internal Loop

$$\Delta G^{\circ}_{\mathrm{loop}}\left(\begin{smallmatrix}5'\mathrm{U}^{\mathrm{GA}}\ \mathrm{U}3'\\3'\mathrm{G}_{\mathrm{AAG}}\mathrm{G}\end{smallmatrix}\right) = \Delta G^{\circ}_{\mathrm{init}}(n_1 + n_2) + \Delta G^{\circ}_{\mathrm{asymm}}(|n_1 - n_2|) + \Delta G^{\circ}_{\mathrm{AU/GUclosure}} + m\Delta G^{\circ}_{\mathrm{GA/UU}}$$

$$= 1.8 + 0.48 \times (3 - 2) + 2 \times 0.2 + 2 \times (-0.8)$$

$$= 1.08\ \ \mathrm{kcal\ \ mol^{-1}}$$

Figure 5 Calculation of predicted free energy increment for an 2 × 3 internal loop in J4/5 of group I intron of *Pneumocystis carinii*.[177] Note that the calculation of free energies for the double helical regions flanking this internal loop would include a penalty of 0.45 kcal mol^{-1}for each of the GU pairs closing the loop.

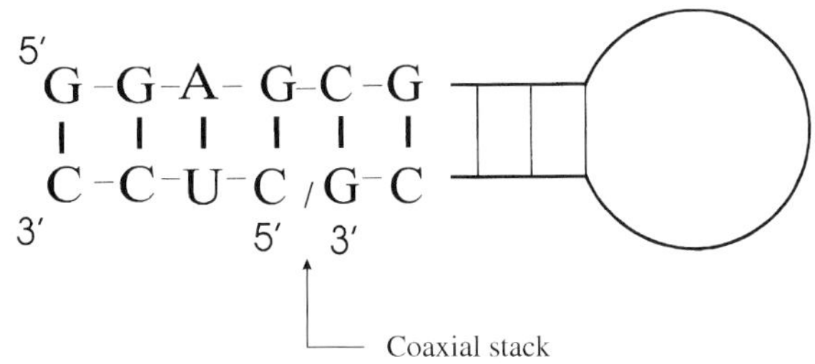

Figure 6 Coaxial stack formed by binding of a short oligonucleotide to the overhang of a hairpin stem. For this case, the free energy of coaxial stacking is determined by $\Delta G^{\circ}_{\text{coax stack}} = \Delta G^{\circ}$ (hairpin/oligomer complex shown) $- \Delta G^{\circ}\left(\begin{smallmatrix}5'\text{GGAG}3'\\3'\text{CCUC}5'\end{smallmatrix}\right)$. Here the free energy changes are for formation of the complex between hairpin and short oligomer, and of the isolated short duplex.

between two helical regions joined by one continuous backbone and a break in the backbone on the other strand (Figure 6). The contribution of coaxial stacking is calculated as the difference in the free energy change between forming the complex and the isolated short-stem duplex:

$$\Delta G^{\circ}_{\text{coaxstack}} = \Delta G^{\circ}\,(\text{hairpin/oligomer}) - \Delta G^{\circ}\,(\text{oligomer duplex}) \tag{33}$$

Since the isolated short oligomer duplexes are usually too short to be experimentally measured, their free energy changes of duplex formation are calculated with nearest-neighbor parameters of the INN–HB model.[75] Note that for interfaces with AU base pairs, the terminal AU term is applied.

The enhanced binding by coaxial stacking of Watson–Crick pairs is found to be independent of the position of the break in the sugar–phosphate backbone, but is dependent on the sequence at the interface, suggesting that the interaction of nearest-neighbor base pairs is the primary determinant of the enhancement.[52,158] These values are collected in Table 8.[48] Compared with the corresponding regular Watson–Crick nearest-neighbor interactions (Table 1), coaxial stacking is 0.8–1.5 kcal mol^{-1} more favorable. Presumably this is because the break in the backbone on one strand offers more flexibility to the base pairs at the interface. Thus, stacking and hydrogen bonding interactions can be optimized. Because the free energy increments for coaxial stacking do not include additional hydrogen bonding interactions, while regular Watson–Crick nearest neighbors do include hydrogen bonding interactions,[75] the real enhancements are even larger. When the chains are extended with an unpaired nucleotide beyond the break, the free energy increments for coaxial stacking are approximately equal to that of a regular Watson–Crick nearest-neighbor interaction.[48,52] This could be because the backbone has to make an unfavorable turn to allow the coaxial stack.

GA mismatches between helices are common in known RNA secondary structures. For example, in yeast phenylalanine tRNA, the D and anticodon arms are separated by a GA mismatch, but coaxially stack with each other. GA mismatches between helices are also found in self-alkylating,[160] and self-ligating[161] ribozymes. Thermodynamic effects of GA mismatches and the much less frequently occurring CC mismatch at the interface have been studied.[159] Despite the difference in abundance, both GA and CC mismatches at the interface contribute roughly 2 kcal mol^{-1}, and are relatively insensitive to the identity of the adjacent base pairs, and to extension beyond the interface.[48]

Multibranch loops or junctions are loops where more than two helices intersect. They usually also contain unpaired nucleotides, and are a major determinant of RNA three-dimensional structure.[120] Coaxial stacking is a major contributor to stabilization of multibranch loops. The thermodynamics of other factors that determine stabilities of multibranch loops have not been studied in detail. In general, the stability of a multibranch loop can depend on the number of helices involved, the number of unpaired nucleotides in the loop, and interactions such as base triples involving nucleotides in and near the loop. The INN–HB model for Watson–Crick nearest neighbors suggests that each helix that ends with an AU

Table 8 Free energy increments of coaxial stacking interaction at 37 °C (kcal mol^{-1}) in 1 M NaCl, pH 7.

Coaxial stacking interface [a]	G-C C/G	G-G C/C	C-C G/G	C-G G/C	C/G G-C	G-A C/U	G-U C/A	A-G U/C	U-G A/C	U-A A/U
Stacking interaction (kcal mol^{-1})	−4.20	−4.01	−4.23	−3.16	−3.46	−3.84	−3.34	−2.95	−2.92	−2.74

[a] A hyphen "-" represents continuation of backbone, while a back slash "/" represents a break in the backbone.

pair should be made less favorable by about 0.5 kcal mol^{-1}.[75] A crude model has been proposed[48,52] that predicts the free energies of multibranch loops based on an unfavorable initiation term and favorable coaxial stacking. These parameters can be estimated by optimizing the accuracy of the folding algorithm for predicting known secondary structures.[48]

2.3.6 Environmental Effects on RNA Secondary Structure Thermodynamics

RNA is a polyanion. In order to form stable structures, it requires counterions to neutralize the repulsions of the negatively charged backbone phosphates. Manning's polyelectrolyte theory predicts that, due to counterion condensation, the local concentration of monovalent cation near a polynucleotide duplex is around 1 M, regardless of the bulk salt concentration if no multivalent cations are present.[162] In solutions containing only monovalent cations, e.g., Na^+ and K^+, the stability of a long duplex increases with salt concentration up to 1 M, with the increase in T_M being linear with log $[Na^+]$ up to about 0.2 M Na^+. At salt concentrations above 1 M, addition of more salt decreases duplex stability. The destabilizing effect at high salt concentrations depends on the anion with $CCl_3COO^- > SCN^- > ClO_4^- > Me_2CO_2^- > Br^-, Cl^-$.[163] The effects depend on RNA length. Less charge will be neutralized in oligonucleotides than in polynucleotides because of the reduced charge density at the ends. The dependence of salt effects on length in DNA oligonucleotides has been analyzed.[164] The length effect is negligible at high salt concentrations.[165,166]

In the presence of multivalent ions such as Mg^{2+}, the monovalent ions around the RNA backbone will be essentially replaced by multivalent ions.[162,167] If saturated with Mg^{2+}, increasing concentrations of Na^+ decrease duplex stability, because saturation with Mg^{2+} leaves the random coil with a higher density of negative charge than the duplex. Thus, a higher monovalent cation concentration favors a random coil resulting in a lower T_M.

The Mg^{2+} ion may provide specific binding to many potential sites in various RNA structural motifs, especially in non-Watson–Crick regions.[18,168] Experimental results show that duplex stability in 1 M NaCl is similar to that observed in the presence of a few millimolar Mg^{2+} plus 0.1 M Na^+ or K^+, which is similar to physiological conditions.[29,32,77,169] Therefore, thermodynamic information derived at 1 M NaCl is useful for predicting RNA properties in biologically relevant environments.

Addition of cosolvents to aqueous solutions of RNA usually destabilizes RNA structures.[170,171] Typically, the T_M of a duplex is a linear function of cosolvent concentration.[171]

The stability of most duplexes is relatively insensitive to pH between 5 and 9.[44] At low or high pH, bases are protonated or deprotonated precluding normal hydrogen bonding and the formation of duplexes. In RNA, CC^+mismatches have been observed in the middle of the duplex $(CGC\underline{CC}GCG)_2$ at pH 5.5.[149]

2.4 APPLICATIONS

Our knowledge of the thermodynamics of RNA secondary structure is steadily increasing. Its applications to calculating the thermodynamic properties of RNA molecules containing specific structural motifs have been shown above. Calculations for RNA molecules with combinations of different motifs can be done in a similar fashion. Other important applications include estimation of tertiary interactions, prediction of RNA structure from sequence, and designing ribozymes to target mRNA.

2.4.1 Estimation of Tertiary Interactions

In addition to the formation of secondary structures, RNA folding usually involves tertiary interactions, about which little is known. One way to detect the effect of tertiary interactions is to compare dissociation constants, K_D, for RNA–RNA association with those predicted based on the thermodynamics for formation of RNA secondary structure.[172] When tertiary interactions appear to be important, their net effect can be determined by comparing K_D values for the complete system with those measured for a model system that can only form secondary structure.[173] The contribution of various groups to tertiary interactions can be determined by measuring the K_D for complexes with functional group substitutions.[108,173–176] Such comparisons provide insight into structure–function relationships, and can suggest ways to target RNA by exploiting tertiary interactions.[177,178]

For example, the predicted K_D for the duplex, $\frac{5'\text{AUGACU3}'}{3'\text{UACUGG5}'}$ is 0.34 mM in 1 M NaCl (Figure 3a), close to the K_D of 0.32 mM measured in 135 mM KCl, 15 mM $MgCl_2$.[177] Binding of 5′AUGACU3′ to a group I ribozyme from *Pneumocystis carinii* involves the same pairing interactions, but the K_D is 5.2 nM. Thus, tertiary interactions must strengthen this binding 60 000-fold.[177] Such binding enhancement by tertiary interactions (BETI) can be used to increase the specificity and binding of short antisense agents to RNA.[178]

2.4.2 RNA Secondary Structure Prediction and Modeling of Three Dimensional Structure

Thermodynamic parameters are crucial for RNA secondary structure predictions because of the large number of possible pairings. The parameters form the basis of algorithms that can predict the most likely foldings.[46,47,113,179–181] Recursive (dynamic programming) algorithms[46,47] use predicted free energies of smaller fragments to predict free energies for larger fragments until the free energy for the whole structure is calculated. Knotted structures are not allowed in this recursive algorithm, but could be allowed in genetic algorithms.[180,181] At this time, little is known about the thermodynamics of knotted structures, and many other motifs require gross approximations, as discussed above. For this reason, folding algorithms have also been designed to calculate suboptimal structures, and to incorporate experimental data such as enzymatic mapping.[47,48,182] These programs facilitate the design of experimental approaches to single out and test the real structure.

Reliable secondary structure is also the basis for modeling three-dimensional structure, and the thermodynamics of RNA secondary structure can help predict three-dimensional structure. For example, as discussed above, sequences with 3′ dangling end stacking more favorable than 0.8 kcal mol^{-1} in model systems are usually stacked in three-dimensional structures.[42,120] Moreover, conserved features in secondary structures often suggest small model systems whose three-dimensional structures can be solved by NMR or X-ray crystallography. Two such examples are illustrated in Figure 7.

2.4.3 Targeting RNA with Ribozymes

Ribozymes have therapeutic potential for targeting RNA either for cleavage[4,183–186] or repair.[187,188] There are several obstacles to realizing this potential. For example, ribozymes may fold into inactive conformations,[189,190] targets may have secondary and tertiary structures that prevent ribozyme binding, and it may be difficult to achieve target specificity.[191–194] In principle, knowledge of the thermodynamics of secondary structure formation can help overcome these obstacles. Secondary structure prediction algorithms can help design ribozymes that have an optimal folding much favored over other foldings.[189] With constraints from chemical and enzymatic mapping, they may also make it possible to predict suitable sites on RNA targets. Specificity may be difficult to achieve because it requires that ribozymes dissociate from mismatched substrate much faster than the rate of the chemical reaction they catalyze.[191] The strength of Watson–Crick base pairing makes this difficult except for short helixes, but sequences that can form short helixes are likely to bind many perfect matches in cellular RNA. One potential solution to this problem is to design ribozymes with enhanced dissociation rates. LeCuyer and Crothers[195] have shown that intramolecular helixes may rearrange much faster than expected, suggesting that the rate-limiting step does not necessarily require disruption of all base pairs in a helix. Another potential solution is to design or evolve a ribozyme that recognizes a three-dimensional shape in the target RNA, rather than simply base sequence. Knowledge of the thermodynamics of secondary structure formation should be useful in implementing these and other strategies.

2.5 FUTURE PERSPECTIVES

Our understanding of the thermodynamics of RNA secondary structure is a powerful tool. The current thermodynamic database provides reasonable prediction of RNA secondary structures. It can also be

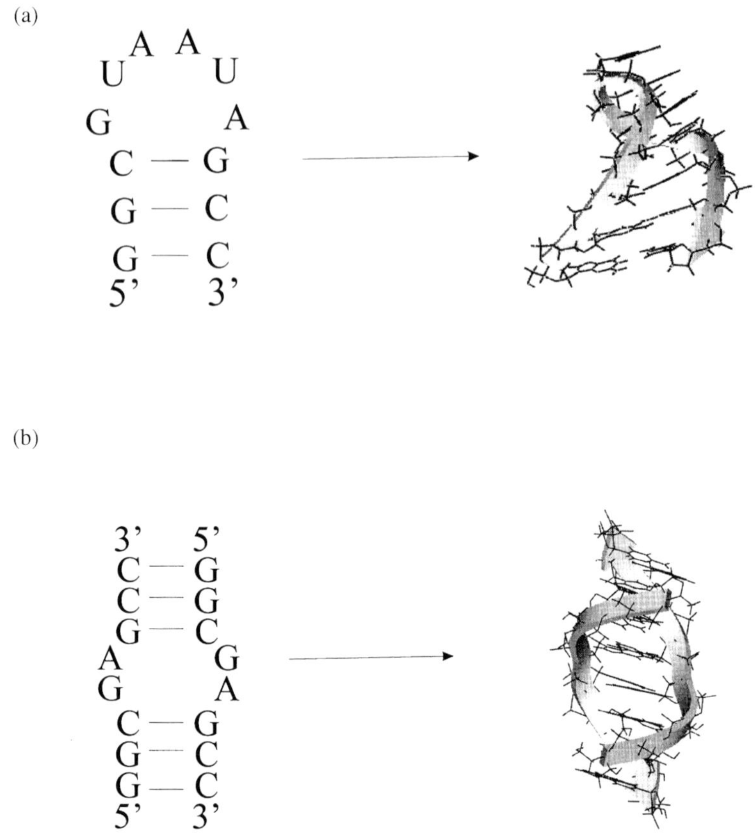

Figure 7 Secondary and three-dimensional structures of two fragments of natural RNAs. Arrows in 3D structures point from 5′ to 3′ along the backbones. (a) Conserved hairpin in the L11 protein binding region of large subunit ribosomal RNA.[27] (b) A tandem mismatch commonly found in RNA.[23]

used to aid the design and interpretation of many experiments. Our knowledge is still crude, however; the stabilities of most motifs have not been experimentally determined. Therefore many approximations have to be made and need to be tested, especially for loop regions. High-density oligonucleotide array technologies,[196–201] may make it possible to speed up the accumulation of this kind of experimental data. Random selection methods will provide a way to identify motifs with unusual properties.[202] Moreover, better understanding of intermolecular interactions should improve approximations for motifs that have not been studied. Thus, secondary structure thermodynamics should become an even more powerful and reliable tool in the future.

2.6 REFERENCES

1. J.D. Watson, N.H. Hopkins, J.W. Roberts, J.A. Steitz and A.M. Weiner, "Molecular Biology of the Gene", Benjamin Cummings Inc, Menlo Park, CA, 1987.
2. T.R. Cech and B.L. Bass, *Annu. Rev. Biochem*, 1986, **55**, 599.
3. T.R. Cech, *Annu. Rev. Biochem.*, 1990, **59**, 543.
4. S. Altman, *Adv. Enzymol.*, 1989, **62**, 1.
5. F. Noller, V. Hoffarth and L. Zimniak, *Science*, 1992, **256**, 1416.
6. P. Nissen, J. Hansen, N. Ban, P.B. Moore and T.A. Steitz, *Science*, 2000, **289**, 920.
7. P. Walter and G. Blobel, *Nature*, 1982, **299**, 691.
8. N. Williams, *Science*, 1997, **275**, 301.
9. D. Gershon, *Nature*, 1997, **389**, 417.
10. (a) J.C. Venter *et al.*, *Science*, 2001, **291**, 1304; (b) International Human Genome Sequencing Consortium, *Nature*, 2001, **409**, 860.
11. S.H. Kim, G.J. Quigley, F.L. Suddath, A. McPherson, D. Sneden, J.J. Kim, J. Weinzierl and A. Rich, *Science*, 1973, **179**, 285.
12. S.H. Kim, F.L. Suddath, G.J. Quigley, A. McPherson, J.L. Sussman, A.H.J. Wang, N.C. Seeman and A. Rich, *Science*, 1974, **185**, 435.

13. J.D. Robertus, J.E. Ladner, J.T. Finch, D. Rhodes, R.S. Brown, B.F.C. Clark and A. Klug, *Nature*, 1974, **250**, 546.
14. H.W. Pley, K.M. Flaherty and D.B. McKay, *Nature*, 1994, **372**, 68.
15. W.G. Scott, J.T. Finch and A. Klug, *Cell*, 1995, **81**, 991.
16. W.G. Scott, J.B. Murray, J.R.P. Arnold, B.L. Stoddard and A. Klug, *Science*, 1996, **274**, 2065.
17. J.B. Murray, D.P. Terwey, L. Maloney, A. Karpeisky, N. Usman, L. Beigelman and W.G. Scott, *Cell*, 1998, **92**, 665.
18. J.H. Cate, A.R. Gooding, E. Podell, K. Zhou, B.L. Golden, C.E. Kundrot, T.R. Cech and J.A. Doudna, *Science*, 1996, **273**, 1678.
19. C. Cheong, G. Varani and I. Tinoco Jr., *Nature*, 1990, **346**, 680.
20. H.A. Heus and A. Pardi, *Science*, 1991, **253**, 191.
21. G. Varani, C. Cheong and I. Tinoco Jr., *Biochemistry*, 1991, **30**, 3280.
22. J.D. Puglisi, R. Tan, B.J. Calnan, A.D. Frankel and J.R. Williamson, *Science*, 1992, **257**, 76.
23. J. SantaLucia Jr. and D.H. Turner, *Biochemistry*, 1993, **32**, 12612.
24. B. Wimberly, G. Varani and I. Tinoco Jr., *Biochemistry*, 1993, **32**, 1078.
25. F.H.-T. Allain and G. Varani, *J. Mol. Biol.*, 1995, **250**, 333.
26. A.A. Szewczak and P.B. Moore, *J. Mol. Biol.*, 1995, **247**, 81.
27. M.A. Fountain, M.J. Serra, T.R. Krugh and D.H. Turner, *Biochemistry*, 1996, **35**, 6539.
28. S. Huang, Y.X. Wang and D.E. Draper, *J. Mol. Biol.*, 1996, **258**, 308.
29. J.A. McDowell and D.H. Turner, *Biochemistry*, 1996, **35**, 14077.
30. S.E. Butcher, T. Dieckmann and J. Feigon, *J. Mol. Biol.*, 1997, **268**, 348.
31. A. Dallas and P.B. Moore, *Structure*, 1997, **5**, 1639.
32. J.A. McDowell, L.Y. He, X.Y. Chen and D.H. Turner, *Biochemistry*, 1997, **36**, 8030.
33. F. Michel and E. Westhof, *J. Mol. Biol.*, 1990, **216**, 585.
34. C. Zwieb, *J. Biol. Chem.*, 1992, **267**, 15650.
35. B.D. James, G.J. Olsen and N.R. Pace, *Methods Enzymol.*, 1989, **180**, 227.
36. R.R. Gutell, N. Larsen and C.R. Woese, *Microbiol. Rev.*, 1994, **58**, 10.
37. D.M. Crothers, P.E. Cole, C.W. Hilbers and R.G. Shulman, *J. Mol. Biol.*, 1974, **87**, 63.
38. C.W. Hilbers, G.T. Robillard, R.G. Shulman, R.D. Blake, P.K. Webb, R. Fresco and D. Riesner, *Biochemistry*, 1976, **15**, 1874.
39. A.R. Banerjee, J.A. Jaeger and D.H. Turner, *Biochemistry*, 1993, **32**, 153.
40. L. Jaeger, E. Westhof and F. Michel, *J. Mol. Biol.*, 1993, **234**, 331.
41. L.G. Laing and D.E. Draper, *J. Mol. Biol.*, 1994, **237**, 560.
42. D.H. Turner, N. Sugimoto and S.M. Freier, *Annu. Rev. Biophys. Chem*, 1988, **17**, 167.
43. D.H. Turner and P.C. Bevilacqua, in "The RNA World", eds. R.F. Gesteland and J.F. Atkins, Cold Spring Harbor Laboratory Press, Cold Spring Harbor, NY, 1993,
44. D.H. Turner, in "Nucleic Acids: Structures, Properties, and Functions", eds. V.A. Bloomfield, D.M. Crothers and I. Tinoco Jr., University Science Books, Sausalito, CA, 2000,
45. I. Tinoco Jr., O.C. Uhlenbeck and M.D. Levine, *Nature*, 1971, **230**, 362.
46. M. Zuker and P. Stiegler, *Nucleic Acids Res.*, 1981, **9**, 133.
47. M. Zuker, *Science*, 1989, **244**, 48.
48. D.H. Mathews, J. Sabina, M. Zuker and D.H. Turner, *J. Mol. Biol.*, 1999, **288**, 911.
49. R. Kierzek, M.H. Caruthers, C.E. Longfellow, D. Swinton, D.H. Turner and S.M. Freier, *Biochemistry*, 1986, **25**, 7840.
50. N. Usman, K.K. Ogilvie, M.-Y. Jiang and R.J. Cedergren, *J. Am. Chem. Soc*, 1987, **109**, 7845.
51. J.A. Jaeger, D.H. Turner and M. Zuker, *Proc. Natl. Acad. Sci. USA*, 1989, **86**, 7706.
52. A.E. Walter, D.H. Turner, J. Kim, M.H. Lyttle, P. Muller, D.H. Mathews and M. Zuker, *Proc. Natl. Acad. Sci. USA*, 1994, **91**, 9218.
53. V.A. Bloomfield, D.M. Crothers and I. Tinoco Jr., "Physical Chemistry of Nucleic Acids", Harper and Row, New York, 1974.
54. C.R. Cantor and P.R. Schimmel, "Biophysical Chemistry, Part III", Freeman, San Francisco, CA, 1980.
55. C.R. Cantor and P.R. Schimmel, "Biophysical Chemistry, Part II", Freeman, San Francisco, CA, 1980.
56. I. Tinoco Jr., *J. Am. Chem. Soc.*, 1960, **82**, 4785.
57. I. Tinoco Jr., *J. Chem. Phys.*, 1960, **33**, 1332.
58. I. Tinoco Jr., *J. Chem. Phys.*, 1961, **34**, 1067.
59. W. Rhodes, *J. Chem. Phys.*, 1961, **83**, 3609.
60. H. DeVoe and I. Tinoco Jr., *J. Mol. Biol*, 1962, **4**, 518.
61. D.F. Bradley, I. Tinoco Jr. and R.W. Woody, *Biopolymers*, 1963, **1**, 239.
62. J.R. Fresco, L.C. Klotz and E.D. Richards, *Cold Spring Harbor Symp. Quant. Biol.*, 1963, **28**, 83.
63. G. Felsenfeld and S.Z. Hirschman, *J. Mol. Biol*, 1965, **13**, 407.
64. J.D. Puglisi and I. Tinoco Jr., *Methods Enzymol.*, 1989, **180**, 304.
65. L.A. Marky and K.J. Breslauer, *Biopolymers*, 1987, **26**, 1601.
66. M. Petersheim and D.H. Turner, *Biochemistry*, 1983, **22**, 256.
67. A.E. Peritz, R. Kierzek, N. Sugimoto and D.H. Turner, *Biochemistry*, 1991, **30**, 6428.
68. C.E. Longfellow, R. Kierzek and D.H. Turner, *Biochemistry*, 1990, **29**, 278.
69. E.G. Richards, in "Handbook of Biochemistry and Molecular Biology: Nucleic Acids", ed. D. Fasman, CRC Press, Cleveland, OH, 1975, 57
70. P.N. Borer, in "Handbook of Biochemistry and Molecular Biology: Nucleic Acids, 3rd edn.", ed. G.D. Fasman, CRC Press, Cleveland, OH, 1975, 589

71. P.R. Bevington and D.K. Robinson, "Data Reduction and Error Analysis for the Physical Sciences, 2nd edn.", McGraw-Hill, Boston, 1992.
72. P.N. Borer, B. Dengler, I. Tinoco Jr. and O.C. Uhlenbeck, *J. Mol. Biol.*, 1974, **86**, 843.
73. D.De P. Albergo, L.A. Marky, K.J. Breslauer and D.H. Turner, *Biochemistry*, 1981, **20**, 1409.
74. S.M. Freier, R. Kierzek, J.A. Jaeger, N. Sugimoto, M.H. Caruthers, T. Neilson and D.H. Turner, *Proc. Natl. Acad. Sci. USA*, 1986, **83**, 9373.
75. T. Xia, J. SantaLucia Jr., M.E. Burkard, R. Kierzek, S.J. Schroeder, C. Cox and D.H. Turner, *Biochemistry*, 1998, **37**, 14719.
76. J. SantaLucia Jr., R. Kierzek and D.H. Turner, *J. Am. Chem. Soc*, 1991, **113**, 4313.
77. T. Xia, J.A. McDowell and D.H. Turner, *Biochemistry*, 1997, **36**, 12486.
78. H.T. Allawi and J. SantaLucia Jr., *Biochemistry*, 1997, **36**, 10581.
79. J.B. Chaires, *Biophys. Chem.*, 1997, **64**, 15.
80. J.M. Sturtevant, *Annu. Rev. Phys. Chem.*, 1987, **38**, 463.
81. P.L. Privalov and S.A. Potekhin, *Methods Enzymol.*, 1986, **131**, 4.
82. K.J. Breslauer, E. Freire and M. Straume, *Methods Enzymol.*, 1992, **211**, 533.
83. H.F. Fisher and N. Singh, *Methods Enzymol.*, 1995, **259**, 194.
84. J. SantaLucia Jr. and D.H. Turner, *Biopolymers*, 1997, **44**, 309.
85. S.M. Law, R. Eritja, M.F. Goodman and K.J. Breslauer, *Biochemistry*, 1996, **35**, 12329.
86. S.M. Freier, K.O. Hill, T.G. Dewey, L.A. Marky, K.J. Breslauer and D.H. Turner, *Biochemistry*, 1981, **20**, 1419.
87. D.E. Draper and T.C. Gluick, *Methods Enzymol.*, 1995, **259**, 281.
88. J. Gralla and D.M. Crothers, *J. Mol. Biol*, 1973, **78**, 301.
89. A. Wada, S. Yubuki and Y. Husimi, *Crit. Rev. Biochem.*, 1980, **9**, 87.
90. O. Gotoh, *Adv. Biophys*, 1983, **16**, 1.
91. R.M. Wartell and A.S. Benight, *Phys. Rep*, 1985, **126**, 67.
92. M. Schmitz and G. Steger, *CABIOS*, 1992, **8**, 389.
93. D. Poland and H. Scheraga, "Theory of Helix-Coil Transitions in Biopolymers", Academic Press, New York, 1970.
94. D. Poland, "Cooperative Equilibria in Physical Biochemistry", Clarendon Press, Oxford, 1978.
95. G. Steger, *Nucleic Acids Res*, 1994, **22**, 2760.
96. D. Porschke and M. Eigen, *J. Mol. Biol.*, 1971, **62**, 361.
97. T.C. Gluick and D.E. Draper, *J. Mol. Biol.*, 1994, **241**, 246.
98. B. Lewin, "Genes", Oxford University Press, Oxford, 1997.
99. R.F. Gesteland and J.F. Atkins, "The RNA World: The Nature of Modern RNA Suggests a Prebiotic RNA World", Cold Spring Harbor Laboratory Press, New York, 1993.
100. I. Tinoco Jr., P.N. Borer, B. Dengler, M.D. Levine, O.C. Uhlenbeck, D.M Crothers and J. Gralla, *Nature New Biol*, 1973, **246**, 40.
101. J. Gralla and D.M. Crothers, *J. Mol. Biol*, 1973, **73**, 497.
102. D. Gray, *Biopolymers*, 1997, **42**, 783.
103. W.F. Bailey and A.S. Monahan, *J. Chem. Ed.*, 1978, **55**, 489.
104. N. Sugimoto, K. Honda and M. Sasaki, *Nucleosides Nucleotides*, 1994, **13**, 1311.
105. N. Sugimoto, S. Nakano, M. Katoh, A. Matsumura, H. Nakamuta, T. Ohmichi, M. Yoneyama and M. Sasaki, *Biochemistry*, 1995, **34**, 11211.
106. P.H. van Knippenberg, L.J. Formenoy and H.A. Heus, *Biochem. Biophys. Acta*, 1990, **1050**, 14.
107. D. Gautheret, D. Konings and R.R. Gutell, *RNA*, 1995, **1**, 807.
108. S.A. Strobel and T.R. Cech, *Science*, 1995, **267**, 675.
109. O.C. Uhlenbeck, F.H. Martin and P. Doty, *J. Mol. Biol.*, 1971, **57**, 217.
110. S.M. Freier, R. Kierzek, M.H. Caruthers, T. Neilson and D.H. Turner, *Biochemistry*, 1986, **25**, 3209.
111. N. Sugimoto, R. Kierzek, S.M. Freier and D.H. Turner, *Biochemistry*, 1986, **25**, 5755.
112. L. He, R. Kierzek, J. SantaLucia Jr., A.E. Walter and D.H. Turner, *Biochemistry*, 1991, **30**, 11–124.
113. M.E. Burkard, R. Kierzek and D.H. Turner, *J. Mol. Biol.*, 1999, **290**, 967.
114. S.M. Freier, B.J. Burger, D. Alkema, T. Neilson and D.H. Turner, *Biochemistry*, 1983, **22**, 6198.
115. D.R. Hickey and D.H. Turner, *Biochemistry*, 1985, **24**, 3987.
116. S.M. Freier, D. Alkema, A. Sinclair, T. Neilson and D.H. Turner, *Biochemistry*, 1985, **24**, 4533.
117. N. Sugimoto, R. Kierzek and D.H. Turner, *Biochemistry*, 1987, **26**, 4554.
118. N. Sugimoto, R. Kierzek and D.H. Turner, *Biochemistry*, 1987, **26**, 4559.
119. M.R. Giese, K. Betschart, T. Dale, C.K. Riley, C. Rowan, K.J. Sprouse and M.J. Serra, *Biochemistry*, 1998, **37**, 1094.
120. M.E. Burkard, D.H. Turner and I. Tinoco Jr., in "The RNA World II", eds. T.R. Cech, R.F. Gesteland and J.F. Atkins, Cold Spring Harbor Laboratory Press, New York, 1998, 233
121. V. Lehnert, L. Jaeger, F. Michel and E. Westhof, *Chem. Biol.*, 1996, **3**, 993.
122. F.M. Jucker and A. Pardi, *Biochemistry*, 1995, **34**, 14416.
123. C.R. Woese, S. Winker and R.R. Gutell, *Proc. Natl. Acad. Sci. USA*, 1990, **87**, 8467.
124. M.J. Serrra, T.J. Axenson and D.H. Turner, *Biochemistry*, 1994, **33**, 14289.
125. M.J. Serra, T.W. Barnes, K. Betschart, M.J. Gutierrez, K.J. Sprouse, C.K. Riley, L. Stewart and R.E. Temel, *Biochemistry*, 1997, **36**, 4844.
126. V.P. Antao and I. Tinoco Jr., *Nucleic Acids Res.*, 1992, **20**, 819.
127. V.P. Antao, S.Y. Lai and I. Tinoco Jr., *Nucleic Acids Res.*, 1991, **19**, 5901.
128. J. SantaLucia Jr., R. Kierzek and D.H. Turner, *Science*, 1992, **256**, 217.

129. C. Tuerk, P. Gauss, C. Thermes, D.R. Groebe, M. Gayle, N. Guild, G. Stormo, Y. D'Aubenton-Carafa, O.C. Uhlenbeck, I. Tinoco Jr., E.N. Brody and L. Gold, *Proc. Natl. Acad. Sci. USA*, 1988, **85**, 1364.
130. M. Orita, F. Nishikawa, T. Shimayama, K. Taira, Y. Endo and S. Nishikawa, *Nucleic Acids Res.*, 1993, **21**, 5670.
131. L. Mueller, P. Legault and A. Pardi, *J. Am. Chem. Soc.*, 1995, **117**, 11043.
132. G. Varani, *Annu. Rev. Biophys. Biomol. Struct.*, 1995, **24**, 379.
133. F.M. Jucker, H.A. Heus, P.F. Yip, E.H.M. Moors and A. Pardi, *J. Mol. Biol.*, 1996, **264**, 968.
134. M. Costa and F. Michel, *EMBO J.*, 1997, **16**, 3289.
135. Y. Endo, Y.L. Chan, A. Lin, K. Tsurugi and I.G. Wool, *J. Biol. Chem*, 1988, **263**, 7917.
136. A. Gluck, Y. Endo and I.G. Wool, *J. Mol. Biol.*, 1992, **226**, 411.
137. D.R. Groebe and O.C. Uhlenbeck, *Nucleic Acids Res.*, 1988, **16**, 11725.
138. P.N. Borer, Y. Lin, S. Wang, M.W. Roggenbuck, J.M. Gott, O.C. Uhlenbeck and I. Pelczer, *Biochemistry*, 1995, **34**, 6488.
139. Y.T. van der Hoogen, A.A. van Beuzekom, E. de Vroom, G.A. van der Marel, J.H. van boom and C. Altona, *Nucleic Acids Res.*, 1988, **16**, 5013.
140. R.S. Tang and D.E. Draper, *Nucleic Acids Res.*, 1994, **22**, 835.
141. D.A. Peattie, S. Southwaite, R.A. Garett and H.F. Noller, *Proc. Natl. Acad. Sci. USA*, 1981, **78**, 7331.
142. P.J. Romaniuk, P. Lowary, H-N. Wu, G. Stormo and O.C. Uhlenbeck, *Biochemistry*, 1987, **26**, 1563.
143. P.J. Flor, J.B. Flanegan and T.R. Cech, *EMBO J.*, 1989, **8**, 3391.
144. K.M. Weeks and D.M. Crothers, *Science*, 1993, **261**, 1574.
145. D.R. Groebe and O.C. Uhlenbeck, *Biochemistry*, 1989, **28**, 742.
146. T.R. Fink and D.M. Crothers, *J. Mol. Biol.*, 1972, **66**, 1.
147. S.E. Morse and D.E. Draper, *Nucleic Acids Res.*, 1995, **23**, 302.
148. R. Kierzek, M.E. Burkard, and H. Turner, *Biochemistry*, 1999, **38**, 14214.
149. J. SantaLucia Jr., R. Kierzek and D.H. Turner, *Biochemistry*, 1991, **30**, 8242.
150. A.E. Walter, M. Wu and D.H. Turner, *Biochemistry*, 1994, **33**, 11349.
151. M. Wu, J.A. McDowell and D.H. Turner, *Biochemistry*, 1995, **34**, 3204.
152. M. Wu and D.H. Turner, *Biochemistry*, 1996, **35**, 9677.
153. M. Wu, J. SantaLucia Jr. and D.H. Turner, *Biochemistry*, 1997, **36**, 4449.
154. S.J. Schroeder, J. Kim and D.H. Turner, *Biochemistry*, 1996, **35**, 16105.
155. C. Papanicolaou, M. Gouy and J. Ninio, *Nucleic Acids Res.*, 1984, **13**, 1717.
156. E. Westhof, P. Dumas and D. Moras, *J. Mol. Biol.*, 1985, **184**, 119.
157. V. Biou, A. Yaremchuk, M. Tukalo and S. Cusack, *Science*, 1994, **263**, 1404.
158. A.E. Walter and D.H. Turner, *Biochemistry*, 1994, **33**, 12715.
159. J. Kim, A.E. Walter and D.H. Turner, *Biochemistry*, 1996, **35**, 13753.
160. C. Wilson and J.W. Szostak, *Nature*, 1995, **374**, 777.
161. E.H. Ekland, J.W. Szostak and D.P. Bartel, *Science*, 1995, **269**, 364.
162. G. Manning, *Q. Rev. Biophys*, 1978, **11**, 179.
163. K. Hamaguchi and E.P. Geiduschek, *J. Am. Chem. Soc.*, 1962, **84**, 1329.
164. J. SantaLucia Jr., *Proc. Natl. Acad. Sci. USA*, 1998, **95**, 1460.
165. M.T. Record Jr. and T.M. Lohman, *Biopolymers*, 1978, **17**, 159.
166. M.C. Olmsted, C.F. Anderson and M.T. Record Jr., *Proc. Natl. Acad. Sci. USA*, 1989, **86**, 7766.
167. M.T. Record Jr., C.F. Anderson and T.M. Lohman, *Q. Rev. Biophys*, 1978, **2**, 103.
168. L.G. Laing, T.C. Gluick and D.E. Draper, *J. Mol. Biol.*, 1994, **237**, 577.
169. A.P. Williams, C.E. Longfellow, S.M. Freier, R. Kierzek and D.H. Turner, *Biochemistry*, 1989, **28**, 4283.
170. D.D. Albergo and D.H. Turner, *Biochemistry*, 1981, **20**, 1413.
171. D.R. Hickey and D.H. Turner, *Biochemistry*, 1985, **24**, 2086.
172. N. Sugimoto, R. Kierzek and D.H. Turner, *Biochemistry*, 1988, **27**, 6384.
173. P.C. Bevilacqua and D.H. Turner, *Biochemistry*, 1991, **30**, 10632.
174. N. Sugimoto, M. Tomka, R. Kierzek, P.C. Bevilacqua and D.H. Turner, *Nucleic Acids Res.*, 1989, **17**, 355.
175. A.M. Pyle and T.R. Cech, *Nature*, 1991, **350**, 628.
176. S.A. Strobel and T.R. Cech, *Biochemistry*, 1993, **32**, 13593.
177. S.M. Testa, C.G. Haidaris, F. Gigliotti and D.H. Turner, *Biochemistry*, 1997, **36**, 15303.
178. S.M. Testa, S.M. Gryaznov and D.H. Turner, *Biochemistry*, 1998, **37**, 9379.
179. D.H. Mathews, A.R. Banerjee, D.D. Luan, T.H. Eickbush and D.H. Turner, *RNA*, 1997, **3**, 1.
180. A.P. Gultyaev, F.H.G. van Batenburg and C.W.A. Pleij, *J. Mol. Biol.*, 1995, **250**, 37.
181. F.H.D. van Batenburg, A.P. Gultyaev and C.W.A. Pleij, *J. Theor. Biol.*, 1995, **174**, 269.
182. J.A. Jaeger, D.H. Turner and M. Zuker, *Methods Enzymol.*, 1990, **183**, 281.
183. O.C. Uhlenbeck, *Nature*, 1987, **328**, 596.
184. J.M. Burke, *Nucleic Acids. Mol. Biol.*, 1994, **8**, 105.
185. J.J. Rossi, E.M. Cantin, N. Sarver and P.F. Chang, *Pharmacol. Ther.*, 1991, **50**, 245.
186. D. Castanotto, H.T. Li, W. Chow, J.J. Rossi and J.O. Deshler, *Antisense Nucleic Acid Drug Dev.*, 1998, **8**, 1.
187. B.A. Sullenger and T.R. Cech, *Nature*, 1994, **371**, 619.
188. N. Lan, R.P. Howrey, S.-W. Lee, C.A. Smith and B.A. Sullenger, *Science*, 1998, **280**, 1593.
189. N. Usman and J.A. McSwiggen, *Annu. Rep. Med. Chem.*, 1992, **30**, 285.
190. O.C. Uhlenbeck, *RNA*, 1995, **1**, 4.
191. D. Herschlag, *Proc. Natl. Acad. Sci. USA*, 1991, **88**, 6921.
192. R.W. Roberts and D.M. Crothers, *Proc. Natl. Acad. Sci. USA*, 1991, **88**, 9397.

193. T.M. Woolf, D.A. Melton and C.G.B. Jennings, *Proc. Natl. Acad. Sci. USA*, 1992, **89**, 7305.
194. K.J. Hertel, D. Herschlag and O.C. Uhlenbeck, *EMBO J.*, 1996, **15**, 3751.
195. K.A. LeCuyer and D.M. Crothers, *Proc. Natl. Acad. Sci. USA*, 1994, **91**, 3373.
196. W. Bains and G.C. Smith, *J. Theor. Biol.*, 1988, **135**, 303.
197. S.P.A. Fodor, J.L. Read, M.C. Pirrung, L. Stryer, A.T. Lu and D. Solas, *Science*, 1991, **270**, 467.
198. U. Maskos and E.M. Southern, *Nucleic Acids Res.*, 1992, **20**, 1679.
199. M.J. O'Donnell-Maloney, C.L. Smith and C.R. Cantor, *TIBTECH*, 1996, **14**, 401.
200. M. Schena, R.A. Heller, T.P. Theriault, K. Konrad, E. Lachenmeier and R.W. Davis, *TIBTECH*, 1998, **16**, 301.
201. A.V. Fotin, A.L. Drobyshev, D.Y. Proudnikov, A.N. Perov and A.D. Mirzabekov, *Nucleic Acids Res.*, 1998, **26**, 1515.
202. J.M. Bevilacqua and P.C. Bevilacqua, *Biochemistry*, 1998, **37**, 15877.
203. M.J. Serra and D.H. Turner, *Methods Enzymol.*, 1995, **259**, 242.
204. H. Jacobson and W.H. Stockmayer, *J. Chem. Phys.*, 1950, **18**, 1600.
205. M.E. Burkard, T. Xia and D.H. Turner, *Biochemistry*, 2001, **40**, 2478.
206. S.J. Schroeder and D.H. Turner, *Biochemistry*, 2000, **39**, 9257.

3

RNA Structures Determined by X-ray Crystallography

JENNIFER A. DOUDNA
Yale University, New Haven, CT, USA

and

JAMIE H. CATE
University of California, Berkeley, CA, USA

3.1 INTRODUCTION

Many RNA molecules have complex three-dimensional structures under physiological conditions, and the chemical basis for their functional properties cannot be understood unless these structures are known. In recent years it has become practical to determine RNA structures by X-ray crystallography, which can provide high-resolution information not only about RNA conformation but also about RNA interactions with ligands such as metal ions. The study of RNA by X-ray crystallography has become technically feasible due to the development of methods for producing milligram quantities of virtually any RNA molecule, and for crystallizing RNA and producing heavy atom derivatives of RNA crystals. This chapter discusses these methods, and reviews the RNA crystal structures that are currently known.

3.2 CRYSTALLIZATION OF RNA

3.2.1 Production of Homogeneous RNA by *In Vitro* Transcription

In vitro transcription with bacteriophage T7 RNA polymerase is the method of choice for obtaining milligram quantities of RNA for crystallization. Its only drawback is that transcription with this enzyme results in molecules that are heterogeneous at their 3′-termini and, depending on template sequence, may also be heterogeneous at their 5′-termini.[1,2] Transcripts containing these extra residues cannot be removed from RNA preparations by preparative purification techniques when chain lengths exceed ~50 nucleotides.

Terminal heterogeneity can be removed from transcripts using ribozymes in *cis* and *trans* geometries. When included as part of an RNA transcript, hammerhead, hairpin, and hepatitis delta virus ribozyme sequences will self-cleave during or after transcription to produce RNA with defined termini. This method has been used by several groups to obtain RNA samples suitable for structure determination.[3,4]

3.2.2 Chemical Synthesis

Short oligoribonucleotides are often conveniently prepared using automated, solid-phase DNA synthesis machines. Chemically protected ribonucleoside phosphoramidites are sequentially coupled to a protected nucleoside attached at its 3′ end to a solid support such as controlled-pore glass or polystyrene. When synthesis of the sequence is complete, base hydrolysis is used to cleave its linkage to the solid support, releasing a 2′-*O*-silyl protected oligomer. Silyl protecting groups are removed using tetrabutylammonium fluoride (TBAF) or similar chemical reagents. Improvements in the 2′-OH protecting groups and deprotection methods, as well as development of effective oligonucleotide purification methods, have made the chemical synthesis of RNA oligonucleotides of up to 40–50 nucleotides routine (for more information, see Chapter 6). With extreme care, RNA oligomers of up to 80 nucleotides in length can be produced in milligram quantities by solid-phase synthesis.[5,6] In practice, however, chemical synthesis of RNA for crystallization is practical for oligonucleotides 30 nucleotides or less in length.

One advantage of this approach is that nucleotide analogues are readily incorporated into synthetic oligoribonucleotides at specific sites. This is useful for the production of heavy atom derivatives (see below) and for investigating ligand–RNA interactions. Enzymatic ligation of short synthetic RNAs and longer RNA molecules prepared by *in vitro* transcription can be used to produce chimeric molecules that contain modified bases at specific locations[7] (see Chapter 15).

3.2.3 Purification of RNA for Crystallization

Prior to crystallization experiments, contaminating salts and chemical reagents must be removed from RNA samples. This is usually accomplished using ion exchange or reverse phase chromatography, for RNA molecules up to ~40 nucleotides long, and by denaturing polyacrylamide gel electrophoresis for larger RNAs. Following purification, the RNA is dialyzed extensively into a low-salt buffer, and often it is then annealed by heating to 60–90 °C and slow cooling in the presence of 1–10 mM magnesium ion.

3.2.4 Establishing the Suitability of RNA Preparations for Crystallization

Whenever possible, purified RNA samples are tested for biological activity prior to crystallization. In the case of a tRNA, this might involve assaying for charging by tRNA synthetase, or, in the case of ribozymes, measuring catalytic activity. Once the activity of a sample is confirmed, it is tested for conformational homogeneity (polydispersity) using native polyacrylamide gel electrophoresis, size exclusion chromatography, or dynamic light scattering.[8] The first two techniques can evaluate polydispersity only under low ionic strength conditions, while light scattering allows the determination of conformational homogeneity of RNA in solutions containing a variety of electrolytes and additives.

3.2.5 Sparse Matrix Approaches to RNA Crystallization

The crystallization of macromolecules is a trial and error process, and it is usually necessary to screen a wide range of conditions to find any that are conducive to crystal nucleation and growth. In the case of RNA, additional factors may complicate crystallization, such as the source and purity of material and the inherent instability of RNA. Furthermore, since some RNA molecules adopt several different conformations in solution, conditions that favor a single conformer must be found and used for crystallization.

To facilitate the search for crystallization conditions, sets of precipitating solutions have been developed that are biased towards conditions that have generated RNA crystals in the past.[9–11] These sets are applied to RNA using approaches based on the incomplete factorial and sparse matrix methods developed for protein crystallization. Satisfactory crystals of RNA duplexes, the hammerhead ribozyme, the P4–P6 domain of the *Tetrahymena* group I intron, and a fragment of 5S ribosomal RNA have all been obtained this way.

3.3 HEAVY ATOM DERIVATIVES OF RNA CRYSTALS

Once satisfactory crystals of a macromolecule are obtained, the phases for structure factors must be determined so that an electron density map can be calculated. For new structures this is usually achieved by making heavy atom derivatives of crystals, measuring diffraction intensities, and calculating phases based on the positions of the heavy atom(s).[12,13] Heavy atom derivatives of tRNA crystals were produced by soaking lanthanides into crystals, or by reacting crystals with osmium pyridine.[14] For the hammerhead ribozyme, crystal derivatives were prepared by lanthanide soaks and by covalent modification of the RNA with bromine. Covalent modification with bromine or iodine has also been used to solve the structures of short RNA duplexes and the loop E fragment of 5S ribosomal RNA. The crystal structure of the P4–P6 domain of the *Tetrahymena* ribozyme was solved by osmium hexammine substitution of magnesium binding sites in the major groove of the RNA.

3.4 DUPLEX STRUCTURES

The A-form helix is the structural unit from which complex, three-dimensional RNA structures are built. Isolated RNA helices often crystallize readily, and their structures can be solved using molecular replacement or covalent modification of the RNA. The high resolution (>2 Å) of some of the structures that have resulted has allowed a detailed look at metal ion binding sites, non-Watson–Crick base pairings, base bulging, helix packing in crystal lattices, and hydration.

3.4.1 Metal Ion Interactions in RNA Duplexes

Most structured RNAs require divalent metal ions for folding, and ribozymes generally need them for catalysis.[15] Several divalent ions have been located in the hammerhead ribozyme crystal structures (see below), but their functional significance remains unclear. One divalent metal ion binding site seen in hammerhead structures has also been found in a duplex containing sheared G · A and asymmetric A · A base pairs.[16] The site occurs at a C · G pair followed by the sheared G · A pair. Interestingly, tandem G · A mismatches have been found near the active sites of a lead-dependent ribozyme and an RNA ligase ribozyme, and they occur frequently in ribosomal RNA.[17–19] Thus, this motif may turn out to be a common way to position divalent metal ions within an RNA structure.

3.4.2 Noncanonical Base Pairs

Noncanonical or mismatch base pairs are common in RNA, and internal loops in rRNA often contain a high proportion of adenosines.[20] How are these nucleotides arranged, and how do they alter helical geometry? One example has been seen in a symmetric duplex, which includes a 5′-GAAA-3′ bulge surrounded by Watson–Crick pairs.[16] In this structure, tandem asymmetric A • A base pairs are sandwiched between sheared G • A pairs. Another common motif in rRNA involves tandem U • U pairs.[20] The three known duplex structures that contain this motif demonstrate that its structure varies depending on flanking sequences.[21–23] Two of these duplexes contain U • U wobble pairs,[21,23] but interestingly, the number of hydrogen bonds between the U • U pairs depends on the flanking base pairs. This result is consistent with effects seen in thermodynamic studies in solution,[24] but crystal packing forces may also affect the base pair geometry. In the third example, the U • U tandems form at the end of a duplex in an intermolecular contact.[22] These U • U tandems form unusual Hoogsteen pairs in which the N-3—H and O-4 of one uridine hydrogen bond to the O-4 and C-5—H of the other. It still is not clear whether tandem Hoogsteen U • U pairs like this can form in the middle of a duplex region, but they certainly might occur at the end of a helix. More importantly, the structure provides clear examples (at 1.4 Å resolution) of CH—O hydrogen bonds in base pairs and provides a model for U • Ψ base pairs in RNA.

3.4.3 RNA Packing and Hydration

RNA packing and hydration play important roles in RNA function, as highlighted in experiments involving large entropic contributions to ΔG.[25,26] While deceptively simple in form, the A-form helix can be greatly distorted, as seen in a structure of an RN–DNA chimeric duplex with a single looped-out adenosine.[27] In addition, the conformation of the extruded adenosine sheds some light on why the backbone of looped-out bases is often susceptible to magnesium-induced hydrolysis. Two high-resolution structures reveal in detail the pattern of hydration of G–C base pairs,[22,28,29] while a lower resolution structure sheds new light on the hydration of A–U pairs.[30,31] In these structures, the backbone plays key roles in the observed hydration patterns: the 2′-OH and the *pro*-R_p phosphate oxygen. As 2′-OH groups play important roles in RNA packing, exemplified by the ribose zipper (see Section 3.7), the heavy involvement of the 2′-OH group in hydration is a major factor to consider in thermodynamic studies of RNA–RNA interactions.

3.5 TRANSFER RNA

The first RNA molecule to be solved by X-ray crystallography that is large enough to have a tertiary structure was transfer RNA (tRNA). They were first because tRNAs are quite small and so abundant that they are readily purified from cells in adequate quantities. Modern methods for RNA production were not required. The crystal structure of $tRNA^{Phe}$, which was determined independently by three groups, became the basis for much of the RNA structural and functional biology that was done for the next 20 years.

tRNA plays a crucial role in protein biosynthesis. It is an adaptor molecule, one end of which interacts with amino acids and the other of which interacts with messenger RNA. Unlike normal double-stranded DNA, tRNA contains short helical elements interspersed with loops and its secondary structure is often drawn as a "cloverleaf" (Figure 1(a)). On the acceptor end, it carries an amino acid that corresponds to the genetic code triplet in its anticodon loop. The anticodon loop forms base pairs with messenger RNA on the ribosome, which then catalyzes peptide bond formation between the amino acid covalently bonded to one tRNA and the growing peptide chain covalently attached to a second one. Our understanding of tRNA structure and function has been reviewed in far more detail than is appropriate here;[32] however, some experiments regarding nucleotide modifications in tRNA deserve mention.

In the crystal structures of $tRNA^{Phe}$, the anticodon loop is ~70 Å away from the acceptor end of the molecule where the amino acid is attached. The four helical stems in that tRNA form an L-shaped molecule, each arm of which consists of a stack of two helices. A network of tertiary interactions between the D and T loops stabilize the assembly (Figure 1(b)),[33,34] a pattern that is conserved in other tRNA crystal structures.[35,36] While some tRNAs fold properly in the absence of divalent ions, Mg^{2+}

stabilizes the tertiary structure of all of them, and the binding sites of some of the Mg^{2+} ions involved have been inferred from tRNA crystal structures.[37,38]

Although tRNAs synthesized *in vitro* from the four naturally occurring nucleotides are active, tRNAs purified from cells always contain at least a few modified nucleotides. Modifications have been found in all of the stems and loops of tRNA.[32] One class of modifications, in the D and T loops, for example, optimizes the tertiary folding of tRNA.[33,34] Other modifications in the acceptor arm (acceptor stem and T stem) play a role in helping the protein-synthesizing machinery distinguish between initiator tRNAs and elongator tRNAs.[36] In the anticodon, modifications often play key roles in specifying codon recognition.[39]

Two advances in the chemical analysis of tRNAs have accelerated our understanding of these modifications. First, mass spectroscopic analysis of nucleotides in natural tRNAs has greatly expanded our knowledge of the kinds of modifications that occur in tRNA. In addition, the extent of modification in different types of organisms can be quickly assessed in the same way. For example, tRNAs from bacteria that grow in cold habitats have a higher abundance of dihydrouridine, which may increase conformational flexibility,[40] while thermophiles have modifications that may stabilize conformation.[41] Second, the chemical synthesis of tRNA-length RNAs may allow for milligram quantities of tRNA to be made that contain single-site modifications.[6] For example, both intra- and interhelical disulfide cross-links have already been incorporated into tRNA for the purpose of biophysical studies.[42] Combined with our increased knowledge of how tRNA interacts with amino-acyl tRNA synthetases and elongation factors, the ability to make designed modifications in tRNA, natural or otherwise, opens many new areas for biochemical and biophysical study.

(a)

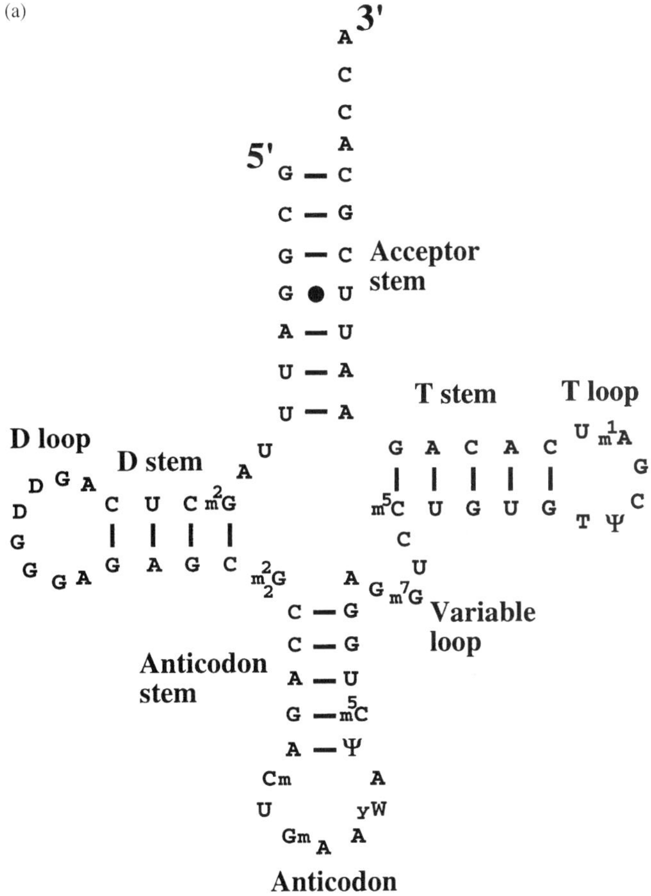

Figure 1 Structure of yeast phenylalanyl tRNA. (a) The secondary structure shows numerous naturally occurring modified nucleosides within the conserved tRNA fold.

(b)

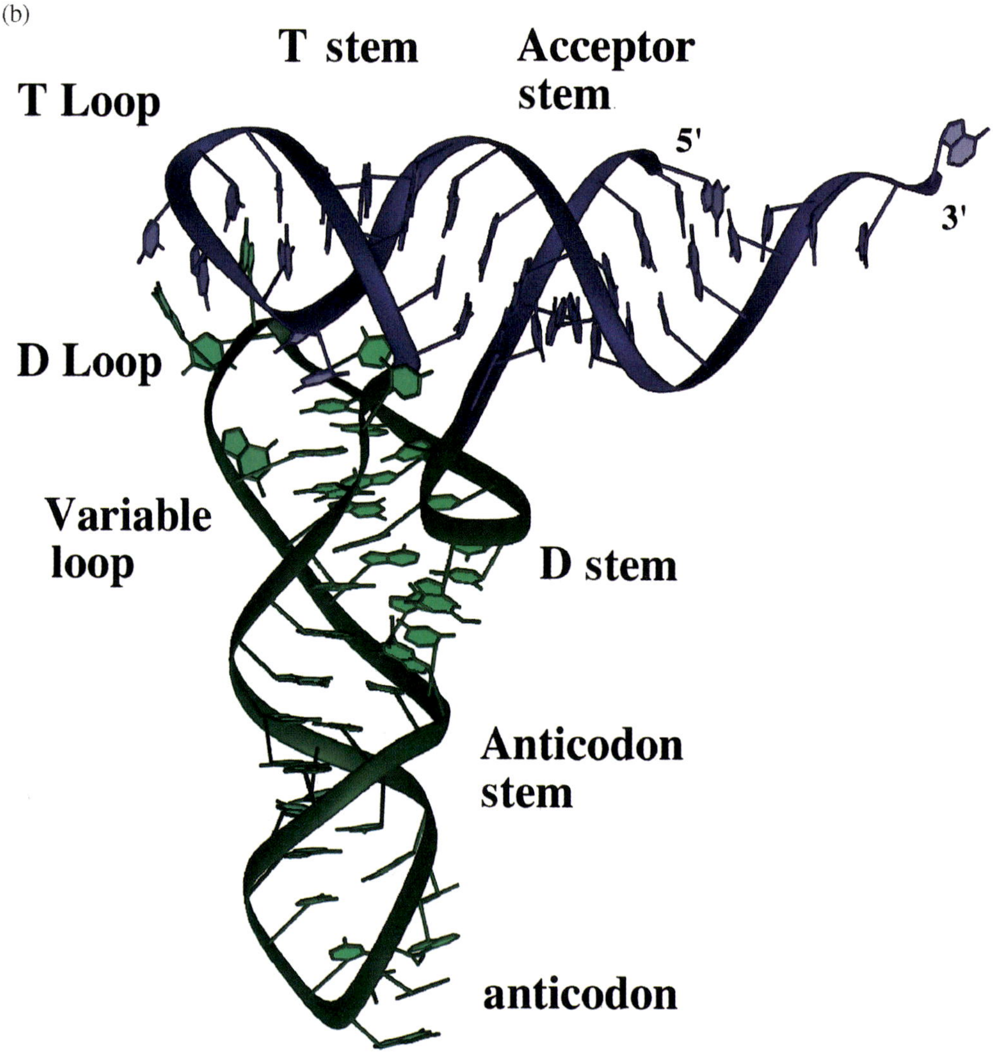

Figure 1 (continued) (b) Ribbon diagram of the crystal structure. Two sets of stacked helices form the L-shaped structure: the T-stem and the acceptor stem are in purple, and the D-stem and anticodon stem are in green.

3.6 THE HAMMERHEAD RIBOZYME

The hammerhead motif is a self-cleaving RNA sequence found in small RNAs that are plant pathogens. They make it possible for the multimeric genomes produced by rolling circle replication to cleave into unit length molecules.[43] Unlike ribozymes such as self-splicing introns and the catalytic RNA subunit of ribonuclease P, the hammerhead domain is small, consisting of three helices that adjoin a core of phylogenetically conserved nucleotides (Figure 2). Cleavage occurs via nucleophilic attack of the 2′-hydroxyl of a specific nucleotide within the core on its adjacent phosphodiester bond to produce a 2′,3′-cyclic phosphate and a 5′-hydroxyl terminus.[44] Normally a single-turnover catalyst, the hammerhead is readily made into a multiple-turnover enzyme by separating the strand containing the cleavage site from the rest of the core.[44,45] Divided molecules like this have proven useful for crystallization because they allow replacement of the substrate strand with an all-DNA strand, or with an RNA strand modified at the cleavage site by a 2′-*O*-methyl group, neither of which can be cleaved. The crystal structures of these hammerhead–inhibitor complexes have revealed the overall geometry of the ribozyme but have raised almost as many new questions concerning the catalytic mechanism as they have answered.

In three dimensions, the hammerhead is shaped like a wishbone or γ, with stems I and II forming the arms, and stem III and the core forming the base.[46,47] This fold is seen in both of the inhibitor complexes solved so far despite differences in RNA backbone connectivities, substrate strand identities,

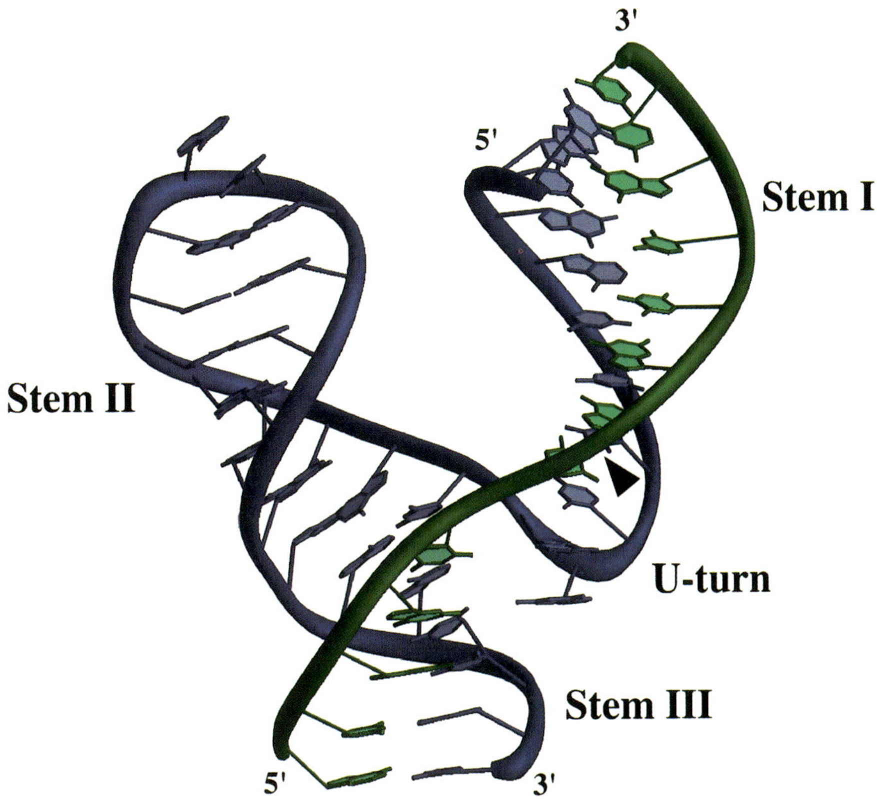

Figure 2 Crystal structure of the hammerhead ribozyme. The substrate strand is in green; the cleavage site is indicated by a black arrow within the U-turn.

crystallization conditions, and crystal packing. Whereas the three stems are all A-form helices, the structure of the central core is created, in part, by noncanonical pairings of the phylogenetically conserved nucleotides. Stems II and III sandwich two sheared G • A base pairs and an A–U base pair to form one long pseudocontinuous helix from which stem I and the catalytic site emanate. The highly conserved sequence CUGA between stems I and II forms a tight turn nearly identical in conformation to the uridine turn previously seen in the X-ray crystal structure of yeast phenylalanine transfer RNA.[34,48] The cytosine at the cleavage site between stems I and III is positioned near the CUGA cleft by interactions with the C and A of that sequence. This proximity led Klug and co-workers[47] to propose that the uridine turn, called domain I by McKay's group,[46] constitutes the catalytic pocket of the ribozyme.

Since the hammerhead-inhibitor complexes do not position the scissile bond correctly for the in-line nucleophilic attack that is believed to be part of the catalytic mechanism, these crystal structures probably represent the ground state of the ribozyme.[46,47] This has led to speculation that the hammerhead ribozyme may have to undergo a conformational change in order for cleavage to occur. Eckstein and co-workers[49] have used fluorescence resonance energy transfer (FRET) data to build a three-dimensional model of the hammerhead ribozyme that is similar to the X-ray models, except that the helical groove of stem I facing stem II differs. To distinguish between the solution and X-ray models, an elegant set of disulfide cross-linking experiments was carried out.[50] When stems I and II are cross-linked in conformations that exclude either the FRET or X-ray models, only the ribozyme cross-linked in a manner consistent with the X-ray structures is active. In addition, gel electrophoresis and transient electric birefringence have shown that the three stems are roughly co-planar and do not rearrange significantly after cleavage.[51,52] On the basis of these data, it is unlikely that the cleavage reaction requires a large change in conformation of the ribozyme.

More recent crystal structures of a hammerhead ribozyme complexed with a cleavable substrate have provided new insight into the rearrangements that occur in the catalytic pocket. Unlike the previous hammerhead ribozymes, the construct examined is active in the crystal lattice, allowing the experimenters to "trap" an intermediate in the reaction pathway.[53,54] The major difference between this structure and its two predecessors is a repositioning of the substrate nucleotides in the catalytic pocket.

In the trapped structures, a divalent metal ion is bound to the *pro*-Rp oxygen of the phosphate involved in the cleavage reaction, as previously proposed. The scissile bond, however, while still not positioned for in-line attack, is rotated closer to the required orientation.

Although Scott *et al.* propose a new model for the transition state based on this structure,[54] some key questions remain unanswered. First, what are the actual positions of the metal ions involved in catalysis? Second, what is the role of G5 in the CUGA U-turn in the catalytic pocket? Biochemical studies have clearly shown that all of the Watson–Crick base functionalities of G5 and its 2′-hydroxyl are critical for catalysis, yet none of the crystal structures reveals a clear role for this nucleotide. In later structures,[55] weak density interpreted as a divalent metal ion appears next to this guanosine, which is consistent with uranium-induced cleavage at that site, but no function for this metal ion has been shown.

Finally, the tandem sheared G•A base pairs seen in the crystal structures seem to be incompatible with the available biochemical data. Functional group modification studies have suggested that the G • A base pairs could not be in the sheared conformation.[56] Besides the clear need to distinguish between ground state and transition state structure stabilities, there are other factors that should be considered when attempting to relate biochemical data to crystal structures. For example, the thermodynamic stability and even the base pairing conformations of tandem G • A pairs are affected dramatically by their context.[57–59] In addition, the chemically modified RNAs used to "trap" the ribozyme in mid-reaction may have many alternate conformations that are not easily detected in the biochemical experiments.[60]

3.7 THE P4–P6 DOMAIN OF THE *TETRAHYMENA* GROUP I SELF-SPLICING INTRON

Group I introns, which are defined by a conserved catalytic core and reaction pathway, splice precursor RNAs so that mature ribosomal, transfer, or messenger RNAs can be formed.[61] Half of the conserved core in the *Tetrahymena thermophila* intron is found in an independently folding domain consisting of the base-paired (P) regions P4 through P6 (P4–P6).[62] By itself, the P4–P6 domain folds into a structure whose chemical protection pattern is very similar to that seen for the P4–P6 region of the intact intron.[62,63] The crystal structure of this domain, a 160-nucleotide RNA, has revealed several new aspects of RNA secondary and tertiary folding, and provides the first example of a kind of helical packing that is thought to occur in large ribozymes and RNA–protein complexes.[64]

In the 2.8 Å crystal structure of the P4–P6 domain, a sharp bend allows stacked helices of the conserved core to pack alongside helices of an extension (helices P5a, P5b, and P5c, or P5abc) that is important for folding and catalytic efficiency (Figure 3).[64,65] Two specific sets of tertiary interactions clamp the two halves of the domain together: an adenosine-rich corkscrew plugs into the minor groove of helix P4, and a GAAA tetraloop binds to a conserved 11-nucleotide internal loop,[66] termed the tetraloop receptor. The A-rich bulge coordinates two magnesium ions via its phosphate oxygens, allowing the backbone to invert and the bases to flip out. The adenosines make numerous tertiary contacts that connect the core helices to the helices in the P5abc extension. From biochemical evidence, these interactions are crucial to the stability of the entire domain.[62–64] The other half of the clamp, equally important to the packing of helices P5abc against the core (although not to the folding of the P5abc region itself), involves a GAAA tetraloop in the same conformation as seen previously.[67,68] The tetraloop receptor, a motif seen in many RNAs, has a widened minor groove that enables it to dock with the tetraloop in a highly specific manner.

The ribose 2′-hydroxyl group is involved in a common motif that occurs in both clamp interactions between the helical stacks. Pairs of riboses form an interhelical "ribose zipper" — a major component of the packing interactions (Figure 4). McKay and co-workers[68] also observed packing that involves pairs of 2′-hydroxyl contacts between a GAAA loop and the stem II minor groove of another hammerhead molecule in the crystal lattice. In a group II intron, riboses likely to be involved in a ribose zipper each contribute 2 kcal mol^{-1} of binding energy via their 2′-OH groups.[69] The number of ribose zippers seen so far suggests that this is a common way to pack RNA helices together.

One unexpected motif seen in the P4–P6 domain structure mediates both intramolecular and intermolecular interactions. At three separate locations in the 160-nucleotide domain, adjacent adenosines in the sequence lie side by side and form a pseudo-base pair within a helix (Figure 5).[70] This AA platform opens the minor groove for base stacking or base pairing with nucleotides from a noncontiguous RNA strand.[70] The platform motif has a distinctive chemical modification signature which may make it possible to detect it in other RNAs chemically.[62,70] The ability of this motif to facilitate higher order

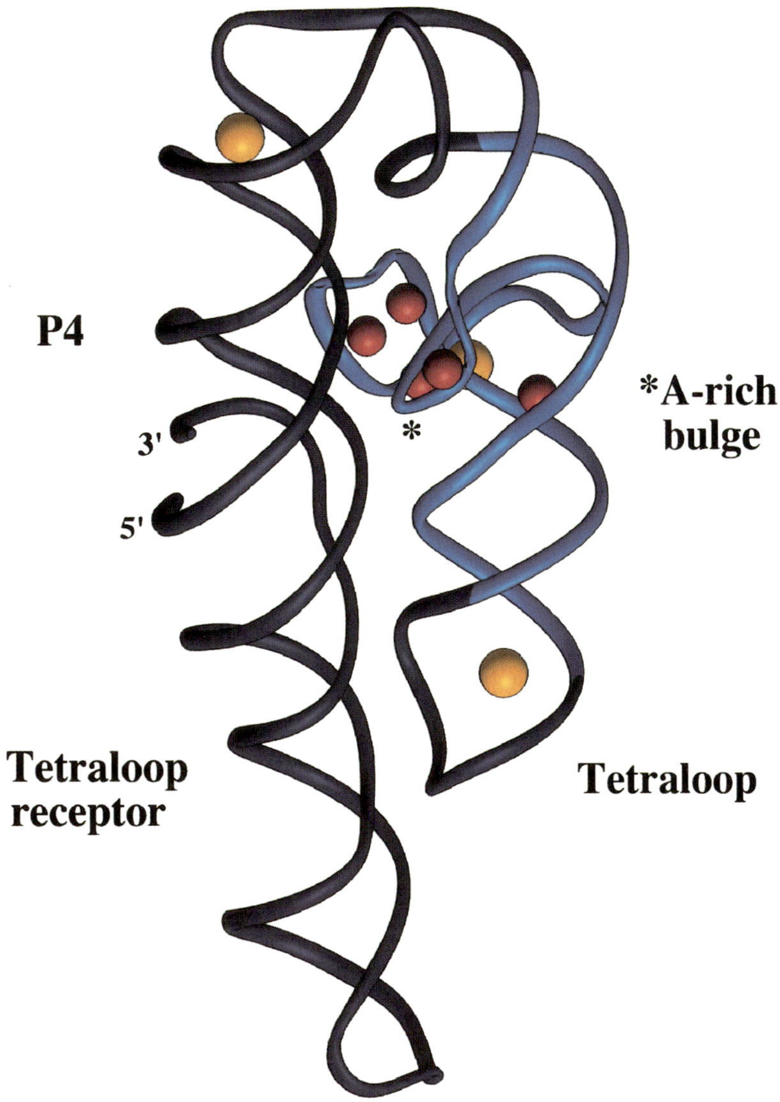

Figure 3 Crystal structure of the P4–P6 domain of the *Tetrahymena* group I intron. In red, magnesium ions in the metal ion core; in gold, osmium hexammine binding sites in the major groove; in light blue, the P5abc helices. The tertiary contacts between the helical stacks are indicated: the A-rich bulge docks into the minor groove of P4, and the tetraloop docks into the minor groove of the tetraloop receptor.

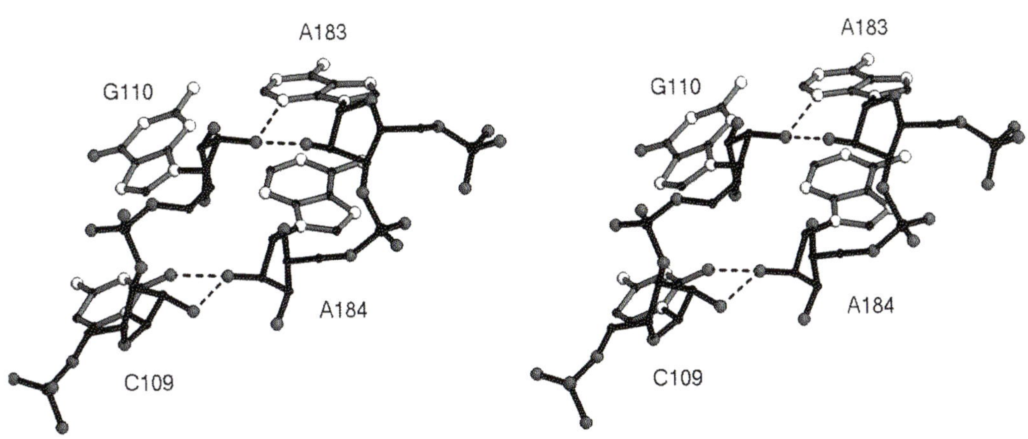

Figure 4 Stereoview of the ribose zipper. Pairs of nucleotides from different strands form a network of directed hydrogen bonds via 6 2′OH groups and the minor-groove functionalities of the bases.

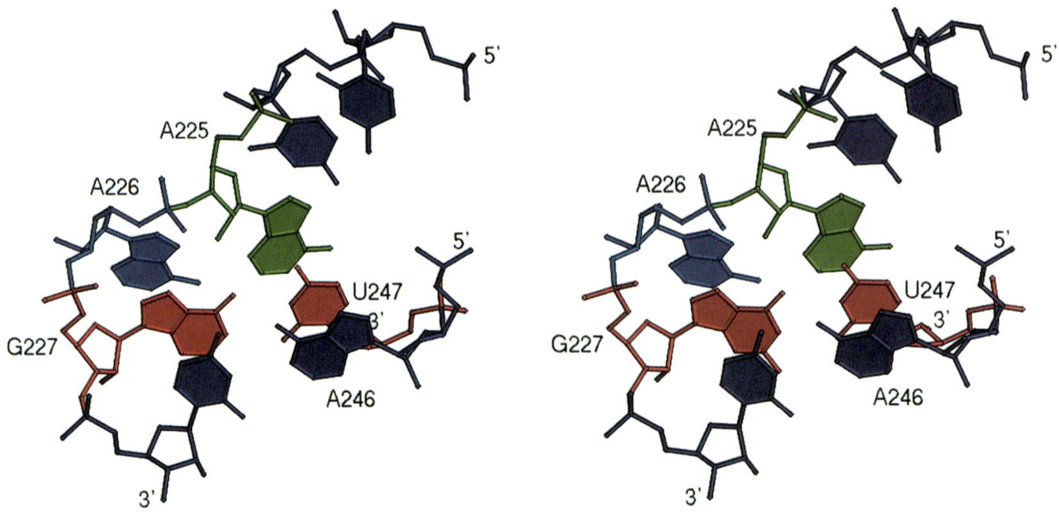

Figure 5 Stereoview of an AA platform. Viewed from the major groove, the AA platform allows cross-strand stacking, opening the minor groove for long-range tertiary contacts. The tertiary contact would be to the upper left of the AA platform as shown.

folding provides at least one explanation for the abundance of adenosine residues in internal loops of many RNAs.

Many of the contacts that stabilize the P4–P6 domain structure, as well as its packing in the crystal lattice, involve the wide and shallow minor groove as opposed to the deep and narrow major groove. However, osmium hexammine, the compound used to determine the RNA structure by multiwavelength anomalous diffraction, binds at three locations in the major groove where nonstandard base pairs create pockets of negative electrostatic potential.[71] In two cases, the heavy atoms occupy sites normally bound by hydrated magnesium ions in the native RNA. One of the motifs involved, tandem G–U wobble pairs, occurs frequently in ribosomal RNAs,[72] suggesting a mechanism for metal binding in the ribosome.

3.8 5S RIBOSOMAL RNA FRAGMENT

E. coli ribosomal 5S RNA (5S rRNA), which contains 120 nucleotides, forms part of the 50S ribosomal subunit and binds three proteins — L25, L18, and L5. Like all structured RNAs, it has internal non-Watson–Crick base paired regions of loops. One of these, loop E, adopts its biologically functional structure only in the presence of millimolar magnesium ion concentrations. Mild nuclease digestion of 5S rRNA yields a 62 nucleotide fragment I that includes helices I and IV and loop E. The ribosomal protein L25 binds to both 5S rRNA and fragment I and protects helix IV and loop E from chemical modification.

Crystals of fragment I were originally obtained in 1983, but the structure proved difficult to solve due to limited diffraction resolution (~4 Å) and the lack of suitable heavy atom derivatives. Heavy atom derivatives were ultimately obtained by incorporation of chemically modified nucleotides into the RNA.[73] The structures of both fragment I and a smaller dodecamer duplex RNA containing the sequence of loop E have been determined at 3 Å and 1.5 Å resolution, respectively.[74]

Together these two structures reveal an interesting helical molecule in which the loop E region is distorted by three "cross-strand purine stacks" (Figure 6).[74] In this motif, the helical backbones are pinched together by stacking of A bases from opposite sides of the helix that are part of a sheared A • G and a Hoogsteen A • U, respectively. Furthermore, four magnesium ions bind in the narrowed major groove of the helix, creating a unique binding surface for the cognate ribosomal protein L25.

3.9 FUTURE DIRECTIONS

While the crystal structures of several RNAs provide new insights into RNA folding and catalysis, exciting challenges lie ahead. The hammerhead catalytic center is now the best understood of numerous

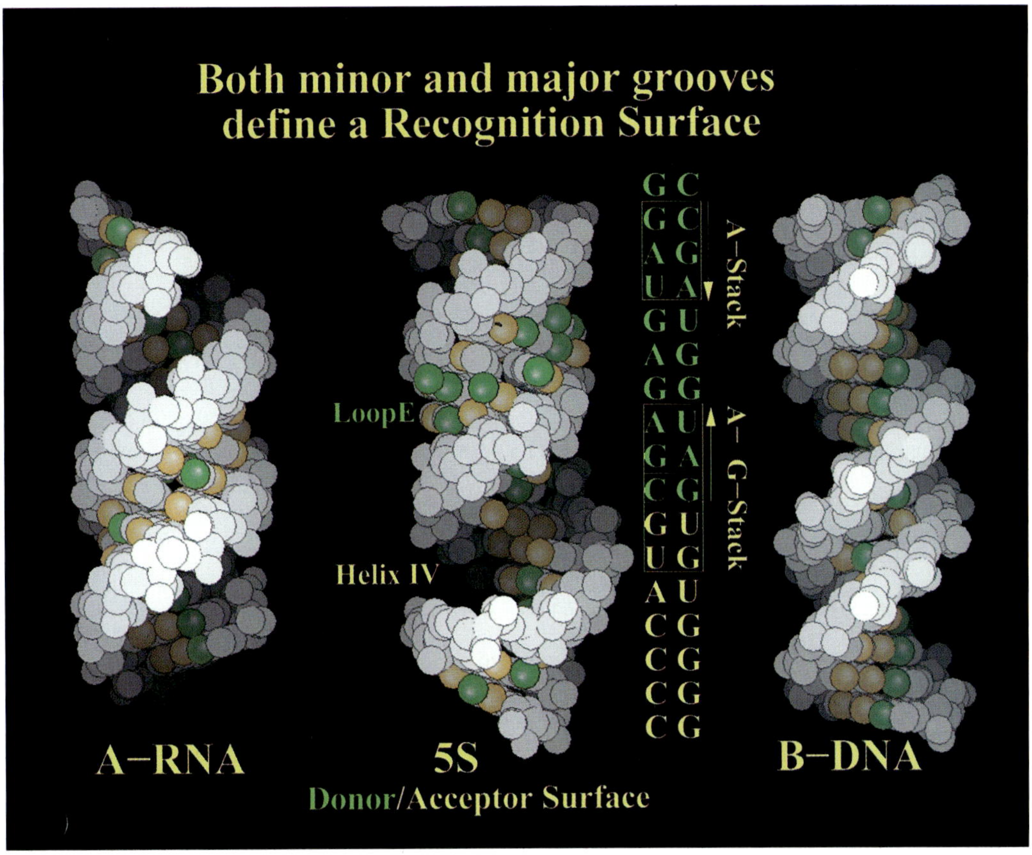

Figure 6 5S rRNA fragment I crystal structure. The irregular helical geometry of the fragment contrasts with A-form RNA (left) or B-form DNA (right). The two cross-strand purine stack motifs in the structure are indicated on the sequence by yellow boxes and arrows.

ribozyme active sites; the others remain mysteries. Structures of the ribosome and its subunits have recently been published, while the structures and roles of RNA in ribonucleoprotein particles including telomerase and the spliceosome remain to be tackled by crystallographers. Chemical and biochemical experiments, critical for solving and understanding the hammerhead and P4–P6 domain structures, will be key to structural studies of these other RNAs also.

3.10 REFERENCES

1. J.F. Milligan, D.R. Groebe, G.W. Witherell and O.C. Uhlenbeck, *Nucleic Acids Res.*, 1987, **15**, 8783.
2. J.F. Milligan and O.C. Uhlenbeck, *Methods Enzymol.*, 1989, **180**, 51.
3. S.R. Price, N. Ito, C. Oubridge, J.M. Avis and K. Nagai, *J. Mol. Biol.*, 1995, **249**, 398.
4. A.R. Ferre-D'Amare and J.A. Doudna, *Nucleic Acids Res.*, 1996, **24**, 977.
5. K.K. Ogilvie, N. Usman, K. Nicoghosian and R.J. Cedergren, *Proc. Natl. Acad. Sci. USA*, 1988, **85**, 5764.
6. J.T. Goodwin, W.A. Stanick and G.D. Glick, *J. Org. Chem.*, 1994, **59**, 7941.
7. M.J. Moore and P.A. Sharp, *Science*, 1992, **256**, 992.
8. A.R. Ferre-d'Amare and J.A. Doudna, *Methods Mol. Biol.*, 1997, **74**, 371.
9. J.A. Doudna, C. Grosshans, A. Gooding and C.E. Kundrot, *Proc. Nat. Acad. Sci. USA*, 1993, **90**, 7829.
10. W.G. Scott, J.T. Finch, R. Grenfell, J. Fogg, T. Smith, M.J. Gait and A. Klug, *J. Mol. Biol.*, 1995, **250**, 327.
11. B.L. Golden, E.R. Podell, A.R. Gooding and T.R. Cech, *J. Mol. Biol.*, 1997, **270**, 711.
12. C.W.J. Carter and R.M. Sweet (eds.), "Macromolecular Crystallography Part A. Methods in Enzymology", Academic Press, New York, 1997,
13. C.W.J. Carter and R.M. Sweet (eds.), "Macromolecular Crystallography Part B. Methods in Enzymology.", Academic Press, New York, 1997,
14. W. Saenger, "Principles of Nucleic Acid Structure", Springer, New York, 1984.
15. A.M. Pyle, *Science,*, 1993, **261**, 709.
16. K.J. Baeyens, H.C. DeBondt, A. Pardi and S.R. Holbrook, *Proc. Natl. Acad. Sci. USA*, 1996, **93**, 12–851.
17. E.H. Ekland, J.W. Szostak and D.P. Bartel, *Science*, 1995, **269**, 364.

18. T. Pan and O.C. Uhlenbeck, *Nature*, 1992, **358**, 560.
19. D. Gautheret, D. Konings and R.R. Gutell, *J. Mol. Biol.*, 1994, **242**, 1.
20. R.R. Gutell, N. Larsen and C.R. Woese, *Microbiol. Rev.*, 1994, **58**, 10.
21. K.J. Baeyens, H.L. DeBondt and S.R. Holbrook, *Nature Struct. Biol.*, 1995, **2**, 56.
22. M.C. Wahl, S.T. Rao and M. Sundaralingam, *Nature Struct. Biol.*, 1996, **3**, 24.
23. S.E. Lietzke, C.L. Barnes, J.A. Berglund and C.E. Kundrot, *Structure*, 1996, **4**, 917.
24. M. Wu, J.A. McDowell and D.H. Turner, *Biochemistry*, 1995, **34**, 3204.
25. Y. Li, P.C. Bevilacqua, D. Mathews and D.H. Turner, *Biochemistry*, 1995, **34**, 14–394.
26. T.S. McConnell and T.R. Cech, *Biochemistry*, 1995, **34**, 4056.
27. S. Portmann, S. Grimm, C. Workman, N. Usman and M. Egli, *Chem. Biol.*, 1996, **3**, 173.
28. S. Portmann, N. Usman and M. Egli, *Biochemistry*, 1995, **34**, 7569.
29. M. Egli, S. Portmann and N. Usman, *Biochemistry*, 1996, **35**, 8489.
30. M.C. Wahl, C. Ban, S. Sekharudu, B. Ramakrishnan and M. Sundaralingam, *Acta Crystallogr., Ser. D*, 1996, **52**, 655.
31. A.C. Dock-Bregeon, B. Chevrier, A. Podjarny, J. Johnson, J.S. Debear, G.R. Gough, P.T. Gilham and D. Moras, *J. Mol. Biol.*, 1989, **209**, 459.
32. D. Söll and U.L. Rajbhandary (eds.), "tRNA: Structure, Biosynthesis and Function", ASM Press, Washington, DC, 1995.
33. F.L. Suddath, G.J. Quigley, A. McPherson, D. Sneden, J.J. Kim, S.H. Kim and A. Rich, *Nature*, 1974, **248**, 20.
34. J.D. Robertus, J.E. Ladner, J.T. Finch, D. Rhodes, R.S. Brown, B.F. Clark and A. Klug, *Nature*, 1974, **250**, 546.
35. D. Moras, M.B. Comarand, J. Fischer, R. Weiss, J.C. Thierry, J.P. Ebel and R. Giege, *Nature*, 1980, **288**, 669.
36. R. Basavappa and P.B. Sigler, *EMBO J.*, 1991, **10**, 3105.
37. A. Jack, J.E. Ladner, D. Rhodes, R.S. Brown and A. Klug, *J. Mol. Biol.*, 1977, **111**, 315.
38. S.R. Holbrook, J.L. Sussman, R.W. Warrant, G.H. Church and S.H. Kim, *Nucleic Acids Res.*, 1977, **4**, 2811.
39. S. Yokoyama and S. Nishimura, in "tRNA: Structure, Function and Recognition", eds. D. Söll and U.L. Rajbhandary, ASM Press, Washington, 1995, 207
40. J.J. Dalluge, T. Hamamoto, K. Horikoshi, R.Y. Morita, K.O. Steller and J.A. McCloskey, *J. Bacteriol.*, 1997, **179**, 1918.
41. J.A. Kowalak, J.J. Dalluge, J.A. McCloskey and K.O. Steller, *Biochemistry*, 1994, **33**, 7869.
42. J.T. Goodwin, S.E. Osborne, E.J. Scholle and G.D. Glick, *J. Am. Chem. Soc.*, 1996, **118**, 5207.
43. G.A. Prody, J.T. Bakos, J.M. Buzayan, I.R. Schneider and G. Bruening, *Science*, 1986, **231**, 1577.
44. O.C. Uhlenbeck, *Nature*, 1987, **328**, 596.
45. J. Haseloff and W.L. Gerlach, *Nature*, 1988, **334**, 585.
46. H.W. Pley, K.M. Flaherty and D.B. McKay, *Nature*, 1994, **372**, 68.
47. W.G. Scott, J.T. Finch and A. Klug, *Cell*, 1995, **81**, 991.
48. S.H. Kim, G.J. Quigley, F.L. Suddath, A. McPherson, D. Sneden, J.J. Kim, J. Weinzierl and A. Rich, *Science*, 1973, **179**, 285.
49. T. Tuschl, C. Gohlke, T.M. Jovin, E. Westhof and F. Eckstein, *Science*, 1994, **266**, 785.
50. S.T. Sigurdsson, T. Tuschl and F. Eckstein, *RNA*, 1995, **1**, 575.
51. K.M. Amiri and P.J. Hagerman, *Biochemistry*, 1994, **33**, 13–172.
52. K.M. Amiri and P.J. Hagerman, *J. Mol. Biol.*, 1996, **261**, 125.
53. J.B. Murray, D.P. Terwey, L. Maloney, A. Karpeisky, N. Usman, L. Beigelman and W.G. Scott, *Cell*, 1998, **92**, 665.
54. W.G. Scott, J.B. Murray, J.R.P. Arnold, B.L. Stoddard and A. Klug, *Science*, 1996, **274**, 2065.
55. A.L. Feig, W.G. Scott and O.C. Uhlenbeck, *Science*, 1998, **279**, 81.
56. T. Tuschl, M.M.P. Ng, W. Pieken, F. Benseler and F. Eckstein, *Biochemistry*, 1993, **32**, 11–658.
57. J. Santalucia Jr. and D.H. Turner, *Biochemistry*, 1993, **32**, 12–612.
58. A.E. Walter, M. Wu and D.H. Turner, *Biochemistry*, 1994, **33**, 11–349.
59. M. Wu and D.H. Turner, *Biochemistry*, 1996, **35**, 9677.
60. O.C. Uhlenbeck, *RNA*, 1995, **1**, 4.
61. T.R. Cech, *Annu. Rev. of Biochem.*, 1990, **59**, 543.
62. F.L. Murphy and T.R. Cech, *Biochemistry*, 1993, **32**, 5291.
63. F.L. Murphy and T.R. Cech, *J. Mol. Biol.*, 1994, **236**, 49.
64. J.H. Cate, A.R. Gooding, E. Podell, K. Zhou, B.L. Golden, G.E. Kundrot, T.R. Cech and J.A. Doudna, *Science*, 1996, **273**, 1678.
65. G. VanderHorst, A. Christian and T. Inoue, *Proc. Natl. Acad. Sci. USA*, 1991, **88**, 184.
66. M. Costa and F. Michel, *EMBO J.*, 1995, **14**, 1276.
67. H.A. Heus and A. Pardi, *Science*, 1991, **253**, 1.
68. H.W. Pley, K.M. Flaherty and D.B. McKay, *Nature*, 1994, **372**, 111.
69. D.L. Abramovitz, R.A. Friedman and A.M. Pyle, *Science*, 1996, **271**, 1410.
70. J.H. Cate, A.R. Gooding, E. Podell, K. Zhou, B.L. Golden, A.A. Szewczak, C.E. Kundrot, T.R. Cech and J.A. Doudna, *Science*, 1996, **273**, 1696.
71. J.H. Cate and J.A. Doudna, *Structure*, 1996, **4**, 1221.
72. D. Gautheret, D. Konings and R.R. Gutell, *RNA*, 1995, **1**, 807.
73. C.C. Correll, ecB. Freeborn, P.B. Moore and T.A. Steitz, *J. Biomol. Struct. Dyn.*, 1997, **15**, 165.
74. C.C. Correll, B. Freeborn, P.B. Moore and T.A. Steitz, *Cell*, 1997, **91**, 705.

4

RNA Conformational Dynamics

DONALD M. CROTHERS
Yale University, New Haven, CT, USA

4.1 INTRODUCTION

4.1.1 The Time Axis

The RNA folding problem is 4-dimensional. The spatial coordinates used to define the structures that an RNA can take on are essential but not sufficient for a full understanding of molecular properties and function. In a living cell there is continuous RNA synthesis, folding, conformational change, and degradation. The time range covered is vast, from pico- and nanoseconds for structural changes in individual nucleotides, to microseconds for forming small hairpin helices, to milliseconds for forming compact tertiary structures such as that in tRNA, to minutes for refolding more complex molecules such as introns. Renaturation, or undoing incorrect RNA folds, can take hours or days or longer.

4.1.2 How RNA and Proteins Differ in the Folding Problem

Consider the following thought-experiment. A protein that forms an αβ dimer is mixed with all the proteins in the cell, and the solution is subjected to thermal or chemical denaturation, causing separation of the subunits, after which it is slowly annealed back to native conditions. What is the probability that a native αβ dimer will emerge from this messy process? Small, indeed! Yet if the same experiment is done with a double helical RNA in the presence of all the RNA in the cell, the chances are quite good of recovering the duplex with native pairing, as verified by the widespread use of hybridization to identify DNA and RNA sequences in complex nucleic acid mixtures. (In most such hybridizations, of course, one adds an excess of strand α in order to drive the kinetics of formation of the αβ duplex.)

The energy needed to stabilize double helical nucleic acids comes largely from the stacking of bases together. However, this interaction is relatively nonspecific, since there are only modest variations in the enthalpy of base pairing, depending on the nature of the dinucleotide step.[1,2] Specificity, of course, comes from the highly polar base pairing interactions. The result of these strong, directional forces is that the secondary structure elements of RNA, local hairpin helices, are thermodynamically stable in the absence of other interactions. This is in contrast to proteins, in which α-helices usually have only marginal stability absent interactions with other parts of the molecule.

Thus, RNAs are capable of rapid formation of stable hairpin helices, followed by slower formation of the non-canonical interactions that together constitute the tertiary structure.[3,4] In these simple cases the folding pathway through the energy "landscape" is well defined, although even apparently simple cases can involve some rearrangement of secondary structure upon tertiary structure formation, as shown by Wu and Tinoco.[5] In more extreme cases, the high potential stability of improper double helices can cause some RNAs to take long detours through metastable states in deep local energy minima that can be characterized as "denatured" by their distinctive mobility on electrophoresis gels. The comparatively large per-residue enthalpy that characterizes double helix formation frequently means that elevated temperature is needed to overcome the activation barrier for melting the incorrect structure to produce the native form.

4.2 THE FAST TIME SCALE: LOCAL INTERNAL DYNAMICS

Infrared spectroscopy reveals vibrational motions of the nucleotide bases and phosphodiester backbone. Typical IR absorbance bands for base vibrations occur around 1600 cm^{-1}, corresponding to frequencies of about 5×10^{13} s^{-1}, or times of the order of 10 femtoseconds (fs). The spectral density function, which describes the frequency components of the motion, is quite densely occupied by internal modes of varying complexity from this time scale out to approximately 1 nanosecond (ns). For example, Raman scattering from DNA crystals shows low frequency vibrational modes in the region from 20 to 120 cm^{-1},[6] corresponding to the time range from about 1 ps to 100 fs. The 100 fs mode was ascribed to concerted librational motions of the bases.

Nuclear spin relaxation studies are effective for examination of internal dynamics in the time scale from about 10 ps to several ns. Motion of one spin relative to another causes fluctuations in the local magnetic field. Frequency components of this motion near the Larmor or resonance frequency enhance the rate of relaxation of magnetization in the z or field direction, measured by the longitudinal relaxation time T1. Slower motions can cause line broadening, reflected in a shortened transverse relaxation time T2 because of incomplete averaging of the chemical shift tensor and dipole–dipole magnetic interactions. The importance of these relaxation effects depends on the correlation time for the internal motion, and the overall tumbling time. Fast overall tumbling of the whole molecule can mask slow internal motions unless the different states have significantly different chemical shifts that are incompletely averaged because of slow interconversion.

Early NMR studies of double helical DNA detected correlation times for sugar proton–aromatic proton and sugar proton–phosphate motions on the order of 1 ns.[7–9] This motion was not quenched by incorporating the DNA into nucleosomes.[9] Selective ^{13}C labeling of the purine bases in an RNA hairpin enabled Hall and Tang[10] to characterize the motion of the bases. Both fast (~10 ps) and slower (~1 ns) motions were reported, which could be interpreted in terms of ~20° rotational fluctuations of the glycosidic torsion angle. The purines in the loop displayed complex motions with ns correlation times as the loop melted at elevated temperature. A similar study with pyrimidines ^{13}C labeled at C1′ in a double helical DNA by Gaudin *et al.*[11] revealed an effective internal correlation time of about 10 ps for motion of the C1′-H1′ vector. Drobney and colleagues[12] and Lane[13] have reviewed the use of magnetic resonance and site-specific labeling to study internal dynamics in DNA.

Another useful probe of nucleic acid internal dynamics is time-dependent fluorescence depolarization of a bound ligand, for which the study of the ethidium–DNA complex by Zewail and colleagues[14] provides an illustration. A polarized laser pulse is used to excite the chromophore, and the fluorescence polarization is recorded as a function of time. An approximately 90° rotation of a molecule whose transition moment was initially aligned with the light polarization is sufficient for randomization of the polarization of the emission fluorescence. The experiment reveals a very rapid, $\leq$10 ps, anisotropy decay, which was ascribed to "wobble" of the drug in its intercalation site. This was followed by a non-exponential decay extending out to hundreds of nanoseconds. The relatively slow character of this relaxation is not a consequence of intranucleotide dynamics that are significantly slower than the ns scale revealed by other experiments. Rather, due to the torsional stiffness of DNA, absorbing twist fluctuations of $\sim$90° requires local twisting motions that cover a hundred or more base pairs. This one-dimensional "diffusion" of the twist fluctuations is the source of the non-exponential character of the fluorescence anisotropy relaxation.

Molecular dynamics simulations provide in principle a rich resource for characterizing internal dynamics on the time scale from ps to many ns, with a longer time limit that steadily increases as computer resources become more powerful. However, most such simulations have been undertaken to characterize the structure rather than the detailed dynamics. Nonetheless, examination of the time-dependence of parameters such as intramolecular distances in simulations of an anticodon hairpin by Auffinger and Westhof[15] shows fluctuations with apparent autocorrelation times from ps to hundreds of ps. In a more recent simulation, the same group reported a simulation in which there are fluctuations in the structure of a base triple in $tRNA^{Asp}$ that occur with a frequency on the order of hundreds of ps.[16] The lifetime of water molecules bound to $tRNA^{Asp}$ has been examined in a series of 500 ps simulations,[17] with a distribution of lifetimes varying from ps up to the maximum 500 ps extent of the simulation. A simulation more explicitly directed at determining dynamic properties, albeit on DNA, is that of Norberg and Nilsson[18], who examined internal correlation times in two small DNA double helices over a 500 ps time scale. Calculated correlation times were typically in the range of 20 to 100 ps, and a sugar pucker alternation was seen for a terminal dG residue, with a lifetime of $\sim$200 ps.

Miller and Kollman[19] simulated a stable RNA tetraloop from two closely related starting structures, and found that each structure remained close to its initial structure for the full length of the 2 ns simulations. They use a simple Arrhenius/Eyring approach as a basis for a useful rule of thumb for estimating the probability that structural transitions can be observed in ns scale simulations: one expects activation barriers of $\sim$5 kcal mol^{-1} to be overcome approximately once in a nanosecond. (The calculation assumes negligible activation entropy. Estimated times should increase by about 10-fold for each additional 1.4 kcal mol^{-1} of activation energy at 300 K.) Since many processes in nucleic acids have activation energies considerably larger than 5 kcal mol^{-1}, much faster computers, or new strategies, will be needed to simulate the dynamics of processes such as opening a base pair in the middle of a double helix at physiological temperature.

4.3 BASE STACKING DYNAMICS

Stacking of adjacent bases is a fundamental source of the free energy of stabilization of ordered structures in nucleic acids. To a first approximation the process in single stranded poly- and oligonucleotides can be analyzed with a localized two-state model, with each dinucleotide either stacked or unstacked. However, both states, especially the unstacked form, represent ensembles of conformations of the nucleotides. Since base stacking is accompanied by a reduction in absorbance in the UV, it has been possible to characterize the dynamics using optical methods. The stacking interaction is energetically favorable, with a negative ΔH of roughly 10 kcal mol^{-1}, so the temperature jump method can be used to induce unstacking. Pörschke[20] used a cable discharge technique to increase the temperature in a $\sim$10 ns interval, allowing study of the 25–100 ns dynamics of the stacking relaxation in CpC and poly C. The stacking rate constant in poly C (at 20 °C) was about 1×10^7 s^{-1}, with a dissociation rate constant of about 2×10^6 s^{-1}. The corresponding activation energies were about 3 kcal mol^{-1} and 12 kcal mol^{-1}, respectively. The rates were faster in CpC by about 2-fold for stacking and 12-fold for unstacking, probably because of the absence of coupling to the other modes of motion present in the polymer. Note that the 6-fold difference in the change of the two rate constants between polymer and dinucleotide implies less stacking in CpC than in the polymer, consistent with a modest degree of cooperativity in the stacking reaction in single strands.

Dewey and Turner[21] used a laser pulse to raise the temperature in a sample of poly A, with results quite similar to those of Pörschke, although the activation energy for unstacking was slightly larger, at 15 kcal mol^{-1}. The stacking rate was found to vary inversely with solution viscosity, implying that the motions required for the conformational search are rate limiting for stacking. Norberg and Nilsson[22] reported simulations of GpU that started from either a stacked or unstacked state. No unstacking of the stacked state was seen in 2 ns, consistent with the slower dynamics measured experimentally, and with simple expectation based on the magnitude of the activation energy. As discussed in Section 4.7 below, more complex reactions that involve a change in stacking pattern can occur on a much slower time scale than the elementary step of unstacking adjacent bases.

4.4 THE RATE OF BASE PAIRING

Formation of a base pair is the elementary reaction step in the transfer of genetic information. Because of the cooperative character of double helix formation, the dynamic and equilibrium properties of base pairs differ greatly depending on whether they are located at the end of a helix or in its interior. Base pairs at the end of a helix open and close rapidly in a process called "fraying", which has an equilibrium constant of roughly 2–10, depending on the nature of the base pair and its neighbor. The equilibrium modestly favors pairing under physiological conditions. In contrast, the equilibrium constant favoring closure of a base pair interior to a double helix is roughly 10^5 to 10^6.

The primary experiments measuring the rate of adding a base pair to a growing helix are more than 20 years old. The first estimates were indirect,[23,24] based on determination of the overall rate of double helix formation by oligonucleotides, which was taken equal to the rate constant for helix growth times an estimated equilibrium constant for forming the critical rate-limiting nucleus, starting from the single strands. Using different ribooligonucleotides, Pörschke and Eigen[23] and Craig *et al.*[24] reached similar conclusions, setting limits of 10^6 to 10^7 s^{-1} on the rate of helix growth. In 1974 Pörschke[25] made a more direct measurement by using the T-jump method to determine the relaxation kinetics of the fraying process in double helices of $A_{14}U_{14}$ and $A_{18}U_{18}$. These were chosen because they are long enough to allow significant fraying at temperatures below the T_m of the duplex. Measured relaxation times in the range of 100–300 ns were used to extract a rate constant of 8×10^6 s^{-1} for helix formation by A • U pairs, with a corresponding activation energy of 3–4 kcal mol^{-1}. Since the fraying reaction is too fast to produce significant NMR line broadening, and too slow to affect the proton T1 relaxation time appreciably, it has not been obvious how to pursue the dynamics of this process in more detail.

In contrast, the rate of opening base pairs interior to the double helix is much slower and therefore quite accessible to NMR studies. The strategy, involving measurement of the buffer-catalyzed rate of exchange of imino protons with solvent H_2O, was pioneered by Guéron, and has been reviewed by Guéron and Leroy.[26] The simplest form of the model is a reversible two-step exchange process, in which the base pair first opens, exposing it to solvent. The imino proton in the opened base pair is then transferred to a proton acceptor, after which there is rapid reprotonation by water, followed by closure of the base pair. The possible proton acceptors are H_2O, OH^-, and the conjugate base of the buffer (B). Imino protons are not sufficiently acidic to transfer protons rapidly to water, and the OH^- concentration is too small to be effective. Therefore, the predominant pathway for exchange is proton transfer to B, typically NH_3 or Tris. The rate of transfer depends on the diffusion-limited rate of collision of B with the imino proton, times a factor that measures the efficiency of proton transfer in a collision. This factor in turn depends on the relative acidity of the imino proton and BH^+; lowering the acidity of BH^+ implies that B becomes more basic, thus increasing the effectiveness of B as a proton acceptor. Typical transfer rates for a water-accessible imino proton are 10^5 to 10^6 s^{-1} in 10 to 100 mM Tris buffer at pH 7.5.[26] It is generally observed that a transiently opened base pair in a double helix closes more rapidly than 10^{-6} s, so a base pair opens and closes a number of times before a proton exchange event occurs. For this reason the opening rate is generally faster than the observed exchange rate. A plot of the reciprocal of the exchange rate versus the reciprocal concentration of B yields the base pair opening time as the intercept at infinite concentration of B. The slope can be used to estimate the equilibrium constant for local opening of the base pair.[26]

Proton exchange in DNA has been studied more extensively than RNA. For example,[26] in the double helical oligomer CGCGAATTCGCG at 15 °C, the G_4 base pair has a lifetime of 36 ms, and an estimated equilibrium constant for opening of about 2×10^{-7}. Taking the equilibrium constant as the ratio of opening and closing rates, one estimates a closing rate constant of about 10^8 s^{-1}. Base pair A_5 has a lifetime of 8 ms, an estimated opening equilibrium constant of 4.5×10^{-6}, and a closing rate constant of about $3 \times 10^{7}\ s^{-1}$. The corresponding numbers for A_6 are 30 ms, 2×10^{-6}, and $2 \times 10^{7}\ s^{-1}$. The rates of opening adjacent internal base pairs appear to be uncorrelated, so the dynamic properties are intrinsic to each base pair, within the equilibrium environment of its neighbors. Typical lifetimes for G • C base pairs in B-DNA are 40–100 ms, with values between 0.5 and 10 ms for A • U pairs. Base pairs at or near the end of a helix, however, open by the fraying process. For example, proton exchange of base pairs C_1 and G_2 in the oligomer above are too fast to measure by the NMR method.

Opening of base pairs in double helical RNA occurs at comparable rates to those observed for DNA. For example, the base pair lifetime in poly A • poly U is about 4 ms at 15 °C. Tertiary structure formation can greatly extend the proton lifetime,[26] as illustrated by the $\sim 10^{3}$ s lifetime for the U_8A_{14} tertiary pair in yeast $tRNA^{AsP}$. Further investigation of imino proton lifetimes in RNA tertiary interactions should provide valuable insight into local base pairing dynamics.

4.5 SECONDARY STRUCTURE FORMATION

Formation of double helix requires a nucleation event to bring the complementary regions of the sequence together, so the overall rate of forming a double helix is much slower than the rate of helix growth. If the reaction is bimolecular, the rate is governed by a corresponding rate constant that is typically in the range of 10^{6} to $10^{7}\ M^{-1}\ s^{-1}$.[23,24] With a strand concentration of 1 μM, the characteristic time for helix formation is thus roughly 100 ms to 1 s.

Hairpin helix formation is substantially faster than the bimolecular reaction, because the local concentration of the two complementary regions is much higher. Typical time constants are 10–100 μs.[27,28] For simple reactions the rate of helix formation can be estimated from the size of the hairpin loop. For example, in the folding–unfolding relaxation kinetics of $tRNA^{fMet}$, it was observed that the anticodon helix, which closes a loop of 7 nucleotides, occurs in 80 μs, whereas the acceptor stem, which closes a 59 nucleotide loop, requires 2 ms.[29] The ratio of the closure rates for these two loops is in close accord with the expectation value of $(7/59)^{-3/2} = 25$, based on the Jacobson–Stockmayer theory for the cyclization of flexible polymer chains:[30] the entropy loss of loop closure slows down the formation of hairpin helices. Nucleation of a helix between more distant parts of the sequence in a large RNA may require times well above the ms scale.

Hairpin helix formation can also be appreciably slowed by the presence of a competing secondary structure that has to be melted before the proper hairpin can form. This circumstance is usually accompanied by an appreciable positive activation energy for the helix formation reaction, resulting from the heat required to melt the competing structure. Helix dissociation reactions generally have activation energies approximately equal to the heat of melting, so the corresponding rate constants are highly temperature dependent. It is useful to recall that at T_m the rates of helix formation and dissociation are equal.

4.6 TERTIARY STRUCTURE DYNAMICS

4.6.1 tRNA

It is a general observation that multivalent ions strongly stabilize tRNA tertiary structure.[31,32] As a consequence, in the absence of Mg^{2+} or other multivalent ions, the tertiary structure is often the least stable part of the native structure. Early work on the dynamics of tRNA tertiary structure by the Crothers[31–34], Maass[35–39] and other[40–42] groups exploited this fact, thus making the tertiary structure unfolding transition an early and experimentally accessible component of the thermal melting curve. At 100–200 mM monovalent counterion concentration, the time constants for tRNA tertiary structure formation are typically in the range of 10–100 ms,[29,31,33,34,37,39] although there is considerable variation from one tRNA to another.

The combination of NMR and relaxation kinetic experiments provides a powerful approach to correlating folding–unfolding relaxation signals with specific structural regions in the molecule.[29,42] The basis for this approach is the observation that imino proton exchange of hairpin helices results in line broadening of the corresponding resonances when the helix lifetime is less than about 5 ms.[29] Using the measured temperature dependence of the helix dissociation rate from relaxation kinetics for a specific relaxation signal, one can extrapolate to lower temperatures to match a 5 ms extrapolated lifetime to a temperature zone for broadening of a group of resonances corresponding to a specific helix.[29] In this way, the structural origin of the relaxation signals can be established.[29,42]

Another important clue for determining the structural origin of relaxation signals derives from the fortuitous presence of 4-thiouridine (s^4U) bases in natural tRNAs. For example, the s^4U residue at position 8 in $tRNA^{fMet}$, between the acceptor stem and DHU helices, shows a large hyperchromism in the tertiary melting transition,[31,33] as is also observed for $tRNA^{Tyr}$,[34] which has s^4U residues at positions 8 and 9. Modern synthetic methods, which permit substitution of modified bases at any position, and much improved NMR methods offer considerable potential for revisiting the problem of tRNA structural dynamics in search of a deeper level of understanding than provided by the experiments of the 1970s.

In most cases examined, some relatively unstable component of the secondary structure is found to melt together with the tertiary structure. For example, in $tRNA^{fMet}$, the dihydrouridine (DHU) stem melts together with the tertiary structure,[29] while in the case of $tRNA^{Gly}$, the tertiary structure melting was accompanied by simultaneous unfolding of the DHU and anticodon stems.[42]

These experiments revealed significant differences in the stability and dynamic properties of the tertiary structure, depending on small changes in base sequence between tRNA isoaccepting species. For example, the replacement of A at position 47 in $tRNA^{fMet}{}_3$ by m^7G in $tRNA^{fMet}{}_1$,[43] leads to an increase in the T_m for tertiary structure unfolding from 30 to 46 °C.[29,33] However, the time constant for folding $tRNA^{fMet}{}_1$ ($\sim$7 ms) is about two-fold slower than that for the less stable $tRNA^{fMet}{}_3$.[33] The activation energy for unfolding the tertiary structure in $tRNA^{fMet}{}_1$ is $\sim$50 kcal mol^{-1}, about two-fold larger than for $tRNA^{fMet}{}_3$.[33] The origin of these effects is not clear, although it is likely that the positive charge contributed by m^7G in a critical region of the folded molecule may be at least partly responsible for the stability increase.

Another example can be found in the contrasting stabilities and dynamics of isoacceptors of $tRNA^{Tyr}$,[34] which differ by two nucleotides in the loop region of the "extra arm" of the cloverleaf structure.[44,45] In spite of their distance from the tertiary interactions, these nucleotides have considerable influence on the folding dynamics of the tertiary structure.[34] The Tyr_{II} isoacceptor is slightly less stable than Tyr_I, and with a relaxation time of 40 ms at T_m, it refolds approximately 4-fold more slowly. Unlike $tRNA^{fMet}$, both Tyr tRNAs have appreciable activation energy for refolding, which is larger for Tyr_{II} than for Tyr_I.

The authors explained this behavior as the result of the formation of two different competing structures in the two tRNAs subsequent to unfolding of the tertiary structure.[34] The extra arm nucleotide differences yield 4-base complementarity with the TψC loop in Tyr_I and with the anticodon loop in Tyr_{II}. In the same vein, Hilbers *et al.*[42] used a combination of NMR and relaxation kinetic measurements to conclude that a non-cloverleaf structure forms after melting the tertiary structure. Another example can be found in the relaxation kinetic studies of Coutts *et al.*[38] on yeast $tRNA^{Phe}$, which led to the conclusion that non-cloverleaf interactions exist (or are formed) after the tertiary structure is unfolded. This is likely to be a common phenomenon in the formation of RNA tertiary interactions, since it is improbable that the final structure contains the only stable interactions of the sequences involved. The general effect is to slow down the folding kinetics, and to contribute a positive activation energy to the folding reaction.

It may turn out to be difficult to formulate simplifying generalizations about the dynamics of even such seemingly related folding processes as formation of tertiary structure in tRNA. For example, Urbanke *et al.*[39] found a substantial dependence of the rate of forming the tertiary structure on ionic strength (at constant temperature) for $tRNA^{Phe}$, whereas Yang and Crothers[34] addressed the same question for $tRNA^{Tyr}$, and found no dependence of the folding rate on salt concentration, again at constant temperature. It is clear that electrostatic repulsion has to be overcome in order to form the tertiary structure, but the question is whether this barrier is crossed before or after the rate-limiting step for the folding reaction. Variability may be the rule rather than the exception.

4.6.2 Folding Dynamics of the Tetrahymena Group I Ribozyme

The group I ribozyme from *Tetrahymena thermophila* has played a pivotal role in the field of RNA catalysis, starting with the report in 1982 by Cech and coworkers[46] that *Tetrahymena* pre-rRNA is capable of excising its own intron. The ribozyme can be split into two domains, one of which, P4–P6, can fold independently, while the other, P3–P9, requires P4–P6 for formation of its native tertiary structure.[47] Both the intrinsic interest of the system and the potential for in-depth analysis of the kinetic measurements has been greatly increased by solution of the crystal structure of the P4–P6 subunit of the ribozyme.[48]

The self-splicing intron is approximately 400 nucleotides long, roughly 5 times the length of tRNAs. Molecules of this size are not suited for NMR spectroscopic studies, and, unlike tRNAs, they contain no naturally occurring modified nucleotides to serve as localized spectroscopic markers. Consequently it was necessary to develop new methods in order to analyze the kinetic properties at nucleotide resolution.

Sclavi *et al.*[49] described the application to RNA folding of a rapid nucleic acid footprinting method that uses a synchrotron source to generate a burst of hydroxyl radicals by radiolysis of water. The X-ray source is combined with a quench-flow device that allows folding to be triggered by mixing with buffer containing Mg^{2+} ions, followed by a variable time delay before application of the X-ray pulse. The initial state contained Na^{+} ions at 10 mM concentration, and in the final state the Mg^{2+} concentration increased from zero to 10 mM. Local kinetic parameters can be determined for any nucleotide whose hydroxyl radical reactivity changes between those two ionic conditions on a time scale substantially slower than the fastest resolvable time, which is about 10 ms.

In applying this technique to the *Tetrahymena* ribozyme, Sclavi *et al.*[49] showed that most parts of the P4–P6 domain fold in a few seconds or less, whereas parts of the P3–P9 domain require times of up to a few min, at 42 °C. This distinction is not unreasonable in view of the ability of the P4–P6 domain to fold independently, nucleating around a Mg^{2+} core,[50] whereas folding of the P3–P9 domain requires P4–P6.

Under the conditions of this assay, the stable hairpin helices are already formed before addition of Mg^{2+}. The first event in folding the tertiary structure was within the P5a–P5c subdomain of P4–P6. This was followed by concerted formation of the P4–P6 tertiary structure in a few seconds.[49] Evidence was found for interdomain contacts between P2 and P9 over the next 10 s, indicating further condensation of the RNA. Organization of the catalytic core of the molecule, including formation of the tertiary structure in P3 and P7, followed on a time scale of minutes. A likely reason for the relatively slow kinetics of this phase of the reaction is presence of a competing structure, requiring a conformational switching event in order to reach the final structure, as discussed in Section 4.7.

Another important technical innovation is represented by the use by Williamson and coworkers of deoxyoligonucleotide hybridization probes and RNAse H cleavage to monitor accessibility of regions of the sequence that are blocked by tertiary structure formation.[51,52] In this method the reaction is initiated by Mg^{2+} addition, followed after a variable time delay by addition of the oligonucleotide probe and RNAse H. The kinetics of processes that occur on a time scale of about 1 min or longer can be resolved. Applying the technique to the *Tetrahymena* ribozyme, Zarrinkar and Williamson[51] showed that oligonucleotides complementary to tertiary structure-forming regions in P3 and P7 become unable to hybridize on a time scale of about 4 min, with an activation energy of 20 kcal mol^{-1}. There was little dependence of the rate on Mg^{2+} concentration around 2 mM Mg^{2+}, but subsequent work has shown that the process has an optimum rate near 2 mM Mg^{2+}, and becomes considerably slower at higher concentrations of Mg^{2+}.[53,54] These features support the view that formation of the final conformation is slowed by the necessity for disrupting a competing structure, as demonstrated persuasively by Woodson and colleagues in a series of experiments using primarily chemical footprinting, site-directed mutagenesis and electrophoretic separation of reaction intermediates and products,[55–63] as discussed in Section 4.7.

4.7 DYNAMICS OF CONFORMATIONAL SWITCHING

It is often proposed that RNA switches between two or more stable conformations in the course of its biological function. Study of such processes probably has more biological significance than *in vitro* examination of the rate of folding an intact transcript, which starts from very different boundary conditions than found in the cell, where folding events accompany RNA synthesis. In addition, cellular proteins are likely to affect the pathway and rate of folding reactions.[64] However, detailed understanding of the dynamics and mechanism of RNA conformational switching reactions has not come easily or rapidly. We examine here some cases that illustrate the mechanisms and wide range of possible time scale on which such processes can occur.

4.7.1 Switch in Stacking in the tRNA Anticodon Loop

Urbanke and Maass[65] studied the dynamics of a conformational switch between states that differ in base stacking. They used the temperature jump method with fluorescence detection to monitor the time dependence of the Y-base fluorescence in yeast $tRNA^{Phe}$. This approach allows one to focus on the dynamics of a local region in the intact RNA, in this case the anticodon loop. The kinetic constants have a substantial activation energy, ~16 kcal mol^{-1}, with time constants varying from about 1 ms at −10 °C to 10 μs at 33 °C. Thus the process, which is thought to involve changes in the stacking pattern in the anticodon loop, remains ~100 times slower than the stacking elementary step even at the high end of this temperature range. The proposed model[65] invokes a switch from stacking of the anticodon loop on the 3′ side of the anticodon helix to stacking on the 5′ side. The cooperative character of this simple switch, in which disfavored unstacked species constitute the transition state, gives rise to a higher activation energy and a slower rate than normally seen for stacking.

4.7.2 A Helix–Helix Switch in a Spliced Leader RNA

The kinetics of a conformational switch that involves two competing secondary structures was studied by LeCuyer and Crothers.[66] The 5′-half of the spliced leader RNA from *Leptomonas collosoma* has two competing, phylogenetically conserved secondary structures that differ only marginally in free energy.[67] According to the model, Form 1 contains 14 bp, and Form 2 has 13. None of the base pairs are common to both structures, which can be forced to interconvert by adding an oligonucleotide that binds to a sequence that is unbonded in only one of the structures. Using stopped flow and T-jump methods, LeCuyer and Crothers[66] showed that the back and forth rate constants for the switch were 2 s^{-1} and 10 s^{-1} at 42 °C. These values correspond to a relaxation time for the equilibrium of about 80 ms, about a thousand-fold slower than the conformational switch in the yeast $tRNA^{Phe}$ anticodon. The estimated activation energy for the SL RNA switch was about 25 kcal mol^{-1}, considerably less than would be expected if the mechanism involved dissociation of 13–14 bp before the alternate structure begins to form. In addition, if that were the mechanism, the process should be much slower, because of the stability of the duplex regions. To explain the results, a branch migration mechanism was proposed, in which the alternate helix is nucleated and grows while the competing structure shrinks.[66]

4.7.3 A Slow Switch in an mRNA Regulatory Region

An example of an even slower conformational switch, whose detailed structural basis is not obvious but probably involves tertiary interactions, comes from the work of Gluick *et al.*[68] They observed two electrophoretically distinguishable conformations of a 112 nucleotide fragment containing the first ribosome initiation site of *E. coli* α mRNA. This element is involved in translational regulation through interaction with ribosomal protein S4. The higher mobility or "fast" conformation is dominant at high Mg^{2+} concentrations, whereas the "slow" conformation is favored in absence of multivalent ions. At 30 °C the formation fate for the fast conformation is about 3.2×10^{-2} min^{-1}, or 5.3×10^{-4} s^{-1}. The corresponding relaxation time for the conformational switch is about 2000 s (about 30 min), or $\sim 2 \times 10^4$-fold slower than the SL RNA switch. Thus the time range spanned by the dynamics of these first three RNA conformational switches is more than 7 orders of magnitude.

4.7.4 Switching and Docking of the P1 Helix in a Ribozyme

As a final example, we consider the RNA conformational switch most intensively studied to date, namely the reorganization of the P1 or substrate (because it contains the splice site) helix, and its docking as part of the tertiary structure of the *Tetrahymena* ribozyme. Involvement of the P1 helix in tertiary interactions is shown, for example, by the requirement for 2′-OH groups in both the internal guide sequence (IGS) and the 5′-intron portions of the helix that pair with the IGS to form P1.[69] The early stopped flow experiments of Turner and associates[70] used fluorescent-labeled substrate oligonucleoticles to form the P1 helix by hybridization in with the IGS in *trans*. Two reaction steps could be resolved kinetically, which were interpreted as formation of the P1 helix, followed by the slower docking reaction.

Narlikar and Herschlag[71] used a crosslinking approach to identify three distinct states of the ribozyme in 10 mM Mg^{2+}, depending on the status of the P1 helix. A reactive azidophenacyl crosslinking group was placed at the 5′-end of the IGS, which was hybridized with a series of substrate oligonucleotides. Some of these, such as a deoxy analog, were able to hybridize but not dock. The alternate crosslinked species formed with these oligonucleotides could be distinguished by their electrophoretic mobilities on denaturing gels. The initial state is that in which the IGS is unpaired, followed by formation of the "open" complex, in which the oligonucleotide has hybridized, but the P1 helix remains undocked. The final, docked state is referred to as the "closed" complex. The formation of distinct crosslinks in all three of these states implies that there are time-dependent specific interactions between the IGS, with or without hybridization to form P1, and the rest of the molecule in 10 mM Mg^{2+}.

Woodson and colleagues adopted the strategy of providing the substrate in *cis* to form the pre-rRNA, and examined the ability of alternate substrate secondary structures to affect the rate of folding.[55–63] A fraction of the RNA usually renatures rapidly to the form active in splicing, but a variable portion of primary transcripts renatured at physiological temperatures is found in inactive states. These slowly interconverting conformers were distinguished by their electrophoretic mobilities on non-denaturing gels. The active form of the pre-rRNA is the highest mobility band, corresponding to a compact structure. Chemical modification of the RNA prior to renaturation allows the nucleotides whose interactions are critical for active and inactive forms to be identified by chemical modification interference experiments on the separate electrophoretic bands.[63] The inactive form is found to have an alternate structure, called alt P3, instead of P3 in the catalytic core of the ribozyme. The presence of this non-native structure is coupled with formation of an alternate hairpin helix, P(−1), in the 5′-exon.[63] The balance between the competing structures can be altered by mutations that stabilize appropriate pairings.[63] The authors propose a model in which the unpaired IGS in P(−1) can pair with P3, thereby favoring the alt P3 structure.

Switching between the P1–P3 and P(−1)–alt P3 structures slows folding of the P3–P9 domain via the pathway through the misfolded intermediate. Many of the earlier experiments on the slow rate of folding of P3–P9 relative to P4–P6 in the ribozyme can be understood on this basis. The P(−1) structure is shown as a 12 bp dumbbell helix, whereas P(1) is a 9 bp hairpin.[63] The activation energy for the folding reaction, about 12 kcal mol^{-1}, is much smaller than would be expected if the mechanism involved full melting of the incorrect structure before initiating formation of the native form. In this case, as in the SL RNA switch,[66] it is likely that some form of branch migration pathway is involved, allowing coexistence of portions of each structure in the transition state. It is notable that mutations that destabilize the P4–P6 domain accelerate the folding of P3–P7,[52] implicating structural fluctuations in P4–P6 in the transition state for folding of P3–P7.

An impressive degree of mechanistic detail at the molecular level has been achieved for *Tetrahymena* pre-rRNA and its corresponding ribozyme. However, it is also striking that a very significant effort by several laboratories was necessary to reach this level of understanding. The methods that have been developed will aid in attacking other RNA folding/switching problems, but present indications are that these systems are idiosyncratic, making each problem a new one and posing a major challenge to formulation of simplifying generalizations.

4.8 REFERENCES

1. K.J. Breslauer, R. Frank, H. Blocker and L.A. Marky, *Proc. Natl. Acad. Sci. USA*, 1986, **83**, 3746–3750.
2. T. Xia, J. SantaLucia Jr., M.E. Burkard, R. Kierzek, S.J. Schroeder, X. Jiao, C. Cox and D.H. Turner, *Biochemistry*, 1998, **37**, 14719–14735.
3. P. Brion and E. Westhof, *Annu. Rev. Biophys. Bimolec. Struct.*, 1997, **26**, 113–137.
4. I. Tinoco Jr. and C. Bustamante, *J. Mol. Biol.*, 1999, **293**, 271–281.
5. M. Wu and I. Tinoco Jr, *Proc. Natl. Acad. Sci. USA*, 1998, **95**, 11555–11560.
6. O.P. Lamda, A.H. Wang and G.J. Thomas Jr., *Biopolymers*, 1989, **28**, 667–678.
7. A. Early and D.R. Kearns, *Proc. Natl. Acad. Sci. USA*, 1979, **76**, 4165–4169.
8. M.E. Hogan and O. Jardetsky, *Proc. Natl. Acad. Sci. USA*, 1979, **76**, 6341–6345.
9. L. Kievan, I.M. Armitage and D.M. Crothers, *Nucl. Acids Res.*, 1979, **6**, 1607–1616.
10. K.B. Hall and C. Tang, *Biochemistry*, 1998, **37**, 9323–9332.
11. F. Gaudin, F. Paquet, L. Chanteloup, J.M. Beau, T.T. Nguyen and G. Lancelot, *J. Biomol. NMR*, 1995, **5**, 49–58.

12. B.H. Robinson, C. Mailer and G. Drobney, *Annu. Rev. Biophys. Bimolec. Struct.*, 1997, **26**, 629–658.
13. A.N. Lane, *Methods Enzymol.*, 1995, **261**, 413–435.
14. D.P. Millar, R.J. Robbins and A.H. Zewail, *Proc. Natl. Acad. Sci. USA*, 1980, **77**, 5593–5597.
15. P. Auffinger and E. Westhof, *Biophys. J.*, 1996, **71**, 940–954.
16. P. Auffinger, S. Louise-May and E. Westhof, *Biophys. J.*, 1999, **76**, 50–64.
17. P. Auffinger and E. Westhof, *J. Mol. Biol.*, 1997, **269**, 326–341.
18. J. Norberg and L. Nilsson, *J. Biomol. NMR*, 1996, **7**, 305–314.
19. J.L. Miller and P.A. Kollman, *J. Mol. Biol.*, 1997, **270**, 436–450.
20. D. Pörschke, *Biochemistry*, 1976, **15**, 1495–1499.
21. T.G. Dewey and D.H. Turner, *Biochemistry*, 1980, **19**, 1681–1685.
22. J. Norberg and L. Nilsson, *Biophys. J.*, 1994, **67**, 812–824.
23. D. Pörschke and M. Eigen, *J. Mol. Biol.*, 1971, **62**, 361–382.
24. M.E. Craig, D.M. Crothers and P. Doty, *J. Mol. Biol.*, 1971, **62**, 383–401.
25. D. Pörschke, *Biophys. Chem.*, 1974, **2**, 97–101.
26. M. Guéron and J.L. Leroy, *Methods Enzymol.*, 1995, **261**.
27. M. Dourlent, J.C. Thrierr, F. Brun and M. Leng, *Biochim. Biophys. Res. Commun.*, 1970, **41**, 1590–1596.
28. S.M. Coutts, *Biochim. Biophys. Acta*, 1971, **232**, 94–106.
29. D.M. Crothers, P.E. Cole, C.W. Hilbers and R.G. Shulman, *J. Mol. Biol.*, 1974, **87**, 63–88.
30. H. Jacobson and W.H. Stockmayer, *J. Chem. Phys.*, 1950, **18**, 1600–1606.
31. P.E. Cole, S.K. Yang and D.M. Crothers, *Biochemistry*, 1972, **11**, 4358–4368.
32. A. Stein and D.M. Crothers, *Biochemistry*, 1976, **15**, 160–168.
33. P.E. Cole and D.M. Crothers, *Biochemistry*, 1972, **11**, 4368–4374.
34. S.K. Yang and D.M. Crothers, *Biochemistry*, 1972, **11**, 4375–4381.
35. R. Römer, D. Riesner, S.M. Coutts and G. Maass, *Eur. J. Biochem.*, 1970, **15**, 7784.
36. D. Riesner, R. Römer and G. Maass, *Eur. J. Biochem.*, 1970, **15**, 85–91.
37. D. Riesner, G. Maass, R. Thiebe, P. Philippsen and H.G. Zachau, *Eur. J. Biochem.*, 1973, **36**, 76–88.
38. S.M. Coutts, D. Riesner, R. Römer, C.R. Rabl and G. Maass, *Biophys. Chem.*, 1975, **3**, 275–289.
39. C. Urbanke, R. Römer and G. Maass, *Eur. J. Biochem.*, 1975, **55**, 439–444.
40. M. Dourlent, M. Yaniv and C. Hélène, *Eur. J. Biochem.*, 1971, **19**, 108–114.
41. S.M. Coutts, J. Gangloff and G. Dirheimer, *Biochemistry*, 1974, **13**, 3938–3948.
42. C.W. Hilbers, G.T. Robilard, R.G. Shulman, R.D. Blake, P.K. Webb, R. Fresco and D. Riesner, *Biochemistry*, 1976, **15**, 1874–1882.
43. S.K. Dube, K.A. Marcker, B.F.-C. Clark and S. Cory, *Nature (London)*, 1968, **218**, 232–233.
44. H.M. Goodman, J. Abelson, A. Landy, S. Brenner and J.D. Smith, *Nature (London)*, 1968, **217**, 1019–1024.
45. U.L. RajBandaary, S.H. Chang, H.J. Gross, F. Harada, F. Kimura and S. Nishimura, *Fed. Proc., Fed. Am. Soc. Exp. Biol.*, 1969, **23**, 409.
46. K. Krueger, P.J. Grabowski, A.J. Zaug, J. Sands, D.E. Gottschling and T.R. Cech, *Cell*, 1982, **32**, 147–157.
47. E.A. Doherty and J.A. Doudna, *Biochemistry*, 1997, **36**, 3159–3169.
48. J.H. Cate, A.R. Gooding, E. Podell, K. Zhou, B.L. Golden, C.E. Kundrot, T.R. Cech and J.A. Doudna, *Science*, 1996, **273**, 1678–1685.
49. B. Sclavi, M. Sullivan, M.R. Chance, M. Brenowitz and S.A. Woodson, *Science*, 1998, **279**, 1940–1943.
50. J.H. Cate, R.L. Hanna and J.A. Doudna, *Nat. Struct. Biol.*, 1997, **4**, 553–558.
51. P.P. Zarrinkar and J.R. Williamson, *Science*, 1994, **265**, 918–924.
52. D.K. Treiber, M.S. Rook, P.P. Zarrinkar and J.R. Williamson, *Science*, 1998, **279**, 1943–1946.
53. M.S. Rook, D.K. Trieber and J.R. Williamson, *Proc. Natl. Acad. Sci. USA*, 1999, **96**, 12471–12476.
54. J. Pan, D. Thirumalai and S.A. Woodson, *Proc. Natl. Acad. Sci. USA*, 1999, **96**, 6149–6154.
55. S.A. Woodson, *Nucleic Acids Res.*, 1992, **20**, 4027–4032.
56. V.L. Emerick and S.A. Woodson, *Biochemistry*, 1993, **32**, 14062–14067.
57. V.L. Emerick and S.A. Woodson, *Proc. Natl. Acad. Sci. USA*, 1994, **91**, 9675–9679.
58. V.L. Emerick, J. Pan and S.A. Woodson, *Biochemistry*, 1996, **35**, 13469–13477.
59. J. Pan, D. Thirumalai and S.A. Woodson, *J. Mol. Biol.*, 1997, **273**, 7–13.
60. J. Pan and S.A. Woodson, *J. Mol. Biol.*, 1998, **280**, 597–609.
61. Y. Cao and S.A. Woodson, *RNA*, 1998, **4**, 901–914.
62. T. Nikolcheva and S.A. Woodson, *J. Mol. Biol.*, 1999, **292**, 557–567.
63. J. Pan and S.A. Woodson, *J. Mol. Biol.*, 1999, **294**, 955–965.
64. K.M. Weeks and T.R. Cech, *Cell*, 1995, **82**, 221–230.
65. C. Urbanke and G. Maass, *Nucleic Acids Res.*, 1978, **5**, 1551–1560.
66. K.A. LeCuyer and D.M. Crothers, *Proc. Natl. Acad. Sci. USA*, 1994, **91**, 3373–3377.
67. K.A. LeCuyer and D.M. Crothers, *Biochemistry*, 1993, **32**, 5301–5311.
68. T.C. Gluick, R.B. Gerstner and D.E. Draper, *J. Mol. Biol.*, 1997, **270**, 451–463.
69. S.A. Strobel and T.R. Cech, *Biochemistry*, 1993, **32**, 13593–13604.
70. P.C. Bevilacqua, R. Kierzek, K.A. Johnson and D.H. Turner, *Science*, 1992, **258**, 1355–1358.
71. G.J. Narlikar and D. Herschlag, *Nat. Struct. Biol.*, 1996, **3**, 701–710.

5

Classical and Novel Chemical Tools for RNA Structure Probing

RICHARD GIEGÉ, MARK HELM and CATHERINE FLORENTZ
Institut de Biologie Moléculaire et Cellulaire, Strasbourg, France

5.1 INTRODUCTION

5.1.1 Historical and Theoretical Background

The chemistry of nucleobases, nucleosides, and nucleotides is a large and well-established field of organic chemistry. The study of their reactivities to alkylating agents has contributed to the

Table 1. Detection methods for modifications of RNA by structural probes

Detection/ revelation method	*RNA*	*Extent of modification*	*Fragmentation*	*Fractionation*	*Corresponding sequencing method*	*Refs.*
A	unlabeled	complete	various RNases	various chromatographic methods	first generation biochemistry	2
B	internally labeled	complete	RNases A and T1	electrophoresis/ TLC	fingerprint	3
C	end labeled	statistical	chemical or enzymatic scission at modified residues	PAGE	Maxam and Gilbert/ Donis-Keller	4,5
D	unlabeled	statistical	reverse transcription	PAGE	primer elongation	6

understanding of mutagenic and carcinogenic mechanisms of the these agents.[1] In addition, today's widely used techniques of sequencing and structural probing of nucleic acids, and related techniques, like footprinting and chemical interference techniques, have their origins in organic chemistry. In particular, alkylating agents are now used in an analytical fashion to deduce structural features of nucleic acids. The basic idea is that RNA structure influences the extent to which parts of it, i.e. nucleotides as a convenient unit, can be modified by a given chemical compound or attacked by a nuclease. If the specificity of the reagent is known, structural features of the RNA can be deduced, given the existence of an efficient detection method for the modification. The chemically modified nucleotides are judged noninvolved in the structure, while nonmodified nucleotides would be protected from the reagent by hydrogen-bonding or other steric encumbrances. Similarly, preferential cuts by nucleases provide information about accessible regions in RNA.

The history of structural probing has always paralleled that of nucleic acid sequencing. Table 1 shows historically important events that mark progress in the field. In a typical early structural probing experiment, a relatively large quantity of unlabeled RNA (classically tRNA) was quantitatively reacted with a probe, then digested with an RNase, like RNase T1 and/or RNase A. The resulting fragments were separated by chromatographic techniques like those used for sequencing of the first tRNAs, and then analyzed for their content of nucleotides modified in the experiment. This early probing approach reflected the then current state of the art of RNA sequencing. With the development of fingerprint techniques it became possible to use internally labeled RNA that had to be purified from radioactive cell cultures. However, the analytical procedure was labor intensive and the approach suffered from the intrinsic problem that a large portion of the target residues had to be modified in order to be detectable. The large body of early work was efficiently validated when the first crystal structure of an RNA became known and, satisfyingly, computed theoretical accessibilities indeed displayed a high degree of correlation with the experimental reactivities.[7,8]

The invention of end-labeling sequencing techniques by Maxam and Gilbert and colleagues[4,5,9] greatly reformed the technique of structural probing. End-labeling enables the use of only small amounts of material and permits sensitive detection.[10] Because of the augmented sensitivity in signal detection, the principle of statistic modification can be used, i.e. only a few percent of the molecules have to be modified to yield a signal. Thus, the number of molecules with multiple modifications is negligible, and a signal results from a molecule that has not been structurally altered by previous interaction with other probe molecules. Intrinsic parts of the technique of Maxam and Gilbert are chemical treatments that induce chain scission of the nucleic acid at the modified sites. In the case of alkylated bases in RNA, the aromaticity of the purine or pyrimidine rings, which contain nitrogen atoms involved in hemiaminal bonding of the base to the sugar, is usually affected. Subsequent treatment with buffered aniline provokes chain scission by β-elimination.[9,11,12] By this mechanism, three nitrogens in RNA (N-3C, N-7G, and N7-A; see Table 2 for atom abbreviations) can be targeted with dimethyl sulfate (DMS) and diethyl pyrocarbonate (DEPC).[10] Soon after, ethylnitrosourea (ENU) was introduced[13] as a useful probe for mapping phosphates, regardless of the corresponding nucleobases and secondary structure.

Paralleling this evolution, the same reagents were used for footprinting studies of RNAs in interaction with proteins or other macromolecules. Here, a crystal structure of an RNA–protein complex allowed an assessment of the usefulness of this technique. Thus, in the case of yeast $tRNA^{Asp}$ complexed to its cognate synthetase, contact points determined in solution[14,15] could be compared to those on the X-ray structure[16] (see Section 5.5.1). Even though it had been shown early on that most RNases cut preferably in the anticodon loops of tRNAs, RNases like RNase V1[17] only became popular in structural probing and footprinting with the development of the end-labeling techniques.[18]

As a result of this progress, many RNAs in addition to tRNAs became appropriate for structural probing. However, the resolution of polyacrylamide gel electrophoresis (PAGE) limits the application of the "direct" detection method to roughly the first 150 residues from either extremity of an RNA molecule. Again, adaptation of state-of-the-art sequencing techniques[6] to RNA structural probing permitted important methodological advances. The use of primer elongation with reverse transcriptase[19] allows the exploration of larger molecules, such as ribosomal RNAs. A further advantage of the primer elongation technique is that probes, which cannot provoke strand scission, but modify bases to efficiently stall the enzyme, can be utilized on a statistical basis (e.g., 1-cyclohexyl-3-(2-morpholino-4-ethyl)carbodiimide methotosylate (CMCT), kethoxal, see Table 2 and Section 5.3.1). This considerably enriches the arsenal of small probes and permits most of the atoms important for the architecture of RNA to be probed.

The growing body of structure–function data indicates that the biological activity of many RNAs is intimately related to the recognition of its tertiary structure and irregular structural features, like base-triples and mismatches (this distinguishes RNA from DNA which, in general, displays fewer three-dimensional peculiarities). As a consequence, this has, in recent times, shifted attention to the development of a second generation of structural probes. These more complex molecules often contain parts specially designed for the recognition of such structures (see Section 5.3.2 and Table 2) and, being of intermediate size, are large enough to possess significant conformational selectivity.

5.1.2 Fields of Application

The techniques outlined above have led to many important biochemical results. Together with computer algorithms for secondary structure prediction[69,70] and sequence alignments, they allow establishment of Watson–Crick interactions and therefore secondary structure models as a basis for further work. In combination with computer modeling and phylogenetic comparisons,[71] structural probing has led to the construction of valuable three-dimensional models of many RNAs, among which are tRNAs with unusual structures,[72–74] tRNA-like viral RNAs,[75,76] tmRNA,[77,78] gRNA,[79] recognition elements in mRNAs,[80] dimerization initiation site of HIV-1 genomic RNA,[81] RNase P,[82,83] and *Tetrahymena* ribozyme.[84,85] These models are of great heuristic value and permit better understanding of the molecular functioning of the RNAs and definition of the domains of importance so that further (e.g., mutational) analyses can be well aimed. Definition of functional domains in this way is also important for other studies, like NMR or crystallogenesis, which cannot (yet) be conducted easily on large or flexible multidomain molecules.

5.2 VALIDATION OF RNA STRUCTURAL PROBING APPROACHES

5.2.1 Probing with Chemical Reagents

When nucleic acids are mapped with chemical reagents, it is often implicitly assumed that the probes map simple steric accessibilities. Strictly speaking, the experimental results, referred to as reactivities, must theoretically, comprise steric and electronic parameters of the targeted residues. For most of the known probes, the correlation of reactivity (experimental value) and accessibility (theoretical value) was first qualitatively validated by comparison with the crystallographic structure of $tRNA^{Phe}$ from yeast. The targeted residues were, by visual inspection of the crystal structure, classified accessible or not, and compared to the reactivities. A more quantitative procedure used computed accessibilities as a product of the "contact surface area" for residues in $tRNA^{Phe}$ and empirically determined effective radii of probes.[8] It accounted for the reactivities of methoxyamine, $TiCl_3/KI$, semicarbazide-bisulfite, monoperphtalic acid, kethoxal, DMS, and DEPC. A more refined concept included electrostatic potentials calculated with an *ab initio* procedure.[7,86] The resulting theoretical values for expected reactivities were termed accessible surface integrated field (ASIF) values.[87] Using yeast $tRNA^{Phe}$ and $tRNA^{Asp}$ as model RNAs,

Table 2. Structural probes for RNAs

Probe	*Size*	*Target*	*Chemistry*			*Refs.*
			Modification extent	*Strand scission*	*Detection method*	
(a) Nucleophiles						
$NaBH_4$	+	D, m^7G, m^1A	C	–	B	20–22
Methoxyamine and hydroxylamines	+	C-4C, C-5C, C-6C	C	–	B	23
Semicarbazide and bisulfite	+	C-4C, C-5C	C,S	–	B,C	24
$KI/TiCl_3$	+	C-5C	C	–	B	25
(b) Electrophiles and alkylating reagents						
Acrylnitrile		s^4U, I, Ψ	C	–		20
Chloroacetaldehyde	+	N-1A, N-3C	C	–	B	6
DMS (DES) #	+	N-7G, N-3C, N-1A	C,S	–*	B,C,D	10,27
DEPC #	+	N-7A	S	–*	B,C,D	10,27
ENU (MNU) #	+	phosphate,	S	–*	C	13
Ketoxal, glyoxal #	+	N-1G, N-2G	C,S	–	B,D	11,20
CMCT #	+	N-3U, N-1G, N-3D	C,S	–	B,D	23,28
Tritiated water	+	HC-8Py	C	–	B	29
1-Fluoro-2,4-dinitro-benzene	+	2′OH, ss A	C	–	B	30
T4MPyP,TMAP, T2MyP/h·υ	++	helix-hinges	S	–*	C,D	31
$BenzN_2^+$ and derivatives / h·υ	++	phosphate	S	–*	C	32
(c) Radical generators/oxidants						
Op-Cu	++	binding pocket	S	+	C	33
$Rh(phen)_2phi^{3+}$/h·υ	++	tert. interactions	S	+	C	34,35
$Rh(DIP)_3{}^{3+}$	++	G–U bp	S	+	C	36
$Ru(tpy)(bpy)O^{2+}$	++	G	S	+	C	37
$Fe^{2+}/EDTA/H_2O_2$ #	+	HC2/4/5Rib	S	+	C,D	38,39
$Fe^{2+}/MPE/H_2O_2$	++	ds RNA	S	+	C,D	40
KONOO	+	HC2/4/5Rib	S	+	C	41
X-rays	+	HC2/4/5Rib	S	+	D	42
Fe-bleomycin	++	specific S + C 43				
NiCR and derivatives/$KHSO_5$ and G–U bp	++	exposed N-7G	S	–*	C,D	44,45
Ozone	+	exposed G			B	46
Monoperphtallic acid	+	N-1A, N-3C	C	–	B	20
Isoalloxazine derivatives/h·υ	++	G–U bp	S		C	47
(d) Hydrolytic cleavages and nuclease mimics						
Water (OH^-)	+	phosphates	S	+	C	48
Metal ion mediated						49,50
Mg^{2+}	+	specific	S	+	C	51,52
Ca^{2+}, Sr^{2+}	+	specific	S	+	D	53
Pb^{2+}	+	specific	C,S	+	C,D	54,55
Lanthanides (Ln^{3+})	+	specific	S	+	C	52,56
Zn^{2+}	+	specific		+		57
Imidazole	+	ss RNA	S	+	C	58
Imidazole-conjugates	++	ss Py-A	S	+	C	57–60
(e) Biological nucleases						
Nuclease S1 #	+++	ss RNA	S	+	C,D	61–63
Nuclease *N. crassa*	+++	ss RNA	S	+	C,D	64
RNase U2	+++	ss A	S	+	C,D	18
RNase T1 #	+++	ss G	S	+	C,D	62
RNase T2 #	+++	ss RNA	S	+	C,D	65
RNase CL3	+++	ss C	S	+	C,D	66
RNase V1 #	+++	ds RNA	S	+	C,D	17,63
Rn nuclease I	+++	ss RNA	S	+	C	67
RNase P1	+++	ss RNA	S	+	C	68

Probes: CMCT, 1-cyclohexyl 3-(2 morpholino (4)-ethyl carbodiimide methotosylate; DMS, dimethyl sulfate; DES, diethylsulfate; DEPC, diethyl pyrocarbonate; ENU or MNU, ethyl- or methylnitrosourea; kethoxal, β-ethoxy-α-ketobutyraldehyde; KONOO, potassium peroxonitrite; Fe^{2+}/MPE, methidiumpropyl-EDTA-iron(II); NiCR, (2,12-dimethyl-3,7,11,17-tetraazabicyclo [11.3.1]heptadeca-1(17),2,11,13,15-pentaenato) nickel(II) perchlorate; Op-Cu, 1,10-phenanthroline-copper; $Rh(phen)_2phi^{3+}$,bis(phenanthroline)(phenanthrene quinone diimine)rhodium(III); $Rh(DIP)_3{}^{3+}$, Tris(4,7-diphenyl-1,10-phenanthroline) rhodium(III); $Ru(tpy)(bpy)O^{2+}$, (2,2′,2″-terpyridine)(2,2′-bipyridine)oxoruthenium(IV); T4MPyP, *meso*-tetrakis-(4-*N*-methylpyridyl) porphine; TMAP, *meso*-tetrakis-(*para*-*N*-trimethyl anilinium)porphine; T2MPyP, *meso*-tetrakis-(2-*N*-methylpyridyl)porphine; BenzN2+, *p*-*N*-dimethyl aminobenzenediazonium cation. Commonly used probes are highlighted by #. →

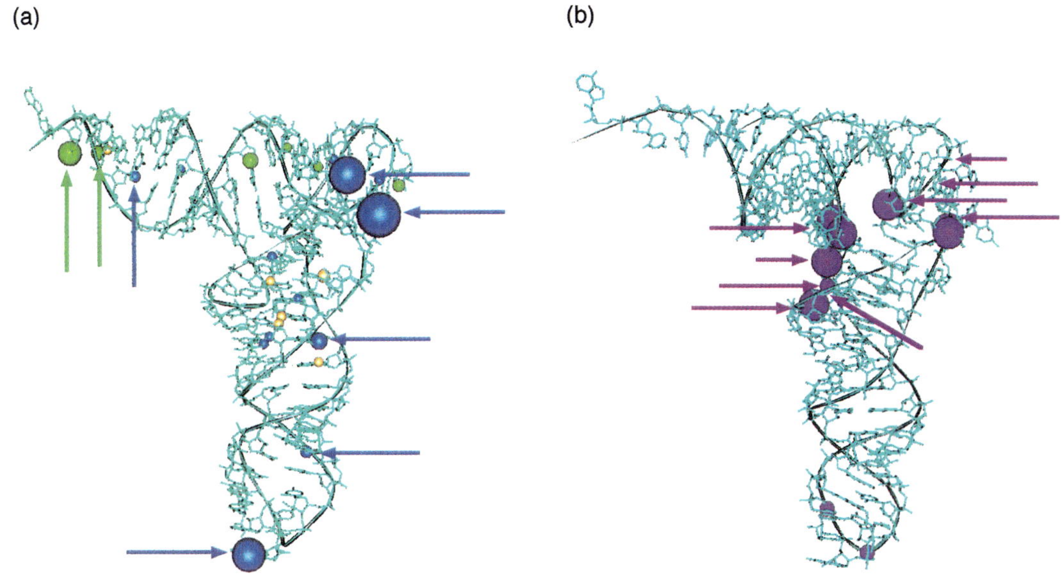

Figure 1. Comparison of calculated and observed reactivities and protections in the crystal structure of yeast $tRNA^{Phe}$ [89,90] as model RNA. (a) Nitrogen atoms are indicated by spheres of sizes proportional to computed reactivities corresponding to their ASIF values for N-3C (green) N-7G (blue), and N-7A (yellow).[86] Experimental results for DMS (N-3C and N-7G) and DEPC (N-7A)[27] are indicated by arrows of corresponding colors. (b) Phosphates protected from alkylation by ENU are in purple, their sizes proportional to calculated protections according to their ASIF values.[86,87] Arrows indicate observed protections.[13,91] Images were made with DRAWNA.[92]

the probes DMS, DEPC[27,86] and ENU[13,88] were validated this way in a thorough comparison of reactivities and ASIF values. It was found that the addition of an electrostatic field to the geometric accessibility leads to a good model in that it yields good predictions of chemical reactivities. This correlation is illustrated in Figure 1. In other words, chemical reactivities represent predominantly architectural features of the RNA. Interestingly, the probes reflect even subtle conformational differences found in the crystal structures of the two tRNAs.[14,27] They are also capable of monitoring tertiary interactions and Mg^{2+} coordination. More recent work has established adequate correlation between the crystal structure of 5S RNA and the reactivities of the probes CMCT, kethoxal, DMS, and DEPC.[93] The crystal structure of the P4–P6 domain from the *Tetrahymena* class I intron allowed to establish a good correlation with data from probing experiments with free radicals.[94] Thus chemical probes represent an analytical tool of good precision and fidelity.

5.2.2 Probing with Nucleases

Nucleases have become popular probes because of easy use and fast results. As for the chemical probes, most of the commonly used nucleases have been validated by studying their specificity on tRNAs of known structure.[61–63,65] However, nucleases are much larger than the commonly used chemical

---

Table notes (continued)

Size: +, small (1–10 Å); ++, intermediate (10–100 Å); +++, large (over 100 Å).

Target: N-1A, nitrogen 1 in adenine; N-7A, nitrogen 7 in adenine; N-3C, nitrogen 3 in cytidine; C-4C, C-5C, C-6C, carbons 4, 5, and 6 in cytidine; N-3D, nitrogen 3 in dihydrouracil; N-1G, nitrogen 1 in guanine; N-2G, nitrogen 1 in guanine; N-7G, nitrogen 7 in guanine; N-3U, nitrogen 3 in uracil; m^1A, 1-methyladenine; m^7G, 7-methylguanine; HC2/4/5Rib, hydrogen at carbons 2, 4, and 5 of ribose; HC-8R, hydrogen on carbon 8 of purines; ss, single-stranded; ds, double-stranded; specific, binding pocket particular for a certain probe.

Chemistry: *modification extent*: C, complete; S, statistical; *strand scission*: −, no strand scission, −*, scission by additional treatment; +, strand scission by probe (nuclease activity).

Detection method: B, fingerprinting; C, end-labeling; D, primer extension (as in Table 1).

Other abbreviation: $h \cdot \upsilon$, requires irradiation.

probes and they do not require, by definition of their enzymatic activity, additional treatment to induce chain scission. The large size of nucleases brings some pitfalls, and thus, some care has to be taken in interpreting the results. Indeed, a nuclease might establish in the process of protein–RNA recognition several interactions at a time with the RNA, and therefore there is a possibility that these interactions induce conformational changes in the RNA and yield artifacts. Also, because the enzyme is much larger in size than the target, the probe can no longer be assumed to be spherical and therefore a validation approach as used for chemical probes is not possible, hence their frequent use in mapping global domains rather than details in RNA structure.

For all probes that directly cleave RNA, the problem of secondary cleavages can, in principle, occur. Primary cleavage of the RNA by the nuclease generates two fragments, which might adopt a structure different from the intact RNA. Upon a second cleavage, such a fragment will yield a signal that does not reflect the intact RNA structure. Because secondary fragments are always shorter than the primary fragments, they can be detected by comparing fragmentation patterns of both 3′- and 5′-labeled RNA.[11,12,17,66] Only cleavages that are found in both experiments are of primary origin.

5.3 PROBES, TARGETS, AND METHODOLOGY

5.3.1 Probes and Targets

Of all the probes described in the literature, only a handful is commonly used (highlighted in Table 2). Among the most popular chemical probes are DEPC and DMS. They were the first to be used with the direct revelation method and their specificity for purine and pyrimidine rings has been well-characterized by a large body of data. As with the equally well-characterized ENU (mapping phosphates), chemical chain scission reactions are specific for one type of modification at a time. Reactive N-7A, acylated by DEPC, can be directly revealed by chain scission with aniline. DMS preferentially alkylates N-7G, followed by N-3C and N-1A. Minor methylation sites are N-3A, N-7A, and even N-3G and O-6G. Interestingly, N-7G modifications can be exclusively revealed by treatment of the RNA with sodium borohydride prior to chain scission with aniline, while treatment with hydrazine reveals methylations at N-3C, the other methylations remaining undetected. This specificity in cleavage is possible because sodium borohydride reduces the N-7-methylated imidazole ring of guanine residues, thereby creating a site for aniline scission. On the other hand, N-3-methylated cytidine residues are rendered susceptible to aniline scission by a nucleophilic attack of their pyrimidine ring by hydrazine.[9–12] Thus, breaking down the different modification sites yields a favorable signal-to-noise ratio for each type of modification. In a similar way, mild alkaline treatment reveals exclusively the phosphates alkylated by ENU, even though, in principle, ENU alkylates all oxygens and some nitrogens in RNA.[1] It is due to the above-mentioned properties that these probes work with high precision. Commonly used with the primer extension revelation method are DMS (mapping N-1A and N-3C, but not N-7G), kethoxal (N-1G and N-2G), and CMCT (N-3U and N-1G). Because these probes map, of each nucleobase, at least one nitrogen atom involved in Watson–Crick interactions, they are an excellent means for determining the secondary structure of a given RNA.

As mentioned above, the use of these probes as revealed by the direct method has been validated with the crystallographic structures of yeast $tRNA^{Phe}$ and $tRNA^{Asp}$ as references. While the direct method, with end-labeling and chain scission specific for one type of modified nucleotide at a time, guarantees reliability, primer extension produces a relatively high amount of background signal, because all modifications at different sites capable of stopping reverse transcription give signals at the same time.[19] The efficiency of stalling the enzyme differs among the different modifications. Usually, the reaction rates of one probe with several different target sites are not identical, e.g., the methylation of N1-A by DMS cannot directly be compared to that of N-3C. Also, strongly structured regions sometimes evade mapping because they hinder the enzyme and create pauses in reverse transcription. These pauses cause regions with a high background in the control experiments. Therefore, the precision of these probes must be judged as slightly lower than those mentioned above. However, their reliability has recently been validated[93] and their usefulness is illustrated by their widespread utilization especially for mapping large RNAs (see Section 5.1.2), but also tRNAs.[74]

5.3.2 Enzyme Mimics, Chemical Nucleases, and Tethered Probes

The phosphates in RNA are several orders of magnitude less resistant to hydrolytic cleavage than those in DNA. Studies on model substrates are consistent with the hydrolysis being subject to general base as well as general acid catalysis.[95] General base catalysis proceeds by deprotonation of the 2′OH, promoting a nucleophilic attack of the 2′-oxygen at the phosphorus to form the often cited 2′–3′ cyclic intermediate. The phosphate can be activated for nucleophilic attack by coordination of an electrophile, e.g., a proton (hence the general acid catalysis[95]) or a divalent metal ion. Therefore, even highly purified RNA in water is subject to some spontaneous hydrolysis, due to trace amounts of hydroxide ions and protons (by definition 10^{-7} M at pH 7). Metal ions can also coordinate water molecules that are then more easily deprotonated. Consequently, divalent metal ions can promote RNA hydrolysis with high efficiency. Because they also favor structuring, especially tertiary interactions, they can be seen as additives that help RNA to adopt its conformation and thus promote cleavage in an indirect manner. Such a role is especially crucial in catalytic RNAs.[96]

In 1976, Werner *et al.* discovered that $tRNA^{Phe}$ from yeast is site-specifically cleaved by Pb^{2+} ions at a rate that assigns catalytic activity to the plombus ions.[54] Lead ions preferably map single-stranded regions unless they coordinate to elements of tertiary interaction, for example in tRNA between the D- and T-loops. They can thus be used as specific probes to verify the structural integrity of a canonical tRNA. Crystallography permitted localization of the coordination pocket[55] and aided the mechanistic studies, which were further supplemented with detailed mutational analyses.[97,98] Other "leadzymes" have been selected, and their properties studied.[49,99] Many other di- and trivalent cations were also found to induce RNA cleavage (see Table 2).[100] As several tRNAs are cleaved by Mg^{2+} ions coordinated near the classical coordination site,[52] such ions, besides their structural role, seem to actively participate in RNA hydrolysis as well (note that, above 60 ºC and/or at slightly alkaline pH, Mg^{2+} ions exhibit strong RNase activity[101]). A proposed model of the active center of a group I intron suggests coordination of different divalent cations to the same binding pocket in which Mg^{2+} is required for cleavage of the natural substrate. Interestingly, Ca^{2+}, Sr^{2+}, and Pb^{2+} induce cleavage of the ribozyme itself, rather than its natural substrate.[53] Not only do these findings render the borders between leadzymes, other metalloenzymes, and ribozymes even more fluent, but they also illustrate the possible use of metal ions to monitor structure in such binding pockets. In view of all the above, plus the fact that some nucleases require divalent cations for functionality (Zn^{2+} for S1 nuclease[102]), metal ions can be regarded as the stripped active center of an RNase or as an "RNase mimic".

With a more detailed knowledge of RNA structures and cleavage mechanisms, it became possible to engineer chemical compounds for structure recognition. Inferred from the crystal structures of RNase A is the knowledge of how imidazole residues, in the form of two histidines, activate phosphates and 3′OH groups of RNA for hydrolysis.[100,103,104] Imidazole is especially suited because its pK_b is 7.0 and thus it is capable of both acid and base catalysis at neutral pH (see above). This situation has been successfully mimicked by supplying one or two imidazole groups tethered to different ligands (e.g., spermine, phenazine) with affinity for RNA structural features.[58–60] Cleavage occurs readily with mimics containing two imidazoles; if only one imidazole group is present, as in the spermine–imidazole conjugate, cleavage is triggered when the reaction medium is supplied by imidazole buffer.[58] These probes exhibit site-specificity and, like RNase A, show preference for pyrimidine/A sequences.

A group of transition metal complexes is capable of cleaving RNA upon irradiation or addition of redox-active co-reagents (e.g., H_2O_2) by mechanisms involving free or metal-coordinated radicals.[105] These compounds, termed "chemical nucleases",[100,105,106] occupy an intermediate position in size between the small electrophiles and the large nucleases, and are therefore potentially valuable for the recognition and mapping of structural features of intermediate size. Indeed, $Rh(DIP)_3{}^{3+}$ (tris(4,7-diphenyl-1,10-phenanthroline)rhodium(III)) selectively cleaves at the 3′-end of the G of G–U mismatches[36] and Fe^{2+}/MPE (methidiumpropyl EDTA-iron(II)) was shown to specifically map double-stranded regions[40] due to the intercalating nature of the methidium moiety. In the latter case, as in the case of RNase A mimics (see above), the cleavage-active center has been tethered to a recognition-active ligand to obtain structural specificity for the cleavage. In a larger view, using a biological macromolecule as the recognition-active ligand is the next logical step. Indeed, Fe^{2+}/EDTA has been tethered to a tRNA,[107] as well as to EF-G[108] and ribosomal protein S20,[109] to probe the corresponding binding sites on the ribosome. Similarly, europium complexes were linked to antisense oligonucleotides to cleave targeted RNA sequences.[110]

5.3.3 Methodology

There are a number of research papers and reviews[10–13,27,111] covering the chemistry and theoretical and experimental aspects of the different probing methods. We will restrict our remarks in this section to a number of technical details and pitfalls concerning the practical aspects of structural probing that demand special attention. Among them are the choice of temperature, buffer, and salt conditions because of their possible influence on RNA structure. Important for the choice of buffer is its inertness towards the probe. For example, Tris, as a popular buffer in the pH range 7.5–8.5, is suitable for probing with nucleases. However, being nucleophile, and its capacity being strongly temperature dependent, it is not recommended for use with electrophiles and in temperature-dependent probing. For the latter applications, it is commonly replaced by cacodylate. Tris is also considered a radical quencher and, although used with chemical nucleases, should be replaced by a phosphate buffer.[43] Even though probing is in most cases performed at 20–25 or 37 °C, it is recommended to conduct UV melting experiments[112] in a way to facilitate the proper choice of parameters for the denaturing/renaturing procedure that usually precedes probing.

A very important requirement for structural probing is a homogeneous population of labeled molecules, be it RNA for the direct method or DNA for primer extension.[18] 5′-Labeling is usually performed using T4 polynucleotide kinase and [γ-^{32}P]ATP. Unless exchange reactions are used, an existing nonradioactive 5′-phosphate has to be removed by bacterial alkaline phosphatase or calf intestine phosphatase. In the case of RNAs with a stable structure near the 5′-extremity which hampers dephosphorylation (typically in native tRNAs), very stringent conditions have to be chosen. 3′-Labeling is usually done by ligation of [5′-^{32}P]pCp to the 3′OH of the RNA by T4-RNA ligase. Purification to the nucleotide level can be a problem when *in vitro* transcribed RNAs are used, because RNA polymerases do not yield homogeneous populations on the 3′-end upon run-off transcription. RNA labeled this way and not properly purified will yield multiple signals for each nucleotide in probing experiments. A different approach can be used for 3′-labeling of tRNAs and tRNA-like molecules. After removal of the 3′-CCA extremity by endonuclease digestion, reconstitution by ATP/CTP-tRNA nucleotidyltransferase is performed in the presence of [α-^{32}P]ATP. Labeled RNA is usually purified by denaturing PAGE, subsequent excision, passive elution, and ethanol precipitation in the presence of carrier RNA. It is recommended to submit the redissolved RNA to gel filtration to exclude contamination by acrylamide residues, residual salt, and ethanol before structure probing.

For direct detection and assignment of cleavage sites, the radioactive RNA is electrophoresed in parallel with: (1) an untreated sample to detect spontaneous and unspecific cleavages in the experimental procedure; (2) a ladder created by degradation of the sample in the denatured state by heating with alkali or imidazole[58] or simply in water or formamide[111] or by P1 digestion; and (3) the products of a partial digestion with RNase T1 (specific for G) or U2 (specific for A). These allow the correct assignment of bands observed in the sample lane to the primary sequence. Note that aniline, radicals, ENU, S1 type nucleases, and T1 type nucleases all leave chemically different chain scission products. This fact is reflected in the different migration behaviors of fragments corresponding to approximately the first 20 nucleotides on the labeled extremity. Signal assignment in these regions can therefore be difficult or ambiguous. In fragments longer than 30 nucleotides, the chemical differences become irrelevant in comparison to other parameters that determine migration behavior, i.e., sterical encumbrances and charge density. Consequently, assignment of these fragments is reliable. For correct assignment with the primer extension method, sequencing reactions with dideoxy chain terminators are run in parallel. When blocked by chemical modification, the reverse transcriptase stops elongation one nucleotide before the modified residue. Thus, its signal is one nucleotide shorter than the corresponding sequencing signal.

5.4 APPROACHING ARCHITECTURAL FEATURES OF RNAs BY STRUCTURAL PROBING

5.4.1 Two- and Three-dimensional Structures

tRNA mimics are nontransfer RNAs that are recognized by tRNA-specific enzymes.[113] They have been among the most prominent models to demonstrate the usefulness of enzymatic and chemical probes

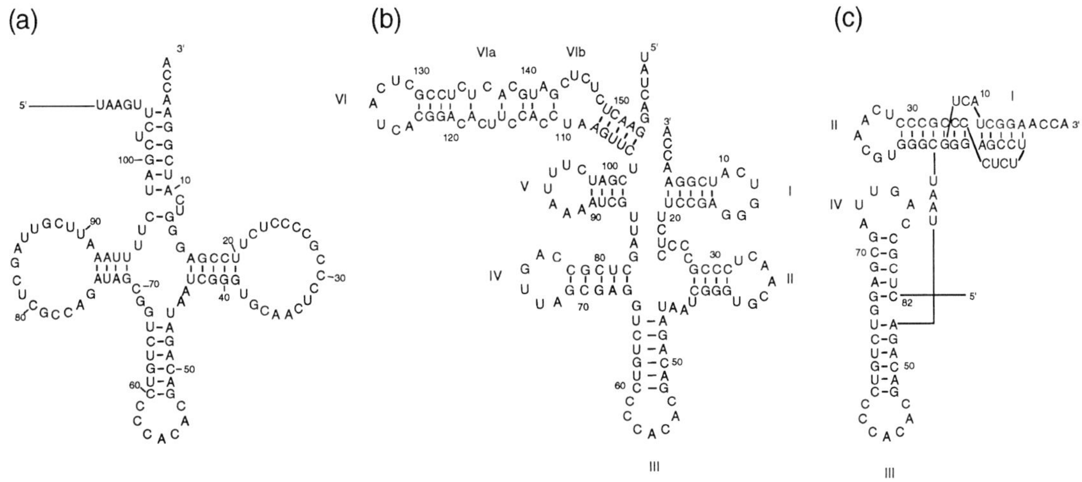

Figure 2. Different steps in the understanding of an RNA structure: the case of the tRNA-like domain of TYMV RNA.[66,114,115] All three displayed models are based on the presence of a CAC valine anticodon located in a loop mimicking a tRNA anticodon loop (bottom of the models). (a) Two-dimensional folding mimicking a cloverleaf established on the sole sequence without additional experimental data (nt ~ 109); (b) experimentally established two-dimensional folding (nt = 159); (c) pseudoknotted fold mimicking three-dimensional features of canonical tRNA, in particular its architectural organization in amino acid accepting (domains I and II) and anticodon (domains III and IV) branches (nt = 82). Noteworthy, the NMR model[117] showed the stacking of the two helical regions in the pseudoknotted fold (domain I) and a conformation of loop L1 (residues 21–24) compatible with functional data.[116]

in unraveling intricate structural features in the RNA world. This relies on the plausible assumption that similar functions are sustained by similar structures. Thus, efforts were made to find in tRNA mimics structural elements recapitulating features of tRNA two-dimensional cloverleaves and/or of their L-shaped three-dimensional architecture. As an example, we take the tRNA-like domain present at the 3′-extremity of turnip yellow mosaic virus (TYMV) RNA. This domain is found within the last 159 nucleotides of the viral RNA and is charged by valyl-tRNA synthetase. Thus it should share structural similarities with $tRNA^{Val}$. Extensive nuclease probing on 5′- and 3′-end-labeled fragments, up to 159 nt long, revealed a folding that partly differs from a cloverleaf[66,114], suggesting that the mimicry with tRNA occurs at the 3D-level (Figure 2). Indeed, the longest fragment is organized into six domains, with the most 5′-domain loosely structured, and likely not involved in the tRNA-like fold. Domains I–V are well-organized stem and loop regions and it is easy to recognize in domains II, III, and IV, the mimics of the T-, anticodon-, and D-arms of canonical tRNAs. In contrast, domains I and V, as well as the joining oligonucleotides between domains I and II, did not show any resemblance to tRNA features, except for the 3′-terminal $-ACCA_{OH}$ sequence characteristic of valine-accepting tRNAs. The understanding of the role of domain I emerged from enzymatic and chemical probing as well as from the observation of compensatory mutations in several strains of TYMV that maintained base pairing between three residues of loop I and the three most 5′-residues in the single-stranded stretch joining stem I to stem II. This was the clue for the proposal of a new RNA folding principle, the pseudoknot[114,115] that allowed construction of a model of the amino acid accepting stem of the tRNA-like molecule involving the sole 3′-end sequence of the viral RNA. Further probing with DMS and DEPC confirmed the pseudoknot fold and ENU probing revealed T-loop features in loop II. A more accurate three-dimensional structure, with correct nucleotide geometries, was then computer modeled, utilizing all the available crystallographic knowledge on tRNAs and the probing data on the isolated RNA fragments. The model showed that the tRNA-like structure is 83 nt long and is thus restricted to domains I–IV. This model, and especially the pseudoknot fold, was of great heuristic value, since it rapidly gave the answer to the folding of other tRNA-like structures, and later allowed prediction of the histidine identity of all plant viral tRNA-like structures since a residue from the pseudoknot loop L1 (residues 21–24, see Figure 2(c)) is stacked 5′ on top of the accepting helix and mimics histidine major identity residue N-1.[116] A NMR structure of a pseudoknot derived from the TYMV tRNA-like structure[117] has confirmed many of the conclusions from the modeling, in particular its overall fold and the conformation of the histidine identity element (U21) in loop L1. It is important to mention, however, that there were fine structural details revealed by NMR that could not be predicted by structural probing.

In general, the problem of how three-dimensional features can be extracted from probing data is a complex procedure always utilizing computer modeling and if possible additional information from sequence comparisons of phylogenetically related molecules and detection of long-range conservations of base pairings.

5.4.2 Comparison of Mutants

Construction and structural analysis of RNA, mutated at strategically important positions, render the technique even more powerful. Thus it is possible to detect, whether a certain mutated position in an RNA is directly responsible for function or rather for structural features. Drastically altered probing patterns indicate an essential role of the mutated position(s) for the RNA architecture. Three-dimensional features are commonly found and verified this way. Typical examples include studies on human mitochondrial $tRNA^{Lys}$ [118] and the RNA binding site of the *Escherichia coli* ribosomal protein S8.[119]

Definition of the precise structural features in the binding site of protein S8 on rRNA from *E. coli* was based on an *in vitro* selection procedure. This powerful mutagenesis method generated RNA aptamers able to bind to the ribosomal protein. Establishment of the consensus sequence of the aptamers and hydroxyl radical probing studies with Fe^{2+}/EDTA revealed a number of noncanonical features in the stem of the large hairpin in 16S rRNA, known to contain the binding site. Thus the core structure of the binding site comprises three interdependent bases (nt 597–641–643) forming a base triple, a conserved A residue at position 642, and above these elements a base pair (598–640) and a bulged residue at position 595.[119] In the case of human mitochondrial $tRNA^{Lys}$, a combined mutational and probing analysis showed the necessity of a methylation at N-1 of A9 for the correct cloverleaf folding of this molecule. In unmodified tRNA transcripts, lack of this methylation that occurs at a Watson–Crick position allows three additional pairings (A8–U65, A9–U64, G10–C63), with the consequence that the transcript folds into an extended hairpin, as deduced from extensive structural probing. In contrast, transcripts with mutations that prevent Watson–Crick pairing between positions 9 and 64 (e.g., replacing the A–U pair by A/A, A/C, or C/U combinations) showed the cloverleaf fold of these molecules, as did the native tRNA with the post-transcriptional modifications.[118]

5.4.3 Defining Domains for Biochemical and Biophysical Studies

Biophysical methods, like NMR and crystallography, yield high-resolution data, but require large amounts of material. If a large RNA molecule is the target of such a study, it has proven useful to first study smaller domains of it because of experimental and financial limitations. To define such domains, and to verify that smaller molecules corresponding to these domains display the same or at least similar structural features, structural probing is an extremely useful if not an essential tool. The case of 5S RNA illustrates well how structural probing in combination with phylogenetic studies provided the basis for two- and three-dimensional models that in return allowed selection of subdomains for further refining studies with different methods. This ubiquitous RNA, discovered in 1963, is part of the large ribosomal subunit, but despite numerous studies, its function could not be unambiguously determined. The first secondary structure models to emerge were based on phylogenetic comparisons and thermodynamic calculations.[120,121] The secondary structure of 5S RNA is highly conserved in evolution and was soundly confirmed in a number of papers featuring thorough structural probing work and mutational analysis.[122–124] Based on these data, models of 5S RNAs from different origins were constructed. Also resulting from this body of work was the discovery that 5S RNA of different origins can exist in several metastable isoforms, a fact that had all the while hampered the studies. The E-loop emerged as a structurally highly interesting region, because it was shown to be highly structured, despite not being a classical Watson–Crick helix. The structure was also shown to be strongly dependent on the Mg^{2+} concentration. Crystals of entire 5S RNA from *Thermus flavus* had been obtained but reportedly diffracted only to 8 Å resolution.[125] Thus, the structure was broken down into domains previously defined by chemical probing and these were studied separately by NMR and crystallographic techniques.[124,126–128]

5.4.4 Detection of Structural Plasticity of Free RNA

Structural probing permits insight into conformational changes of RNA that can be provoked by a variety of parameters including pH, temperature, salt conditions, and even interaction with proteins (see Section 5.5.1). Not being prone to effects that take place in crystals, probing reflects the status of a given RNA in solution. Thus, given the choice of an adequate probe, the conditions can be chosen freely and are not unique as for crystallization. Under crystallization conditions, yeast $tRNA^{Asp}$ forms anticodon–anticodon interactions with a second $tRNA^{Asp}$ molecule, thus representing a model of a tRNA in interaction with an mRNA. The tertiary interaction G18–C56 is disrupted in both molecules. Under milder structural probing conditions, on the other hand, this tertiary interaction is present. However, it is also absent in transcripts devoid of modified bases.[129] Other effects might be induced by agents added for crystallization that would not usually be chosen when an RNA structure is to be monitored.

Temperature-dependent chemical probing can be an alternative to UV melting measurements. In addition to information concerning the melting temperature, it gives indication of the structural domains that melt in a transition. Probing of $tRNA^{Phe}$ [91] and $tRNA^{Asp}$ [14] with ENU allowed the direct demonstration of thermal denaturation beginning in the D-stem and continuing in the T-stem. From an opposite point of view, the structuring process of RNA can be monitored. Starting from denatured RNA molecules at high temperatures, the successive formation of structural domains can be followed by disappearing reactivities at lowered temperatures. This allows a more definite assignment of interactions between different regions of the molecule, because residues involved in the same interaction simultaneously become protected below the melting temperature. One of the first three-dimensional RNA models to be constructed according to probing data was that of bovine mt $tRNA^{Ser}$. This model is based on sequence comparison and temperature-dependent structural probing of the N-7G, N-7A, and N-3C.[72,130] As is now typical, experiments were carried out in a comparative fashion in the absence and in the presence of Mg^{2+}. The results indicated a stabilization of the tertiary structure by the divalent cations, an effect that had already been observed on canonical tRNAs[131] and through further experiments on other RNAs, has proven to be of general importance.[49,96] For ribozymes in particular, divalent cations and salt conditions have proven to be essential for proper structure and function (see Section 5.3.2), as evidenced for the *sunY* group I ribozyme by temperature-dependent structural probing.[132,133]

5.5 PROBING RNAs IN COMPLEXES

5.5.1 RNA–Protein Complexes

Since structure formation can protect a nucleic acid from being attacked by enzymatic or chemical probes, it is logical that any other molecule that binds to the nucleic acid can also give protection. These protected regions on the nucleic acid then indicate the binding site. RNA–protein complexes were investigated early on by digestion with nucleases (such as nuclease S1 or RNase V1).[17] This approach, however, suffers from the size of the probes (see Section 5.2.2) which is often close to the size of the investigated protein and thus might suffer from sterical hindrances.

In vitro footprinting with chemical probes is restricted to complexes where the protein withstands treatment with the probe without substantial damage affecting its binding or functional properties. The chemical probes listed below have been shown to be harmless to various proteins under the experimental conditions used for footprinting. Artifactual data can be minimized if experimental conditions are optimized so that a maximum of RNA molecules are complexed to the protein, and if the absence of protein-induced cleavages of the RNA has been excluded in control experiments. The minimal amount of protein can be estimated if K_D or K_m values are available. If not, footprinting should be performed in the presence of increasing amounts of protein. The smaller chemical probes offer a more detailed insight into the interaction pattern between RNA and protein. Thus DMS, ENU, hydroxyl radicals, and iodine are frequently employed probes. The specificities of ENU (phosphates), hydroxyradicals (riboses), and iodine (thiophosphates) allow probing of all nucleotides of an RNA chain. DMS has slightly less specificity since it probes the base components of A, C, and G residues. These probes have been applied for the investigation of most if not all RNA families and their interacting proteins. Selected examples include tRNAs interacting with aminoacyl-tRNA synthetases,[134] elongation factor,[135] RNase P,[136] modification enzyme,[137] and ribosome.[138] Other examples include rRNAs interacting with ribosomal proteins,[139]

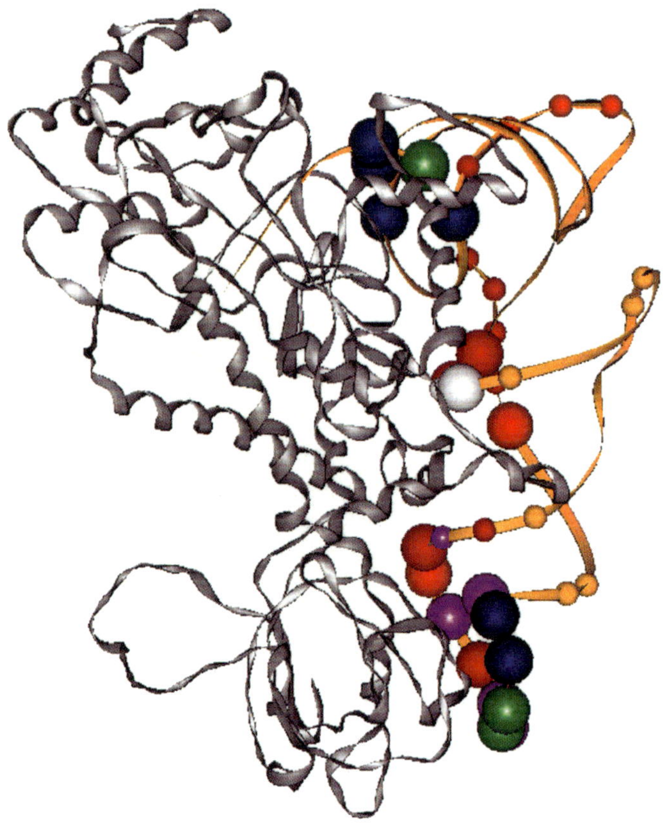

Figure 3. Crystal structure of yeast tRNA^{Asp} complexed to its cognate synthetase.[16,145] For simplicity, the figure displays only one monomer of the $\alpha 2$ dimeric complex. Spheres indicate contacts of nucleotides with the protein found either by crystallography (blue) or by phosphate mapping as monitored with ENU[14] (red) and iodine[146] (yellow). Coincidences of ENU and iodine protections are marked in orange. Coincidence of crystal structure contacts with probing data are indicated in purple (ENU), green (iodine), and white (ENU and iodine).

mRNAs binding to the ribosome,[140] or to translation regulator factors,[141] snRNP RNAs and proteins,[142] as well as viral RNAs interacting with coat proteins[143] or with other viral proteins.[144]

In a few instances, the interaction domains of a protein on an RNA have not only been determined by chemical probing, but also by X-ray crystallography of the complex. Thus, a comparative study of both approaches is feasible. The structure of yeast tRNA^{Asp} complexed to its cognate aspartyl-tRNA synthetase has been established at high resolution.[16,145] It has the great advantage of defining at once hydrogen bonds between any type of atom of the RNA and, in addition, to reveal the partners in the amino acids of the protein. This complex has also been investigated by several probes including the very sensitive and phosphate-specific probes ENU[14] and iodine.[146] All three approaches lead to very similar data, namely that the enzyme binds to the tRNA along the side containing the variable region and the T-loop (Figure 3). A detailed analysis reveals several fine differences between the two chemical probes as well as between solution and crystallographic data. Indeed, some phosphates appear either protected by the enzyme or not, depending on the probe. Also, some moderately protected phosphates appear far from the protein in the crystal structure (these protections likely reflect unspecific interactions between the tRNA and the synthetase[146]). Again, as already pointed out, reactivities do not reflect simple geometrical accessibilities, and the intervention of proteins potentially adds more significance to electrostatic and quantum chemical effects in the modification. Apart from differences arising from the sizes of the probes, the differential protection patterns would also be linked to the fact that ENU can attack either of the two nonesterified oxygens of the phosphate, while iodine only attacks the sulfur that substitutes one of the two oxygens. Whereas the resolution of chemical probes to determine points of contact is clearly higher than that of nucleases, interpretation of footprinting data must be done with some care.

Whereas most footprinting experiments are performed *in vitro* with purified couples of macromolecules, investigations *in vivo* have become popular. They consist of a short treatment of growing cells with the probe, followed by quenching, RNA extraction, and searching for modified positions in target

RNAs by reverse transcription.[147] Classically, DMS is used because it can readily pass cell membranes and can be quenched by mercaptoethanol. Other probes, such as CMCT, are also used. Protection of nucleotides are deduced by comparison with *in vitro* modification of free target RNA extracted from untreated cells. Examples of *in vivo* footprinting concern the *E. coli* rRNA[148] or U3A snoRNP from *Saccharomyces cerevisiae*.[142]

5.5.2 RNA–RNA Complexes

Whereas proteins recognize practically all functional groups of RNA with comparable frequency, RNA–RNA interactions are strongly biased towards Watson–Crick interactions of complementary strands, although tertiary interactions are also frequently encountered. Intermolecular RNA–RNA association can be studied in much detail by chemical probing. A recent example is the reverse transcription initiation complex of HIV-1. Reverse transcription of the genomic viral RNA into cDNA requires a short RNA primer to initiate transcription. HIV reverse transcriptase uses $tRNA^{Lys}{}_3$ (anticodon s^4UUU) which hybridizes to the primer binding site (PBS) of the viral RNA. The formation of this complex can be effected *in vitro* and has been monitored by structural probing with chemical (DMS, DEPC, kethoxal, CMCT) and enzymatic probes (S1, V1, T_2). Both RNAs have been probed in their free and complexed state.[149] In the process of primer annealing, the tRNA structure melts and completely rearranges upon hybridization of an 18 nt sequence on the 3′-end to a complementary stretch of the PBS loop. A second domain of functionally essential interaction is the s^4UUU tRNA anticodon hybridized to an A-rich loop of PBS. Interestingly, this second interaction requires thiolation at position 34 for an appreciable efficiency.[149,150] A second interesting example of an RNA–RNA complex is an antisense–sense RNA association in *E. coli*. The secondary structures of the antisense RNA (micF), its binding site on the mRNA (ompF mRNA), and of both molecules when complexed, have been established with both chemical and enzymatic probes.[151] A third example concerns RNA editing of pre-mRNAs in kinetoplastid organisms, which is specified by guide RNAs. Editing involves a covalent binding between both RNAs converting intermolecular to intramolecular interactions. Advantage has been taken of this fact to investigate the secondary structures of the interacting guide complexed to the mRNA as one single molecule with kethoxal, CMCT, DEPC, and DMS.[152]

5.5.3 RNA–Antibiotic Complexes

RNAs are primary targets for interaction with many antibiotics. Sensitivity, and thus localization of binding sites for these molecules on various RNAs, has been widely investigated by mutagenic approaches of the RNA but also by footprinting with chemical probes. DMS, kethoxal, and lead have been preferentially used for the investigation of the interaction between antibiotics and either rRNAs (both 16S and 23S), ribozymes, or a catalytic group I intron. To be noted is the pioneering work on the interaction of several families of antibiotics with 16S rRNA, that allowed to define concise sets of highly conserved nucleotides as strategic binding domains of the antibiotics.[153] Other examples concern aminoglycoside binding sites.[154,155]

5.6 OTHER APPLICATIONS OF NUCLEOTIDE MODIFICATIONS

Alternative strategies to those described above, that also make use of base modifications in RNA, have been successfully applied to investigate both structural and functional aspects of RNA. These include interference experiments, site-selected insertion of defined modified residues and exploiting the presence of enzyme-catalyzed modified nucleotides in natural RNAs.

5.6.1 Interference Experiments

Interference experiments are aimed to identify at once those elements within an RNA that are involved in a functional process (e.g., binding to a protein or another RNA) and those which are not.

Conceptually, these experiments resemble classical probing as described above in that they are based on a statistical modification of the RNA in its native conformation. However, the resulting mixed population of molecules is then complexed with the partner molecule and submitted to a selection procedure (e.g., band shift on gel). RNA molecules modified at residues important for interaction with the target molecule are discriminated since they are no longer able to bind. Depending on the experimental strategy and circumstances, either the negatively selected molecules are analyzed for their modified nucleotides or the positively selected molecules are analyzed for all nucleotides that were not modified, i.e., those important for the interaction. All the formerly discussed detection methods can, in principle, be used. RNA can be modified by external chemicals or during *in vitro* transcription by statistical insertion of internal probes such as phosphorothioates or deoxyribonucleotides.

A straightforward example was the determination of nucleotides in tRNA important for interaction with nucleotidyl-transferases from different organisms.[156] Different tRNAs from *E. coli* were first modified by DEPC, then end-labeled by 3′-terminal incorporation of [α-^{32}P]ATP by the transferase, then submitted to chain scission. Note the elegance of performing labeling and selection procedures in a single step. An example of application to RNA–RNA interaction refers to the dimerization initiation site of HIV-1 genomic RNA as identified by means of treatment with CMCT, DMS, DEPC, and kethoxal prior to dimerization/hybridization.[157] Monomeric RNA was separated from dimeric RNA by agarose gel electrophoresis under nondenaturing conditions, and the modified residues determined by primer extension. A three-dimensional model of this part of the HIV genomic RNA, based on probing data has been published.[81]

A major recent innovation in the field is the use of base analogues in combination with thiophosphates. After the interference procedure, iodine cleavage of the thiophosphates provides a means to easily identify the interference sites.[158] This approach, first pioneered on defining the chemical groups essential for *Tetrahymena* group I intron function, has found other applications such as the identification of adenosine functional groups involved in substrate binding by the RNase P ribozyme.[159]

5.6.2 Site-Selected Modification of Nucleotides: Chemical Applications

Detailed understanding of RNA structure and function may require analysis exceeding simple base mutagenesis. Base analogues may differ from the natural bases by as little as one atom, which is why the technique of chemical and enzymatic incorporation of these modified bases can be seen as atomic mutagenesis. Incorporation by RNA polymerases requires triphosphates of modified nucleotides compatible with the enzymatic machinery.[160] 2′-Deoxynucleotides were incorporated into tRNAs to study the impact of 2′OH-groups in structure and function[161] and replacement of U by 4-thiouridines in tRNA^{Phe} during *in vitro* transcription was used to define conjunction sites for attachment of the cleavage reagent 5-iodoacetamido-1,10-*o*-phenanthroline (IoP). This reagent will bind to other RNA, such as ribosomal 16S RNA, in a reducing environment.[162] Alternatively, modified nucleotides can be incorporated by recombinant RNA techniques based on the preparation of discrete RNA fragments and their association *via* enzymatic ligations.[160]

Progress in chemical RNA synthesis, combined with enzymatic cut-and-paste techniques allows site-selected insertion of modified nucleotides at virtually any desired position of an RNA, thus presenting a powerful tool allowing one to address specific questions on structure–function relationships of these macromolecules. A large number of phosphoramidite derivatives of modified nucleotides, required for chemical synthesis of RNA, have been created and allow insertion of minimally modified heterocyclic bases, ribose rings, and phosphates.[160,163,164] To be mentioned are inosine, isoguanosine, deazaguanosine, deazaadenosines, 2-aminopurines, 2′-fluororibose, 2′-aminoribose, phorphorothioates, and methylphosphates. Much larger modifications can also be introduced, as is the case for example with uridine derivatives including photoreactive components.[165]

Utilization of such derivatives has contributed to the understanding of tRNAs,[166,167] ribozymes[168,169] as well as anti-HIV[170,171] and antisense strategies.[172] Internal insertion of modified elements have found applications in cross-linking experiments, e.g., insertion of disulfide derivatives that allow intrahelical cross-linking of RNA chains.[172,174]

5.6.3 Targeting Modified Residues

Natural RNA modification is an enzyme-catalyzed process that takes place in all organisms. Interest in these modified bases steadily increases as they are discovered in more and more different types of RNAs. Also, the number of different modifications of ribonucleotides is on a steady climb.[175] Modified nucleotides stand out among the four major nucleotides thanks to their peculiar properties which make them preferential targets for investigators. Many examples in the older literature within the tRNA field, the RNA family where most modifications are found, concern bases such as wybutine or m^7G[176,177] for specific strand scission, or modified riboses which hinder enzymatic cleavage. Alternatively, fluorescent dyes have been bound to tRNA *via* N-ribosylic linkages upon removal of modified residues, such as wybutine or dihydrouracil.[178] Also, the photoreactive nucleotide s^4U and its derivative s^4T have been used for cross-linking studies. Many natural modifications are also used as insertion points for additional modifications. Thus, for example, the undermodified *E. coli* $tRNA^{Tyr}$ possessing a 7-(aminomethyl)-7-deazaguanosine instead of Q, can be modified chemically at this specific position by dansylchloride, a fluorescent compound.[179] *E. coli* $tRNA^{fMet}$ has been derivatized with side chains containing a disulfide bond capable of reaction with cysteine residues of a protein and with *N*-hydroxysuccinimide ester groups capable of coupling to lysine. This tRNA has been used for cross-linking to methionyl-tRNA synthetase.[180] Along a same view, t^6A and acp^3U are target sites for introduction of photoreactive azidointrophenyl probes which allow cross-linking at 20 Å distance.[181]

5.6.4 Site-Selected Modification of RNA Residues: Biophysical Applications

Introduction of fluorescent markers to RNA, either internally or at terminal positions, has been a main issue over the years.[178,182] Chemical insertion of fluorophores (a donor and acceptor, respectively) at two distant positions within the RNA lead to the most interesting fluorescence resonance energy transfer (FRET) experiments which allow the relative spatial orientation of RNA domains to be established. Donor and acceptor fluorophores are classically attached to the extremities of domains and the energy transfer resulting from a dipolar coupling between the transition moments of the two fluorophores measured. This energy transfer is inversely dependent on the distance between the fluorophores. The method was first used to measure intramolecular distances in tRNAs.[183] More recently, a three-dimensional model for the hammerhead ribozyme has been proposed on the basis of such an analysis where 5-carboxyfluorescein was the donor and 5-carboxytetramethylrhodamine the acceptor fluorophore.[184] Also, the ion-induced folding of this ribozyme has been evaluated[185] using fluorescein and cyanine-3 as donor and acceptor fluorophores, respectively. As a last example, intra-tRNA distances have been measured for nucleocapsid protein-dependent tRNA unwinding during priming of HIV reverse transcription.[186]

5.7 PERSPECTIVES

The field of structural probing is in constant motion and the new trends that are emerging concern both novel fields of applications and development of new methodological tools. Trends towards developing a deeper understanding of RNA structure in solution point to finding novel nonclassical structural features, like noncanonical base pairs and tertiary interactions, to approaching the dynamics of RNA as well as its folding and unfolding behavior, and to searching for conformational changes and alternate structures. The development of new RNA cleaving reagents will continue to be a challenge for chemists. Even though research will take manifold directions, the authors believe that two principal directions will be favored. First, for the design of more efficient cleavers and, second, for that of structure-specific ligands on which they will be tethered. In other words, chemical constructs will be engineered that, like enzymes, comprise an active site and a ligand binding domain. Because of their size which can be small, the artificial RNases will have recognition specificities very different from those of the protein RNases. An interesting related issue is the use of antisense oligonucleotides as ligands to confer sequence-specificity to the artificial RNases. Finally, the recent innovation of time-resolved footprinting[42] and probing methods[187] bears the potential to develop into an all new domain by itself.

5.8 REFERENCES

1. B. Singer, *Nature*, 1976, **264**, 333.
2. R.W. Holley, J. Apgar, G.A. Everett, J.T. Madison, M. Marquise, S.H. Merril, J.R. Penswick and R. Zamir, *Science*, 1965, **147**, 1462.
3. A.D. Branch, B.J. Benenfeld and H.D. Robertson, *Methods Enzymol.*, 1989, **180**, 130.
4. H. Donis-Keller, A.M. Maxam and W. Gilbert, *Nucleic Acids Res.*, 1977, **4**, 2527.
5. A.M. Maxam and W. Gilbert, *Proc. Natl. Acad. Sci. USA*, 1977, **74**, 560.
6. F. Sanger, S. Nicklen and A.R. Coulson, *Proc. Natl. Acad. Sci. USA*, 1977, **74**, 5463.
7. R. Lavery, A. Pullman, B. Pullman and M.d. Oliveira, *Nucleic Acids Res.*, 1980, **8**, 5095.
8. S.R. Holbrook and S.H. Kim, *Biopolymers*, 1983, **22**, 1145.
9. D.A. Peattie, *Proc. Natl. Acad. Sci. USA*, 1979, **76**, 1760.
10. D.A. Peattie and W. Gilbert, *Proc. Natl. Acad. Sci. USA*, 1980, **77**, 4679.
11. C. Ehresmann, F. Baudin, M. Mougel, P. Romby, J.-P. Ebel and B. Ehresmann, *Nucleic Acids Res.*, 1987, **15**, 9109.
12. N.A. Kolchanov, I.I. Titov, I.E. Vlassova and V.V. Vlassov, *Prog. Nucleic Acid Res. Mol. Biol.*, 1996, **53**, 131.
13. V.V. Vlassov, R. Giegé and J.-P. Ebel, *FEBS Lett.*, 1980, **120**, 12.
14. P. Romby, D. Moras, M. Bergdoll, P. Dumas, V.V. Vlassov, E. Westhof, J.-P. Ebel and R. Giegé, *J. Mol. Biol.*, 1985, **184**, 455.
15. A. Garcia and R. Giegé, *Biochem. Biophys. Res. Commun.*, 1992, **186**, 956.
16. M. Ruff, S. Krishnaswamy, M. Boeglin, A. Poterszman, A. Mitschler, A. Podjarny, B. Rees, J.-C. Thierry and D. Moras, *Science*, 1991, **252**, 1682.
17. O.O. Favorova, F. Fasiolo, G. Keith, S.K. Vassilenko and J.-P. Ebel, *Biochemistry*, 1981, **20**, 1006.
18. G. Knapp, *Methods Enzymol.*, 1989, **180**, 192.
19. L. Lempereur, M. Nicoloso, N. Riehl, C. Ehresmann, B. Ehresmann and J.-P. Bachellerie, *Nucleic Acids Res.*, 1985, **13**, 8339.
20. F. Cramer, *Prog. Nucleic Acid Res. Mol. Biol.*, 1971, **11**, 391.
21. W. Wintermeyer and H.G. Zachau, *FEBS Lett.*, 1970, **11**, 160.
22. W. Wintermeyer and H.G. Zachau, *FEBS Lett.*, 1975, **58**, 306.
23. D. Rhodes, *J. Mol. Biol.*, 1975, **94**, 449.
24. H. Hayatsu, *J. Biochem. (Tokyo)*, 1996, **119**, 391.
25. I.L. Batey and D.M. Brown, *Biochim. Biophys. Acta*, 1977, **474**, 378.
26. W.J. Krzyzosiak and J. Ciesolka, *Nucleic Acids Res.*, 1983, **11**, 6913.
27. P. Romby, D. Moras, P. Dumas, J.-P. Ebel and R. Giegé, *J. Mol. Biol.*, 1987, **195**, 193.
28. M. Litt, *Biochemistry*, 1969, **8**, 3249.
29. R.C. Gable and P.R. Schimmel, *Proc. Natl. Acad. Sci. USA*, 1974, **71**, 1356.
30. K. Watanabe and F. Cramer, *Eur. J. Biochem.*, 1978, **89**, 425.
31. D.W. Celander and J.M. Nussbaum, *Biochemistry*, 1996, **35**, 12061.
32. A. Garcia, R. Giegé and J.-P. Behr, *Nucleic Acids Res.*, 1990, **18**, 89.
33. T. Hermann and H. Heumann, *RNA*, 1995, **1**, 1009.
34. C.S. Chow, L.S. Behlen, O.C. Uhlenbeck and J.K. Barton, *Biochemistry*, 1992, **31**, 972.
35. A.C. Lim and J.K. Barton, *Biochemistry*, 1998, **37**, 9138.
36. C.S. Chow and J.K. Barton, *Biochemistry*, 1992, **31**, 5423.
37. S.A. Ciftan, E.C. Theil and H.H. Thorp, *Chem. Biol.*, 1998, **5**, 679.
38. J.A. Latham and T.R. Cech, *Science*, 1989, **245**, 276.
39. B. Balasubramanian, W.K. Pogozelski and T.D. Tullius, *Proc. Natl. Acad. Sci. USA*, 1998, **95**, 9738.
40. C.P.H. Vary and J.N. Vournakis, *Proc. Natl. Acad. Sci. USA*, 1984, **81**, 6978.
41. M. Götte, R. Marquet, C. Isel, V.E. Anderson, G. Keith, H.J. Gross, C. Ehresmann, B. Ehresmann and H. Heumann, *FEBS Lett.*, 1996, **390**, 226.
42. B. Sclavi, S. Woodson, M. Sullivan, M.R. Chance and M. Brenowitz, *J. Mol. Biol.*, 1997, **266**, 144.
43. C.E. Holmes, A.T. Abraham, S.M. Hecht, C. Florentz and R. Giegé, *Nucleic Acids Res.*, 1996, **24**, 3399.
44. C. Burrows and S.E. Rokita, *Accounts Chem. Res.*, 1994, **27**, 295.
45. R.P. Hickerson, C.D. Watkins-Sims, C.J. Burrows, J.F. Atkins, R.F. Gesteland and B. Felden, *J. Mol. Biol.*, 1998, **279**, 577.
46. N. Shinriki, K. Ishizaki, A. Ikehata, K. Miura, T. Ueda, N. Kato and F. Harada, *Symposium Series Nucleic Acids Res.*, 1981, **10**, 211.
47. P. Burgstaller, T. Hermann, C. Huber, E. Westhof and M. Famulok, *Nucleic Acids Res.*, 1997, **25**, 4018.
48. A.-C. Dock-Bregeon and D. Moras, *Cold Spring Harbor Symp. Quant. Biol.*, 1987, **52**, 113.
49. T. Pan, D.M. Long and O.C. Uhlenbeck, in "The RNA World", eds. R.F. Gesteland and J.F. Atkins, Cold Spring Harbor Laboratory Press, New York, 1993.
50. J.R. Morrow, in "Metal Ions in Biological Systems", ed. E. Sigel, vol. 33, Dekker, New York, 1996, ch. 19.
51. M. Matsuo, T. Yokogawa, K. Nishikawa, K. Watanabe and N. Okada, *J. Biol. Chem.*, 1995, **270**, 10097.
52. T. Marciniec, J. Ciesiolka, J. Wrzesindki and W.J. Krzyzosiak, *FEBS Lett.*, 1989, **243**, 293.
53. B. Streicher, E. Westhof and R. Schroeder, *EMBO J.*, 1996, **15**, 2556.
54. C. Werner, B. Krebs, G. Keith and G. Dirheimer, *Biochim. Biophys. Acta*, 1976, **432**, 161.
55. R.S. Brown, J.C. Dewan and A. Klug, *Biochemistry*, 1985, **24**, 4785.
56. M. Komiyama, *J. Biochem (Tokyo)*, 1995, **118**, 665.
57. R. Breslow, D.-L. Huang and E. Anslyn, *Proc. Natl. Acad. Sci. USA*, 1989, **86**, 1746.

58. V.V. Vlassov, G. Zuber, B. Felden, J.-P. Behr and R. Giegé, *Nucleic Acids Res.*, 1995, **23**, 3161.
59. M.A. Podyminogin, V.V. Vlassov and R. Giegé, *Nucleic Acids Res.*, 1993, **21**, 5950.
60. R. Giegé, B. Felden, M.A. Zenkova, V.N. Sil'nikov and V.V. Vlassov, *Methods Enzymol.*, in press.
61. R.M. Wurst, J.N. Vournakis and A.M. Maxam, *Biochemistry*, 1978, **17**, 4493.
62. P. Wrede, R. Wurst, J. Vournakis and A. Rich, *J. Biol. Chem.*, 1979, **254**, 9608.
63. P. Auron, L. Weber and A. Rich, *Biochemistry*, 1982, **21**, 4700.
64. M. Garret, P. Romby, R. Giegé and S. Litvak, *Nucleic Acids Res.*, 1984, **12**, 2259.
65. C.P.H. Vary and J.N. Vournakis, *Nucleic Acids Res.*, 1984, **12**, 6763.
66. C. Florentz, J.-P. Briand, P. Romby, L. Hirth, J.-P. Ebel and R. Giegé, *EMBO J.*, 1982, **1**, 269.
67. A. Przykorska, C. el Adlouni, G. Keith, J.W. Szarkowsli and G. Dirheimer, *Nucleic Acids Res.*, 1992, **20**, 659.
68. K.S. Aultmann and S.H. Chang, *Eur. J. Biochem.*, 1982, **124**, 471.
69. M. Zucker, *Science*, 1989, **244**, 48.
70. C. Gaspin and E. Westhof, *J. Mol. Biol.*, 1995, **254**, 163.
71. D. Gautheret and R.R. Gutell, *Nucleic Acids Res.*, 1997, **25**, 1559.
72. M.H.L. de Bruijn and A. Klug, *EMBO J.*, 1983, **2**, 1309.
73. B.A. Dock, E. Westhof, R. Giegé and D. Moras, *J. Mol. Biol.*, 1989, **206**, 707.
74. C. Baron, E. Westhof, A. Böck and R. Giegé, *J. Mol. Biol.*, 1993, **231**, 274.
75. P. Dumas, D. Moras, C. Florentz, R. Giegé, P. Verlaan, A.V. Belkum and C.W.A. Pleij, *J. Biomol. Struct. Dyn.*, 1987, **4**, 707.
76. B. Felden, C. Florentz, R. Giegé and E. Westhof, *J. Mol. Biol.*, 1994, **235**, 508.
77. B. Felden, J.F. Atkins and R.F. Gesteland, *Nature Struct. Biol.*, 1996, **3**, 494.
78. B. Felden, H. Himeno, A. Muto, J.P. McCutcheon, J.F. Atkins and R.F. Gesteland, *RNA*, 1997, **3**, 89.
79. T. Hermann, B. Schmid, H. Heumann and H.U. Göringer, *Nucleic Acids Res.*, 1997, **25**, 2311.
80. R. Walczak, E. Westhof, P. Carbon and A. Krol, *RNA*, 1996, **2**, 367.
81. J.-C. Paillart, E. Westhof, C. Ehresmann, B. Ehresmann and R. Marquet, *J. Mol. Biol.*, 1997, **270**, 36.
82. M.E. Harris, J.M. Nolan, A. Malhotra, J.W. Brown, S.C. Harvey and N.R. Pace, *EMBO J.*, 1994, **13**, 3953.
83. E. Westhof, D. Wesolowski and S. Altman, *J. Mol. Biol.*, 1996, **258**, 600.
84. V. Lehnert, L. Jaeger, F. Michel and E. Westhof, *Chem. Biol.*, 1996, **3**, 993.
85. L. Jaeger, F. Michel and E. Westhof, in "Nucleic Acids and Molecular Biology", ed. F. Eckstein, Springer, Berlin, 1996.
86. S. Furois-Corbin and A. Pullman, *Biophys. Chem.*, 1985, **22**, 1.
87. R. Lavery and A. Pullman, *Biophys. Chem.*, 1984, **19**, 171.
88. R. Lavery, A. Pullman and B. Pullman, *Nucleic Acids Res.*, 1980, **8**, 1061.
89. J.D. Robertus, J.E. Ladner, J.T. Finch, D. Rhodes, R.S. Brown, B.F.C. Clark and A. Klug, *Nature*, 1974, **250**, 546.
90. S.H. Kim, F.L. Suddath, G.J. Quigley, A. McPherson, J.L. Sussman, A.H.J. Wang, N.C. Seeman and A. Rich, *Science*, 1974, **185**, 435.
91. V.V. Vlassov, R. Giegé and J.-P. Ebel, *Eur. J. Biochem.*, 1981, **119**, 51.
92. C. Massire, C. Gaspin and E. Westhof, *J. Mol. Graphics*, 1994, **12**, 201.
93. N.B. Leontis and E. Westhof, *RNA*, 1998, **4**, 1134.
94. J.H. Cate, A.R. Gooding, E. Podell, K. Zhou, B.L. Golden, C.E. Kundroth, T.R. Cech and J.A. Doudna, *Science*, 1996, **273**, 1678.
95. A. J. Chandler and A. J. Kirby, *J. Chem. Soc. Chem. Commun.* 1992, 1769.
96. P. Brion and E. Westhof, *Annu. Rev. Biophys. Biomol. Struct.*, 1997, **26**, 113.
97. L.S. Behlen, J.R. Sampson, A.B. DiRenzo and O.C. Uhlenbeck, *Biochemistry*, 1990, **29**, 2515.
98. D. Michalowski, J. Wrzesinski and W. Krzyzosiak, *Biochemistry*, 1996, **35**, 10727.
99. T. Pan and O.C. Uhlenbeck, *Biochemistry*, 1992, **31**, 3887.
100. J.R. Morrow, in "Advances in Inorganic Chemistry", eds. G.L. Eichhorn and L.G. Marzolli, vol. 9, Prentice Hall, Englewood Cliffs, NJ, 1994, ch. 2.
101. W. Wintermeyer and H.G. Zachau, *Biochim. Biophys. Acta*, 1973, **299**, 82.
102. A.E. Oleson and M. Sasakuma, *Arch. Biochem. Biophys.*, 1980, **204**, 361.
103. R. Breslow and M. Labelle, *J. Am. Chem. Soc.*, 1986, **108**, 2655.
104. R. Breslow, *Acc. Chem. Res.*, 1991, **24**, 317.
105. P.W. Huber, *FASEB J.*, 1993, **7**, 1367.
106. D.S. Sigman and C.-h. Chen, *Annu. Rev. Biochem.*, 1990, **59**, 207.
107. S. Joseph and H.F. Noller, *EMBO J.*, 1996, **15**, 910.
108. K.F. Wilson and H.F. Noller, *Cell*, 1998, **92**, 131.
109. G.M. Culver and H.F. Noller, *RNA*, 1998, **4**, 1471.
110. J. Hall, D. Hüsken, U. Pieles, H.E. Moser and R. Häner, *Chem. Biol.*, 1994, **1**, 185.
111. A. Krol and P. Carbon, *Methods Enzymol.*, 1989, **180**, 212.
112. J.D. Puglisi and I.J. Tinoco, *Methods Enzymol.*, 1989, **180**, 304.
113. R. Giegé, M. Frugier and J. Rudinger, *Curr. Opin. Struct. Biol.*, 1998, **8**, 286.
114. K. Rietveld, R. Van Poelgeest, C.W.A. Pleij, J.H. Van Boom and L. Bosch, *Nucleic Acids Res.*, 1982, **10**, 1929.
115. C.W.A. Pleij, *Curr. Opin. Struct. Biol.*, 1994, **4**, 337.
116. J. Rudinger, B. Felden, C. Florentz and R. Giegé, *Bioorg. Med. Chem.*, 1997, **5**, 1001.
117. M.H. Kolk, M. van der Graaf, S.S. Wijmenga, C.W. Pleij, H.A. Heus and C. Hilbers, *Science*, 1998, **280**, 434.
118. M. Helm, H. Brulé, F. Degoul, C. Cepanec, J.-P. Leroux, R. Giegé and C. Florentz, *Nucleic Acids Res.*, 1998, **26**, 1636.
119. H. Moine, C. Cachia, E. Weshof, B. Ehresmann and C. Ehresmann, *RNA*, 1997, **3**, 255.

120. K. Nishikawa and S. Takemura, *J. Biochem. (Tokyo)*, 1974, **76**, 935.
121. G.W. Fox and C.R. Woese, *Nature*, 1975, **256**, 505.
122. P.J. Romaniuk, I. Leal de Stevenson, C. Ehresmann, P. Romby and B. Ehresmann, *Nucleic Acids Res.*, 1988, **16**, 2295.
123. C. Brunel, P. Romby, E. Westhof, C. Ehresmann and B. Ehresmann, *J. Mol. Biol.*, 1991, **221**, 293.
124. H. Moine, B. Ehresmann, C. Ehresmann and P. Romby, in "RNA Structure and Function", eds. R.W. Simons and M. Grunberg-Manago, Cold Spring Harbor Laboratory Press, New York, 1998.
125. S. Lorenz, C. Betzel, E. Raderschall, Z. Dauter, K.S. Wilson and V.A. Erdmann, *J. Mol. Biol.*, 1991, **219**, 399.
126. C. Betzel, S. Lorenz, J.P. Furste, R. Bald, M. Zhang, T.R. Schneider, K.S. Wilson and V.A. Erdmann, *FEBS Lett.*, 1994, **351**, 159.
127. S.A. White, M. Nilges, A. Huang, A.T. Brünger and P.B. Moore, *Biochemistry*, 1992, **31**, 1610.
128. C.C. Correll, B. Freeborn, P.B. Moore and T.A. Steitz, *Cell*, 1997, **91**, 705.
129. V. Perret, A. Garcia, J.D. Puglisi, H. Grosjean, J.-P. Ebel, C. Florentz and R. Giegé, *Biochimie*, 1990, **72**, 735.
130. M.H.L. de Bruijn, P.H. Schreier, I.C. Eperon and B.G. Barell, *Nucleic Acids Res.*, 1980, **8**, 5213.
131. D. Rhodes, *Eur. J. Biochem.*, 1977, **81**, 91.
132. A.R. Bannerjee, J.A. Jaeger and D.H. Turner, *Biochemistry*, 1993, **32**, 153.
133. L. Jaeger, E. Westhof and F. Michel, *J. Mol. Biol.*, 1993, **234**, 331.
134. A. Théobald, M. Springer, M. Grunberg-Manago, J.-P. Ebel and R. Giegé, *Eur. J. Biochem.*, 1988, **175**, 511.
135. D. Otzen, J. Barciszewski and B.F.C. Clark, *Biochem. Mol. Biol. Int.*, 1993, **31**, 95.
136. T.E. LaGrandeur, A. Huttenhofer, H.F. Noller and N.R. Pace, *EMBO J.*, 1994, **13**, 3945.
137. J. Gabryszuk and W.M. Holmes, *RNA*, 1997, **3**, 1327.
138. A. Hüttenhofer and H.F. Noller, *Proc. Natl. Acad. Sci. USA*, 1992, **89**, 7851.
139. T. Uchiumi, *J. Biol. Chem.*, 1997, **272**, 3302.
140. A. Hüttenhofer and H.F. Noller, *EMBO J.*, 1994, **13**, 3892.
141. J. Schlegl, V. Gegout, B. Schlager, M. Hentze, E. Westhof, C. Ehresmann, B. Ehresmann and P. Romby, *RNA*, 1997, **3**, 1159.
142. A. Mereau, R. Fournier, A. Gregoire, A. Mougin, P. Fabrizio, R. Luhrmann and C. Branlant, *J. Mol. Biol.*, 1997, **273**, 552.
143. L. Pearson, C.B. Chen, R.P. Gaynor and D.S. Sigman, *Nucleic Acids Res.*, 1994, **22**, 2255.
144. D. Brown, J. Brown, C. Kang, L. Gold and P. Allen, *J. Biol. Chem.*, 1997, **272**, 14969.
145. J. Cavarelli, B. Rees, M. Ruff, J.-C. Thierry and D. Moras, *Nature*, 1993, **362**, 181.
146. J. Rudinger, J.D. Puglisi, J. Pütz, D. Schatz, F. Eckstein, C. Florentz and R. Giegé, *Proc. Natl. Acad. Sci. USA*, 1992, **89**, 5882.
147. P. Ares and A. Igel, *Genes Dev.*, 1990, **4**, 2132.
148. M. Laughrea and J. Tam, *Biochemistry*, 1992, **31**, 12035.
149. C. Isel, J.-M. Lanchy, S.F.J. Le Grice, C. Ehresmann, B. Ehresmann and R. Marquet, *EMBO J.*, 1996, **15**, 917.
150. C. Isel, C. Ehresmann, G. Keith, B. Ehresmann and R. Marquet, *J. Mol. Biol.*, 1995, **247**, 236.
151. M. Schmidt, P. Zheng and N. Delihas, *Biochemistry*, 1995, **34**, 3621.
152. B. Schmid, L. Read, K. Stuart and H. Goringer, *Eur. J. Biochem.*, 1996, **240**, 721.
153. D. Moazed and H.F. Noller, *Nature*, 1987, **327**, 389.
154. M.I. Recht, D. Fourmy, S.C. Blanchard, K.D. Dahlquist and J.D. Puglisi, *J. Mol. Biol.*, 1996, **262**, 421.
155. M.I. Recht, S. Douthwaite, K.D. Dahlquist and J.D. Puglisi, *J. Mol. Biol.*, 1999, **86**, 33.
156. P. Spacciapoli, L. Doviken, J.J. Mulero and D.L. Thurlow, *J. Biol. Chem.*, 1989, **264**, 3799.
157. E. Skripin, J.-C. Paillart, R. Marquet, B. Ehresmann and C. Ehresmann, *Proc. Natl. Acad. Sci. USA*, 1994, **91**, 4945.
158. S.A. Strobel and K. Sketty, *Proc. Natl. Acad. Sci. USA*, 1997, **94**, 2903.
159. D. Siew, N.H. Wahler, A.G. Cassano, S.A. Strobel and M.E. Harris, *Biochemistry*, 1999, **38**, 1873.
160. R.A. Zimmermann, M.J. Gait and M.J. Moore, in "Modification and Editing of RNA", eds. H. Grosjean and R. Benne, American Society of Microbiology Press, Washington, DC, 1998.
161. R. Aphasizhev, A. Théobald-Dietrich, D. Kostyuk, S.N. Kochetkov, L. Kisselev, R. Giegé and F. Fasiolo, *RNA*, 1997, **3**, 893.
162. J. Bullard, M. van Waes, D. Bucklin, M. Rice and W. Hill, *Biochemistry*, 1998, **37**, 1350.
163. N. Usman and R. Cedergren, *Trends Biochem.*, 1992, **17**, 334.
164. P.F. Agris, A. Malkiewicz, A. Krszewski, K. Everett, B. Nawrot, E. Sochaka, J. Jankowska and R. Guenther, *Biochimie*, 1995, **77**, 125.
165. B.E. Eaton and W.A. Pieken, *Annu. Rev. Biochem.*, 1995, **64**, 837.
166. K. Musier-Forsyth and P. Schimmel, *Nature*, 1992, **357**, 513.
167. K. Musier-Forsyth, N. Usman, S. Scaringe, J. Doudna, R. Green and P. Schimmel, *Science*, 1991, **253**, 784.
168. L. Beigelman, A. Karpeisky, J. Matulic-Adamic, P. Haeberli, D. Sweedler and N. Usman, *Nucleic Acids Res.*, 1995, **23**, 4434.
169. F. Seela, *Nucleic Acids Res.*, 1998, **15**, 1010.
170. H. Aurup, A. Siebert, F. Benseler, D. Williams and F. Eckstein, *Nucleic Acids Res.* 1994, **22**.
171. K. Shah, *Bioconj. Chem*, 1996, **7**, 283.
172. R. Griffey, *J. Med. Chem.*, 1996, **39**, 5100.
173. S. Sigurdsson, T. Tuschl and F. Eckstein, *RNA*, 1995, **1**, 575.
174. D. Earnshaw, B. Masquida, S. Muller, S. Sigurdsson, F. Eckstein, E. Westhof and M. Gait, *J. Mol. Biol.*, 1997, **274**, 197.
175. J.A. McCloskey and P.F. Crain, *Nucleic Acids Res.*, 1998, **26**, 196.
176. K. Harbers, R. Thiebe and H.G. Zachau, *Eur. J. Biochem.*, 1972, **26**, 132.
177. V.S. Zueva, A.S. Mankin, A.A. Bogdanov and L.A. Baratova, *Eur. J. Biochem.*, 1985, **146**, 679.

178. W. Wintermeyer and H.G. Zachau, *Eur. J. Biochem.*, 1979, **98**, 465.
179. H. Kasai, N. Shindo-Okada, S. Noguchi and S. Nishimura, *Nucleic Acids Res.*, 1979, **7**, 231.
180. D. Valenzuela, O. Leon and L.H. Schulman, *Biochem. Biophys. Res. Commun.*, 1984, **119**, 677.
181. J. Podkowinski, T. Dymarek-Babs and P. Gornicki, *Acta Biochem. Pol.*, 1989, **36**, 235.
182. Y. Kinoshita, K. Nishigaki and Y. Husimi, *Nucleic Acids Res.*, 1997, **25**, 3747.
183. C.H. Yang and D. Söll, *Proc. Natl. Acad. Sci. USA*, 1974, **71**, 2838.
184. T. Tuschl, C. Gohlke, T.M. Jovin, E. Westhof and F. Eckstein, *Science*, 1994, **266**, 785.
185. G. Bassi, A. Murchie, F. Walter, R. Clegg and D. Lilley, *EMBO J.*, 1997, **16**, 7481.
186. B. Chan, K. Weidemaier, W.T. Yip, P.F. Narbara and K. Musier-Forsyth, *Proc. Natl. Acad. Sci. USA*, 1999, **96**, 459.
187. B. Sclavi, M. Sullivan, M.R. Chance, M. Brenowitz and S.A. Woodson, *Science*, 1998, **279**, 1940.

6

Chemical RNA Synthesis (Including RNA with Unusual Constituents)

YASUO KOMATSU and EIKO OHTSUKA
Hokkaido University, Sapporo, Japan

6.1 INTRODUCTION

The chemical synthesis of RNA has now been developed to the stage of automated large scale production providing the substantial amounts needed for physicochemical and structural studies of RNA. As these synthetic methods are often taken for granted by today's researchers the authors give a brief overview of the methods currently used and discuss their scope and limitations. No effort is made to provide a complete coverage of all the synthetic achievements; rather the authors would like to present an instructive chapter to scientists in related areas.

Although the principle of the coupling reaction is the same in both DNA and RNA syntheses, obtaining a quantity of RNA is more difficult compared to the DNA synthesis because of lower yields probably due to less accessibility of the 3′- 5′-functional group of 2′-*O*-protected ribo-derivatives during condensation. The yields of oligoribonucleotides depend also on the efficiency in the deprotection and separation from the side products. A variety of combinations of protection methods for the 2′-hydroxyl and 5′-hydroxyl groups have been employed for RNA syntheses. Larger molecules such as tRNAs consisting of approximately 80 nucleotides have been obtained by solid-phase synthesis using the phosphoramidite approach, although the overall yield of the product is much lower than that of DNA of similar size. Practical sizes of RNA for physicochemical studies are still limited to between 10 and 20 nucleotides. Enzymatic syntheses of RNA by transcription or enzymatic ligation of oligoribonucleotides have been used for larger oligoribonucleotides to compensate for the limitations of RNA synthesis.

The two approaches used at present are the solid-phase synthesis of RNA and synthesis in solution. Solid-phase synthesis has become the method of choice as will be described later. Solution-phase synthesis of RNA has some advantages for developing new protecting groups or catalysts for the reaction. The new strategies are important for the synthesis of unusual RNA molecules. There are several approaches to incorporating modified nucleotides into RNA. Suitably protected synthons for the phosphoramidite method are the most straightforward approach, if the modified nucleosides are stable under the conditions used in the synthesis. Although postsynthetic modifications of deoxyribooligonucleotides have been studied, and a few experiments have been reported for the ribo-series, this method needs to be developed further. The replacement of modified RNA fragments is another method of incorporating unusual nucleosides in defined positions, provided that methods for specific cleavage and joining of the RNA fragments are available.

This review will discuss some protecting groups and coupling reagents used in RNA synthesis, and then describe methods for the synthesis of RNA derivatives that are useful for studies of the biological and structural functions of RNA. A few examples of the applications of these RNA derivatives are also included. For the enzymatic incorporation of modified nucleotides, see Chapter 15.

6.2 SYNTHETIC METHODS FOR OLIGONUCLEOTIDES

In the late 1950s, Khorana and co-workers developed phosphodiester methods, which involved the 5′-phosphomonoester of one nucleoside and the 3′-OH of another, using dicyclohexylcarbodiimide (DCC).[1] This approach was used in the synthesis of the 64 ribo-triplets for elucidation for the genetic code.[2] In the 1970s, as has been described elsewhere,[3] oligonucleotides were synthesized by the phosphotriester method (Scheme 1). Nucleoside phosphodiesters (**1**) were coupled with a 5′-free nucleoside (**2** in the presence of a condensing agent, such as 1-(mesitylene-2-sulfonyl)-3-nitro-1,2,4-triazole (**3**) (MSNBT).[4] Letsinger and co-workers introduced the phosphite coupling approach using phosphorodichloridite,[5] and several oligoribonucleotides were also synthesized. However, the instability of the intermediate prohibited further application to larger molecules. The phosphoramidite method was developed by Caruthers and co-workers to solve this problem, by replacing one of the chlorines on the phosphorodichloridite with a secondary amine[6] (Scheme 2). The amidite unit (**5**) is condensed

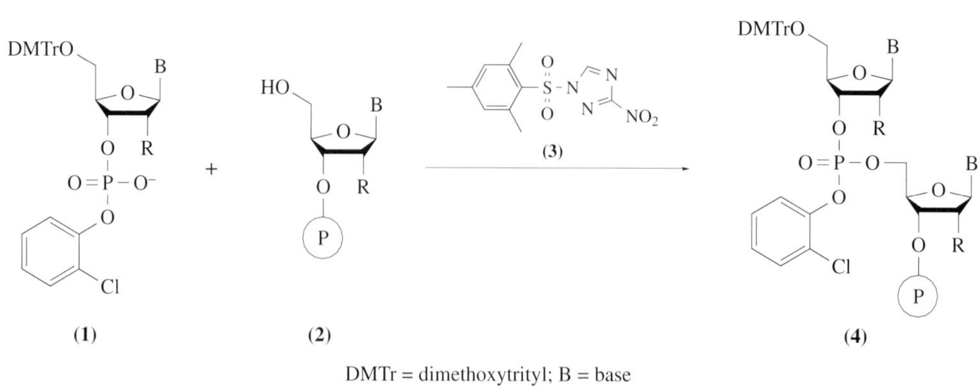

DMTr = dimethoxytrityl; B = base

Scheme 1

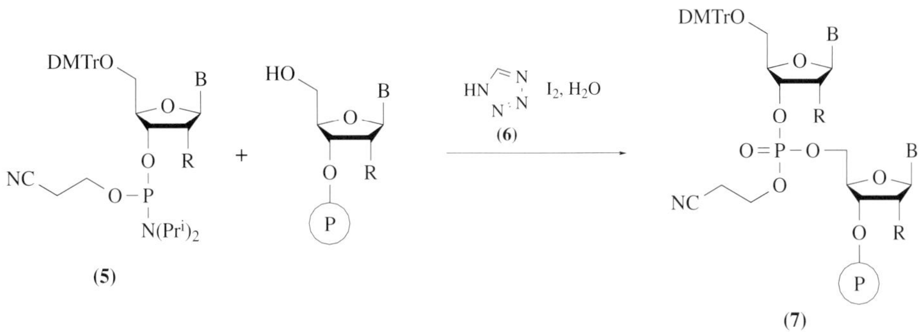

Scheme 2

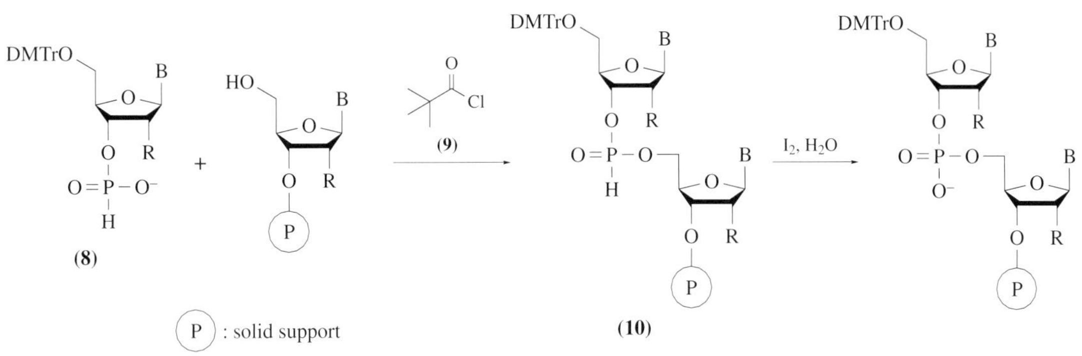

Scheme 3

with (**2**) in the presence of tetrazole (**6**). This method has been widely used to synthesize oligodeoxyribonucleotides by solid-phase synthesis, and more recently for the ribo series (see Section 6.3.2 for 2′-*O*-protecting groups). The H-phosphonate method has been improved and found to be suitable for modifying the internucleotidic phosphate.[7,8] As shown in Scheme 3, the 3′-*O*-(H-phosphonate) (**8**) is coupled with the 5′-hydroxyl group using a hindered acyl chloride (**9**). Since the diester (**10**) is stable under the conditions of the further condensation, the oxidation can be performed at the end of the synthesis. This approach has been applied to the synthesis of biologically active RNA fragments.[9]

The chemical synthesis of oligoribonucleotides requires extra steps, due to the presence of the 2′-hydroxyl group. The 2′-OH of the ribonucleoside has to be continuously protected until the end of the condensation, because a free 2′-OH easily breaks the RNA strand under alkaline conditions, and the 3′-5′-phosphodiester linkage migrates to a 2′-5′ linkage under acidic conditions. The 2′-protecting group must be stable during the condensing reactions of the nucleoside units, as well as during the deprotection of the 5′-OH. In the following section, the protecting groups that are frequently used in amidite chemistry are described.

6.3 PROTECTING GROUPS

6.3.1 5′-OH Protection

Figure 1 shows some protecting groups for the hydroxyl groups. The trityl derivatives (dimethoxytrityl, DMTr (**11**);[10] 9-phenylxanthen-9-yl, (**12**); and 9-*p*-methoxyphenylxanthen-9-yl, Mox (**13**)[11–13]) are widely used to protect the 5′-hydroxyl group. Trityl chlorides alkylate the primary 5′-OH preferentially, compared to the secondary hydroxyl groups, because of steric hindrance. These protecting groups are removed by acids. Groups (**12**) and (**13**) are more acid-labile than (**11**). The lipophilicity of the 5′-*O*-terminal dimethoxytrityl derivatives allows rapid purification of the crude tritylated oligonucleotides by reversed-phase chromatography on C-18 silica gel.

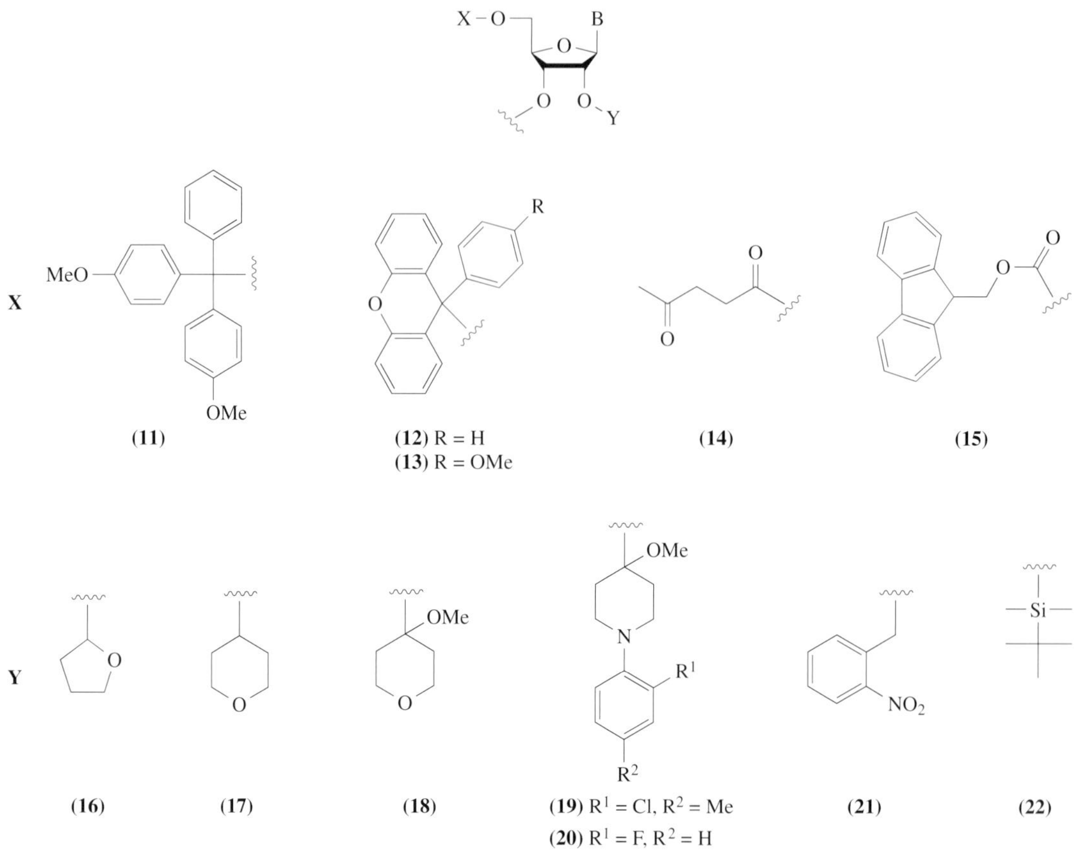

Figure 1 Protecting groups for 5′ and 2′-OH groups.

When ketal/acetal groups are used to protect the 2′-OH, they are partially cleaved during the detritylation steps, and the yield of the oligoribonucleotides is decreased. To overcome this problem, base-labile groups (levulinyl (**14**); 9-fluorenylmethyl-oxycarbonyl, Fmoc (**15**) are used for the 5′-OH protection. van Boom and co-workers developed the use of 5′-*O*-levulinyl protection for RNA synthesis.[14] 5′-*O*-Levulinyl and 2′-*O*-terahydrofuranyl groups have been combined (X = (**14**), Y = (**16**)).[15] The levulinyl group can be removed after each coupling step by treatment with 0.5 M hydrazine monohydrate in pyridine-acetic acid (3:2 v/v). Gait and co-workers utilized (**15**) for the 5′-*O*-protection and 4-methoxytetrahydropyran-4-yl (**18**) for the 2′-OH (X = (**15**), Y = (**18**)).[16] The Fmoc (**15**) was removed after each coupling step by treatment with 0.1 M DBU in acetonitrile. Since exocyclic amino groups and phosphate are usually protected by base-labile protecting groups, the basic conditions for the deprotection of these groups should be carefully optimized.

6.3.2 2′-OH Protection

6.3.2.1 Acetal/ketal groups

Acetal and ketal groups have been used as protecting groups for the 2′-OH. These functions are removed under acidic conditions. The ketal groups are deprotected more rapidly than the acetal type and 4-methoxytetrahydropyran-4-yl (**18**) which is removed in milder conditions than tetrahydropyranyl (**17**) has the advantage of no chirality.[17]

Although the 5′-*O*-dimethoxytrityl group can be deprotected under mild acidic conditions, the acetal and ketal groups are partially removed during the solid-phase synthesis of RNA, which leads to the breakage of the internucleotidic bonds by the following treatment with concentrated ammonia. It is important to prevent the loss of the 2′-OH protecting groups during the deprotection of the 5′-OH. New derivatives 1-(2-chloro-4-methylphenyl)-4-methoxypiperidin-4-yl (**19**) and 1-[(2-fluoro)

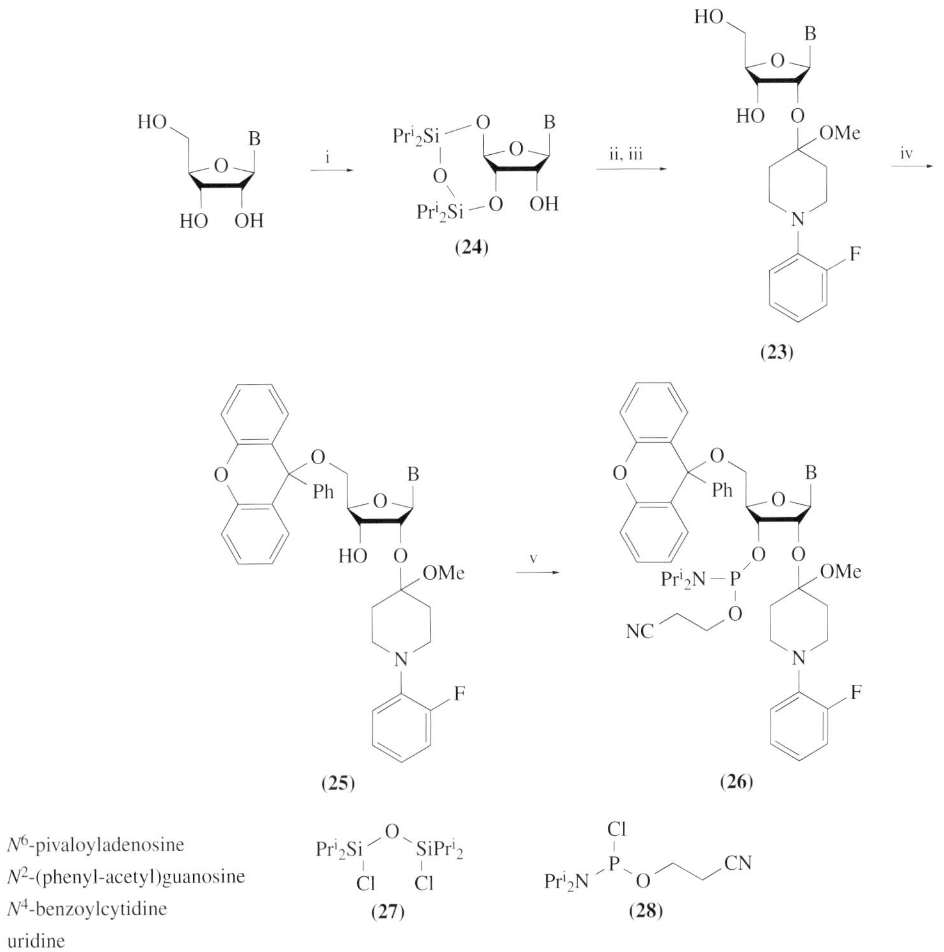

i, **(27)**, imidazole, MeCN; ii, 1-(2-fluorophenyl)-4-methoxy-1,2,3,6-tetrahydropyridine, CF_3CO_2H, CH_2Cl_2; iii, Et_4NF, MeCN; iv, 9-chloro-9-phenyl-xanthene, pyridine; v, **(28)**, THF, $EtN(Pr^i)_2$

Scheme 4

phenyl]-4-methoxypiperidin-4-yl (Fpmp, (**20**)) were proposed as 2′-OH protecting groups by Reese and co-workers.[18,19] Under strong acidic conditions, the nitrogen atom of the piperidine ring of these protecting groups is protonated, which inhibits cleavage of the 2′-*O*-ketal function. At a lower proton activity (pH2), the nitrogen atom is largely unprotonated, and then the 2′-*O*-ketal function is cleaved easily. Since the dihydro derivative of (**19**) is difficult to prepare, (**20**) is frequently used to protect the 2′-OH. Scheme 4 shows the synthesis of 5′-*O*-Px-2′-*O*-Fpmp amidite units (**26**). The exocyclic amino-protected nucleoside is protected with a bifunctional reagent, 1,3-dichloro-1,1,3,3-tetraisopropyldisiloxane (**27**) (Markiewicz reagent), to give the 3′,5′-TBDMS-nucleoside (**23**).[20,21] The 2′-OH of (**23**) is protected to yield (**20**), and the 3′- and 5′-protecting groups are removed by fluoride anions to yield a 2-*O*-Fpmp nucleoside (**24**). After the 5′-OH of (**24**) is protected by Px-chloride, the 3′-OH of (**25**) is subjected to phosphitylation by 2-cyanoethyl-*N*,*N*-diisopropylchloro-phosphoramidite (**28**).

6.3.2.2 o-*Nitrobenzyl group*

A physical analysis suggested that the *o*-nitrobenzyl ethers of nucleosides, ONB (**21**), are stable under acidic and alkaline conditions, and can be removed by UV light (280 nm) under slightly acidic conditions. Ikehara and co-workers[22] developed synthetic procedures for 2′-*O*-(ONB) nucleosides, using *o*-nitrobenzyl chloride or the diazomethane. Scheme 5 shows the reactions to pre-

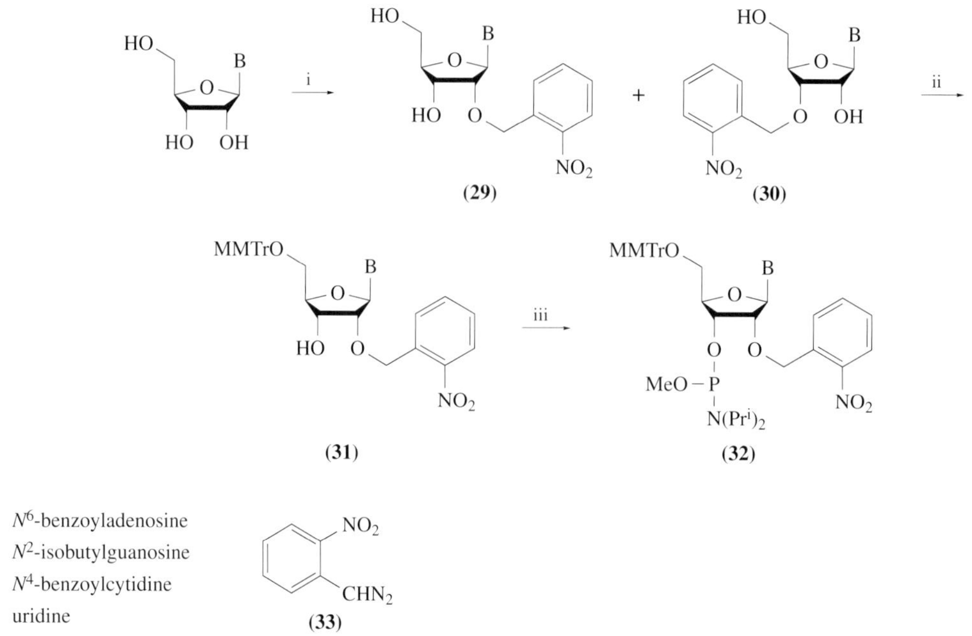

i, (**33**), $SnCl_2$•$2H_2O$; ii, MMTrCl, pyridine; iii, $MeOP[N(Pr^i)_2]_2$, diisopropylamine-hydrtetrazole, CH_2Cl_2

Scheme 5

pare 5′-*O*-monomethoxytrityl-2′-*O*-(ONB)-amidite units (**32**). *N*-Benzoylcytidine, *N*-benzoyladenosine, and *N*-isobutyrylguanosine were each treated with *o*-nitrophenyldiazomethane (**33**), obtained from *o*-nitrobenzaldehyde and tosyl hydrazide, to give a mixture of 2′- and 3′-*O*-(*o*-nitrobenzyl) nucleosides (**29**) and (**30**).[22] The isomers can be separated by either precipitation or silica gel chromatography. A physical analysis suggested that the base in the 2′-isomer (**29**) was stacked with *o*-nitrobenzyl more than the other isomer (**30**). After the 5′-hydroxyl group being protected using monomethoxytrityl chloride (MMTr-Cl) the 3′-hydroxyl group was phosphitylated for the oligonucleotide synthesis.[23] The *o*-nitrobenzyl groups on the oligonucleotides are deprotected after the removal of the other protecting groups. Although ONB is chemically stable, UV light has to be avoided during the synthesis, and the removal by UV light requires stringent conditions to reduce side reactions.[23–25]

6.3.2.3 O-t-*Butyldimethylsilyl group*

Ogilvie and co-workers applied a *t*-butyldimethylsilyl group, TBDMS (**22**), to the 2′-OH protection in oligoribonucleotide syntheses. They succeeded in synthesizing a 77 ribonucleotide-long RNA with a tRNA sequence by using the solid-phase phosphoramidite method.[26] This protecting group is widely used in RNA synthesis, and protected monomer units are commercially available. Scheme 6 shows the preparation of the 5′-*O*-DMTr-2′-*O*-TBDMS amidite units (**37**). The *N*-protected nucleosides are converted to the 5′-*O*-DMTr-nucleosides (**34**), which can be silylated with *t*-butyldimethylsilyl chloride (**38**) in the presence of imidazole/DMF. A mixture of 2′- and 3′-*O*-TBDMS derivatives (**35**), (**36**) is obtained, and (**35**) is separated by either silica gel chromatography or selective precipitation.[27] The ratio of (**35**) to (**36**) depends on the reaction conditions. Silver nitrate enhances the reaction rate and increases the yield of the 2′-isomer.[28] Since (**35**) migrates to (**36**) under basic conditions, phosphitylation of the 3′-OH should be performed carefully using tertiary bases, 2,4,6-collidine and *N*-methylimidazole, as catalysts.[29] The purity of the phosphoramidite should be monitored by ^{31}P NMR to detect contamination by side products. After the coupling reactions, the fully protected oligoribonucleotides are cleaved from the solid support, and deblocking of the phosphate and the exocyclic amino groups is carried out with ethanolic ammonia or concentrated ammonia in ethanol. Concentrated ammonia alone results in the loss of the TBDMS and degradation of the RNA. Ethanol increases the solubility of the

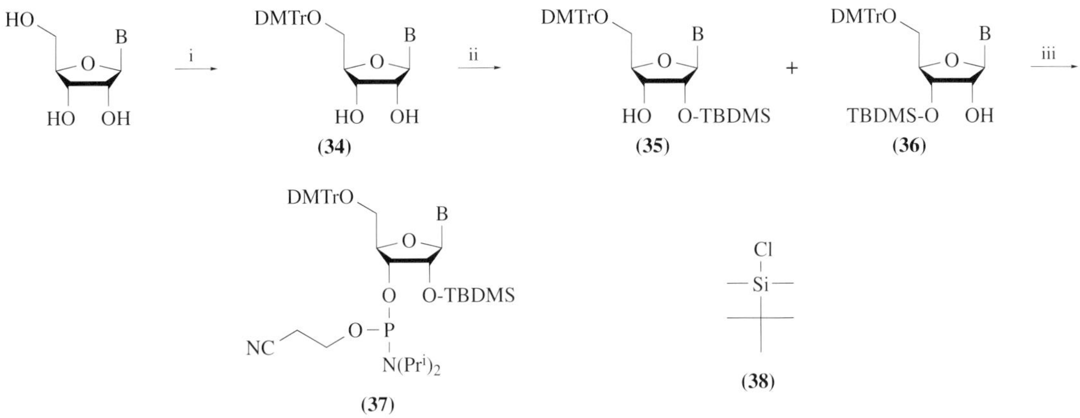

i, DMTr-Cl, THF, pyridine; ii, **(38)**, $AgNO_3$; iii, **(28)**, 2,4,6-collidine, *N*-methylimidazole, THF

Scheme 6

oligoribonucleotides and suppresses the hydrolysis of the silyl ethers. The TBDMS groups are removed from the 2′-OH by treatment with tetra-*n*-butylammonium fluoride (TBAF) in tetrahydrofuran. However, the deprotection of the TBDMS groups by TBAF is time consuming (16 h) and an improved method using triethylamine/hydrogen fluoride ($Et_3N \cdot 3HF$) has been introduced for the deprotection of silyl groups.[30,31] A mixture of $ET_3N \cdot 3HF$ and *N*-methylpyrrolidinone, which can remove the 2′-*O*-TBDMS at 65 °C in 1.5 h, was also shown to be a convenient reagent.[32]

Although the deblocking conditions for the *N*-protecting groups are different, depending on the stability (see the next section), partial desilylation occurs during the deblocking. The resulting shorter side products can be separated by reversed-phase chromatography from the dimethoxytritylated product. Alternatively, the completely deblocked oligonucleotide can be purified by either anion-exchange chromatography or gel electrophoresis. Since deblocked RNA fragments are susceptible to ribonucleases and to alkaline hydrolysis, water and glassware have to be sterilized and cleaned completely.

6.3.3 Protecting Groups for Exocyclic Aminos

The exocyclic amino groups of nucleosides are usually protected by acyl groups during the condensing reactions to avoid the formation of P-N linkages, as well as to augment the solubility. The *N*-acyl groups are removed by treatment with bases. Since strong basic conditions cause partial removal of the 2′-*O*-silyl ether and result in chain cleavage, more easily removed *N*-protecting groups, such as phenoxy acetyl derivatives (**41–43**), have been introduced. These protecting groups are especially useful for the incorporation of base-labile nucleoside analogues into oligonucleotides by the phosphoramidite method.While groups such as benzoyl (**39**) on N^6-adenosine and N^4-cytidine and isobutyryl (**40**) on N^2-guanosine require 16 h in concentrated ammonium hydroxide : ethanol (3:1) for removal, the phenoxy derivatives require only 4 h.[32,33] Dimethylformamidine (**45**) is also used as a labile *N*-protecting group for N^6-adenosine, as well as N^2-guanosine, in the phosphoramidite approach, and can be removed with concentrated ammonium hydroxide : ethanol (3:1) at room temperature for 8 h.[34] Dimethylaminomethylene-protected purine nucleosides have been used in the synthesis by the H-phosphonate method.[9,35]

6.4 SYNTHESIS OF OLIGONUCLEOTIDE ANALOGUES

6.4.1 2′-O-Modified Oligonucleotides as Antisense Nucleic Acids

The 2′-OH of ribonucleotides has various functions in RNA, such as hydrogen bonding, metal ion binding, and protein interactions. The presence of the 2′-OH affects the sugar puckering, and

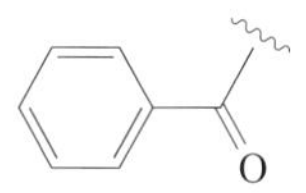

(39)

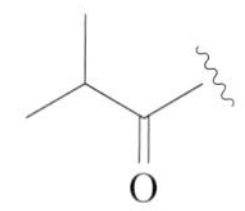

(40)

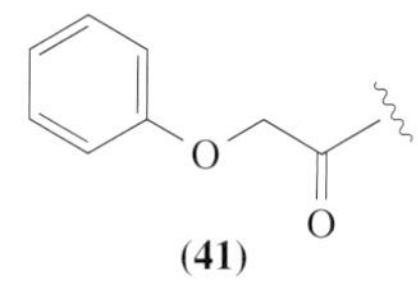

(41)

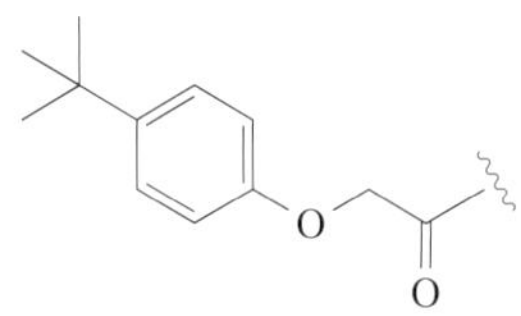

(42)

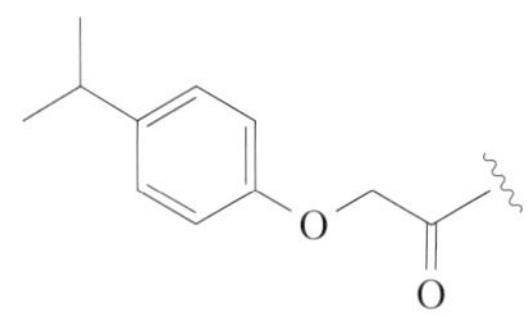

(43)

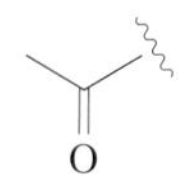

(44)

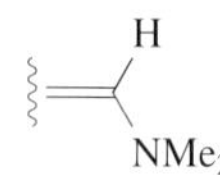

(45)

substitutions at the 2′-position yield a wide variety of RNA analogues, which are useful to investigate RNA functions.

6.4.1.1 2′-O-Alkylnucleotides

Among 2′-substituted RNA's, 2′-*O*-alkylderivatives have been synthesized and utilized most frequently as alkaline stable analogues. These analogues maintain the 3′-*endo* puckering and are resistant to ribonucleases. 2′-*O*-methyl (**46**) and allyl (**47**) RNA derivatives, which have been used to hybridize to DNA and RNA, are illustrated. The methylated nucleosides are also found in tRNAs, rRNA and snRNA. Chemical syntheses of 2′-*O*-methyloligoribonucleotides were reported using the phosphotriester[36] and phosphoramidite methods.[37] Inoue *et al.* found 2′-*O*-methyl oligonucleotides hybrids with RNA showed stability similar to RNA • RNA and greater stability compared with RNA • DNA.[36,38] This stability assists in hybrid formation with chimeric DNAs 3 or 5 mer deoxyoligonucleotides with 2′-*O*-methyl oligomers on the 3′- and 5′-sides, d(UmGmAATGGAmCm)). In an RNase H reaction, 5′r(ACUUA↓CCUG)3′ was found to be cleaved at the allowed site (**50**) in Figure 2(a).[39] These chimeric antisense oligonucleotides have been applied to the site-specific cleavage of RNA.[40] Figure 2(b) shows the specific cleavage of *E. coli* $tRNA^{Gln}$ for replacement of the 5′-fragment with an inosine-containing oligonucleotide. This experiment proved the participation of the 2-amino group of guanosine at position 3 in the reaction with the glutaminyl tRNA synthetase, because the rate of the amino acylation of the modified tRNA was substantially reduced.[41] The 2′-*O*-methylnucleosides were also incorporated into antisense DNA, and these chimeric antisense oligonucleotides were shown to be efficient, nuclease resistant reagents from the induction of RNase H reactivity against the target RNA.[42]

2′-*O*-Alkyloligoribonucleotides serve as good biochemical probes, due to their ability to form highly stable duplexes. 2′-*O*-Methyl and 2′-*O*-allyloligoribonucleotides, complementary to the RNA components of snRNPs, inhibited pre-mRNA splicing.[43,44] Biotinylated 2′-*O*-alkyloligoribonucleotides were used to isolate a snRNP from crude extracts by antisense chromatography.[38,45] 2′-*O*-Allyl probes were shown to have higher specificity to the target RNA than the 2′-*O*-methyl and 3,3-dimethylallyl probes.[46]

Homopyrimidine oligonucleotides recognize homopurine-homopyrimidine double stranded DNA by triple helix formation. It was also reported that 2′-*O*-methyl oligoribonucleotides could form triple helices through binding to double stranded DNA.[47,48] Shimizu *et al* reported that 2′-*O*-methyl pyrimidine oligonucleotides show the highest thermal stability among the various 2′-substituents (2′-deoxy, 2′-fluoro,

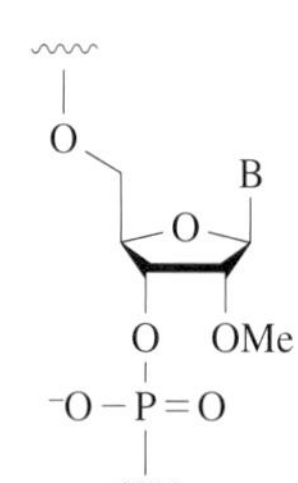

(46)

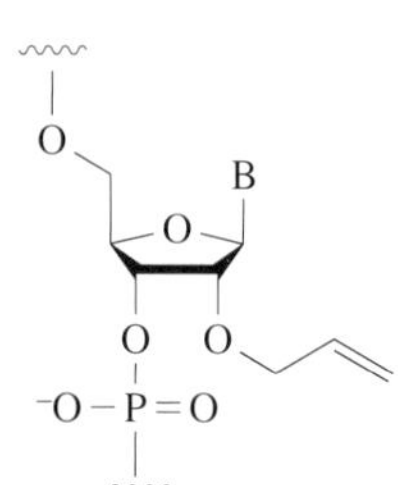

(47)

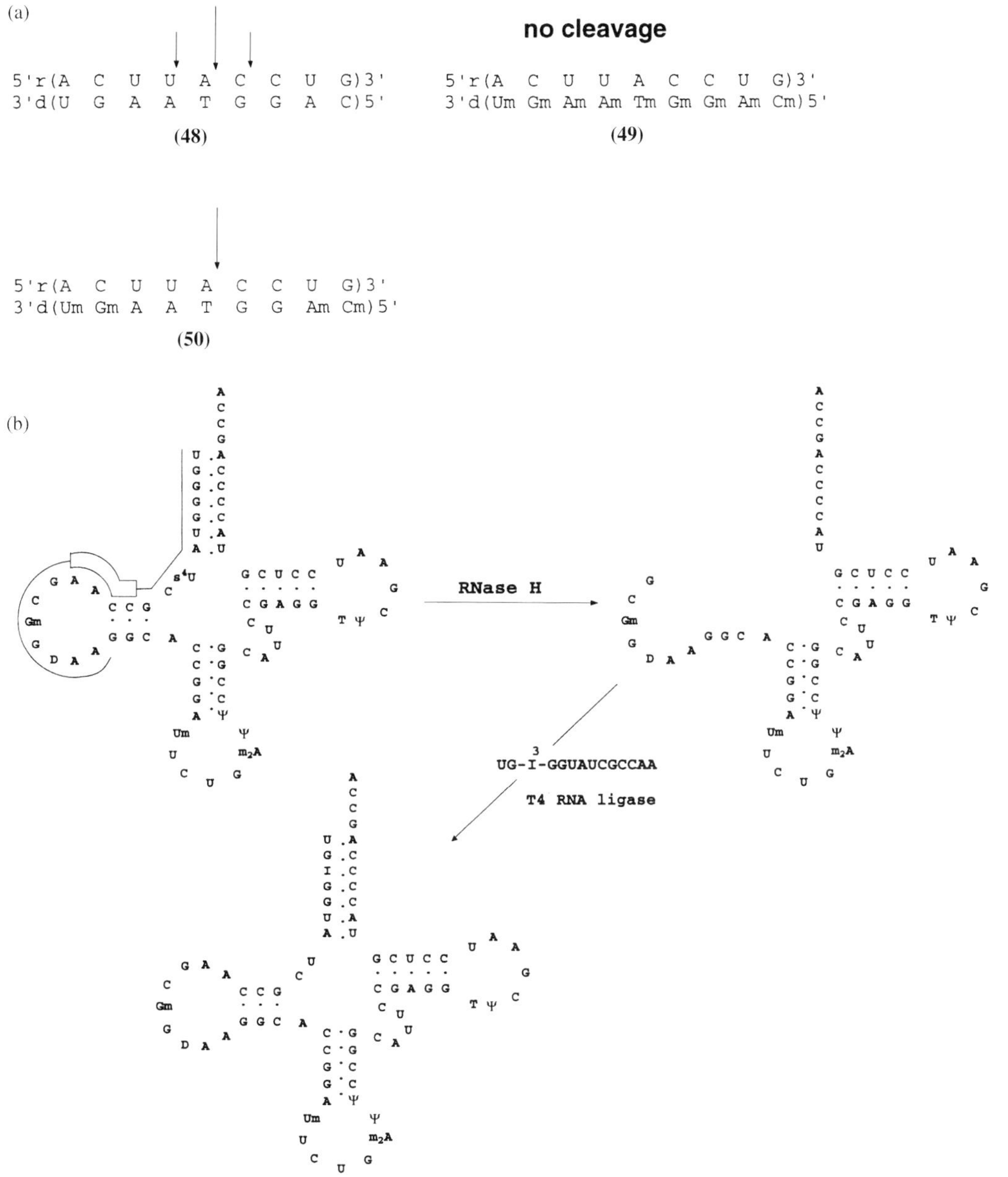

Figure 2 Site specific cleavage of RNA by RNase H using chimeric oligodeoxynucleotides. (a) RNase H cleavage sites in hybrid duplexes. (b) Substitution of guanosine with inosine in the acceptor stem of $tRNA^{Gln}$.

and RNA). 1,10-Phenanthroline-linked 2′-*O*-methyl oligonucleotides were able to cleave double-stranded DNA at the expected sites.[49]

The biological functions of RNA were examined by partial incorporation of 2′-*O*-alkylnucleotides into the RNA. The role of the 2′-OH function in hammerhead ribozymes has been studied by the substitution of ribonucleotides with 2′-*O*-alkylnucleotides (see Chapter 15).

6.4.1.2 Other 2′-O-substituted analogues

The 2′-fluoronucleosides (**51**) have the 3′-endo puckering like ribonucleosides,[50] and have been used in functional RNA, such as hammerhead ribozymes, to prove the importance of the conformation in the catalytic site.[51,52] The 2′-aminonucleosides (**52**) have been incorporated in RNA and further functionalized for cross-linking experiments to investigate higher structures of RNA. Since the 2′-position is located on the surface of the minor groove, intra- and interstrand cross-linking by disulfide bond formation have been carried out in hammerhead (**53**)[53] and *Tetrahymena* ribozymes (**54**).[54]

6.4.2 Phosphate Modification

Phosphodiester linkages give acidic characteristics to nucleic acids and play important roles in the interactions with proteins and metal ions. The oxygen atoms of the phosphate can be substituted with sulfur atoms to alter the chemical or steric character. The nonbridging oxygen can be derivatized at the oxidation step during the condensing reaction, using phosphite chemistry. In addition to sulfurization of the phosphodiester, phosphoramidates and methylphosphonates are obtained. These derivatives have been extensively studied in the deoxy series, including the stereospecific synthesis. However, in the ribo chemistry, nonbonding phosphorothioates have been synthesized by chemical and enzymatic methods, and are extensively used in mechanistic studies of enzymes as well as ribozymes (see Chapter 6.15).

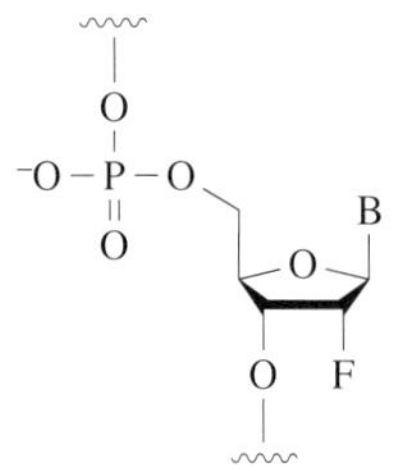

(51)

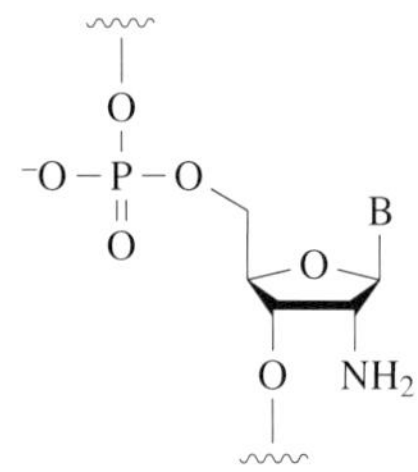

(52)

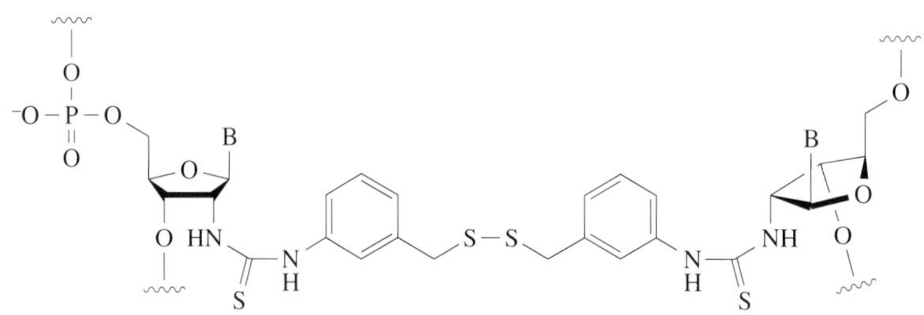

(53)

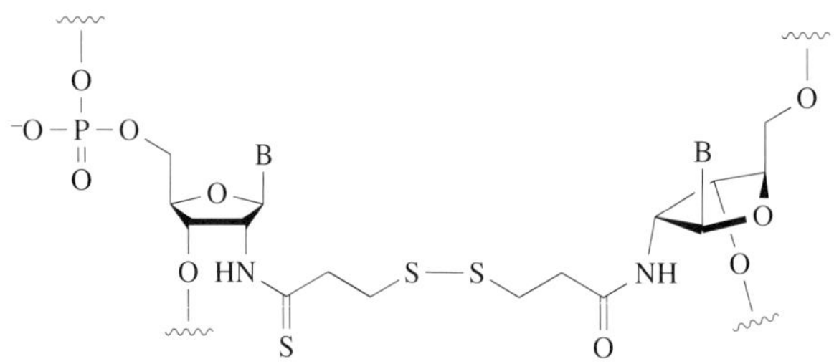

(54)

6.4.2.1 *Substitution of nonbridging oxygen*

RNA with a phosphorothioate linkage can be prepared by chemical and enzymatic syntheses. While the enzymatic synthesis gives the phosphorothioate with the Rp-configuration, chemical syntheses yield both Sp (**55**) and RP (**56**) isomers. The nonbridging oxygen of the internucleotidic linkage can be substituted during the oxidation step in the phosphoramidite or H-phosphonate solid-phase synthesis by using sulfurizing agents, such as elemental sulfur in carbon disulfide/2,6-lutidine[55] or tetraethylthiuram disulfide (TETD) in acetonitrile.[56] Since elemental sulfur tends to precipitate in the delivery line, other soluble reagents are preferred for mechanical synthesis. The Rp and Sp isomers can be separated by reverse phase chromatography depending on the sequences and the lengths.

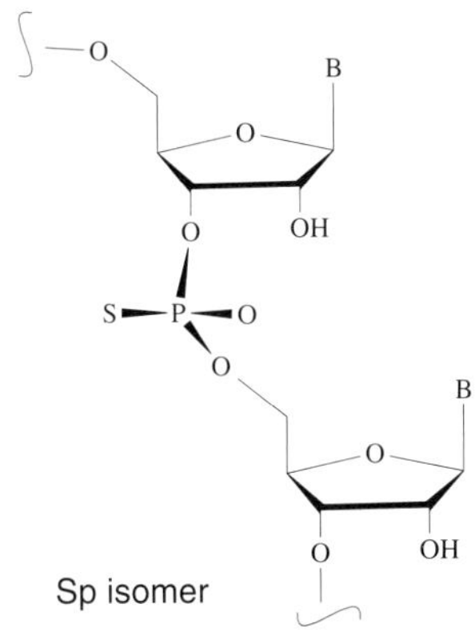

(55)

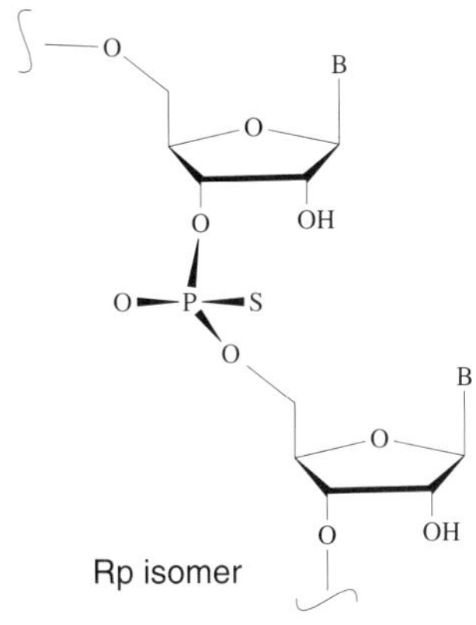

(56)

6.4.2.2 Substitution of 5′- or 3′-bridging oxygen

The bridging phosphorothioate analogues of DNA have been used to elucidate the cleavage mechanism of nucleases.[57] However, it was difficult to synthesize RNA containing bridging phosphorothioate linkages, due to the reactions for the removal of the protecting groups of the ribose 2′-OH. The deprotecting agents for TBDMS, such as triethylamine/trihydrofluoride or tetrabutyl ammonium fluoride, cause substantial chain cleavage of the thio-RNA. MacLaughlin and co-workers synthesized a chimeric oligonucleotide containing a single thioribonucleotide in an otherwise all-deoxyribonucleotide, using phosphoramidite chemistry.[58,59] They prepared a 2′-*O*-chloroethoxyethyl ribonucleotide amidite (**57**) and a 5′-*S*-triphenylmethyl ribonucleotide amidite (**58**), and then synthesized an oligonucleotide (**59**) containing the 5′-thioribonucleotide. The 5′-triphenylmethyl on the thio-compound was removed by treatment with silver nitrate, and the other trityl derivatives were removed with dichloroacetic acid. It was shown that (**59**), with a 5′-bridging phosphorothioate, was labile under alkaline conditions and in the presence of soft metal ions.[59] These chimeric nucleotides have been used to investigate the cleavage mechanisms of hammerhead ribozymes.[60] Taira and co-workers also proposed a mechanism for the hammerhead ribozyme. They prepared an all ribonucleotide sequence containing one 5′-bridging thiophosphate linkage.[61] Reese *et al.* synthesized uridyl-(3′-5′)-(5′-thiouridine) (**60**) by the H-phosphonate method,[62] using tetrahydropyranyl as a 2′-OH protecting group. They also synthesized a diribonucleotide containing a 3′-bridging phosphorothioate (**61**) by a similar procedure.[63]

6.4.3 Base-modified Nucleotides

Tfhe nucleobases are directly involved in the hydrogen bonds and play the key role in the functions of nucleic acids. The exocyclic functional groups on nucleobases can be replaced by the incorporation of properly protected monomer units during solid-phase synthesis. Alternatively, those functions can be modified by postmodification methods, provided that suitable leaving groups are incorporated in the oligonucleotides. For applications of base-modified RNA in studies of ribozyme functions, see Chapter 15.

6.4.3.1 Incorporation of modified pyrimidine

(i) 5-Substituted pyrimidines

RNA containing 5-bromouridine (**62**) was shown to form a stable complex with MS2 coat protein through a transient covalent bond between a cysteine residue via Michael addition.[64] Iwai *et al.* synthesized a Rev-responsive element (RRE) containing (**62**) at the bubble structure to investigate the interaction between RRE and Rev protein.[65] The 5-bromouridine was incorporated by the phosphoramidite method, using TBDMS as the protecting group for the 2′-OH. It has been shown that the recognition of the RRE by Rev requires hydrogen bonds between the protein and the functional groups in the major groove of the distorted RRE.

Bradly and Hanna synthesized 5-thiocyanato-uridine (**63**) phosphoramidites and incorporated them into oligoribonucleotides.[66] After standard deprotection of the acyl and 2′-*O*-TBDMS groups, the thiocyanate group was reduced to a 5-mercapto function, to react with either *p*-azidophenacyl bromide or 5-iodoacetamidofluorescein.

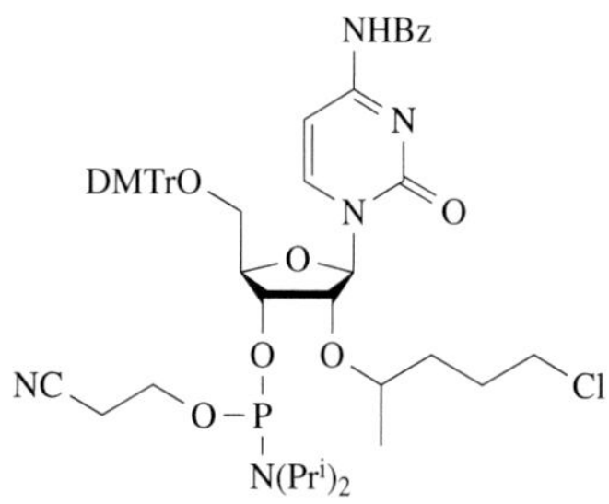

(57)

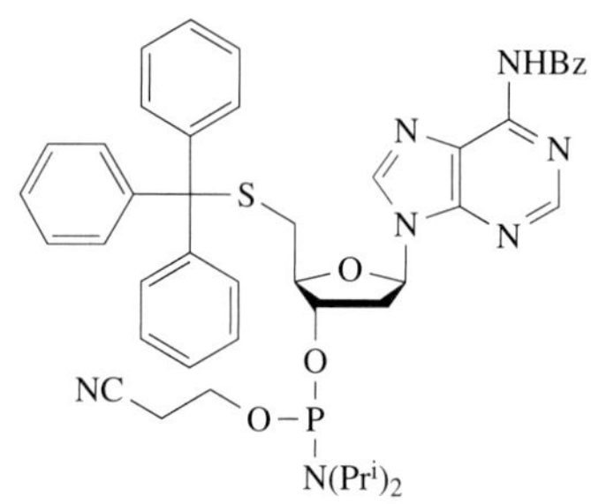

(58)

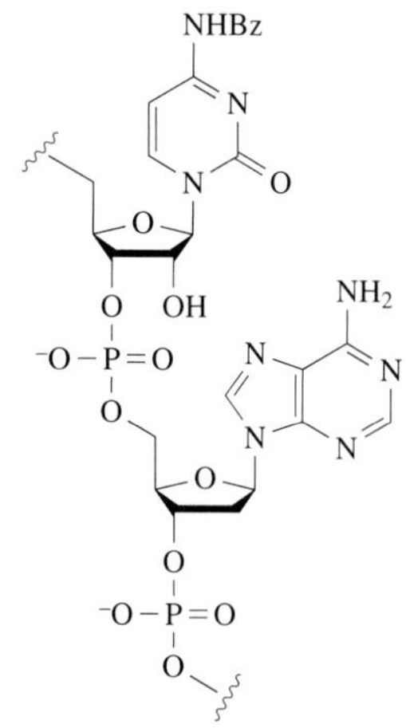

(59)

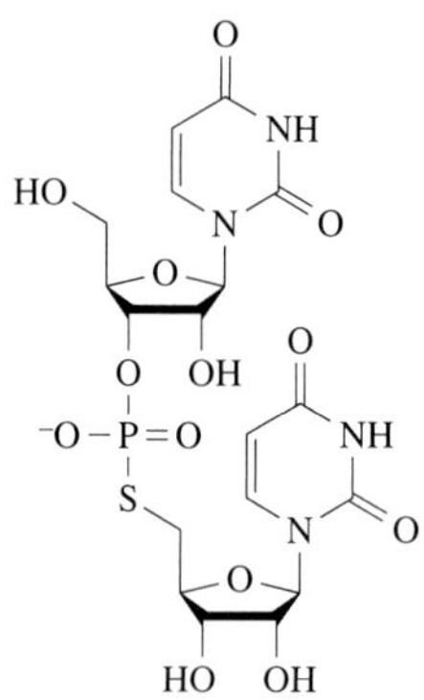

(60)

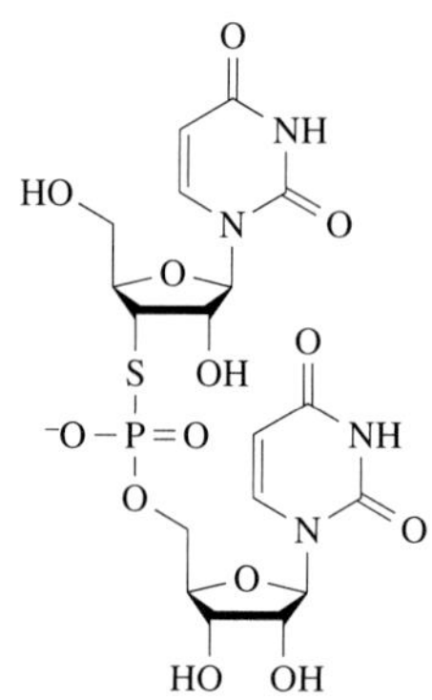

(61)

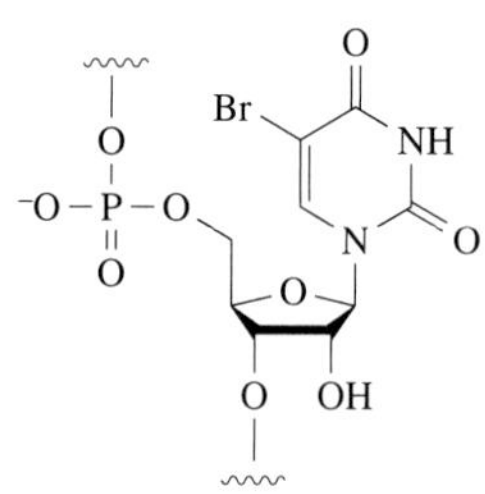

(62)

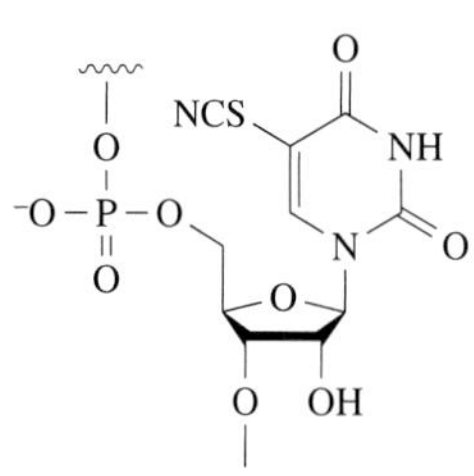

(63)

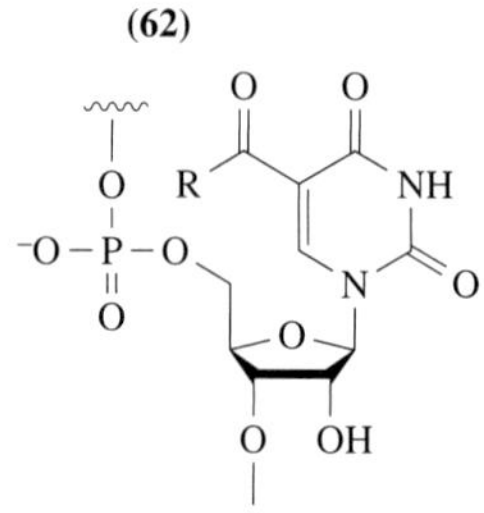

(64) R = 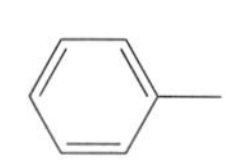

(65) R = 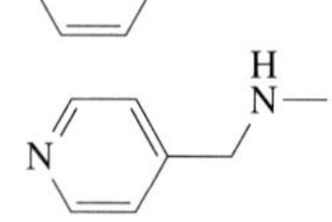

Eaton and co-workers have developed a convenient route that yields 5-carbonyluridine derivatives.[67] 2′,3′-Isopropylidene-5-iodouridine was converted to the 5-carbonyl uridine derivative by using palladium. The 5-carbonyl uridine derivative was converted to the corresponding 5′-triphosphate, and the analogue was found to be a good substrate for T7 RNA polymerase in the transcription of a dsDNA template. The transcripts containing uridine derivatives, (**64**) and (**65**), served as templates for avian myeloblastosis virus (AMV) reverse transcriptase to obtain the cDNA. They also applied this method to incorporate 5-pyridylmethylaminocarbonyl-uridine (**65**) into transcripts for *in vitro* selection of catalytic RNA, and succeeded in the generation of a catalyst for a Diels–Alder cycloaddition.[68] The RNA molecules accelerated the reaction rate by a factor of up to 800 relative to the uncatalyzed reaction. They prepared various 5-position derivatives to select catalytic RNA or the aptamer.[69]

(ii) 4-Substituted pyrimidines

4-Thiouridine (**67**) is known to be a photoreactive compound, and has been used in cross-linking reactions as a probe to investigate the interactions of nucleic acids with proteins and nucleic acids with nucleic acids. Stockley and co-workers prepared a 4-*S*-cyanoethyl-4-thiouridine phosphoramidite unit (**66**) and incorporated it into oligoribonucleotides (Figure 3).[70] Oligoribonucleotides with the *S*-cyanoethyl-4-thiouridine were treated with DBU to remove the *S*- and *O*-cyanoethyl groups, and deprotection of the exocyclic amines was achieved with methanolic ammonia. The 2′-*O*-TBDMS was deprotected using either $Et_3N \cdot 3HF$ or TBAF in THF. For the synthesis of the RRE containing *S*-cyanoethyl-4-thiouridine, Fpmp (**20**) was used as the 2′-OH protecting group.[71] Using this probe, specific cross-linking was demonstrated between the Rev protein of HIV-1 and the RRE by irradiation with 350 nm light. 4-Thiouridine could also be incorporated into oligoribonucleotides by postsynthetic modification (see postmodification section). Since 4-thiouridine absorbs 340 nm light, which is a characteristic of the thione chromophore, the presence of 4-thiouridine was detected from the UV-spectrum.

(iii) 2-Substituted pyrimidines

2-Thiouridine is found predominantly in the wobble position of tRNAs. It has been shown that 2-thiouridine favors the 3′-*endo* sugar conformation at the mono- and dinucleotide levels. Kumar and Davis have synthesized a 2-thiouridine phosphoramidite unit (**68**) (Figure 3) and have incorporated it into a pentamer, an anticodon sequence mimic of $tRNA^{Lys}$.[72] They modified the oxidation step in the solid-phase synthesis such that *t*-butyl hydroperoxide was used instead of I_2/H_2O. The duplex containing 2-thiouridine (**69**) showed higher stability than RNA containing either normal uridine or 4-thiouridine.

6.4.3.2 *Incorporation of modified purines*

Stockley and co-workers synthesized an oligoribonucleotide containing 6-thioguanosine (**71**), using an amidite unit (**70**) (Figure 3).[73] The thio group on the *N*-benzoyl-6-thioguanosine was protected with 3-bromopropionitrile, and the 2′-OH was protected with TBDMS. To deprotect the 2′-*O*-TBDMS, $Et_3N \cdot 3HF$/DMSO (1:1) was used, because TBAF in THF resulted in the degradation of the thiocarbonyl group. They incorporated the 6-thioguanosine into a hammerhead ribozyme strand, and compared the effect of the substitution with that of inosine and 2-aminopurine. Rana and co-workers synthesized a Rev-responsive element containing 6-thiouridine in a specific site, and carried out photo cross-linking experiments in the presence or absence of Rev.[74] Both RNA-RNA and RNA-protein cross-linking reactions were observed.

Other purine analogues, such as isoguanosine and deazapurine nucleosides, have been derivatized and incorporated into ribozymes to investigate the roles of the heterocyclic rings in the catalytic reactions.

6.4.4 Postsynthetic Modifications of RNA

Postsynthetic modification methods consist of the conversion of nucleosides with leaving groups to various derivatives on oligonucleotides. Although postsynthetic modifications have been used to obtain DNA derivatives,[75–77] the approach seemed to be difficult to apply to RNA, because of the presence of

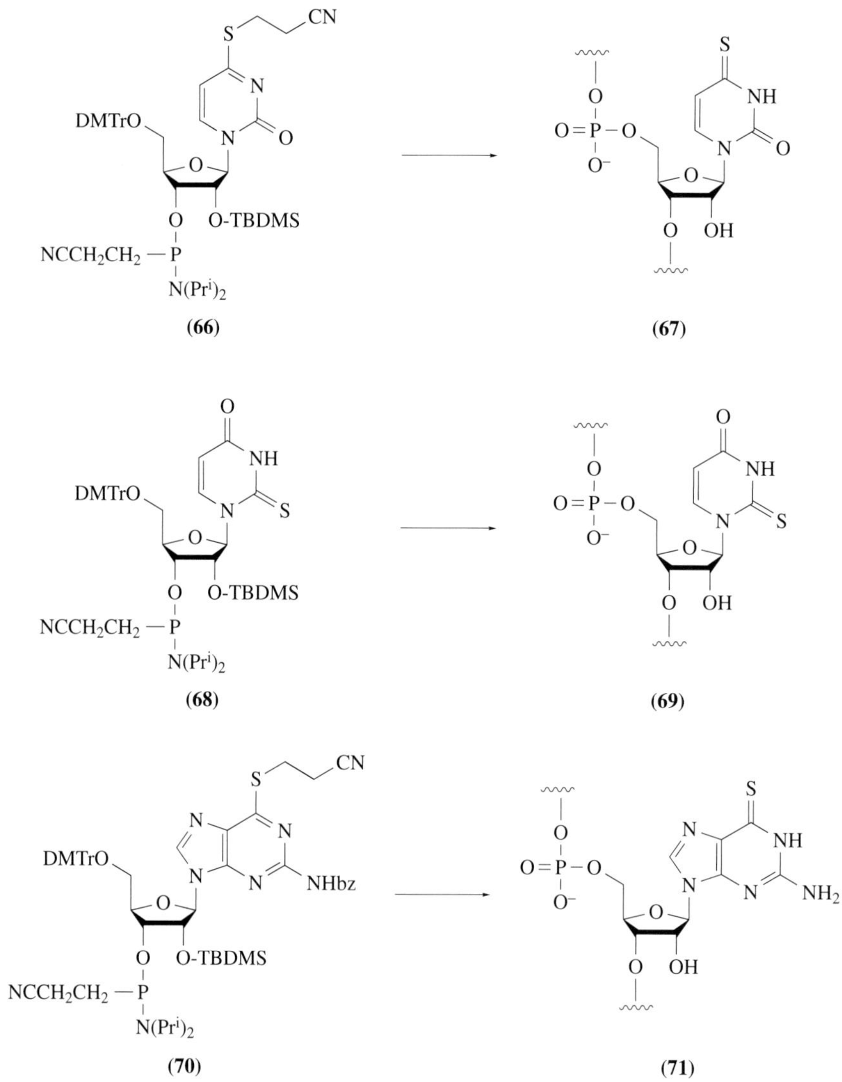

Figure 3 Preparation of 4-thiouridine, 2-thiouridine, and 6-thioguanosine.

the 2′-hydroxyl groups. For the postsynthetic modification of the C-4 of pyrimidine, triazol was used as the leaving group.[78] They synthesized a dimer and a heptamer containing 4-triazolylpyrimidione by the triester method. The 4-triazoylylpyridine derivative was converted to 4-dimethylaminopyrimidine and cytidine by treatment with dimethylamine and ammonia, respectively. The phosphoramidite unit of the 4-triazoylylpyridine was prepared from the amidite of uridine by treatment with triazole in the presence of $POCl_3$,[79] and was used for the synthesis of oligoribonucleotides. The 4-triazoylpyridine derivative could be converted to 4-thiouridine by treatment with thiolacetic acid. It has been shown that the 4-*O*-aryloxy group ribonucleoside can act as a leaving group.[80]

Verdine and co-workers have reported the postsynthetic modification of C, A, and G in RNA. 4-*O*-(4-Chlorophenyl)uridine and 6-*O*-(4-chlorophenyl)inosine were synthesized as the reactive derivatives.[81] For the functionalization of G, 2-fluoro-6-*O*-(4-nitrophenethyl)inosine was prepared. The protected nucleosides with DMTr and TBDMS, were incorporated into oligoribonucleotides by the phosphoramidite chemistry. The resin-bound oligoribonucleotides, which contained one convertible nucleoside, were treated with alkylamines to give the N^2-alkyl-G, N^4-alkyl-C, and N^6-alkyl-A derivatives.

A 5′-*O*-phosphoramidite can be incorporated into the 5′-ends of synthetic oligodeoxyribonucleotides, and these were converted into oligonucleotides with a (5′-5′) internucleotidic linkage. Oxidation of the 5′-end nucleoside with periodate yielded dialdehydes, which can be condensed with biotin hydrazide to prepare 5′-biotinylated oligonucleotides by reduction with sodium borohydride.[82] As the length of an oligoribonucleotide increases, the yield decreases in chemical syntheses, so it is difficult to prepare long RNA fragments in a high yield. Wincott and co-workers synthesized two strands of a hammerhead ribozyme separately, and then conjugated the two strands covalently at loop II.[83] The 3′-end of the 5′-half ribozyme strand was oxidized by sodium periodate, and was allowed to react with the 5′-aminohexyl-3′-half ribozyme in the presence of $NaBH_3CN$. Interestingly, a cyanoborane adduct of the ribozyme was also obtained from the reductive ligation. The postsynthetic ligated ribozymes exhibited activities similar to that of the control ribozyme.

6.4.5 Other Analogues

6.4.5.1 Transition state analogues

The generation of catalytic antibodies that possess novel phosphodiesterase activities has been a focus of much effort. Janda and co-workers succeeded in eliciting an antibody catalyst that cleaves phosphodiester linkages by using an oxorhenium(V) complex of uridine as a transition state analogue.[84] The screened monoclonal antibodies catalyzed the transesterification of uridine 3′-(*p*-nitrophenyl phosphate) in the presence of EDTA. Catalytic antibodies that cleave RNA strands at specific sites may be elicited by using transition state analogues with a defined higher structure.

6.4.5.2 Peptide nucleic acid

A nucleic acid analogue that contains a peptide backbone, called a "peptide nucleic acid" (PNA), was synthesized by Neilsen *et al.* and was shown to bind to either single or double stranded DNA. The PNA • DNA duplex had a much higher melting temperature than the DNA • DNA duplex.[85] One reason for the extraordinarily high stability of PNA • DNA could be the neutral nature of the PNA backbone. The loss of the negatively charged phosphate backbone decreases the repulsion existing between the polyanionic phosphodiesters of the nucleic acids. The high stability of PNA • DNA was also revealed in triplex (PNA • dsDNA) formation by strand displacement.[86] PNAs have been used as probes for nucleic acid functions and may become potential antisense or antigene drugs.

6.5 REFERENCES

1. H.G. Khorana, *Pure Appl. Chem.*, 1968, **17**, 349.
2. R. Lohrmann, D. Soll, H. Hayatsu, E. Ohtsuka and H.G. Khorana, *J. Am. Chem. Soc.*, 1966, **88**, 819.
3. G.M. Blackburn and M.J. Gait, "Nucleic Acids in Chemistry and Biology, 2nd edn.", Oxford University Press, Oxford, 1996.
4. S.S. Jones, B. Rayner, C.B. Reese, A. Ubasawa and M. Ubasawa, *Tetrahedron*, 1980, **36**, 3075.
5. R.L. Letsinger and W.B. Lunsford, *J. Am. Chem. Soc.*, 1976, **98**, 3655.
6. S.L. Beaucage and H. Caruthers, *Tetrahedron Lett.*, 1981, **22**, 1859.
7. B.C. Froehler, P.G. Ng and M.D. Matteucci, *Nucleic Acids Res.*, 1986, **14**, 5399.
8. P.J. Garegg, I. Lindh, T. Regberg, J. Stawinski, R. Stromberg and C. Henrichson, *Tetrahedron Lett.*, 1986, **27**, 4051.
9. L. Arnold, J. Smrt, J. Zajicek, G. Ott, M. Schiesswohl and M. Sprinzl, *Collect. Czech. Chem. Commun.*, 1991, **56**, 1948.
10. M. Smith, D.H. Rammler, I.H. Goldberg and H.G. Khorana, *J. Am. Chem. Soc.*, 1962, **84**, 430.
11. J.B. Chattopadhyaya and C.B. Reese, *J. Chem. Soc. Chem. Commun.*, 1978, **1978**, 639.
12. M. Kwiatkowski and J. Chattopadhyaya, *Acta Chem. Scand.*, 1984, **38**, 657.
13. E. Sonveaux, in "Protocols for oligonucleotide conjugates", ed. S. Agrawal, Humana Press, Totowa, NJ, 1994, p.1
14. J.H. van Boom and P.M. Burgers, *Tetrahedron Lett.*, 1976, **17**, 4875.
15. S. Iwai and E. Ohtsuka, *Nucleic Acids Res.*, 1988, **16**, 9443.
16. C. Lehmann, Y.-Z. Xu, C. Christodoulou, Z.-K. Tan and M.J. Gait, *Nucleic Acids Res.*, 1989, **17**, 2379.
17. C.B. Reese, R. Saffhill and J.E. Sulston, *J., Am. Chem. Soc.*, 1967, **89**, 3366.
18. C.B. Reese, H.T. Serafinowska and G. Zappia, *Tetrahedron Lett.*, 1986, **27**, 2291.
19. M.V. Rao, C.B. Reese, V. Schelhmann and P.S. Yu, *J. Chem. Soc., Perkin Trans*, 1993, **1**, 43.
20. E. Ohtsuka, M. Ohkubo, A. Yamane and M. Ikehara, *Chem. Pharm. Bull.*, 1983, **31**, 1910.
21. W.T. Markiewicz, E. Biala and R. Kierzek, *Bull. Pol. Acad. Sci. Chem.*, 1984, **32**, 433.
22. E. Ohtsuka, T. Wakabayashi, S. Tanaka, T. Tanaka, K. Oshie, A. Hasegawa and M. Ikehara, *Chem. Pharm. Bull.*, 1981, **29**, 318.
23. T. Tanaka, S. Tamatsukuri and M. Ikehara, *Nucleic Acids Res.*, 1986, **14**, 6265.

24. E. Ohtsuka, T. Tanaka, S. Tanaka and M. Ikehara, *J. Am. Chem. Soc.*, 1978, **100**, 4580.
25. J.A. Hayes, M.J. Brunden, P.T. Gilham and G.R. Gough, *Tetrahedron Lett.*, 1985, **26**, 2407.
26. K.K. Ogilvie, N. Usman, K. Nicoghosian and R.J. Cedergren, *Proc. Natl. Acad. Sci. USA*, 1988, **85**, 5764.
27. N. Usman, K.K. Ogilvie, M.Y. Jiang and R.J. Cedergren, *J. Am. Chem. Soc.*, 1987, **109**, 7845.
28. G.H. Hakimelahi, Z.A. Proba and K.K. Ogilvie, *Can. J. Chem.*, 1982, **60**, 1106.
29. S.A. Scaringe, C. Francklyn and N. Usman, *Nucleic Acids Res.*, 1990, **18**, 5433.
30. D. Gasparutto, T. Livache, H. Bazin, A.-M. Duplaa, A. Guy, A. Khorlin, D. Molko, A. Roget and R. Teoule, *Nucleic Acids Res.*, 1992, **20**, 5159.
31. E. Westman and R. Stromberg, *Nucleic Acids Res.*, 1994, **22**, 2430.
32. F. Wincott, A. DiRenzo, C. Shaffer, S. Grimm, D. Tracz, C. Workman, D Sweedler, C. Gonzalez, S. Scaringe and N. Usman, *Nucleic Acids Res.*, 1995, **23**, 2677.
33. T.F. Wu, K.K. Ogilvie and R.T. Pon, *Nucleic Acids Res.*, 1989, **17**, 3501.
34. R. Vinayak, P. Anderson, C. McCollum and A. Hampel, *Nucleic Acids Res.*, 1992, **20**, 1265.
35. G. Ott, L. Arnold, J. Smrt, M. Sobkowski, S. Limmer, H.-P. Hofmann and M. Sprinzl, *Nucleosides Nucleotides*, 1994, **13**, 1069.
36. H. Inoue, Y. Hayase, S. Iwai, K. Miura and E. Ohtsuka, *Nucleic Acids Res.*, 1987, **15**, 6131.
37. B.S. Sproat and A.I. Lamond, in "Oligonucleotides and Analogues", ed. F. Eckstein, Oxford University Press, New York, 1991, p49
38. A.I. Lamond and B.S. Sproat, *FEBS Lett.*, 1993, **325**, 123.
39. H. Inoue, Y. Hayase, S. Iwai and E. Ohtsuka, *FEBS Lett.*, 1987, **215**, 327.
40. Y. Hayase, H. Inoue and E. Ohtsuka, *Biochemistry*, 1990, **29**, 8793.
41. Y. Hayase, M. Jahn, M.J. Rogers, L.A. Silvers, M. Koizumi, H. Inoue, E. Ohtsuka and D. Soll, *EMBO J.*, 1992, **11**, 4159.
42. B.P. Monia, E.A. Lesnik, C. Gonzalez, W.F. Lima, D. McGee, C.J. Guinosso, A.M. Kawasaki, P.D. Cook and S.M. Freier, *J. Biol. Chem.*, 1993, **268**, 14514.
43. A.I. Lamond, B. Sproat, U. Ryder and J. Hamm, *Cell*, 1989, **58**, 383.
44. A. Mayeda, Y. Hayase, H. Inoue, E. Ohtsuka and Y. Ohshima, *J. Biochem. (Tokyo)*, 1990, **108**, 399.
45. B.J. Blencowe, B.S. Sprat, U. Ryder, S. Barabino and A.I. Lamond, *Cell*, 1989, **59**, 531.
46. A.M. Iribarren, B.S. Sproat, P. Neuner, I. Sulston, U. Ryder and A.I. Lamond, *Proc. Natl. Acad. Sci. USA*, 1990, **87**, 7747.
47. M. Shimizu, A. Konishi, Y. Shimada, H. Inoue and E. Ohtsuka, *FEBS Lett.*, 1992, **302**, 155.
48. S.H. Wang and E.T. Kool, *Nucleic Acids Res.*, 1995, **23**, 1157.
49. M. Shimizu, H. Morioka, H. Inoue and E. Ohtsuka, *FEBS Lett.*, 1996, **384**, 207.
50. D.M. Williams, F. Benseler and F. Eckstein, *Biochemistry*, 1991, **30**, 4001.
51. W.A. Pieken, D.B. Olsen, F. Benseler, H. Aurup and F. Eckstein, *Science*, 1991, **253**, 314.
52. D.B. Olsen, F. Benseler, H. Aurup, W.A. Pieken and F. Eckstein, *Biochemistry*, 1991, **30**, 9735.
53. S.T. Sigurdsson, T. Tuschl and F. Eckstein, *RNA*, 1995, **1**, 575.
54. S.B. Cohen and T.R. Cech, *J. Am. Chem. Soc.*, 1997, **119**, 6259.
55. J. Ott and F. Eckstein, *Biochemistry*, 1987, **26**, 8237.
56. H. Vu and B.L. Hirschbein, *Tetrahedron Letters*, 1991, **32**, 3005.
57. M. Mag, S. Luking and J.W. Engels, *Nucleic Acids Res.*, 1991, **19**, 1437.
58. R.G. Kuimelis and L.W. McLaughlin, *Nucleic Acids Res.*, 1995, **23**, 4753.
59. R.G. Kuimelis and L.W. McLaughlin, *Bioorg. Med. Chem.*, 1997, **5**, 1051.
60. R.G. Kuimelis and L.W. Mclaughlin, *J. Am. Chem. Soc.*, 1995, **117**, 11–019.
61. D.-M. Zhou, N. Usman, F.E. Wincott, J. Matulic-Adamic, M. Orita, L.-H. Zhang, M. Komiyama, P.K.R. Kumar and K. Taira, *J. Am. Chem. Soc.*, 1996, **118**, 5862.
62. X.H. Lui and C.B. Reese, *Tetrahedron Lett.*, 1995, **36**, 3413.
63. X.H. Liu and C.B. Reese, *Tetrahedron Lett.*, 1996, **37**, 925.
64. S.J. Talbot, S. Goodman, S.R.E. Bates, C.W.G. Fishwick and P.G. Stockley, *Nucleic Acids Res.*, 1990, **18**, 3521.
65. S. Iwai, C. Pritchard, D.A. Mann, J. Karn and M.J. Gait, *Nucleic Acids Res.*, 1992, **20**, 6465.
66. D.H. Bradley and M.M. Hanna, *Tetrahedron Lett.*, 1992, **33**, 6223.
67. T.M. Dewey, A.A. Mundt, G.J. Crouch, M.C. Zyzniewski and B.E. Eaton, *J. Am. Chem. Soc.*, 1995, **117**, 8474.
68. T.M. Tarasow, S.L. Tarasow and B.E. Eaton, *Nature*, 1997, **389**, 54.
69. B.E. Eaton, L. Gold, B.J. Hicke, N. Janjic, F.M. Jucker, D.P. Sebesta, T.M. Tarasow, M.C. Willis and D.A. Zichi, *Bioorg. Med. Chem.*, 1997, **5**, 1087.
70. C.J. Adams, J.B. Murray, J.R.P. Arnold and P.G. Stockley, *Tetrahedron Lett.*, 1994, **35**, 765.
71. A. McGregor, M.V. Rao, G. Duckworth, P.G. Stockley and B.A. Connolly, *Nucleic Acids Res.*, 1996, **24**, 3173.
72. R.K. Kumar and D.R. Davis, *Nucleic Acids Res.*, 1997, **25**, 1272.
73. C.J. Adams, J.B. Murray, M.A. Farrow, J.R.P. Arnold and P.G. Stockley, *Tetrahedron Lett.*, 1995, **36**, 5421.
74. Y.-H. Ping, Y.P. Liu, X.L. Wang, H.R. Neenhold and T.M. Rana, *RNA*, 1997, **3**, 850.
75. A.M. MacMillan and G.L. Verdine, *J. Org. Chem.*, 1990, **55**, 5931.
76. D.A. Erlanson, L. Chen and G.L. Verdine, *J. Am. Chem. Soc.*, 1993, **115**, 12–583.
77. Y.-Z. Xu, Q.G. Zheng and P.F. Swann, *J. Org. Chem.*, 1992, **57**, 3839.
78. W.L. Sung, *J. Org. Chem.*, 1982, **47**, 3623.
79. K. Shah, H.Y. Wu and T.M. Rana, *Bioconjugate Chem.*, 1994, **5**, 508.
80. A. Miah, C.B. Reese and Q.L. Song, *Nucleosides Nucleotides*, 1997, **16**, 53.
81. C.R. Allerson, S.L. Chen and G.L. Verdine, *J. Am. Chem. Soc.*, 1997, **119**, 7423.
82. S. Agrawal, C. Christodoulou and M.J. Gait, *Nucleic Acids Res.*, 1986, **14**, 6227.

83. L. Bellon, C. Workman, J. Scherrer, N. Usman and F. Wincott, *J. Am. Chem. Soc.*, 1996, **118**, 3771.
84. D.P. Weiner, T. Weimann, M.M. Wolfe, P. Wentworth and K.D. Janda, *J. Am. Chem. Soc.*, 1997, **119**, 4088.
85. M. Egholm, O. Buchardt, P.E. Nielsen and R.H. Berg, *J. Am. Chem. Soc.*, 1992, **114**, 1895.
86. P.E. Nielsen, M. Egholm, R.H. Berg and O. Buchardt, *Science*, 1991, **254**, 1497.

7

The Phosphoryl Transfer Reactions in Pre-Messenger RNA Splicing

MARIANO A. GARCIA-BLANCO, LAURA A. LINDSEY-BOLTZ[1] and SAGARMOY GHOSH
Duke University Medical Center, Durham, NC, USA

7.1 INTRODUCTION

Much has been learned about introns found in precursors to messenger RNAs (pre-mRNAs) since their discovery more than twenty years ago.[1,2] Pre-mRNA introns[2] are found in plants, animals, and fungi, and variant forms of these are encountered in some protists.[3] These introns are found in almost every RNA polymerase II gene in mammals, and most genes contain many introns.[4] In the majority of cases, these introns are precisely removed in the cell nucleus in a process known as

[1] Laura A. Lindsey-Boltz is now in the Department of Biochemistry, University of North Carolina at Chapel Hill, NC.

[2] In this chapter the term intron refers exclusively to a pre-mRNA intron excised by a spliceosome; other types of introns will be specifically identified (e.g., group I introns).

```
5' splice site                    branch site               3' splice site

                    Major class or U2 dependent introns

AG/GURAGUNNNN..............CURAYNNN-YYYYYYYYYYYY-NYAG/GN
RG/AUAAGUNNNN..............??????NNNNNN[N]NNNNNNNNYAC/NN

                    ATAC class or U12 dependent introns

NN/AUAUCCUUUN .............UUCCUURACYCYNNN[N]NNNNNYAC/NN
NU/GUATCCUUUN .............UUCCUUAACYNYNNN[N]NNNNNYAG/AN
```

Figure 1. Consensus sequences of major and ATAC introns, adapted from Sharp and Burge[6]. The putative branch site is underlined, (?) indicates that the branch is unknown.

pre-mRNA splicing[3]. The primary sequence characteristics that define the borders of introns and the corresponding exons have been recognized for years, although new surprising variants have recently emerged (Figure 1).[6,7] The enzymes that catalyze the precise removal of introns, the spliceosomes, have been identified and many of their components, small nuclear RNAs (snRNAs) and proteins, have been isolated.[4,8,9] Its bewildering size and complexity notwithstanding, the spliceosome has surrendered its secrets to a powerful combination of biochemistry and genetics. The biochemical details of the two transesterification reactions that lead to intron removal and exon ligation have been understood for some time, and a consensus has emerged in favor of RNA-based catalysis.[4,9,10] This has been supported by the concurrent and equally exciting discovery of ribozymes, such as RNAse P and group I introns, which can catalyze phosphodiester bond cleavage in the absence of proteins.[11,12] Among these ribozymes or ribonucleoprotein enzymes (RNPzymes), group II introns are particularly interesting because they carry out two transesterification reactions that are remarkably similar to those seen in pre-mRNA splicing.

This chapter will cover the current knowledge of the chemistry of the transesterification reactions and will present speculative ideas on the structure of the active site(s) within the spliceosome. Because knowledge of the chemical details of the transesterification reactions is far from complete, it is necessary to make a few assumptions in order to organize the available information.

The first assumption posits that the two chemical reactions of pre-mRNA splicing, known as step one and two, are catalyzed by spliceosomal snRNAs[4]. Two arguments compel us to embrace this assumption. First, the remarkable discoveries of RNA-based catalysis in group I introns, RNase P, group II introns, small ribozymes, and ribosomes provide a solid foundation for RNA-based chemical catalysis in the spliceosome (see chapters in this volume by Strobel and Vioque, and Dahlberg and Altman). Second, designing a model with proteins in the reaction center seems hopeless at this time, given the large and growing number of protein components in the spliceosome and the lack of structural information.

The second assumption, which should not raise many objections, excludes U1 and U4 snRNAs from chemical involvement. The evidence that rules out these two snRNAs is not absolutely unassailable, but it is convincing. U1 and U4 snRNAs have been noted to dissociate or to be only loosely associated with the fully functional spliceosome.[13–15] Depletion or inactivation of U1 snRNA/snRNP, which normally leads to inhibition of splicing, can be compensated by addition of excess SR proteins, a family of splicing factors.[16,17] Using an elegant *trans*-splicing system, Konforti and Konarska[18] showed that cleavage at a 5′ splice site proceeds better in the absence of functional U1 snRNA. Staley and Guthrie[15] further demonstrated that displacement of U1 is required for U6 binding at the 5′ splice site and that the mutually exclusive interactions with the intron may be in equilibrium[5]. U4 plays an indirect role by extensively base pairing to U6 and in so doing repressing its activity until the enzyme is fully assembled. This discussion does not intend to suggest that the remaining spliceosomal snRNAs, namely, U2, U5 and U6, are all necessarily active participants in the reaction center of the spliceosome. It is important to note here that there is no definitive knowledge of the structure of the active site(s) at this time.

---

[3] Many pre-mRNAs can be alternatively spliced to produce different messenger RNAs (mRNAs) leading to the production of many different protein isoforms from one gene.[5]

[4] A more moderate posture compatible with the conclusions of this chapter is that catalysis of the chemical steps is primarily the role of snRNAs with some assistance from protein factors.

[5] See Murray and Jarrell[19] for an interesting discussion of these findings.

7.2 THE ASSEMBLY OF THE ENZYME · SUBSTRATE COMPLEX

The spliceosome, like many macromolecular machines, is not pre-assembled as an active enzyme. A simple enzyme must, as any catalyst, be unchanged at the end of the reaction it catalyzes.[20] The spliceosome is an enzyme in pieces, and thus to fit the strict definition of an enzyme, these pieces must be regenerated so that they may catalyze multiple rounds of pre-mRNA splicing. A consideration of the number of introns spliced and the maximum number of spliceosomes potentially active at any one time leads to the conclusion that spliceosomal components are involved in multiple splicing reactions.[9,14] The whole system involved in the assembly of the spliceosome on a pre-mRNA and in the regeneration of active pieces [6] can, therefore, be considered to be an enzyme.

The *in vitro* pathway of enzyme–substrate assembly can be diagrammed as follows:

$$S + U1 + P \rightarrow \{S \cdot E\}_{CC} \tag{1}$$

$$\{S \cdot E\}_{CC} + U2 + P \rightarrow \{S \cdot E\}_{A} \tag{2}$$

$$\{S \cdot E\}_{A} + U5 \cdot U4/U6 + P \rightarrow \{S \cdot E\}_{B} \tag{3}$$

$$\{S \cdot E\}_{B}\ (+P?) \rightarrow \{S \cdot E\}_{C} \tag{4}$$

The splicing substrate (S), U1 snRNP (U1), protein factors (e.g., SR proteins) (P), the commitment complex ($\{S \cdot E\}_{CC}$), U2 snRNP (U2), the pre-spliceosome or complex A ($\{S \cdot E\}_{A}$), U5·U4/U6 tri-snRNP complex (U5 · U4/U6), the immature spliceosome or complex B ($\{S \cdot E\}_{B}$), and the spliceosome or complex C ($\{S \cdot E\}_{C}$) are indicated. The enzyme is designated as E.

The formation of $\{S \cdot E\}_{C}$ occurs concomitantly with the appearance of the products of the first chemical step. Clearly, the assembly of complexes $\{S \cdot E\}_{CC}$ through $\{S \cdot E\}_{B}$ modifies the structure of the substrate to juxtapose splice sites that may otherwise be thousands of nucleotides apart in a linear pre-mRNA [7]. Moreover, it is very likely that these complexes also constrain the number of possible substrate conformations, thus contributing to catalysis.[20,22]

The substrate, a functional pre-mRNA, is thought to first interact with U1 snRNP and protein factors in a binding reaction that does not require ATP hydrolysis.[4,9,10] The interaction is determined by RNA–RNA base pairing between the 5′ end of U1 snRNA and the consensus sequence at the 5′ splice site, and by interactions mediated by protein factors (ibid.). Protein factors also recognize the branchpoint sequence, thus bridging the two groups that subsequently will be involved in the first transesterification reaction. This leads to the formation of the commitment complex (CC) [8], which is designated $\{S \cdot E\}_{CC}$ in the scheme above. The CC and the U2 snRNP interact to yield $\{S \cdot E\}_{A}$, and the branchpoint sequence is recognized again, albeit differently in this complex. The U2 snRNA base pairs with this sequence leaving one adenosine residue unpaired or "bulged".[24–27] $\{S \cdot E\}_{A}$ interacts with a preformed U5 · U4/U6 tri-snRNP to form the immature spliceosome $\{S \cdot E\}_{B}$ which then undergoes rearrangements that result in the formation of a fully competent enzyme, $\{S \cdot E\}_{C}$. The most significant rearrangements occurring between $\{S \cdot E\}_{B}$ and $\{S \cdot E\}_{C}$ are the displacement of U1 snRNP from the 5′ splice site and its replacement with U6 snRNP, the melting of the extensive U4–U6 [9] base pairing interaction, the formation of U2–U6 base pairing interactions, and the juxtaposition of the reactants for the first transesterification reaction.[10,14,19] The formation of early splicing complexes and the rearrangements observed after formation of the immature spliceosome can together account for the lag in the appearance of splicing intermediates and products [10].

---

[6] See, for example, the role of Prp24p.[21]

[7] This may be less important *in vivo* where synthesis and processing of pre-mRNAs are temporally coupled, and the RNA polymerase II elongation complexes probably sequester the reactants.

[8] The commitment complex is among the U1 snRNA complexes that have been defined as Complex E,[23] and the terms commitment complex and Complex E have been used interchangeably in the literature.

[9] U snRNPs will be so indicated, U snRNAs will be labeled as such or will be indicated as U1, U2, etc.

[10] The rate of formation of the $\{S \cdot E\}_{CC}$ has not been adequately measured using pure components; however, experiments in HeLa nuclear extract suggest that the rate of the reaction may be slow relative to $\{S \cdot E\}_{A}$ formation (Jamison and Garcia-Blanco, unpublished data). The appearance of $\{S \cdot E\}_{C}$ coincides with that of the products of the first chemical reaction, an event that lags significantly behind $\{S \cdot E\}_{B}$ formation.[28] Thus it is reasonable to assume that the rearrangements which take place in the spliceosome represent a slow step in the splicing reaction.

7.3 THE FIRST PHOSPHORYL TRANSFER REACTION

7.3.1 Introduction and Assumptions

The first chemical step in pre-mRNA splicing is a transesterification reaction, or in more general chemical terms, a phosphoryl transfer reaction.[29,30] A general schematic for the reaction is as follows:

$$E^1\text{–O–P}(O_2^-)\text{–O–I} + A^{2'}\text{–OH} \rightarrow A^{2'}\text{–O–P}(O_2^-)\text{–O–I} + E^1\text{–OH} \qquad (5)$$

In the case of the first phosphoryl transfer for conventional pre-mRNA splicing E^1–O–P(O_2^-)–O–I is the phosphodiester bond between the 3′-most nucleoside of exon 1 (E^1) and the 5′-most nucleoside of the intron (I) (see Figure 2 for the same schematic drawn in the context of an intron). $A^{2'}$ is the branchpoint adenosine with the 2′OH group indicated. $A^{2'}$–O–P(O_2^-)–O–I is the newly formed 2′–5′ phosphodiester bond between the 2′OH of the branchpoint adenosine and the 5′ phosphate of the first nucleoside of the intron. This molecule is the product of an intramolecular transesterification, which results in a "lariat" structure.[11] The adenosine is said to be branched because this residue connects to three other residues via phosphodiester bonds (Fig. 2). The leaving group, E^1–OH, is exon 1, and the hydroxyl represented is

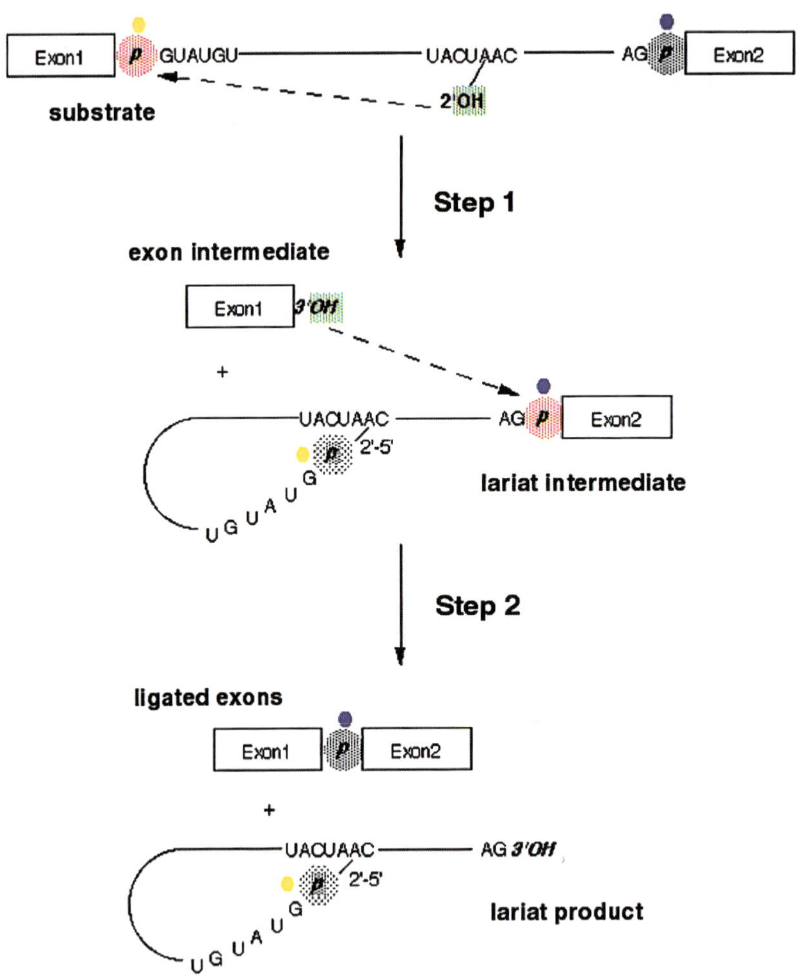

Figure 2. The two phosphoryl transfer reactions of pre-mRNA splicing. The nucleophile in each step is colored green, and the scissile phosphate is red. The two active phosphates at the 5′ splice site and the 3′ splice site are indicated with a yellow and a blue dot, respectively.

[11] In *trans*-splicing reactions the equivalent product is a "Y"-shaped molecule.

the 3′OH of the last nucleotide of the exon (Figure 2). The first step of group II intron splicing proceeds via the same reaction [12].[32,33]

The simplest biochemical schematic for this reaction is as follows:

$$\{S \cdot E\}_C \rightarrow [S \cdot E]^{ts} \rightarrow \{I_E \cdot E \cdot I_L\} \quad (6)$$

The transition state(s) is represented by $[S \cdot E]^{ts}$, and the products of the first step are the I_E and I_L, for exon 1 and the lariat intermediate (Figure 2). [13]

In order to evaluate the groups involved in the catalysis of the first transesterification reaction several assumptions have been made. In addition to the two mentioned above, RNA-based chemistry and the exclusion of U1 and U4, two other assumptions have been made specifically for this reaction. The first, and possibly more controversial one, is that U5 snRNA is not directly involved in this chemical step. Even though there is physical evidence for an interaction between the highly conserved loop 1 of U5 snRNA and exon nucleotides at the 5′ splice site prior to the first transesterification reaction,[34,35] functional studies in yeast and human nuclear extracts do not support an essential role for loop 1 in this reaction [14].[36,37] There are no compelling data to implicate other regions of the U5 snRNA in this reaction, and thus it is reasonable to model the reaction in the absence of this RNA [15]. The second assumption, which is more widely accepted, is that sequences at the 3′ splice site are not essential for the first step. Many introns can efficiently undergo step one in the absence of a 3′ splice site and have thus been defined as AG independent.[39–42] This suggests that these sequences are not critical components of the active site of the enzyme and that introns that are AG dependent for the first step may be so because of requirements for spliceosome assembly rather than chemistry. Similar conclusions can be surmised from studies of group II intron splicing.[32]

7.3.2 The Structure of the Nucleophile

The nucleophile in the first phosphoryl transfer reaction is a 2′OH, which in the overwhelming majority of pre-mRNA introns is on an adenosine present within the more or less conserved branchpoint sequence. The unique three-dimensional structure [16] of the nucleophile should reveal how this specific 2′OH is activated above all others in the intron and selected to attack the 5′ splice site. The branchpoint sequence is highly conserved in *S. cerevisiae* introns (5′-UACUAA̲C-3′) and mammalian ATAC introns [17] (5′-UCCUUAA̲C-3′), but less so in the major class of mammalian introns (5′-CURA̲Y-3′) [18].[8] The underlined A̲, the most 3′ A in the sequence, has the reactive 2′OH. In the major spliceosome, the branchpoint sequence forms Watson–Crick base pairs with the sequence 5′-GUAGUA-3′ in U2 snRNA (Figure 3(a)).[24,26,27] A similar structure can form between the ATAC branchpoint and a sequence in U12 snRNA (Figure 3(b)).[7,44] Duplex formation with U2 snRNA, or U12 for ATAC introns, is absolutely critical for selection of the nucleophile within the branchpoint and for activation of this group. In group II introns the nucleophile is also a bulged A near the base of a long stem loop known as domain VI (Figure 3(c)).[32] A similar structure, again with a bulged A, can also be drawn for the mysterious group III introns.[45] Phylogenetic conservation of adenosine as the bulged base in all of these cases, particularly in light of variation among 5′ splice site sequences (to be discussed below), strongly suggests that unique

[12] See Weiner[31] for a discussion of the evolutionary relationship between group II and pre-mRNA splicing.

[13] The products of the first step have traditionally been referred to as intermediates in the splicing reaction and thus have been designated with the letter "I", to distinguish them from the products of the second reaction that are designated with the letter "P".

[14] McConnell and Steitz have recent data that implicate proximity of the invariant loop and intronic residues at the 5′ splice site (T. McConnell, personal communication).

[15] U5 snRNP has many unique proteins, and one or more of these could contribute to the catalysis of the chemical step. Prp8 (U5-220K in mammals) has always been a favorite candidate because it is the most highly conserved splicing factor and has been shown to cross-link to the branchpoint, 5′ splice site, and 3′ splice site (Luo et al.[38], and references therein). Nonetheless, in keeping with the assumptions made in the Introduction, which at this stage in the chapter become dogmatic, protein players will be ignored.

[16] In this instance three-dimensional structure includes the influence of groups in the intron or in the snRNAs that are in close spatial proximity and can thus influence the chemical properties of the 2′OH of the branchpoint adenosine.

[17] In this chapter we will designate U12 snRNA dependent introns as ATAC introns.[6]

[18] Single letter abbreviations for nucleotides are by convention, Y is a pyrimidine, R is a purine, and N is any nucleotide.[43]

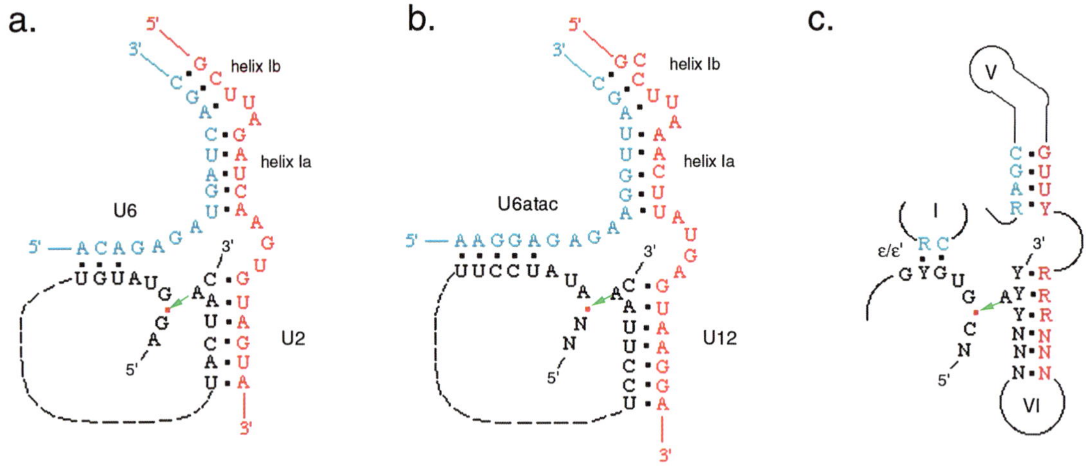

Figure 3. Network of RNA interactions prior to the first catalytic step for (a) the yeast spliceosome, (b) the ATAC spliceosome, and (c) group II introns. Sequences corresponding to, or are the equivalent of, pre-mRNA sequences are black, U2 are red, and U6 are blue.

chemical properties of this nucleotide determine this choice. Indirectly, this argues for recognition of the adenine base during catalysis of the first chemical step; this is also supported by experimental data.[46]

The reactive A is "bulged", which in molecular terms means that it does not have a base opposite it in the RNA A-type double helix assumed for the short U2 snRNA–branchpoint duplex (Figure 3(a)).[25] In studies where two consecutive purines were used within a strong branchpoint sequence, it was concluded that the reactive nucleoside has to be bulged (ibid.) and strongly suggested that this bulged helix was the structure of the nucleophile during the first chemical step. Even though in many cases there are two alternative bulged structures that can form, for instance the duplex formed between U2 snRNA and the branchpoint in yeast (Figure 3(a)), the 3′ adenosine is heavily favored to branch. More detailed structural data may explain this preference.

The three-dimensional structure of the branchpoint bulge has not been established; however, several studies permit an educated guess at what this structure may be. The closest structure solved to date is that of the GA coat protein binding site containing two adenosine residues opposite one uridine in an A-type double helix.[47] Using heteronuclear NMR, Smith and Nikonowicz[47] showed that the most likely structure had the 3′-most A in a bulged configuration, whereas the 5′-most A formed a base pair with the uridine. The 3′-most A, which can be considered analogous to the branching A in the UACUAAC sequence, was found to be stacked within the helix (ibid.). This structure is not static as discerned from the broadening of the NMR signals for the imino protons involved in base pairing (ibid.)[19]. It is likely that the preference to stack the bulged A rather than to displace it into the minor groove is dictated by the neighboring bases.[48] In other cases, bulged As have been seen or modeled to be extruded into the minor groove.[49–51] Activation of the nucleophile very likely involves "flipping" of the adenosine so that the base adopts an extrahelical configuration; this makes the 2′OH of the ribose more accessible to the solvent and less encumbered by the electrostatic influence of the phosphate backbone.[47] Chu *et al.*[52] have argued that in group II introns the essential property of the bulged A is its capacity to "flip" into an extrahelical configuration. Thus two alternative structures for a distorted type-A helix between the *S. cerevisiae* U2 snRNA 5′-GUAGUA-3′ sequence and the conserved branchpoint 5′-UACUAAC-3′ sequence are envisioned: one with a stacked A (Figure 4) and the other the extrahelical A (not shown).

7.3.3 The Structure of the Scissile Phosphodiester Bond

The scissile phosphodiester bond at the 5′ splice site is the site of attack by the nucleophile; it must be selectively recognized and activated for the first chemical step. In the major class of introns, the sequence at the 5′ splice site is usually 5′-GpGURAGU-3′, where the phosphate to be attacked is underlined. It is

[19] See chapter by P. Moore in this volume.

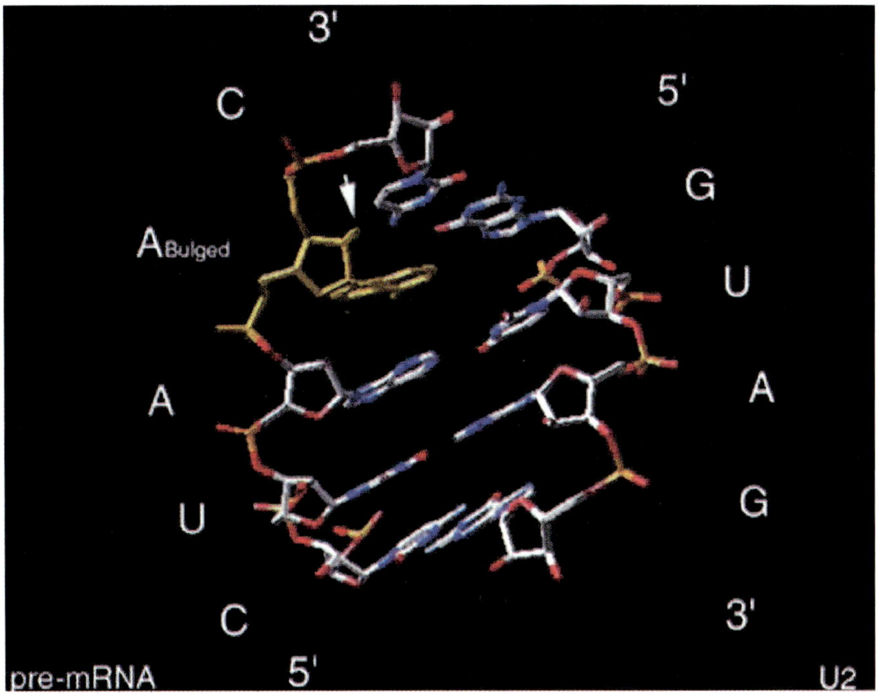

Figure 4. Model of the distorted A-type helix formed between the pre-mRNA branchpoint sequence (only 5′CUAAC3′ is shown) and the U2 snRNA sequence (only 5′GUAG3′ is shown). The bulged A is shown in yellow and is shown in the stacked configuration. This view faces the minor groove and the 2′OH of the bulged A is indicated with an arrow.

clear that selection of the correct 5′ splice site is mediated by multiple components of the spliceosome beginning with early recognition by U1 and SR proteins. In the mature yeast spliceosome, $\{S \cdot E\}_C$, the sequence 5′-ACA-3′ in the U6 snRNA forms base pairs with nucleotides +4 to +6 at the 5′ splice site (5′-GpGUAUGU-3′) (Figure 3(a)).[53,54] This U6 sequence is the beginning of the highly conserved 5′-ACAGAGA-3′ sequence.[55] An even longer helix is predicted for equivalent sequence at the 5′ splice site of ATAC introns and the analogous sequence in U6atac snRNA (Figure 3(b)).[7]

The interactions between U6 snRNA and the splice site may be more complex than this short duplex. The intronic residue at position +3 is almost always a purine in mammals and is an A in *S. cerevisiae* and in ATAC introns. At this time it is not clear whether or not this residue makes contacts with residues in U6 snRNA. The U_{+2} at the 5′ splice site is conserved in both types of pre-mRNA introns, whereas the first position of the intron is not.[20] The remarkable conservation of this residue may be determined by an interaction with U6 snRNA, or alternatively this residue may also form an important interaction with the bulged adenosine, maybe stabilizing an extrahelical configuration for the base. The conservation of this residue may also reflect a requirement subsequent to step one and may be connected to the equally striking conservation of an adenosine at the penultimate position[21].[8] The +1 nucleotide is a G in the great majority of U2-dependent introns (major class) and an A in the majority of the ATAC class of U12-dependent introns (Figure 1). Nonetheless, it seems that the major spliceosome can process introns starting with AU, and the ATAC spliceosome can process introns starting with GU. It is the branchpoint sequence that determines which spliceosome will remove a specific intron.[6] This makes it unlikely that the G_{+1} is uniquely recognized during the first chemical step in a way that is essential for catalysis. The conservation of this nucleotide may reflect a requirement in the second step, and is discussed later. The 3′-most residue of the exon (G_{-1}), which is not highly conserved in yeast or in ATAC introns, is also unlikely to be essential for catalysis of the first step. Therefore, the sequence specificity at the 5′ splice site that can be attributed to contacts required for the presentation and activation of the scissile phosphodiester bond is limited to the region that base pairs with U6 snRNA and possibly U_{+2}[22]. In

[20] This U residue at the +2 position is also conserved in group II introns and the rare group III introns.

[21] Recent data suggest a genetic interaction between the U_{+2} and the penultimate intronic residue (see footnote 38).

[22] The U_{+2} has been shown to cross-link to p220, again suggesting that this protein may have to included among the components

agreement with that discussed above, the first nucleotide of group II introns does not play a direct role in catalysis of the first phosphoryl transfer reaction.[32]

The analysis of the chemical groups required at and around the 5′ splice site is not complete. Replacing the 2′OH of G_{-1} with 2′H does not affect the rate of the first step, suggesting that this hydroxyl group is not involved in the chemistry of the reaction[23].[58] A 2′OCH_3 at this position is equally innocuous suggesting that the ribose at position G_{-1} is not recognized for cleavage of the adjacent phosphodiester bond and that the inclusion of this bulkier group does not sterically hinder the spliceosome (ibid.). These results have to be viewed with some degree of caution given that the rate-limiting step in the splicing reaction has not been determined. Nonetheless, it is likely that the removal of a critical group would significantly decrease the rate of the reaction, and therefore these data indicate that the 2′OH at the scissile bond is not essential for cleavage.

Group II introns may also recognize and activate the phosphodiester bond at the 5′ splice site in a manner analogous to the spliceosomes. The 5′ splice site consensus sequence for group II introns is 5′-NpG$\underline{U}$GYG-3′, where the scissile phosphodiester bond is indicated (p), and the highly conserved U_{+2} is underlined.[32] The GY residues at positions +3 and +4 are involved in the ε–ε' long-distance interaction with two nucleotides within Domain I.[32,33] This may be analogous to the U6 snRNA–5′ splice site interaction described above for pre-mRNA introns (Figure 3(c)). As with its pre-mRNA intron cousin, group II introns also do not require a 2′OH at the 5′ splice site[24].[59]

Hydrolytic cleavage at the 5′ splice site of group II introns can be relatively efficient in the absence of Domain VI, which suggests that the bulged A nucleophile is not directly involved in activation of the phosphate at the 5′ splice site.[32,33] Hydrolysis has not been documented for pre-mRNA splicing, but there are indications that such a pathway may be operative during *in vitro* splicing of permuted intron–exon RNAs that yield circular exon products (Pasman and Garcia-Blanco, unpublished results). By analogy with group II introns, it is reasonable to speculate that the branchpoint–U2 snRNA duplex does not directly activate the 5′ splice site. Moreover, based on these observations it is reasonable to assume that a component of the reaction center prevents the access of water by sequestering the activated scissile phosphodiester bond within a cavity in the enzyme–substrate complex.[22]

7.3.4 The Formation of the Active Site

The bulged adenosine nucleophile and the reactive phosphate group of the first intronic residue must be juxtaposed by conformational changes during formation of the active spliceosome. A critical result of the rearrangement is the formation of extensive base pairing between highly conserved regions of U6 and U2 snRNAs.[10,14] Among the helices formed between these snRNAs, helix Ia and Ib are conserved between the major class of spliceosome from mammals to yeast, and the ATAC spliceosome (Figure 3(a) and (b)).[7] The network of RNA–RNA interactions brings the nucleophile (and the U2 snRNA–branchpoint helix) to the same neighborhood as the scissile phosphate bond (constrained within 5′–splice site sequences that interact with U6 snRNA) (Figure 3(a) and (b)). It is also likely that the base pairing forms some of the active components involved in the catalysis of the chemical step; a molecular understanding of this will require a three-dimensional structure of these components. Extensive remodeling of intramolecular interactions within U6 snRNA and the formation of an extended stem-loop are also thought to be important in creating the active site for the first transesterification,[14] and this stem-loop is also conserved in the ATAC spliceosome.[7]

Helices Ia and Ib, formed by U2 and U6, and the intramolecular U6 stem-loop are probably analogous to Domain V in group II introns (Figure 3).[32,33] This domain is thought to have all of the necessary groups to catalyze the first phosphoryl transfer reaction either with the 2′OH of the bulged adenosine in the branching reaction or with water during hydrolysis (ibid.). Much like helices Ia and Ib in the spliceosome, Domain V has an asymmetric internal loop that may provide flexibility. It is possible to draw a two-dimensional representation of the presumed reaction center for the first step of group II splicing that resembles the depictions shown for the spliceosomal sites (Figure 3(c)).

---

of the active site of the spliceosome.[56]

[23] The 2′OH group of U_{+2} is also not required for this transesterification.[57]

[24] Griffin *et al.* [59] showed only a 16-fold reduction in the k_{chem} with a 2′H substitution at the −1 position, thus concluding that group II introns had little preference for RNA vs. DNA at the scissile bond.

7.3.5 The Stereochemistry of the First Phosphoryl Transfer Reaction

Knowledge of the stereochemistry of a reaction can lead to significant insights into the structure of the transition state. The reactive phosphorus at the 5′ splice site can be made chiral by the substitution of one of the non-bridging oxygens with sulfur. The resulting phosphorothioate diester will be either the R_P or the S_P diastereomer and can be distinguished by its sensitivity to certain nucleases.[60] Random incorporation of the R_P diastereomer in a pre-mRNA precursor resulted in inhibition of the first step of splicing when the phosphorothioate was found at the 5′ splice site, suggesting stereoselectivity for the first reaction.[61] This was confirmed using splicing precursors that contained a single R_P or S_P phosphorothioate at the scissile bond.[62,63] As expected from the above results, the R_P diastereomer inhibited the first transesterification, whereas the S_P diastereomer behaved as the phosphate containing pre-mRNA (ibid.). This clearly supports stereoselective recognition of the pro-R_P oxygen at the 5′ splice site.

Enzymes that carry out phosphoryl transfer reactions can be divided into two general categories, those that carry out a direct transfer and those that utilize a covalent E∼P intermediate.[30] The stereochemistry of the transfer reaction is used to distinguish between these two categories: inversion of configuration is predicted for a direct S_N2 ("in line") displacement not involving a covalent intermediate, whereas retention of the stereochemical configuration is expected for a mechanism that involves a covalent E∼P intermediate and thus two S_N2 ("in line") displacement reactions[25]. Racemization would be expected for an S_N1 reaction without steric bias, and this is not expected for enzymatic phosphoryl transfer reactions.[26] In elegant work, Moore and Sharp[63] showed that the first transesterification reaction in pre-mRNA splicing proceeds with inversion of stereochemical configuration. Splicing of a pre-mRNA with a single S_P phosphorothioate at the 5′ splice site produced a lariat product with a R_P phosphorothioate at the 2′–5′ phosphodiester linkage between the bulged adenosine and the first nucleoside of the intron (ibid.). Thus, the first chemical step of pre-mRNA splicing is believed to proceed via an S_N2 ("in line") displacement reaction and not to involve a covalent intermediate between the phosphate at the 5′ splice site and a component of the spliceosome. Subsequently, the same preference for the S_P phosphorothioate and inversion of configuration have been shown for the first chemical step of group II intron splicing[27].[64]

The S_N2 ("in line") displacement reaction at the 5′ splice site is modeled to proceed via a bipyramidal transition state (Figure 5). When designing a model of the spliceosome active site, data about the relative positions of the 2′OH of the bulged adenosine and the phosphate at the 5′ splice site, and data about groups activating the pro-Rp oxygen at the reactive center have to be taken into consideration.

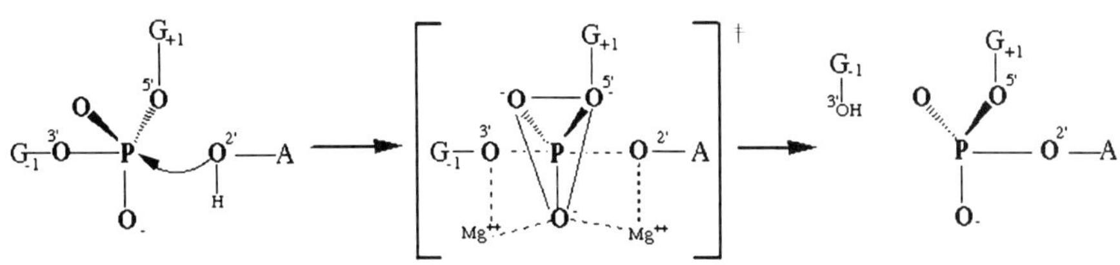

Figure 5. The first step of splicing proceeds through a bipyramidal transition state. The bulged adenosine 2′-hydroxyl attacks through an in-line S_N2 nucleophilic displacement mechanism, releasing the 3′-hydroxyl leaving group of exon 1. The G_{-1} represents the last base of exon 1, and G_{+1} represents the first base of the intron. Two metal ions are shown,[65] but at this time there is only evidence for metal ion stabilization of the oxyanion leaving group during first transesterification.[57]

[25] Inversion of stereochemical configuration provides evidence in favor of an odd number of S_N2 ("in line") displacements; the most parsimonious interpretation is that one transfer has taken place. Retention of configuration provides evidence for an even number of displacements. Again, this is usually interpreted as two transfers.

[26] Attributed to W.P. Jencks in Knowles.[30]

[27] In contrast, the first transesterification in the splicing of group I introns proceeds only with an R_P diastereomer phosphorothioate at the 5′ splice site, through a mechanism that involves inversion of configuration.

7.3.6 A Metal Ion Stabilizes the Leaving Group of the First Transesterification Reaction

Most ribozymes have been thought to be metalloenzymes [28]; the argument has been made that the limited chemical repertoire provided by ribonucleic acid had to be supplemented with metal ions [29].[67] Steitz and Steitz[65] proposed a two-metal ion mechanism for catalysis of pre-mRNA based on biochemical data on protein-based and RNA-based enzymes that carry out phosphoryl transfer reactions, and on the three-dimensional structure of the active site of *E. coli* DNA polymerase I. Some aspects of this model have received experimental support, and will be discussed here. A splicing precursor was synthesized with a 2′-deoxy-3′-thio-uridine as the last nucleoside of the 5′ exon, so as to make a sulfur bridging atom at the 5′ splice site and thus the leaving group of the S_N2 displacement reaction (Figure 5) [30].[57] When the splicing reaction with this modified nucleoside was carried out with Mg^{2+} as the only divalent cation, no products of the first transesterification reactions were observed (ibid.). When Mn^{2+}, a thiophilic metal that can interact with both sulfur and oxygen, was added to the reaction, the first phosphoryl transfer was now efficient (ibid.). The Mn^{2+} "rescued" the modified pre-mRNA. This type of "thio" effect, which was pioneered by Jaffe and Cohn[68], strongly suggests that a Mg^{2+} ion stabilizes the 3′-oxyanion predicted to be the leaving group for the reaction. The metal may also stabilize the transition state. These data prove one aspect of the model proposed by Steitz and Steitz[65]: the spliceosome uses a metal ion for catalysis of the first chemical step.

7.4 THE SECOND PHOSPHORYL TRANSFER REACTION

7.4.1 Introduction and Assumptions

The second chemical step in pre-mRNA splicing is also a transesterification, where a phosphoryl is transferred without net gain or loss of high energy bonds. The general schematic for the reaction is the same as described for the first step:

$$I_L\text{–O–P}(O_2^-)\text{–O–}E^2 + E^1\text{–OH} \rightarrow E^1\text{–O–P}(O_2^-)\text{–O–}E^2 + P_L\text{–OH} \tag{7}$$

For the second step the groups are the following: I_L–O–P(O_2^-)–O–E^2 is the phosphodiester bond between the 3′-most nucleoside of the intron (I_L) and the 5′-most nucleoside of the downstream exon (E^2) (see Figure 2 for the same schematic drawn in the context of the lariat intermediate). E^1 is the 3′-most nucleotide of exon 1, and the 3′OH group is indicated. E^1–O–P(O_2^-)–O–E^2 is the newly formed 3′–5′ phosphodiester bond between the 3′OH upstream exon and the 5′ phosphate of the first nucleoside of the downstream exon. The leaving group, P_L–OH, is a lariat product, and the hydroxyl represented is the 3′OH of the last nucleotide of the intron (Figure 2). The simplest biochemical schematic for this reaction is as follows:

$$\{I_E \cdot E \cdot I_L\} \rightarrow \{I_E \cdot E \cdot I_L\}^* \rightarrow [I_E \cdot E \cdot I_L]^{*ts} \rightarrow \{P_E \cdot E \cdot P_L\} \tag{8}$$

the transition state(s) is represented by $[I_E \cdot E \cdot I_L]^{*ts}$, and the products of the second step are the P_E and P_L, the product of exon ligation and the lariat product respectively (see Figure 2). [31]

It is clear that the enzyme–substrate complex must undergo conformational changes, $\{I_E \cdot E' \cdot I_L\} \rightarrow \{I_E \cdot E' \cdot I_L\}^*$, between the two chemical steps. Some of these changes will be described in detail in the sections that address the structure of the nucleophile and the scissile bond for second phosphoryl transfer reaction (Figure 6(b)). Following the first step, the invariant loop of U5 snRNA makes contacts with nucleotides immediately downstream of the 3′ splice site.[34,69] The U_{+2} of the intron makes contacts with a conserved nucleotide of U6 snRNA.[34,69,70] Finally, several new interactions are observed between U2 snRNA and residues in the substrate and in U6 snRNA (ibid.). The implications of these conformational

[28] See chapters by Vioque and Altman (Chapter 9) and Strauss-Soukup and Strobel (Chapter 11) in this volume.

[29] Recent data from M. Been and colleagues clearly show that acid–base catalysis is another modality that ribozymes share with protein enzymes.[66]

[30] A 2′-deoxy nucleoside at position −1 was shown by Moore and Sharp[58] to be functional for the first step.

[31] The products of the second reaction are designated with the letter "P".

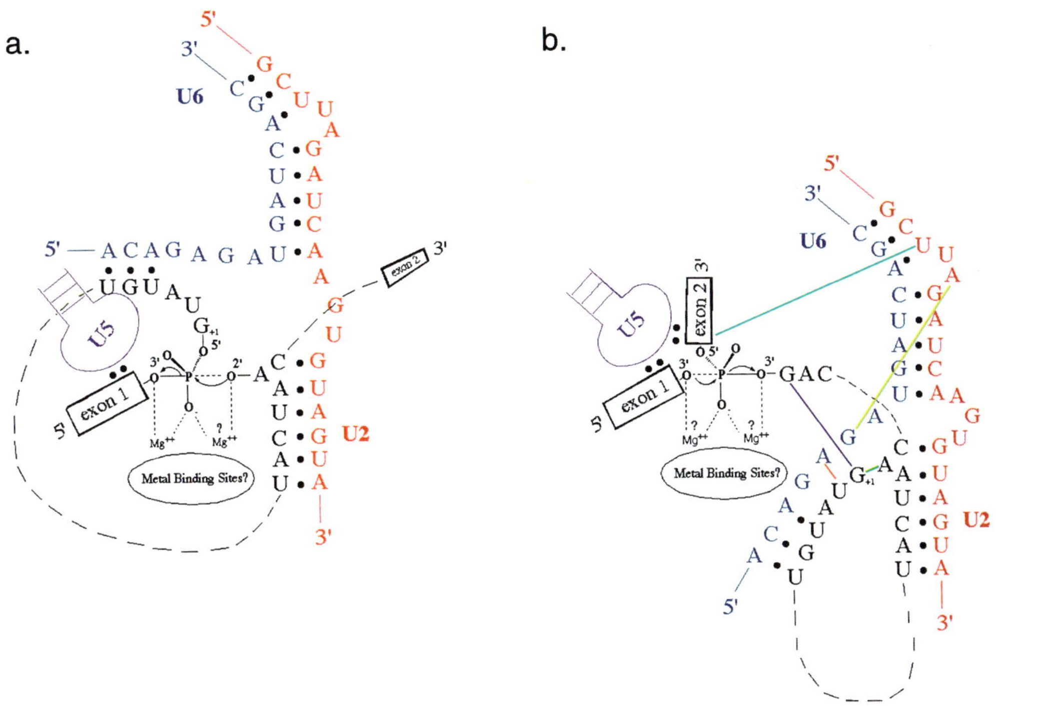

Figure 6. Active sites and transition states during (a) the first step, and (b) the second step of pre-mRNA splicing. The new interactions in the spliceosome which occur after the first step are: U5 interaction with exon 2; the non-canonical G–G interaction between the first and last intron nucleotides (pink line); the tertiary interaction between U2 and U6 (yellow line); cross-links between U6 and intron +2 (orange line) and between U2 and exon 2 (light blue line); and the new 2′–5′ A–G bond at the lariat branch (green line).

changes *vis-a-vis* the structure of the active site will be discussed below. The conformational changes are designated $\{I_E \cdot E' \cdot I_L\} \rightarrow \{I_E \cdot E' \cdot I_L\}^*$, and very likely involve the activity of proteins, especially RNA–RNA helices.[14] The distance between the branchpoint and the 3′ splice site can be longer than fifty nucleotides,[71] and this suggests that translocation of the 3′ splice site over considerable distances may be a requisite function of this conformational change.[32]

Several other assumptions can be made regarding the second transesterification reaction. It is safe to assume that U1 and U4 snRNAs are not involved in the chemistry of the reaction[33]. The role of U5 snRNA is harder to assess. Although it is clear in *S. cerevisiae* that the invariant loop of U5 is required for the second step,[36] the same cannot be said for mammalian reactions.[37] Even if required for proper alignment of the exons to be ligated in the second transesterification, U5 snRNA is not likely to participate directly in chemical activation of the reactive groups[34]. It is our view, however, that the invariant loop of U5 may participate in approximating and aligning the reactants to achieve the appropriate geometry for the transition state (Figure 6(b)).

In conventional *in vitro* splicing reactions, splicing intermediates transiently accumulate, suggesting that under these conditions the rate of the second step is slower than the rate for the first step. Two observations suggest that this may be an artifact of this *in vitro* system. First, splicing intermediates (the products of the first step) are not easily detectable among nuclear RNA. Second, in an *in vitro* system where RNA polymerase II transcription and splicing are coupled, the intermediates of the reaction do not accumulate (Ghosh and Garcia-Blanco, unpublished results).

---

[32] It is not clear which component moves, and this statement is simply meant to imply a relative translocation with respect to spliceosomal components.

[33] See discussion above.

[34] As previously discussed above (footnotes 15 and 22), genetic and biochemical data suggest that the U5 snRNP protein, Prp8p, may play a role in chemical catalysis for step two.

7.4.2 Formation of the Active Site

The rearrangements leading to a new enzyme substrate complex, $\{I_E \cdot E_{'} \cdot I_L\} \rightarrow \{I_E \cdot E_{'} \cdot I_L\}^*$, are likely to be rate-limiting for the second step rather than the chemistry of the second phosphoryl transfer reaction [35]. This is also the case with group II.[73,74] The translocation of the 3′ splice site relative to the reaction center(s) [36] of the spliceosome must be required for the second step of splicing because there is evidence against dissociation of the enzyme–intermediates and subsequent reassociation [37]. From the point of view of the enzyme the scissile phosphodiester bond at the 3′ splice site must be brought into the active site. The most elegant data suggest that the sequences downstream of the branched adenosine "thread" through the enzyme until the first pyAG is recognized [38].[39,71] There is no evidence for directed 5′ to 3′ scanning, and the short distances (an average of 18 nucleotides) usually found between the branched adenosine and the 3′ splice site suggest that two-dimensional bidirectional diffusion on the RNA can explain the available data.

It is absolutely clear that the branched adenosine is recognized during translocation, given the strong preference for the most 3′ proximal pyAG (ibid.). Whether or not the branched adenosine is part of the active site for the second step is less clear. The existence of a hydrolytic pathway in group II splicing, which is independent of branching, argues against an essential requirement for the branched adenosine in the chemistry of the second transesterification reaction.[32,33] By analogy, it is assumed here that the branch adenosine of a pre-mRNA is not an integral component of the reactive center for the second step. However, the G_{+1} at the 5′ splice site, which is linked via a 2′–5′ phosphodiester bond to the branch adenosine, is required for the second step of splicing,[77,78] and will be discussed later in the section on the structure of the scissile phosphodiester bond. [39]

7.4.3 The Structure of the Nucleophile

The nucleophile for the second phosphoryl transfer reaction is the 3′OH of the 5′ exon, the leaving group of the first reaction. The vicinal 2′OH is not required for activation of the nucleophile given that a 2′H (2′deoxy) sugar at the last position of the exon results in a fully functional splicing precursor [40]. An important interaction, however, is suggested by the fact that a 2′OMe group interferes with the second transesterification reaction.[58] A group normally involved in the activation of the 3′OH is likely to be sterically hindered by the bulkier OMe group. The base composition at the end of the exon does not seem to be absolutely critical suggesting interactions that are either redundant or flexible. It is likely that specific activation of the nucleophile is driven by interactions with intronic sequences that exist prior to the first step; in other words, the leaving group of the first reaction is pre-determined to be the attacking group of the second reaction [41]. Nonetheless, the second step is not a reversal of the first step.

7.4.4 The Structure of the Scissile Phosphodiester Bond

The electrophile in the reaction is the phosphate atom at the 3′ splice site. While it has been difficult to assess the requirements for activation, it is clear that the 2′OH of the last nucleoside of the intron plays a role in this reaction. Both 2′H and 2′OMe substitutions impair the rate of this step, nineteen and

---

[35] See Umen and Guthrie[72]; M. Moore, personal communication.

[36] Much debate has concentrated on whether the two transesterification reactions are carried out in different active sites. However meritorious this debate may be, this chapter will merely describe the reactive centers. A decision on whether these are the same or different rest with the reader.

[37] See, however, Anderson and Moore[39] and Chanfreau *et al.*[75].

[38] This type of scanning is unlikely to explain the juxtaposition of the reactants for the first transesterification (see Pasman *et al.*[76], and references therein).

[39] Recent data from the laboratory of C. Guthrie suggest that nucleotides at the 5′ splice, 3′ splice site, and U6 snRNA interact with each other and with Prp8p at the active site of the second transesterification.

[40] Given recent data of Sontheimer *et al.*[74], which suggest that effects on the chemistry of the second step could be masked by the very slow rearrangement, one should consider this conclusion tentative. Note added in proof: see Gordon *et al.*[79].

[41] See, however, Chanfreau *et al.*[75].

seven times, respectively.[58] Given the modest effect in the apparent rate of the second step, the authors concluded that the 2′OH at the 3′ splice site is not absolutely required for the chemistry of the reaction (ibid). Nonetheless, the greater effect of the 2′H substitution and the uncertainty of rate measurements in very crude systems suggests that a role for the 2′OH in chemistry is viable and should be considered[42].

The first nucleotide of the intron is involved in a non-Watson–Crick interaction with the last nucleotide of the intron, which is a G in the major class of introns.[77,78,80] This normal G–G pairing can be substituted with nucleotides that can form similar configurations, A–C or I–I (ibid.). Furthermore, this critical interaction may be why the nucleotides at the ends of both classes of introns are either G–G or A–C.[7] In group II introns, there is a similar interaction between the first and penultimate positions of the intron.[32] The penultimate intronic position is almost always an A in both major and ATAC introns,[8] and group II introns,[32] but not necessarily in group III introns,[45] and may be important for catalysis of the second transesterification.[38]

7.4.5 The Stereochemistry of the Second Phosphoryl Transfer Reaction

The stereochemistry of the second phosphoryl transfer reaction was studied as described above for the first reaction. A randomly incorporated R_P phosphorothioate diastereomer resulted in inhibition of the second step of splicing when incorporated at the 3′ splice site.[61] This was confirmed using splicing substrates that contained a single R_P or S_P phosphorothioate at the scissile bond.[63] As expected from the above results, the R_P diastereomer inhibited the second transesterification, whereas the S_P diastereomer behaved as the phosphate containing pre-mRNA (ibid.). This clearly supports stereoselective recognition of the pro-R_P oxygen at the 3′ splice site. As shown for the first step, the second phosphoryl transfer reaction proceeds with inversion of configuration, implying an "in line" attack (ibid.).[43] This, together with marked differences in the nucleophiles and in the requirement for a 2′OH at the splice sites, argues that the two chemical steps of pre-mRNA splicing are not forward and reverse reactions within a catalytic center (ibid.).[44] Similar conclusions have been reached for group II introns.[64] Group I introns on the other hand, very likely proceed via two transesterifications that are the forward and reverse reactions catalyzed within one active site (see discussion in Moore and Sharp[63]). The S_N2 ("in line") displacement reaction at the 3′ splice site is modeled to proceed via a bipyramidal transition state (Figure 5). Clearly the pro-R_P nonbridging oxygen at the reactive phosphate must be activated by a chemical group incapable of interacting with a sulfur atom.

7.4.6 Metal Ion Requirements of the Reaction

In order to address the metal ion requirement for leaving group stabilization during the second step, a pre-mRNA was synthesized with a 3′-thioinosine as the last nucleoside of the intron (Figure 2)[45].[57] Substrates with guanosine, inosine or 3′-thiosine at this position were able to proceed through both the first and the second step of splicing in buffers where the only divalent cation was Mg^{2+} (ibid.). Thus, Sontheimer *et al.*[57] concluded that either the leaving group in the second phosphoryl transfer reaction is not stabilized by a metal ion or that the disruption of such a stabilization is masked by a rate-limiting binding or conformation step. Recent results with group II introns suggest that the latter may be correct. Sontheimer *et al.*[74] showed that a 3′thio substitution at the 3′ splice site has little or no effect on the second step of a conventional (*cis*) group II splicing reaction. If the second step is carried out in *trans*, however, a metal ion effect is noted for the leaving group (ibid.). The authors also have recently reported that a spliceosome-dependent *trans*-splicing reaction, which was achieved using the method of Anderson and Moore[39], showed that the spliceosome may also have a metal ion requirement during the second step. In both cases, slow conformational changes between the first and second steps of *cis*-splicing reactions mask the "thio" effect.

---

[42] See footnote 39.

[43] See discussion above.

[44] See Steitz and Steitz[65] for a slightly different interpretation.

[45] An inosine at this position had been shown to be compatible with splicing.[80]

7.5 SUMMARY

Given the state of current knowledge, it is only possible to develop crude models of the reactive centers and transition states for the two phosphoryl transfer reactions. Nonetheless, these can be useful in designing better experiments. In this spirit, the schematics shown in Figure 6(a) and (b) summarize the authors' views of the active sites and transition states of the chemical steps of pre-mRNA splicing. Powerful computer algorithms, increasing knowledge of RNA structure–function relationships[46], and rapidly advancing methods for structure determination should soon refine these models and lead to an understanding of the chemical principles that guide the reactions of intron removal.

7.6 REFERENCES

1. S.M. Berget, C. Moore and P.A. Sharp, *Proc. Natl. Acad. Sci. USA*, 1977, **74**, 3171–3175.
2. L.T. Chow, R.E. Gelinas, T.R. Broker and R.J. Roberts, *Cell*, 1977, **12**, 1–8.
3. T.W. Nilsen, in "Eukaryotic mRNA processing", ed. A. Krainer, Oxford University Press, Oxford, 1997, pp. 310–344.
4. A. Kramer, *Annu. Rev. Biochem.*, 1996, **65**, 367–409.
5. J. Wang and J.L. Manley, *Curr. Opin. Genet. Dev.*, 1997, **7**, 205–211.
6. P.A. Sharp and C.B. Burge, *Cell*, 1997, **91**, 875–879.
7. W.Y. Tarn and J.A. Steitz, *Trends Biochem. Sci.*, 1997, **22**, 132–137.
8. C.B. Burge, T.H. Tuschl and P.A. Sharp, in "The RNA World", eds. R.F. Gesteland, T. Cech and J.F. Atkins, Cold Spring Harbor Laboratory Press, Cold Spring Harbor, NY, 1999, pp. 525–560.
9. J.M. Moore, C.C. Query and P.A. Sharp, in "The RNA World ", eds. R.F. Gesteland and J.F. Atkins, Cold Spring Harbor Laboratory Press, Cold Spring Harbor, NY, 1993, pp. 303–357.
10. T.W. Nilsen, in "RNA Structure and Function", eds. R.W. Simons and M. Grunberg-Manago, Cold Spring Harbor Laboratory Press, Cold Spring Harbor, NY, 1998, pp. 303–357.
11. C. Guerrier-Takada, K. Gardiner, T.L. Marsh, N.R. Pace and S. Altman, *Cell*, 1983, **35**, 849–857.
12. K. Kruger, P.J. Grabowski, A.J. Zaug, J. Sands, D.E. Gottschling and T.R. Cech, *Cell*, 1982, **31**, 147–157.
13. A.N. Kuhn, Z. Li and D.A. Brow, *Mol. Cell*, 1999, **3**, 65–75.
14. J.P. Staley and C. Guthrie, *Cell*, 1998, **92**, 315–326.
15. J.P. Staley and C. Guthrie, *Mol. Cell*, 1999, **3**, 55–64.
16. J.D. Crispino, B.J. Blencowe and P.A. Sharp, *Science*, 1994, **265**, 1866–1869.
17. W.Y. Tarn and J.A. Steitz, *Genes Dev.*, 1994, **8**, 2704–2717.
18. B.B. Konforti and M.M. Konarska, *RNA*, 1995, **1**, 815–827.
19. H.L. Murray and K.A. Jarrell, *Cell*, 1999, **96**, 599–602.
20. W.P. Jencks, "Catalysis in Chemistry and Enzymology", McGraw-Hill, New York, 1969.
21. P.L. Raghunathan and C. Guthrie, *Science*, 1998, **279**, 857–860.
22. J.R. Knowles, *Nature*, 1991, **350**, 121–124.
23. S. Michaud and R. Reed, *Genes Dev.*, 1991, **5**, 2534–2546.
24. R. Parker, P.G. Siliciano and C. Guthrie, *Cell*, 1987, **49**, 229–239.
25. C.C. Query, M.J. Moore and P.A. Sharp, *Genes Dev.*, 1994, **8**, 587–597.
26. J. Wu and J.L. Manley, *Genes Dev.*, 1989, **3**, 1553–1561.
27. Y. Zhuang and A.M. Weiner, *Genes Dev.*, 1989, **3**, 1545–1552.
28. S.F. Jamison, A. Crow and M.A. Garcia-Blanco, *Mol. Cell. Biol.*, 1992, **12**, 4279–4287.
29. W.W. Cleland and A.C. Hengge, *FASEB J.*, 1995, **9**, 1585–1594.
30. J.R. Knowles, *Annu. Rev. Biochem.*, 1980, **49**, 877–919.
31. A.M. Weiner, *Cell*, 1993, **72**, 161–164.
32. A. Jacquier, *Biochimie*, 1996, **78**, 474–487.
33. A.M. Pyle, in "Nucleic Acids and Molecular Biology", eds. F. Eckstein and D.M.J. Lilley, Springer, Berlin, Vol. 10, 1996, pp. 75–107.
34. A.J. Newman, S. Teigelkamp and J.D. Beggs, *RNA*, 1995, **1**, 968–980.
35. J.R. Wyatt, E.J. Sontheimer and J.A. Steitz, *Genes Dev.*, 1992, **6**, 2542–2553.
36. R.T. O'Keefe, C. Norman and A.J. Newman, *Cell*, 1996, **86**, 679–689.
37. V. Segault, C.L. Will, M. Polycarpou-Schwarz, I.W. Mattaj, C. Branlant and R. Luhrmann, *Mol. Cell. Biol.*, 1999, **19**, 2782–2790.
38. H.R. Luo, G.A. Moreau, N. Levin and M.J. Moore, *RNA*, 1999, **5**, 893–908.
39. K. Anderson and M.J. Moore, *Science*, 1997, **276**, 1712–1716.
40. R. Reed, *Genes Dev.*, 1989, **3**, 2113–2123.
41. B. Ruskin and M.R. Green, *Nature*, 1985, **317**, 732–734.
42. B.C. Rymond and M. Rosbash, *Nature*, 1985, **317**, 735–737.
43. W. Saenger, "Principles of Nucleic Acid Structure", Springer, New York, 1984.

[46] See Chapters 1–5 in this volume.

44. S.L. Hall and R.A. Padgett, *Science*, 1996, **271**, 1716–1718.
45. D.W. Copertino and R.B. Hallick, *Trends Biochem. Sci.*, 1993, **18**, 467–471.
46. C.C. Query, S.A. Strobel and P.A. Sharp, *EMBO J.*, 1996, **15**, 1392–1402.
47. J.S. Smith and E.P. Nikonowicz, *Biochemistry*, 1998, **37**, 13486–13498.
48. M. Chastain and I.J. Tinoco, *Prog. Nucleic Acid Res. Mol. Biol.*, 1991, **41**, 131–177.
49. N.L. Greenbaum, I. Radhakrishnan, D.J. Patel and D. Hirsh, *Structure*, 1996, **4**, 725–733.
50. K. Kalurachchi, K. Uma, R.A. Zimmermann and E.P. Nikonowicz, *Proc. Natl. Acad. Sci. USA*, 1997, **94**, 2139–2144.
51. S. Portmann, S. Grimm, C. Workman, N. Usman and M. Egli, *Chem. Biol.*, 1996, **3**, 173–184.
52. V.T. Chu, Q. Liu, M. Podar, P.S. Perlman and A.M. Pyle, *RNA*, 1998, **4**, 1186–1202.
53. S. Kandels-Lewis and B. Seraphin, *Science*, 1993, **262**, 2035–2039.
54. C.F. Lesser and C. Guthrie, *Science*, 1993, **262**, 1982–1988.
55. H.D. Madhani and C. Guthrie, *Genes Dev.*, 1994, **8**, 1071–1086.
56. J.L. Reyes, P. Kois, B.B. Konforti and M.M. Konarska, *RNA*, 1996, **2**, 213–225.
57. E.J. Sontheimer, S. Sun and J.A. Piccirilli, *Nature*, 1997, **388**, 801–805.
58. M.J. Moore and P.A. Sharp, *Science*, 1992, **256**, 992–997.
59. E.A. Griffin Jr., Z. Qin, W.J. Michels Jr. and A.M. Pyle, *Chem. Biol.*, 1995, **2**, 761–770.
60. F. Eckstein, *Annu. Rev. Biochem.*, 1985, **54**, 367–402.
61. K.L. Maschhoff and R.A. Padgett, *Nucleic Acids Res.*, 1992, **20**, 1949–1957.
62. K.L. Maschhoff and R.A. Padgett, *Nucleic Acids Res.*, 1993, **21**, 5456–5462.
63. M.J. Moore and P.A. Sharp, *Nature*, 1993, **365**, 364–368.
64. M. Podar, P.S. Perlman and R.A. Padgett, *Mol. Cell. Biol.*, 1995, **15**, 4466–4478.
65. T.A. Steitz and J.A. Steitz, *Proc. Natl. Acad. Sci. USA*, 1993, **90**, 6498–6502.
66. A.T. Perrotta, I. Shih and M.D. Been *Science*, 1999, **286**, 123–126.
67. G.J. Narlikar and D. Herschlag, *Annu. Rev. Biochem.*, 1997, **66**, 19–59.
68. E.K. Jaffe and M. Cohn, *J. Biol. Chem.*, 1978, **253**, 4823–4825.
69. E.J. Sontheimer and J.A. Steitz, *Science*, 1993, **262**, 1989–1996.
70. C.H. Kim and J. Abelson, *RNA*, 1996, **2**, 995–1010.
71. C.W. Smith, E.B. Porro, J.G. Patton and G.B. Nadal, *Nature*, 1989, **342**, 243–247.
72. J.G. Umen and C. Guthrie, *RNA*, 1995, **1**, 869–885.
73. G. Chanfreau and A. Jacquier, *EMBO J.*, 1996, **15**, 3466–3476.
74. E.J. Sontheimer, P.M. Gordon and J.A. Piccirilli, *Genes Dev.*, 1999, **13**, 1729–1741.
75. G. Chanfreau, C. Gouyette, B. Schwer and A. Jacquier, *RNA*, 1999, **5**, 876–882.
76. Z. Pasman, M.D. Been and M.A. Garcia-Blanco, *RNA*, 1996, **2**, 603–610.
77. A. Deirdre, J. Scadden and C.W. Smith, *EMBO J.*, 1995, **14**, 3236–3246.
78. R. Parker and P.G. Siliciano, *Nature*, 1993, **361**, 660–662.
79. P.M. Gordon, E.J. Sontheimer and J.A. Piccirilli, *RNA*, 2000 **6**, 199–205.
80. W.Y. Tarn, *Biochimie*, 1996, **78**, 1057–1065.

8

RNA Editing

MARIE ÖHMAN
Stockholm University, Stockholm, Sweden

and

BRENDA L. BASS
University of Utah, Salt Lake City, UT, USA

8.1 INTRODUCTION TO RNA EDITING

RNA editing, a type of RNA processing, was first discovered by Benne and co-workers in a mitochondrion-encoded mRNA of a kinetoplastid trypanosome.[1] The term RNA editing initially referred only to the process as it occurs in trypanosomes, which involves the post-transcriptional insertion and deletion of uridylate (UMP) within nascent transcripts. The discovery of additional examples of such post-transcriptional sequence alterations led to a broader use of the term. RNA editing is now used to describe the insertion and deletion of nucleotides other than UMP, base deamination, and the cotranscriptional insertion of nonencoded nucleotides. RNA editing has been observed in mRNAs, tRNAs, and rRNAs, in mitochondrial and chloroplast encoded RNAs, as well as in nuclear encoded RNAs. Examples of RNA editing have been found in many metazoa, unicellar eukaryotes such as trypanosomes, and in plants. To date, RNA editing has not been observed in a prokaryote.

Typically, RNA editing reactions are put into two broad catagories based on their reaction mechanisms. One type, insertion/deletion RNA editing, involves the insertion or deletion of nucleotides and actually changes the length of the target RNA. The second type, RNA editing by base modification, changes an encoded nucleotide into a different nucleotide, without changing the overall length of the RNA. In this chapter the authors focus on the most well-characterized examples of these two types of editing, and particularly on those examples where a model for the catalytic mechanism has been described. Details on other examples of RNA editing can be found in several reviews.[2,3]

8.2 EDITING BY BASE MODIFICATION

RNA modifications have long been known to occur in tRNAs, rRNAs, and mRNAs.[4,5] Some of these modifications are simple, involving merely the methylation of an existing nucleotide, while others are quite complex, involving multiple enzymes for their synthesis. As additional types of RNA editing are discovered, it becomes harder to distinguish the difference between what has traditionally been called an RNA modification, from what is now categorized as RNA editing. At one point in time the term RNA editing was reserved for processes that involved canonical bases, that is, the insertion or deletion of Watson–Crick nucleotides, or the conversion of one Watson–Crick nucleotide into another. Except for one adenosine modification (see below), for the most part this nomenclature rule still applies. However, some scientists now prefer to use the term RNA editing for any post-transcriptional modification of an RNA.

The two types of base modification that are always categorized as RNA editing both involve deamination reactions. In one case, deamination of cytidine produces uridine[6] (see Figure 1), while in the other case, deamination of adenosine produces inosine[7] (see Figure 3). RNA editing by base modification can change one sense codon to that of a different sense, as well as create (C to U editing) or remove (A to I editing) stop codons. In most cases, a protein with new unique features is produced, and a single genomically-encoded RNA gives rise to a variety of transcripts. Despite their mechanistic similarities, as discussed below, RNA editing by cytidine and adenosine deamination involve very different mechanisms for recognizing their RNA substrates.

8.2.1 Editing by Cytidine Deamination

Soon after the first example of editing in kinetoplastid protozoa was discovered, editing was observed within the mammalian apolipoprotein B (apoB) mRNA.[8,9] ApoB mRNA editing was the first example of RNA editing by base modification, as well as the first example of editing of a nuclear-encoded RNA (for review see references 10 and 11). The conversion of a C to a U within apoB mRNA is thought to involve hydrolytic deamination at the C-4 position of cytidine[6] (Figure 1). The cytidine deaminase activity involved in apoB mRNA editing derives from the enzyme APOBEC-1 (apolipoprotein B mRNA editing

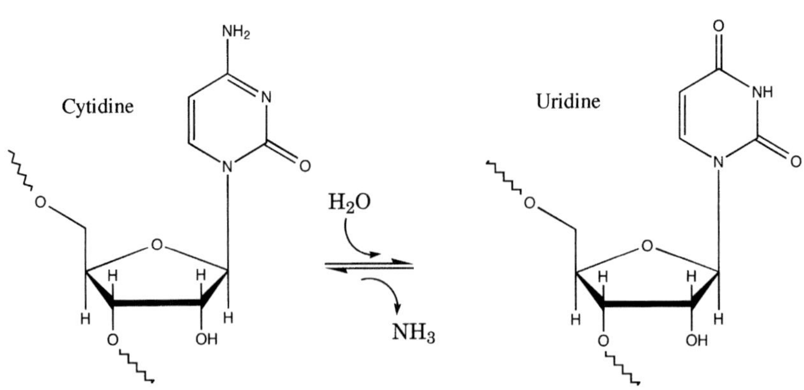

Figure 1 RNA editing of apoB mRNA involves a single cytidine (C) to uridine (U) conversion. As shown, the reaction is thought to proceed by the hydrolytic deamination of the N-4 amino group of cytidine. The deamination is catalyzed by the enzyme APOBEC-1, although additional factors are required for the RNA editing reaction.

enzyme, catalytic polypeptide #1),[12,13] but additional factors are required to edit apoB mRNA. However, interestingly, APOBEC-1 can deaminate cytidine mononucleoside in the absence of other factors,[14,15] consistent with the idea that this polypeptide contains the catalytic active site.

Other natural substrates for APOBEC-1 have been looked for, but so far, they have not been found, and the enzyme's only known biological target is apoB mRNA. However, when APOBEC-1 is overexpressed, aberrant C to U editing is observed in additional RNAs.[16] Although not covered here, C to U editing has also been observed in mitochondria and chloroplasts of plants, as well as in the mitochondria of *Physarum polycephalum* (for reviews see reference 2). Interestingly, plant organelles also exhibit U to C editing, the reverse of the deamination reaction. The endogenous factors that catalyze RNA editing in plants and *Physarum* have not yet been identified.

8.2.1.1 Apolipoprotein B mRNA

Apolipoproteins are required in mammals for the assembly, secretion, and transport of complex lipids; among the various apolipoproteins, apoB is particularly important (reviewed in references 10 and 11). There are two apoB proteins, B-100 and B-48, and each has a different biological function in the metabolism of lipoproteins. Prior to the discovery that the apo-B mRNA was subject to RNA editing, it was suggested that the two proteins were expressed from the same gene because apoB-48 was found to be identical to the amino-terminal half of apoB-100.[17] However, the mechanism of generating the two distinct proteins from a single gene was an enigma.

The mystery was finally solved when cDNA and genomic sequences were compared. In humans, apoB-100 is synthesized mainly in the liver, while apoB-48 is synthesized in the intestine. Comparison of the human genomic sequence of apoB, with human liver and intestinal cDNAs, led to the conclusion that RNA editing was the basis for the production of apoB-48 mRNA. While both the genomic and liver cDNA sequences encoded glutamine (CAA) at codon 2153, the intestinal cDNA had a TAA stop at this codon.[8,9] Thus, a single C to U change at nucleotide 6666 (numbered according to the human sequence) converts the glutamine-2153 codon (CAA) in apoB-100 mRNA to an in-frame stop codon (UAA) in apoB-48 mRNA (Figure 2). The apoB-48 protein is therefore a truncation of apoB-100.

So far, editing of apoB mRNA has been found in human, rat, mouse and rabbit, although there is a difference in the extent of editing in liver and intestine between the various organisms. In the mammalian small intestine, the majority (70–95%) of the steady-state mRNA population corresponds to the edited apoB-48 mRNA (reviewed in reference 10). Apo-B mRNA editing occurs in the nucleus and is a

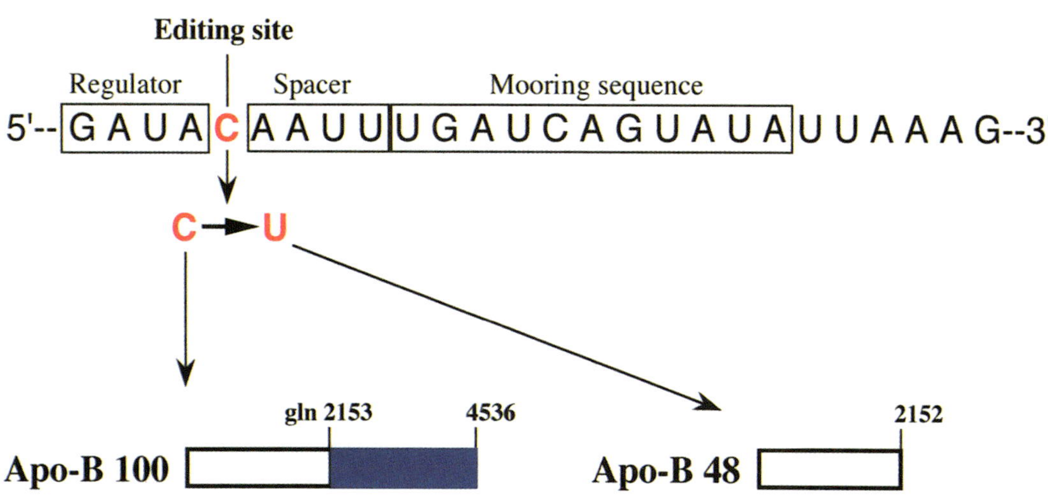

Figure 2 The apoB mRNA editing reaction. The upper part of the figure shows the editing site cytidine (red) in the context of the sequences immediately surrounding the editing site that are important for the reaction. The three different regions that have been determined to be important for editing are boxed (regulator, spacer, and mooring sequence, see text for details). The lower section of the figure illustrates the consequences of editing apoB mRNA. A glutamine codon is changed to a stop codon so that a long (ApoB 100) and a short (ApoB 48) protein can be made from a single open reading frame. The blue rectangle shows sequences in ApoB 100 that are absent in ApoB 48.

post-transcriptional process. This conclusion derives from the observation that a fully processed apoB mRNA isolated from the nucleus is edited to about the same extent as the cytoplasmic apoB mRNA.[18]

RNA editing of apoB mRNA occurs at a single site in the middle of the 14-kilobase mature message. Mutagenesis studies show that a sequence of 22–26 nucleotides surrounding the editing site is important for apoB mRNA editing.[19,20] Within these 22–26 nucleotides, three subdomains have been defined: a *spacer* sequence of 4–6 nucleotides immediately downstream of the editing site; an 11 nucleotide "mooring sequence" (UGAUCAGUAUA) following the spacer, and a "regulator" region immediately upstream of C-6666 (Figure 2; reviewed in references 10 and 11). The mooring sequence is both sufficient and required for editing of the upstream C, as shown by ligating the mooring sequence to an unrelated RNA.[21] Editing is site specific, the hypothesis being that once the editing enzyme (or enzyme complex) binds the mooring sequence, a cytidine at a fixed distance upstream is precisely targeted. The AU-rich regulator region immediately 5′ of the target site appears to stimulate editing, although editing will occur in its absence. Maximal editing appears to require even other, more distal, sequences. For example, a study showed that 139 nucleotides surrounding the rabbit apoB mRNA were required for optimal editing *in vitro*.[22]

8.2.1.2 APOBEC-1 and its catalytic mechanism

RNA editing of apolipoprotein-B is thought to require multiple proteins, and so far, only one of these, APOBEC-1, has been conclusively identified.[12] APOBEC-1 is the catalytic subunit of the holoenzyme, while one or more subunits are proposed to provide the RNA binding functions. When the APOBEC-1 gene is inactivated in mice, editing of apoB mRNA does not occur, and the apoB-48 protein cannot be detected;[23,24] this indicates that there is no genetic redundancy for APOBEC-1 activity.

The APOBEC-1 protein is a 27 kDa protein that contains an N-terminal bipartite nuclear localization signal, a centrally located deamination domain, a possible RNA binding domain that overlaps this catalytic domain, and finally, a C-terminal leucine-rich domain that is presumed to be involved in protein–protein interactions (reviewed in references 10 and 11). The extensive sequence similarities between APOBEC-1 and the *Escherichia coli* cytidine deaminase (ECCDA) suggest that the two enzymes use similar catalytic mechanisms.[25] Thus, since ECCDA requires zinc for catalysis, it has been proposed that APOBEC-1 also requires a catalytic zinc, and *in vitro* experiments support this idea.[15] In addition, certain regions of the APOBEC-1 sequence can be aligned with the amino acids of ECCDA known to be involved in coordinating the catalytic zinc.

Like ECCDA,[26] APOBEC-1 forms a homodimer.[27] Although the function of dimerization is not known, it has been suggested that one of the active sites binds a U within the AU-rich mooring sequence, while the other active site binds and deaminates the target C.[25] This hypothesis is based on the sequence similarity of APOBEC-1 to ECCDA. However, the substrate specificity of the two enzymes differs considerably. ECCDA is specific for nucleoside substrates, while APOBEC-1 recognizes a single C in a specific RNA context.

The missing factor(s) that acts in concert with APOBEC-1 to edit apoB mRNA has long been searched for. A 65 kDa complementing protein has been shown to interact with APOBEC-1 *in vitro* to form an enzymatically active complex.[28] More recently, this factor, named APOBEC-1 complementation factor (ACF) has been cloned and shown to bind apoB mRNA in a mooring sequence dependent manner.[29] Further, when combined with APOBEC-1, ACF is sufficient to reconstitute efficient and specific editing of apo-B mRNA. However, it remains possible that there are other accessory factors that modulate activity, or other factors, perhaps in other tissues, that can replace ACF.

8.2.2 Editing by Adenosine Deamination

The second example of RNA editing by base modification involves adenosine deamination. This type of RNA editing requires a specific structure in the RNA substrate, rather than a specific sequence, and has been observed in a number of different RNAs. Adenosine to inosine conversion occurs by a hydrolytic deamination that removes the C-6 amino group from adenosine[7] (Figure 3). A single enzyme called an ADAR (adenosine deaminase that acts on RNA; Figure 3) can catalyze the reaction, without the addition of cofactors.

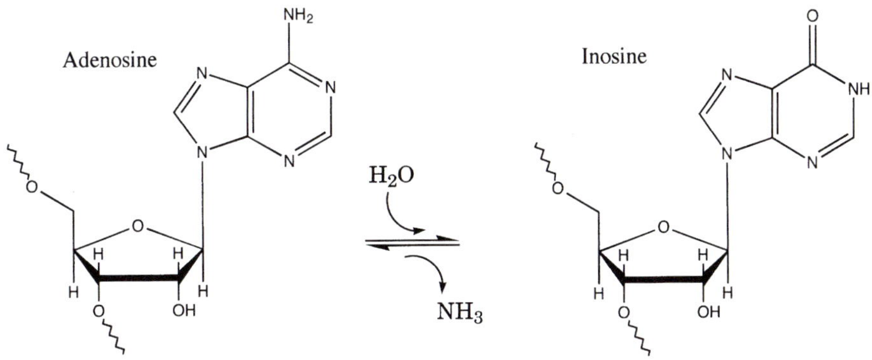

Figure 3 Many RNAs are edited by adenosine (A) to inosine (I) conversion. Enzymes that catalyze the reaction are called ADARs and studies of ADAR1 demonstrate that the reaction occurs by a simple hydrolytic deamination. As shown, the hydrolytic deamination of the amino group at C-6 yields inosine.

The natural RNA substrates for ADARs have been discovered by noticing sequence discrepancies between genes and their cDNAs. Since inosine, like guanosine, prefers to pair with cytidine, adenosine deamination events are indicated when a genomically encoded adenosine appears as a guanosine in a cDNA. Based on such A to G changes, two types of ADAR substrates have been recognized: those that are hypermutated, containing up to 50% A to G changes in a single cDNA, and those that are selectively deaminated, containing less than 10% A to G changes in their cDNAs (reviewed in reference 30). Most of the hypermutated substrates are viral RNAs, and here the function of the adenosine modification is unclear. However, in the selectively deaminated substrates it is clear that adenosine deamination functions to change one codon into another. In this way, ADARs act to regulate gene expression by allowing multiple proteins to be synthesized from a single encoded sequence.

8.2.2.1 ADAR proteins and their catalytic mechanism

ADAR activity was first detected when double-stranded RNA (dsRNA) was microinjected into *Xenopus laevis* embryos.[31,32] Since then, the activity has been found in every metazoan assayed (reviewed in reference 30). Two different enzymes with ADAR activity, ADAR1 and ADAR2, have been discovered in mammals (for review see references 2 and 3). Further, cDNAs have been discovered that, based on sequence similarities, are thought to encode additional ADARs, and ADARs have now been identified in flies and worms as well.

Formerly, ADAR1 was called dsRAD or DRADA, while ADAR2 was called RED1.[33] In contrast to the C to U editing that occurs within apo-B mRNA, both ADAR1 and ADAR2 can catalyze deamination within their substrates without additional factors. The ADAR1 protein contains three dsRNA binding motifs (dsRBMs) upstream of a C-terminal catalytic domain that is highly conserved among ADARs (Figure 4); ADAR2 lacks the amino terminal half of ADAR1, and contains only two

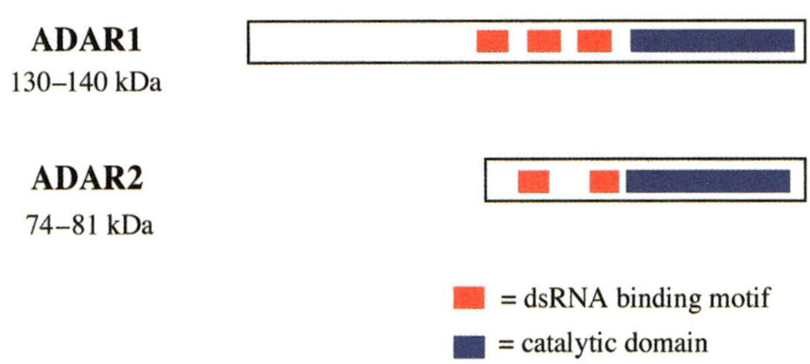

Figure 4 The open reading frame structures of ADAR1 and ADAR2. ADAR1 has been cloned from human, rat, and frog, and ADAR2 from human and rat; the proteins vary slightly from organism to organism so the diagrams are not exact. Red boxes indicate dsRNA binding motifs, and the highly conserved catalytic domain is illustrated with a blue rectangle. Numbers below ADAR1 and ADAR2 correspond to the range of molecular weights observed for the proteins of different organisms.

dsRBMs.[34] The transcripts for both ADAR1 and ADAR2 are alternatively spliced to generate multiple isoforms.[35–37]

ADARs catalyze a nucleophilic attack at the C-6 atom of adenine. Since the C-6 atom is buried deep within the narrow major groove of dsRNA, it has been proposed that ADAR gains access to its target by flipping the adenine base out of the RNA helix and into the active site.[38,39] Such base-flipping mechanisms have been observed in DNA methyltransferases (reviewed in references 40 and 41).

8.2.2.2 ADAR specificity

The deamination specificity of the founding member of the ADAR family, ADAR1, has been characterized *in vitro*.[38] The enzyme does not have strict sequence requirements, and multiple adenosines can be deaminated in a single RNA. However, deaminations do not occur randomly. ADAR1 has a 5′ nearest-neighbor preference such that adenosines with a 5′ neighbor of A, U, or C are more likely to be deaminated than those with a 5′ G. ADAR1 also disfavors adenosines near 3′ termini (within ~8 nucleotides).

The structure of the RNA substrate is also important for ADAR deamination specificity and dictates whether the RNA is promiscuously deaminated or selectively deaminated. In promiscuously deaminated substrates, ~50% of the adenosines are deaminated at complete reaction, while in selectively deaminated substrates, <10% of the adenosines are deaminated at complete reaction. *In vitro* studies, as well as analyses of endogenous substrates, indicate that the promiscuous type of deamination occurs in the context of long, completely base-paired dsRNA, while the more selective deamination occurs in molecules where base-paired regions are periodically interrupted by mismatches, bulges, and loops (e.g., see Figure 5).

Although the deamination specificity of ADAR2 has not been extensively studied, data gathered so far clearly show that it is similar to, but distinct from, that of ADAR1. Like ADAR1, ADAR2 can act with high or low selectivity, depending on the structure of the RNA substrate. For example, ~50% of the adenosines in completely base-paired dsRNA are deaminated,[34,35,37] while base-paired regions of biological substrates, such as glutamate receptor (gluR) B mRNA (see below), are deaminated more selectively. Various laboratories have analyzed the ability of ADAR2 to deaminate editing sites within gluR-B pre-mRNA.[34,35,37,42,43] ADAR2 is clearly more efficient at editing the Q/R site, while both enzymes edit the R/G site (see Figure 5(a) and text below). Thus, the two enzymes have distinct but overlapping specificities. Because of this, at present, it is hard to say which enzyme is responsible for which editing event *in vivo*.

8.2.2.3 Mammalian glutamate receptor mRNA

Based on pharmacological studies, glutamate receptors in the mammalian central nervous system are classified into three distinct classes: *N*-methyl-d-aspartate (NMDA) receptors, α-amino-3 hydroxyl-5-methyl-isoxazole-4-propionate (AMPA) receptors, and kainate (KA) receptors. Editing by adenosine deamination has been demonstrated in the pre-mRNA of the AMPA receptors, gluR-B, -C, and -D, as well as in the KA receptor subunits gluR-5 and gluR-6. As might be expected, the editing sites are found within sequences that can form base-paired structures. So far, all of the structures required for gluR RNA editing involve base-pairing between exons and introns (see Figure 5(a)).

Editing within gluR-B pre-mRNA has been extensively studied, and two biologically important editing events have been defined within its coding sequence. Since inosine is translated as guanosine,[44] the A to I conversions are synonymous with A to G changes. Editing within exon 11 of gluR-B pre-mRNA converts a glutamine codon (CAG) to an arginine codon (CIG). ADAR editing sites are named according to the amino acid change they produce, and thus, this site is called the Q/R editing site. The Q/R site is edited in 99% of gluR-B pre-mRNAs, at all stages of development. In contrast, gluR-5 and gluR-6 are edited at the Q/R site to 50% and 70%, respectively, in adult rat brain, but to a much lower extent early in embryogenesis (for review see reference 45).

Another editing site has been found in exon 13 of gluR-B, -C, and -D. Here editing converts an arginine codon (AGA) to a glycine codon (IGA) and accordingly, the site is called the R/G editing site. Editing at the R/G site also varies during early development, reaching a maximum in the adult brain of

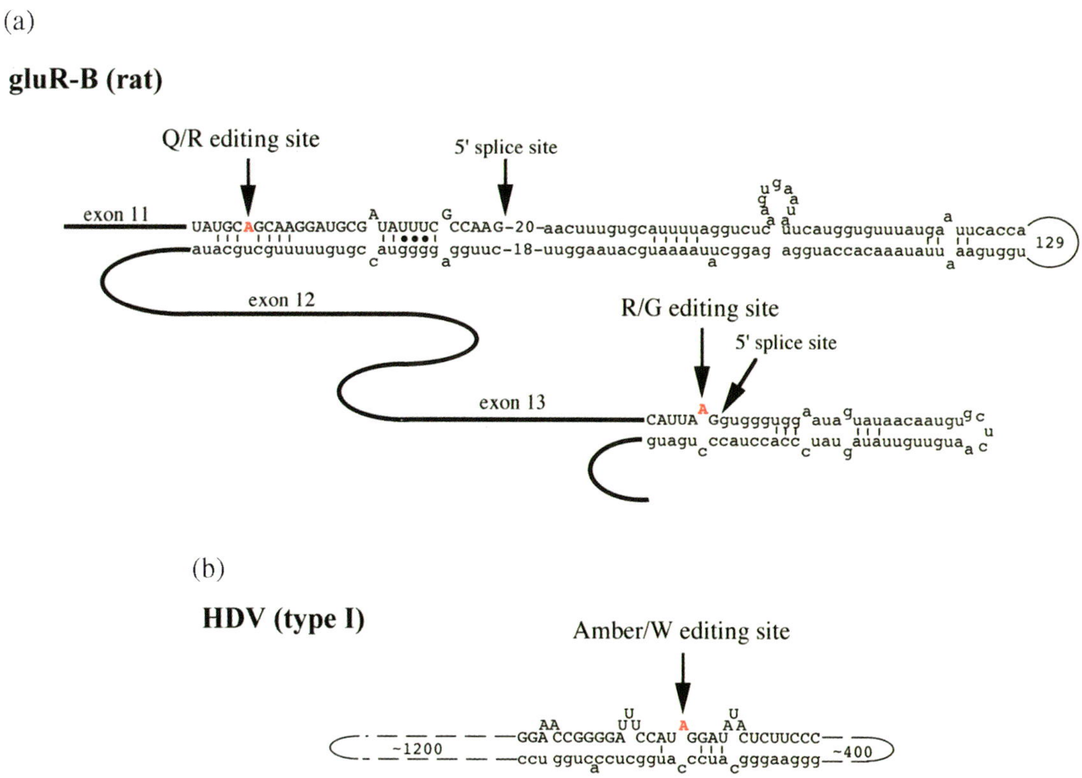

Figure 5 Examples of the base-paired structures that surround ADAR editing sites. (a) The structures required for editing at the Q/R site in exon 11 of gluR-B, and the R/G site in exon 13 of gluR-B, are formed by base pairing between the exon (upper case sequences) and downstream introns (lower case sequences). Editing sites with biological consequences are shown as red capital letters. Only base pairs supported by compensatory mutations are shown (as cited in reference 30). Numbers indicate the length of sequences excluded from the diagram. (b) The sequences immediately surrounding the editing site within the antigenomic RNA of hepatitis delta virus are shown, with symbols and labels as in (a) (reproduced by permission of Elsevier Science Ltd from *Trends Biochem. Sci.*, 1997, 22, 157).

80–90% for gluR-B and -C.[46] Additional editing sites have also been detected within the coding region of gluR-6, in particular, the I/V and the Y/C site.

Additional studies are required to fully understand the functional consequences of each of the editing events within the various glutamate receptors. However, editing at the Q/R site of gluR-B is particularly well characterized and clearly has a striking effect on function. Specifically, ion channels that contain gluR-B with arginine instead of glutamine at the Q/R site are much less permeable to calcium.[47] Mice that have lower than normal levels of editing at the Q/R site of gluR-B mRNA suffer epilepsy and die about 12 days after birth.[48]

8.2.2.4 Hepatitis delta virus antigenomic RNA

Hepatitis delta virus (HDV) is a closed, circular RNA of about 1700 nucleotides. The genome is replicated via an RNA intermediate, the antigenome (for review see reference 49). Both the genome and the antigenome can be folded into rod-shaped, highly base-paired structures, similar to those of viroids. Although the virus expresses only a single open reading frame (ORF), RNA editing by adenosine deamination allows the synthesis of two essential proteins from this single ORF (Figure 6). The shorter protein, HDAg-p24, is synthesized from the encoded ORF, which terminates at an amber stop codon (UAG). The longer protein, HDAg-p27, is synthesized after RNA editing converts the amber stop codon (UAG) to a tryptophan codon (UGG), extending the open reading frame by 19 amino acids. The shorter form of the protein, HDAg-p24, is necessary for replication, while the longer version, HDAg-p27, is required for virus packaging. Editing occurs on the antigenomic RNA at an adenosine referred to as the amber/W editing site[50] and the sequence change is passed to the genome during replication. HDV

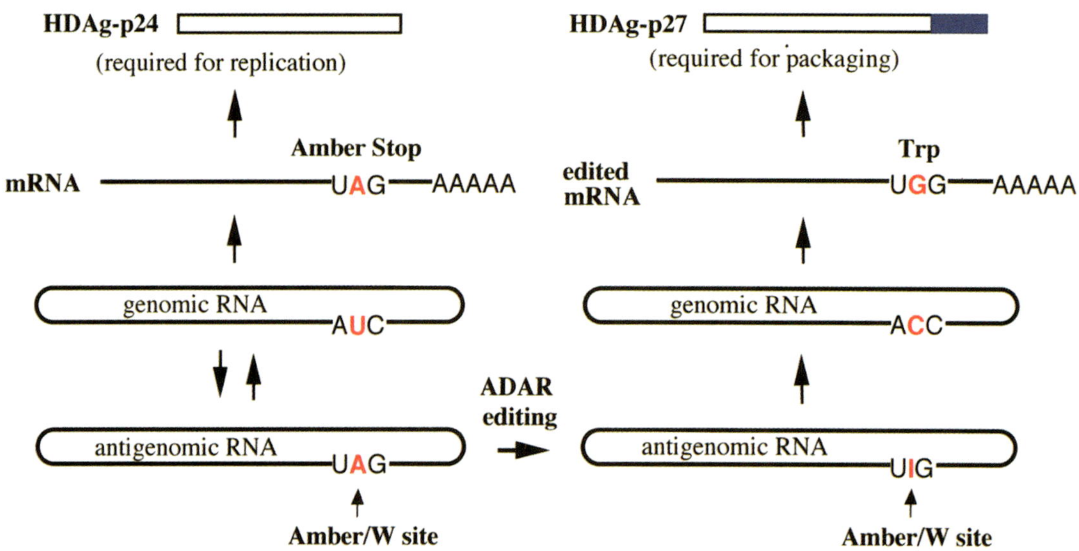

Figure 6 A model for RNA editing of hepatitis delta virus (HDV) antigenomic RNA. The ADAR enzyme acts on the antigenomic RNA at the amber/W site to convert an adenosine (A) to an inosine (I). The sequence change is passed to the genomic RNA during replication. Transcription of the "unedited" genome yields an mRNA that encodes HDAg-p24, a protein required for HDV replication. Transcription of the "edited" genome yields an mRNA that encodes HDAg-p27, a protein required for packaging of the HDV genome and which contains 19 additional amino acids (blue) (reproduced by permission of Macmillan Magazines Ltd from *Nature*, 1996, **380**, 454).

antigenomic RNA is specifically and efficiently edited *in vitro* at the amber/W site by *X. laevis* ADAR1, consistent with the idea that the ADAR enzyme is responsible for the editing event.[51]

8.2.2.5 Other transcripts edited by adenosine deamination

Although only a few examples of ADAR substrates have been described here, other hypermutated as well as selectively deaminated RNAs have been described (as cited in reference 30). So far, the examples of hypermutated RNAs are predominantly viral RNAs, while the selectively deaminated RNAs mainly encode proteins involved in neurotransmission. In addition to the gluR mRNAs, these include mRNAs encoding serotonin receptors[52] as well as potassium and sodium channels.[53,54] As with other ADAR substrates, in these cases, selective deamination functions to allow multiple proteins to be derived from a single mRNA, while as yet, the function of the hypermutations is unclear.

In contrast to other types of mRNA editing described in this chapter, which involve the conversion of one Watson–Crick nucleotide into another, or the insertion or deletion of Watson–Crick nucleotides, adenosine deamination within an mRNA creates a nucleotide that is not normally found in nascent transcripts. This has allowed a direct measurement of the amount of inosine contained within mRNA of various mammalian tissues.[55] The amounts are quite astonishing and suggest that there are many more inosine-containing RNAs yet to be discovered.

8.3 INSERTION/DELETION EDITING

The editing events discussed so far involve the covalent modification of an encoded nucleotide. In this section the second major category of editing is presented, which involves the insertion or deletion of nucleotides that are not genomically encoded. The most well-characterized type of insertion/deletion editing occurs in mitochondrial pre-mRNAs of trypanosomes, a protozoan, and involves the insertion and deletion of UMP. In the current model for this type of editing, several protein enzymes catalyze the requisite cleavage and ligation events, while small RNAs, called guide RNAs (gRNAs), interact with the pre-mRNA to specify where the UMPs are to be added or removed. Other insertion/deletion types of editing, in some cases involving nucleotides other than uridine, have been observed in mitochondria of other organisms as well (reviewed in reference 2). At present it is not clear if these other examples will have mechanistic similarities to the insertion/deletion editing that occurs in trypanosomes.

8.3.1 Editing in Trypanosomatid Protozoa

Many mitochondrial-encoded mRNAs of kinetoplastid trypanosomes are edited by insertions or deletions of UMPs (reviewed in references 56 and 57). The number of editing sites depends on the particular RNA, with some RNAs requiring very few insertions or deletions to create the correct ORF, and others requiring editing at hundreds of sites to create the functional ORF. At the extreme are the pan-edited RNAs, where more than 50% of the nucleotides derive from post-transcriptional U insertions. Overall, deletions are found less frequently, about one-tenth as often as insertions. Editing leads to the creation of new initiation and stop codons, as well as the conversion of one codon into another. Despite the complexity of RNA editing in trypanosomes, functional RNAs are produced. The parasitic trypanosome depends on RNA editing, at least in the stages of the life cycle requiring aerobic respiration.

8.3.1.1 The catalytic mechanism of U insertion/deletion RNA editing

Overall, editing by U insertion and deletion occurs in a 3′ to 5′ direction along the pre-mRNA. The gRNAs base pair with the unedited pre-mRNA, using GU as well as canonical base pairs, and designate where the uridines should be inserted or deleted. The gRNAs are about 70 nucleotides long and consist of three domains: the *anchor*, a 5′ sequence that is perfectly complementary to the pre-mRNA downstream of the region to be edited; a central domain that is partially complementary to the region to be edited and a 3′ U-tail of 5–24 residues that may stabilize the initial gRNA–mRNA interaction[58–60] (Figure 7). It is the central domain, which is only partially complementary to the unedited pre-mRNA, that actually designates where UMPs are to be inserted or deleted. As shown in Figure 7, positions of mismatched bases mark sites where uridines are to be inserted or deleted.

Over the years, two models have been considered for how insertion/deletion editing is catalyzed. One model, the transesterification model, proposes that insertion/deletion is an RNA-catalyzed reaction, while the other model invokes a series of protein enzymes to catalyze the requisite cleavage and ligation events. Recent evidence all but proves that protein enzymes are involved, and thus, here the authors will focus on the cleavage–ligation model which assumes catalysis by proteins.

An important step in understanding the mechanism of U-insertion/deletion RNA editing was the development of *in vitro* editing systems. Although inefficient, *in vitro* systems for both U-insertion and U-deletion have now been established.[59,61–63] The mechanism for U-insertion, based on the cleavage–ligation model, involves four distinct enzymatic activities: an RNA endonuclease, a uridylyl transferase, an exonuclease, and an RNA ligase[59,61,62] (Figure 7, left panel). All of these enzyme activities have been detected in trypanosomes.[64–67] Initially, the mismatches formed between the central domain of the gRNA and the unedited pre-mRNA direct cleavage by an endonuclease. The anchor sequence of the gRNA binds to the mRNA and cleavage occurs immediately 5′ of the gRNA–mRNA duplex (with respect to the mRNA). The nuclease generates a 3′ hydroxyl on the 5′ cleavage product and a 5′ phosphate on the 3′ cleavage product of the mRNA. Subsequently, terminal uridylyl transferase (TUTase) adds U residues to the 3′-end of the mRNA, using free UTP to incorporate 5′-UMP. Although in the original model,[58] UMP residues were added one at a time, recent evidence suggests that multiple UMPs are added during each round of editing;[68] those that do not base pair with the gRNA are subsequently trimmed by an 3′–5′ exonuclease. The last step is a rejoining of the mRNA by an RNA ligase.

The mechanism for U-deletion resembles U-insertion but probably differs at the initial cleavage step.[63] U-deletion is also initiated by an endonucleolytic cleavage event (Figure 7, right panel), but in contrast to U insertion, requires a U residue immediately upstream of the duplex formed between the pre-mRNA and gRNA anchor sequence. U-deletion requires high concentrations of adenosine nucleotides and therefore differs from U-insertional editing which is inhibited by high ATP or ADP concentrations.[63] The second step of U-deletion editing involves a 3′–5′ exonuclease that removes one or several U residue(s) from the 3′ end of the 5′ half of the pre-mRNA. The U residues are thought to be removed from the RNA individually as UMP. Finally, the RNA is ligated in a manner similar

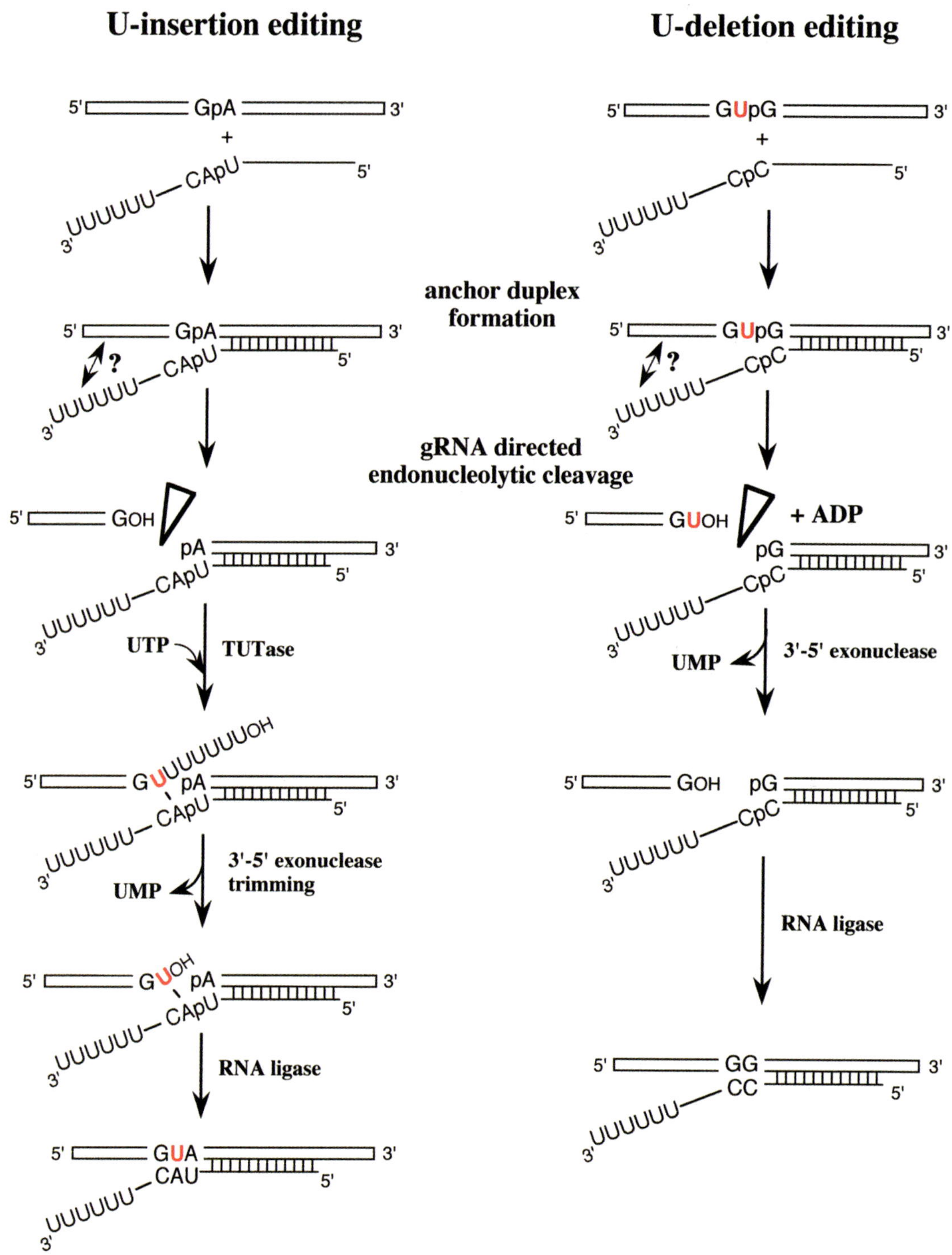

Figure 7 Current models for the mechanism of RNA editing in kinetoplastid mitochondria. The left panel shows a model for U-insertion, and the right panel shows a model for U-deletion. Pre-mRNAs are shown as open boxes, guide RNAs are indicated as lines, and vertical lines indicate base pairs. Open arrowheads indicate cleavage sites, and only sequences immediately surrounding the editing sites are shown. The double arrow indicates a possible interaction between the 3′ U-tail of the gRNA and the pre-edited sequence. Uridines to be inserted or deleted are shown in red. See text for further details.

to the last step in U-insertional editing.[60,63,66] The models presented here, for both U-deletion and U-insertion, are based on *in vitro* studies of sequences from natural substrates, such as the pre-mRNA of ATPase subunit 6 in *Trypanosoma brucei* and that of NADH dehydrogenase 7 in *Leishmania tarentolae*.

8.4 SUMMARY AND PERSPECTIVES

The discovery that nuclear-encoded mRNAs of metazoa were subject to RNA editing altered the prevailing view that editing was relevant only to those studying mitochondrial gene expression in single-celled eukaryotes. Two types of RNA editing have been found in nuclear-encoded mRNAs, and one type has been observed ubiquitously in every metazoan assayed. The RNA editing observed in nuclear-encoded RNAs involves deamination of encoded nucleotides, rather than the insertion and deletion of nucleotides common in mitochondrial RNA editing; in one case deamination of C creates U, and in the other case, deamination of A creates I. While C to U editing of a nuclear-encoded mRNA has only been observed in a few mammalian mRNA, A to I editing has been observed in many RNAs, in many different metazoa. While a single enzyme called an ADAR is sufficient for catalysis of A to I editing, the C to U editing of apoB mRNA requires the enzyme APOBEC-1, as well as additional, as yet unconfirmed, factors.

In the examples of RNA editing within nuclear-encoded mRNAs, editing changes a functional transcript so as to allow a different and additional function. Thus, these deamination reactions are similar to alternative RNA splicing and frameshifting, that is, RNA processing reactions which allow multiple proteins to be generated from a single-encoded sequence. In contrast, the various types of RNA editing that occur in mitochondria can largely be considered as types of repair, that is, processes which create functional RNAs from garbled nascent transcripts. While most of the RNA editing that occurs in mitochondria involves insertion/deletion mechanisms, C to U as well as U to C editing has been observed.

In most cases, the mechanism and macromolecular machinery required for RNA editing in mitochondrial-encoded transcripts has not been defined. However, a great deal of progress has been made in characterizing the editing reaction that occurs in the mitochondria of trypanosomes, the organism where insertion/deletion editing was discovered. Editing in trypanosomal mitochondria requires both protein and RNAs, but recent results indicate the catalytic active sites are contained in the proteins, thus settling a long-standing debate. At least four different enzymatic activities are involved in catalyzing the requisite cleavage and ligation reactions, while the small RNAs, the guide RNAs, designate where the mRNA should be edited by base pairing with the unedited pre-mRNA. To date, the proteins responsible for the enzymatic activities have not been purified to homogeneity.

The focus of future studies on RNA editing will undoubtedly differ for the various types of RNA editing. For most of the mitochondrial systems, studies will be directed towards the definitive identification and characterization of the macromolecules that catalyze the reactions. In these cases, many different RNA substrates have been identified, and once the reactions can be reconstituted *in vitro*, the details of substrate recognition can be worked out. For the RNA editing systems in the nucleus, some of the proteins that catalyze the reactions have been characterized, and thus, here studies will instead focus on the identification of additional substrates, since few are known. Of course, even though some of the proteins that catalyze these reactions have been identified, little is known about how the reactions are regulated *in vivo*, and future studies will also explore the possibility that regulatory factors exist.

8.5 REFERENCES

1. R. Benne, J. VandenBurg, J.P.J. Brakenhoff, P. Sloof, J.H. van Boom and M.C. Tromp, *Cell*, 1986, **46**, 819.
2. J.M. Gott and R.B. Emeson, *Annu. Rev. Genet.*, 2000, **34**, 499
3. S. Maas and A. Rich, *BioEssays*, 2000, **22**, 790.
4. J.A. McCloskey and P.F. Crain, *Nucleic Acids Res.*, 1998, **26**, 196.
5. P.A. Limbach, P.F. Crain and J.A. McCloskey, *Nucleic Acids Res.*, 1994, **22**, 2183.
6. D.F. Johnson, K.S. Poksay and T.L. Innerarity, *Biochem. Biophys. Res. Commun.*, 1993, **195**, 1204.
7. A.G. Polson, P.F. Crain, S.C. Pomerantz, J.A. McCloskey and B.L. Bass, *Biochemistry*, 1991, **30**, 11–507.
8. L.M. Powell, S.C. Wallis, R.J. Pease, Y.H. Edwards, T.J. Knott and J. Scott, *Cell*, 1987, **50**, 831.
9. S.-H. Chen, G. Habib, C.-Y. Yang, Z.-W. Gu, B.R. Lo, S.-A. Weng, S.R. Silberman, S.-J. Cai, J.P. Deslypere and M. Rosseneu, *Science*, 1987, **328**, 363.
10. L. Chan, B.H. Chang, M. Nakamuta, W.H. Li and L.C. Smith, *Biochim. Biophys. Acta*, 1997, **1345**, 11.
11. A. Chester, J. Scott, S. Anant and N. Navaratnam, *Biochem. Biophys. Acta*, 2000, **1494**, 1.
12. B. Teng, C.F. Burant and N.O. Davidson, *Science*, 1993, **260**, 1816.
13. N.O. Davidson, T.L. Innerarity, J. Scott, H. Smith, D.M. Driscoll, B. Teng and L. Chan, *RNA*, 1995, **1**, 3.
14. A.J. MacGinnitie, S. Anant and N.O. Davidson, *J. Biol. Chem.*, 1995, **270**, 14–768.

15. N. Navaratnam, J.R. Morrison, S. Bhattacharya, D. Patel, T. Funihashi, F. Giannoni, B.B. Teng, N.O. Davidson and J. Scott, *J. Biol. Chem.*, 1993, **268**, 20–709.
16. S. Yamanaka, K.S. Poksay, K.S. Arnold and T.L. Innerarity, *Genes Dev.*, 1997, **11**, 321.
17. S.G. Young, S.J. Bertics, T.M. Scott, B.W. Dubois, L.K. Curtiss and J.L. Witztum, *J. Biol. Chem.*, 1986, **261**, 2995.
18. P.P. Lau, W.J. Xiong, H.J. Zhu, S.-H. Chen and L. Chan, *J. Biol. Chem.*, 1991, **266**, 20–550.
19. J.W. Backus and H.C. Smith, *Nucleic Acids Res.*, 1992, **20**, 6007.
20. R.R. Shah, T.J. Knott, J.E. Legros, N. Navaratnum, J.C. Greeve and J. Scott, *J. Biol. Chem.*, 1991, **266**, 16–301.
21. D.M. Driscoll, S. Lakhe-Reddy, L.M. Oleksa and D. Martinez, *Mol. Cell. Biol.*, 1993, **13**, 7288.
22. M. Hersberger and T.L. Innerarity, *J. Biol. Chem.*, 1998, **273**, 9435.
23. K. Hirano, S.G. Young, R.V. Farese Jr., J. Ng, E. Sande, C. Warburton, L.M. Powell-Braxton and N.O. Davidson, *J. Biol. Chem.*, 1996, **271**, 9887.
24. J.R. Morrison, C. Paszty, M.E. Stevens, S.D. Hughes, T. Forte, J. Scott and E.M. Rubin, *Proc. Natl. Acad. Sci. USA*, 1996, **93**, 7154.
25. N. Navaratnam, T. Fujino, J. Bayliss, A. Jarmuz, A. How, N. Richardson, A. Samasekaram, S. Bhattacharya, C. Carter and J. Scott, *J. Mol. Biol.*, 1998, **275**, 695.
26. L. Betts, S. Xiang, S.A. Short, R. Wolfenden and C.W. Carter Jr., *J. Mol. Biol.*, 1994, **235**, 635.
27. P.P. Lau, H.-J. Zhu, A. Baldini, C. Charnsangavej and L. Chan, *Proc. Natl. Acad. Sci USA*, 1994, **91**, 8522.
28. A. Mehta, S. Banerjee and D.M. Driscoll, *J. Biol. Chem.*, 1996, **271**, 28–294.
29. A. Mehta, M.T. Kinter, N.E. Sherman and D.M. Driscoll, *Mol. Cell. Biol.*, 2000, **20**, 1846.
30. B.L. Bass, *Trends Biochem. Sci.*, 1997, **22**, 157.
31. B.L. Bass and H. Weintraub, *Cell*, 1987, **48**, 607.
32. M.R. Rebagliati and D.A. Melton, *Cell*, 1987, **48**, 599.
33. B.L. Bass, K. Nishikura, W. Keller, P.H. Seeburg, R.B. Emeson, M.A. O'Connell, C.E. Samuel and A. Herbert, *RNA*, 1997, **3**, 947.
34. T. Melcher, S. Maas, A. Herb, R. Sprengel, P.H. Seeburg and M. Higuchi, *Nature*, 1996, **379**, 460.
35. F. Lai, C.X. Chen, K.C. Carter and K. Nishikura, *Mol. Cell. Biol.*, 1997, **17**, 2413.
36. Y. Liu, C.X. George, J.B. Patterson and C.E. Samuel, *J. Biol. Chem.*, 1997, **272**, 4419.
37. A. Gerber, M.A. O'Connell and W. Keller, *RNA*, 1997, **3**, 453.
38. A.G. Polson and B.L. Bass, *EMBO J.*, 1994, **13**, 5701.
39. R.F. Hough and B.L. Bass, *RNA*, 1997, **3**, 356.
40. R.J. Roberts, *Cell*, 1995, **82**, 9.
41. X.D. Cheng and R.M. Blumenthal, *Structure*, 1996, **4**, 639.
42. S. Maas, T. Melcher, A. Herb, P.H. Seeburg, W. Keller, S. Krause, M. Higuchi and M.A. O'Connell, *J. Biol. Chem.*, 1996, **271**, 12–221.
43. T. Melcher, S. Maas, A. Herb, R. Sprengel, M. Higuchi and P.H. Seeburg, *J. Biol. Chem.*, 1996, **271**, 31–795.
44. C. Basilio, A.J. Wahba, P. Lengyel, J.F. Speyer and S. Ochoa, *Proc. Natl. Acad. Sci. USA*, 1962, **48**, 613.
45. P.H. Seeburg, *J. Neurochem.*, 1996, **66**, 1.
46. H. Lomeli, J. Mosbacher, T. Melcher, T. Hoger, J.R.P. Geiger, T. Kuner, H. Monyer, M. Higuchi, A. Bach and P.H. Seeburg, *Science*, 1994, **266**, 1709.
47. M. Hollman, M. Hartley and S. Heinemann, *Science*, 1991, **252**, 851.
48. R. Brusa, F. Zimmermann, D.-S. Koh, D. Feldmeyer, P. Gass, P.H. Seeburg and R. Sprengel, *Science*, 1995, **270**, 1677.
49. M.M. Lai, *Annu. Rev. Biochem.*, 1995, **64**, 259.
50. J.L. Casey and J.L. Gerin, *J. Virol.*, 1995, **69**, 7593.
51. A.G. Polson, L.B. Bass and J.L. Casey, *Nature*, 1996, **380**, 454.
52. C.M. Burns, H. Chu, S.M. Reuter, L.K. Hutchinson, H. Canton, E. Sanders-Bush and R.B. Emeson, *Nature*, 1997, **387**, 303.
53. D.E. Patton, T. Silva and F. Bezanilla, *Neuron*, 1997, **19**, 711.
54. C.J. Hanrahan, M.J. Palladino, L.J. Bonneau and R.A. Reenan, *Ann. N.Y. Acad. Sci.*, 1999, **868**, 51.
55. M.S. Paul and B.L. Bass, *EMBO J.*, 1998, **17**, 1120.
56. A. M. Estévez and L. Simpson, *Gene*, 1999, **240**, 247
57. M.L. Kable, S. Heidmann and K.D. Stuart, *Trends Biochem. Sci.*, 1997, **22**, 162.
58. B. Blum, N. Bakalara and L. Simpson, *Cell*, 1990, **60**, 189.
59. M.L. Kable, S.D. Seiwert, S. Heidmann and K. Stuart, *Science*, 1996, **273**, 1189.
60. S.D. Seiwert, S. Heidmann and K. Stuart, *Cell*, 1996, **84**, 831.
61. E.M. Byrne, G.J. Connell and L. Simpson, *EMBO J.*, 1996, **15**, 6758.
62. G.J. Connell, E.M. Byrne and L. Simpson, *J. Biol. Chem.*, 1997, **272**, 4212.
63. J. Cruz-Reyes, L.N. Rusche, K.J. Piller and B. Sollner-Webb, *Mol. Cell*, 1998, **1**, 401.
64. K.J. Piller, L.N. Rusche, J. Cruz-Reyes and B. Sollner-Webb, *RNA*, 1997, **3**, 279.
65. N. Bakalara, A.M. Simpson and L. Simpson, *J. Biol. Chem.*, 1989, **264**, 18679.
66. J. Cruz-Reyes and B. Sollner-Webb, *Proc. Natl. Acad. Sci. USA*, 1996, **93**, 8901.
67. R. Sabatini and S.L. Hajduk, *J. Biol. Chem.*, 1995, **270**, 7233.
68. J.D. Alfonzo, O. Thiemann and L. Simpson, *Nucleic Acids Res.*, 1997, **25**, 3751.

9

Ribonuclease P

AGUSTÍN VIOQUE
Universidad de Sevilla, Sevilla, Spain

and

SIDNEY ALTMAN
Yale University, New Haven, CT, USA

9.1 INTRODUCTION

tRNA biosynthesis is a complex process. tRNAs are transcribed as precursors that contain extra sequences at the 3′ and 5′ ends. They can also contain introns. Several enzymes are required to remove the extra sequences at the 5′ and 3′ ends, remove the intron when present, add the 3′-terminal CCA when required, and, finally, introduce the nucleotide modifications that are present in all tRNAs.[18] In eukaryotes, additional steps are required to transport the tRNA from the nucleus to the cytoplasm and to check for accuracy of processing.[102]

5′-end processing occurs by a single endonucleolytic cleavage that removes the leader sequence. This reaction is catalyzed by ribonuclease P (RNase P). RNase P is, therefore, an endonuclease responsible for generating the mature 5′ end of tRNAs. RNase P generates a 5′-phosphate group and a 3′-OH (Figure 1). In this article, we will summarize the properties of RNase P and its catalytic RNA with

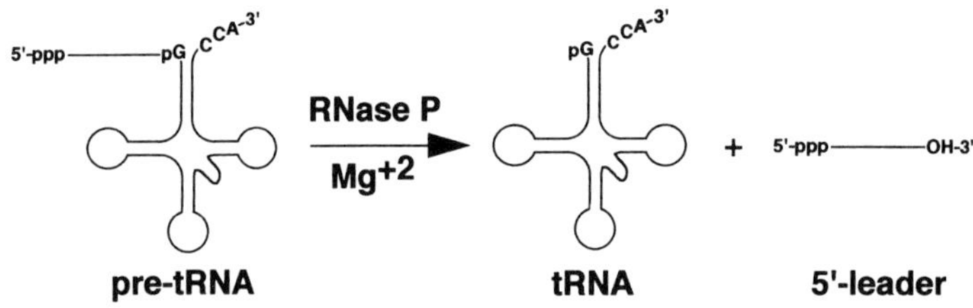

Figure 1. Reaction catalyzed by RNase P. Schematic diagram of the reaction catalyzed by RNase P with a pre-tRNA substrate. RNase P cleavage leaves 3′-OH and 5′-phosphate end groups.

special emphasis on the bacterial system that is the best known. Three main points of interest will be addressed: structure and evolution of the RNA subunit, substrate recognition, and the role of the protein subunit. Several recent reviews have been published in which these and other topics are treated in a more comprehensive way.[3,24]

RNase P activity was demonstrated in extracts of *Escherichia coli* in the early 70s, shortly after a radioactively pure precursor tRNA was available.[80] Very soon after the initial characterization, it was evident that RNase P was an enzyme with peculiar properties due to the presence of an RNA molecule that was essential for activity. The enzyme was a ribonucleoprotein that could be dissociated in separate inactive RNA and protein fractions. An active enzyme was reconstituted by mixing again both fractions.[46] In 1983, it was found that the RNA subunit alone, in the absence of protein, had RNase P activity *in vitro* provided that a high ionic strength and high magnesium concentration were used in the reaction buffer.[30] At that time the RNA subunit of RNase P was the only RNA known to behave as a true enzyme, in the sense that it was not modified in the course of the reaction and each molecule could catalyze more than one round of cleavage (multiple turnover). Another catalytic RNA had been identified, the autocatalytic group I intron,[48] but cleaves itself and therefore is modified in the course of the reaction. Subsequently it was shown that, *in vitro*, the group I intron could also behave as a true enzyme.[104]

Some relevant properties of RNase P from diverse sources are summarized in Table 1. *E. coli* RNase P is composed of a catalytic RNA subunit of 377 nucleotides (M1 RNA) and a protein cofactor of 119 amino acids (C5 protein). The buoyant density of the holoenzyme reflects this composition, being

Table 1. Examples of the diverse types of RNase P.

Organism/organelle	*Buoyant density (g/ml)*	*Micrococcal nuclease sensitivity*	*Protein subunits*	*RNA subunits*	*Active ribozyme*
Bacteria:					
Escherichia coli	1.71 (CsCl)[86]	+	14 kDa (*rnpA*)	377 nt (*rnpB*)	+
Escherichia coli	1.55 ($CsSO_4$)[1]				
Bacillus subtilis	1.7 (CsCl)[25]	+	14 kDa	401	+
Archaea:					
Sulfolobus acidocaldarius	1.27 (Cs_2SO_4)[17]	−	+	315 nt	−
Haloferax volcanii	1.61 (Cs_2SO_4)[54]	+	+	435 nt	(+)[a]
Eukaryotes (nuclear):					
Schizosaccharomyces pombe	1.40 (Cs_2SO_4)[45]	+	100 kDa	285 nt	−
Saccharomyces cerevisiae	n.d.	+	Several proteins from 23 to 100 kDa	369 nt	−
Homo sapiens	1.28 (Cs_2SO_4)[54]	+	Several proteins from 30 kDa to 115 kDa	340 nt	−
Mitochondria:					
Saccharomyces cerevisiae	1.28 (Cs_2SO_4)[63]	+	105 kDa (RPM2)	490 nt (RPM1)	−
Chloroplasts:					
Spinach	1.28 (CsCl)[99]	−	+	−	n.a.
Cyanophora paradoxa	n.d.	+	+	350 nt	−

n.d. = not determined; n.a =. not applicable.
[a] The *Haloferax volcanii* RNase P RNA has a low level of activity under extreme ionic conditions. References to buoyant density calculations are given; see text for references to other data in the table.

intermediate in CsCl or $CsSO_4$ between that of pure RNA and pure protein. A similar enzyme structure is found in all Bacteria, with an RNA subunit that ranges in size from 300 to 450 nucleotides and a small basic protein of around 120 amino acids. The RNA subunit of bacterial RNase P is the only one that has been shown to be catalytic *in vitro* in the absence of protein. The enzyme in Archaea, the eukaryotic nucleus and mitochondria also contains an RNA subunit. Except for some recent observations of catalytic activity of the RNase P RNA of some Archaea[70], the enzyme from these sources is not active *in vitro* in the absence of protein. These enzymes are larger than the bacterial RNase P, with a more complex protein composition. The *Sulfolobus* enzyme has a low buoyant density, close to that of pure protein and is not sensitive to micrococcal nuclease. These data suggested initially that RNase P from *Sulfolobus* might lack an RNA subunit.[17] However, it was later identified as a 315 nt RNA.[52] Although the protein subunits of *Sulfolobus* RNase P have not been characterized, it is implied that this enzyme has a high proportion of protein that protects the RNA from micrococcal nuclease action. The nuclear enzyme from eukaryotes has also a higher protein content. The best characterized is the *Saccharomyces cerevisiae* RNase P: nine proteins have been identified as subunits of the enzyme.[10] The human enzyme has also a complex protein composition and at least eight subunits have been identified.[38,39] The situation in chloroplasts is very interesting because there may be two structurally different types of RNase P, depending on the type of chloroplast.

RNase P from higher plant chloroplasts seems to lack RNA. The enzyme is resistant to micrococcal nuclease and no RNA subunit has yet been detected in highly purified preparations of the enzyme.[99] Furthermore, differences in susceptibility to thiosubstitution at the scissile bond indicate that the higher plant chloroplast enzyme has a catalytic mechanism different from other RNase P enzymes.[95] The *E. coli* RNase P RNA absolutely requires Mg^{2+} or Mn^{2+} coordinated to the *pro-*R_P oxygen of the scissile phosphodiester bond for cleavage.[14] When the *pro-*R_P non-bridging oxygen is replaced with sulphur, which binds Mg^{2+} about 30,000-fold less tightly than oxygen, cleavage occurred with a rate constant reduced 15,000 to 20,000 with respect to the unsubstituted pre-tRNA. In contrast, chloroplast RNase P is little affected by the R_P-thiosubstitution at the scissile bond. This experiment suggests that in the higher plant chloroplast RNase P, there is no Mg^{2+} ion coordinated to the *pro-*R_P oxygen of the scissile bond and positioning of the attacking hydroxide is accomplished by an amino acid side chain.

A second class of chloroplast RNase P is present in the green algae *Nephroselmis olivacea*[96], the red algae *Porphyra purpurea*[79] and the Glaucocystophyceae algae *Cyanophora paradoxa*.[82] The genome of those chloroplasts code for an RNase P RNA similar to the bacterial RNA. Only the *Cyanophora paradoxa* enzyme has been studied biochemically.[4] It is sensitive to micrococcal nuclease and the cyanobacterial-like RNase P RNA copurifies with the enzyme. In spite of its similarity to bacterial RNAse P RNA, the *Cyanophora* RNA is not catalytically active *in vitro* in the absence of protein. However, a functional enzyme can be reconstituted with this RNA and the protein subunit from a cyanobacterium RNase P.[72] Chloroplasts originated from cyanobacteria through endosymbiosis. There are three lines of descent from the original endosymbiosis event represented by green algae and plants, red algae, and Glaucocystophyceae, respectively. While it is possible that in the lineage of plant chloroplasts the cyanobacterial RNase P has been replaced by an all-protein enzyme, some algae have retained the cyanobacterial-like RNA, although this RNA has no catalytic ability. A complete understanding of the evolution of RNase P in plastids would require analysis of a larger sample of the algal radiation and identification of both the protein and RNA component(s).

9.2 STRUCTURE AND EVOLUTION OF THE RNA SUBUNIT

The RNA subunit of RNase P from Bacteria has been characterized in many strains. The RNase P database,[7] maintained by Dr. James Brown (North Carolina State University; URL: http://www.mbio.ncsu.edu/RnaseP/home.html), contains almost 200 sequences of bacterial RNase P RNAs in addition to representatives of all other main phylogenetic groups. Once the first RNase P RNA genes were identified, additional ones were isolated by heterologous hybridization, analysis of genome databases and PCR amplification. PCR has been particularly helpful in the rapid identification of many RNase P genes, even from natural populations without prior cultivation of the organisms.[8] This large data set has allowed a very precise definition of the secondary structure of this RNA through a comparative approach. Secondary structure elements are identified through the analysis of compensatory base changes in the sequence of homologous RNAs. In this way, basically all the base pairs present in the secondary structure

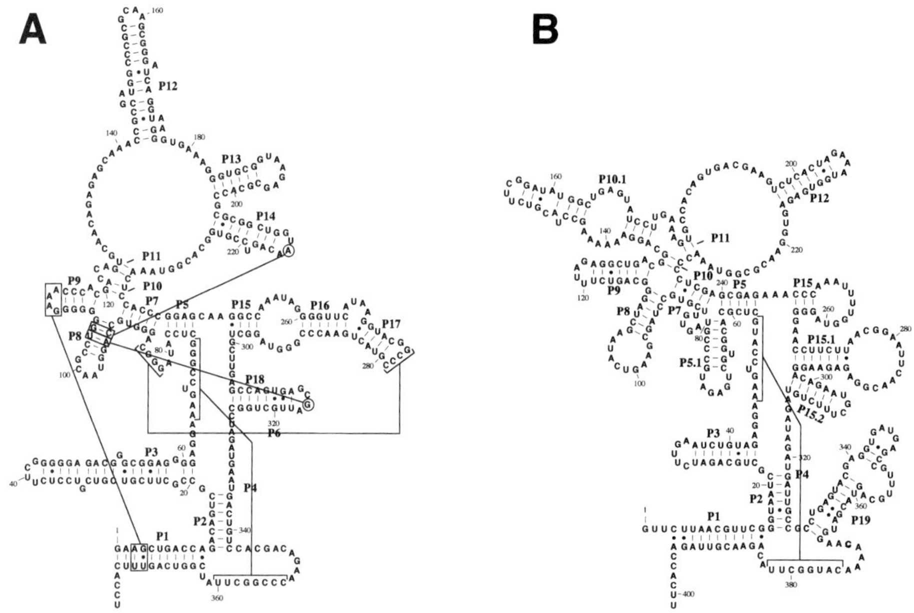

Figure 2. Secondary structure of RNase P RNA from *E. coli* (A) and *B. subtilis* (B). The structures shown are adapted from the RNase P database.[7] Tertiary interactions and pseudoknots are shown by lines connecting the interacting regions.

of RNase P RNA have been proved. The analysis of covariation has allowed also the identification of several possible tertiary interactions[8,61] and some of these predictions have been verified by functional tests *in vivo*.[76] Tests of function *in vitro* are not correlated as well with the phylogenetic predictions.

The proposed secondary structures of *E. coli* and *Bacillus subtilis* RNase P RNA are shown in Figure 2. RNase P RNA can be divided into two distinct structural classes: type A, represented by *E. coli*, which is the ancestral type found in most bacteria, and type B, represented by *B. subtilis*, which is found in the low GC content Gram-positive bacteria. However, the different secondary structure organization of part of the RNA structure in type B RNAs can be modelled in a three-dimensional structure equivalent to the alternative structure in type A RNAs.[31] This illustrates a common phenomenon in RNA architecture, where different, non-homologous structural elements are used for long-range interactions that result in functionally equivalent structures.

When the secondary structures of bacterial RNase P RNAs are compared, it is evident that they have both a conserved core structure and variable elements. Some helices are restricted to certain groups of bacteria. The length of particular helices can also be highly variable. The "optional" variable structures are unlikely to be involved in the basic functions such as substrate binding and catalysis. A synthetic RNA molecule that contains only the common structural elements, a "minimal RNA" containing only 211 nucleotides, corresponding to the core conserved elements, has been shown to have catalytic activity *in vitro*[83,100] although greatly reduced in cleavage ability compared to RNase P RNAs found in nature. This minimal RNA requires very high ionic strength for activity. Actually, the RNase P RNA from *Mycoplasma fermentans* is the smallest natural RNase P RNA known with activity and is similar in structure to the synthetic minimal RNA.[83] Figure 3 shows a diagram of the universally conserved sequence and structure of bacterial RNase P.

Progress in understanding the function of RNase P RNA requires knowledge of its three-dimensional structure. RNase P RNA has not been crystallized yet. No RNA of this size has been crystallized but the objective is probably within reach of current technology as some large ribozymes, such as the 247 nucleotide catalytic core of the *Tetrahymena* intron, has been crystallized recently and the structure solved at 5 Å resolution.[27] The structure of a small, autonomously folding domain of M1 RNA, comprising helices P15 and P16 has been solved by nuclear magnetic resonance and computer modelling.[19,26] This domain is critically involved in substrate interactions and will be discussed below.

We mentioned above that analysis of covariation can provide evidence of interactions between helices. In addition, site-specific crosslinking with photoactivable crosslinking reagents can be used to identify interacting regions of the molecule placed at long distances from each other in the secondary

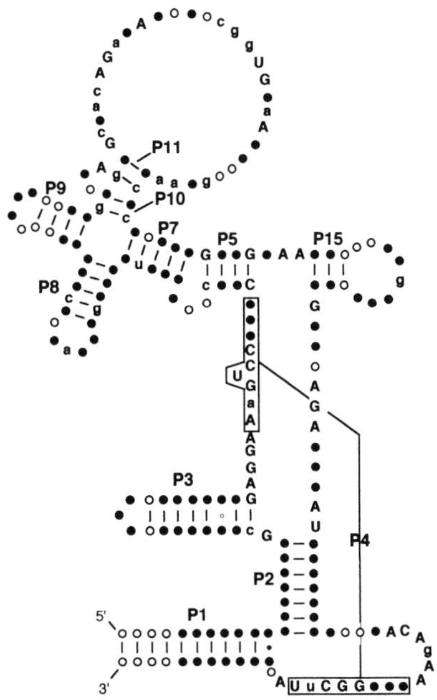

Figure 3. Minimum bacterial consensus RNase P RNA as determined by phylogenetic analysis. Only the structural elements conserved in all known bacterial RNase P RNAs are present in the structure shown. Universally conserved nucleotides are in uppercase letters; nucleotides conserved in at least 80% of RNAs are in lowercase letters; nucleotides universally present although not conserved in identity are shown as filled circles; nucleotides present in at least 80% of RNAs are indicated by open circles. Adapted from Siegel *et al.*[83].

structure. This technique requires the introduction of a photoactive agent at specific positions in RNase P RNA or pre-tRNA. The RNA–RNA complex is irradiated with UV light and the position of the crosslink identified by primer extension. The introduction of the photoactive agent at specific positions is facilitated because the RNase P RNA can be circularly permutated without loss of activity.[23] Therefore, the 5′-end of the RNA can be placed at different specific positions facilitating the incorporation of the photoactive agent. These kinds of data provide a set of constraints on distances that can be used to model the RNA structure. In addition, binding of pre-tRNA can also be modelled on the RNase P RNA structure using information provided by crosslinking and footprinting experiments (see below). The reliability of this approach is supported by the fact that two recent models developed independently[12,60] have a similar overall architecture for the RNase P RNA of *E. coli* and *B. subtilis*, although they differ in the details. Figure 4 shows these two models of the structure of *E. coli* RNase P RNA bound to substrate. The compactness of the structure is achieved by GNRA tetraloop–helix (i.e. L14–P8; L18–P8; L9–P1), loop–loop contacts and coaxial stacking of helical stems. These long-range RNA–RNA interactions, as well as the hierarchical assembly of the RNA, are reminiscent of similar interactions demonstrated in other highly organized RNAs such as group I introns. In particular, interactions between GNRA tetraloops and the shallow groove of RNA helices are found in all three classes of large catalytic RNAs. All the conserved GNRA tetraloops in *E. coli* RNase P RNA are involved in such interactions. The proposed, low resolution, structures for RNase P RNA, shown in Figure 4, are useful as a framework for the design of experiments to test different hypotheses about interactions within the RNase P RNA and its interactions with the substrate.

9.2.1 Conserved Domains

The most highly conserved part of bacterial RNase P RNA is helix P4. This helix is universal and, in contrast with other helices where the sequence differs but the structure is maintained by compensatory substitutions, its sequence is mostly invariant. A homolog to helix P4 has been identified even in archaeal and eukaryotic enzymes.[13] These data point to an important function for this helix. Base substitutions in P4 are lethal for activity even when the helical structure is not affected. An experiment in which the sequence of helix P4 was randomized, and then catalytically active RNA molecules were selected

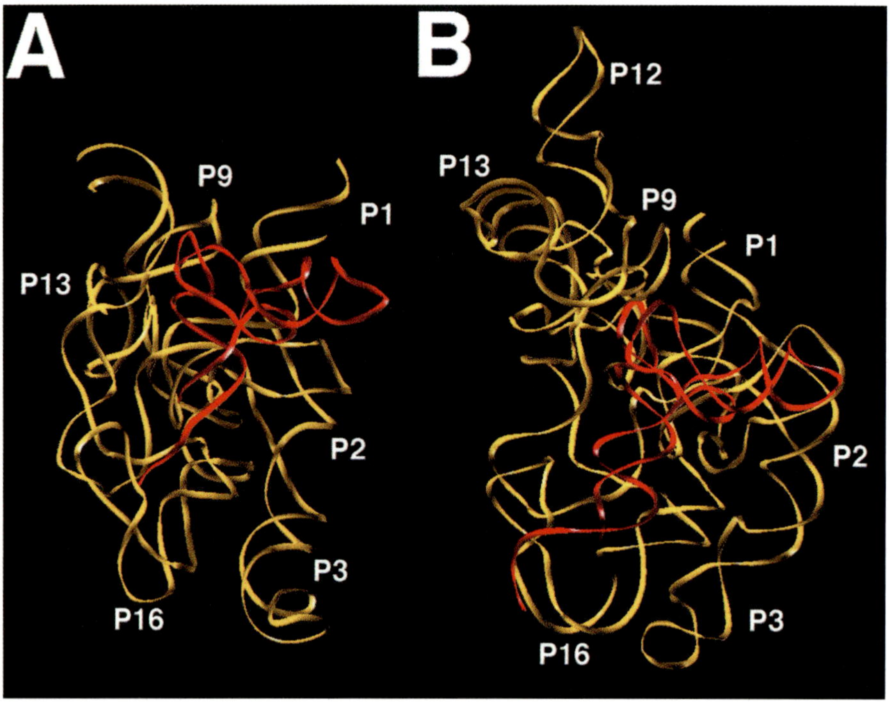

Figure 4. Three-dimensional models of RNase P RNA from *E. coli*. Ribbon representations of the model of Chen *et al.*[12] (A) and Massire *et al.*[60] (B) of RNase P RNA (yellow) complexed with a tRNA (red). The structures were drawn with SYBYL on a SGI Workstation with roughly the same orientation.

from the random sequence pool, resulted in the wild-type sequence in all the selected molecules.[22] This experiment suggests that the nucleotide sequence of P4 is optimal for catalytic efficiency and changes in that sequence, even if helix structure is maintained, are detrimental for activity. A number of biochemical experiments indicate that P4 contains the active site nucleotides. Four phosphate oxygens were identified in *E. coli* RNase P RNA where substitution by sulphur reduces substantially the catalytic rate.[36] Those phosphate oxygens are placed 5′ to nucleotides A67, G68, U69 and A352. These four nucleotides are adjacent or within helix P4 (Figure 2) and are universally conserved. It is assumed that substitution by sulphur inhibits binding of magnesium ions without perturbing secondary structure. Therefore, the phosphate oxygens identified are probably involved in binding of the magnesium ions required for catalysis. In the three-dimensional models of RNase P RNA, helix P4 is placed within the core of the structure in close proximity to the cleavable bond in pre-tRNA (Figure 4).

All extant RNase P RNAs have a common ancestry: therefore, they should share some basic structural elements. However, it is difficult to align RNase P RNAs from Bacteria, Archaea, and Eucarya, because of the low sequence conservation. Despite the low sequence conservation, it has been possible to identify several conserved regions and helical elements present in all these RNAs.[13] The recognition of these universally conserved elements allows also a definition of a structure for the eucaryal RNA similar to the minimal bacterial consensus (Figure 3). As expected, two of the conserved sequence elements correspond to helix P4 and adjacent sequences, which, as mentioned above, is an essential part of the catalytic center of the bacterial enzyme.

The fact that there is a common consensus structural basis for all RNAse P RNAs raises the question of why only the bacterial RNAs (and some archaeal RNAs) are catalytically active in the absence of protein. The catalytic RNase P RNAs must obviously be able to fold into a functional conformation in the absence of protein. Elements in the RNA sequence provide all the requirements for folding. It is possible that the non-catalytic RNase P RNAs are missing some structural elements (e.g. the internal loop joining P15 and P16) necessary to make the required molecular contacts and cannot fold properly in an active conformation. In these cases, the protein(s) subunit would replace the missing structural elements in the RNAs.

Optimal catalytic activity of the RNA-alone reaction requires high ionic strength in the absence of protein. The high concentration of ions facilitates folding of the RNA. The *Mycoplasma fermentans*

RNase P RNA requires higher ionic strength than the *E. coli* RNase P RNA for optimal activity.[83] We can presume that because the *Mycoplasma* RNA is missing several of the secondary structure elements found in *E. coli*, its folding is less stable and, therefore, is more dependent on the high ionic strength.

9.3 SUBSTRATE RECOGNITION AND CATALYTIC MECHANISM

In *E. coli*, RNase P processes not only pre-tRNAs but also a number of other substrates (Figure 5) such as the precursors to 4.5S RNA [5] and tmRNA,[47] and some small phage RNAs.[37] The polycistronic *his* operon mRNA has a cleavage site for RNase P but this site is only accessible after a previous cleavage of the mRNA by RNase E.[2] It is possible that there are other natural substrates for bacterial RNase P yet to be identified. Those hypothetical substrates might be difficult to identify if they are in low abundance, are present only under specific metabolic conditions, or are cryptic and only accessible to RNase P after some previous modification of the RNA. The construction of strains in which expression of RNase P can be shut down under controlled experimental conditions will help in identifying new substrates.

In contrast to bacteria, in human cells only pre-tRNAs have been identified as substrates for RNase P. Although human RNase P appears as part of a complex with RNase MRP, an enzyme that is involved in the cleavage of pre-rRNAs,[39,58] highly purified RNase P cannot carry out the cleavage of pre-rRNA. In yeast, however, there is both genetic and biochemical evidence that, in addition to pre-tRNAs, RNase P subunits, which may be shared with RNase MRP, are necessary for the processing of pre-rRNA.[11,88]

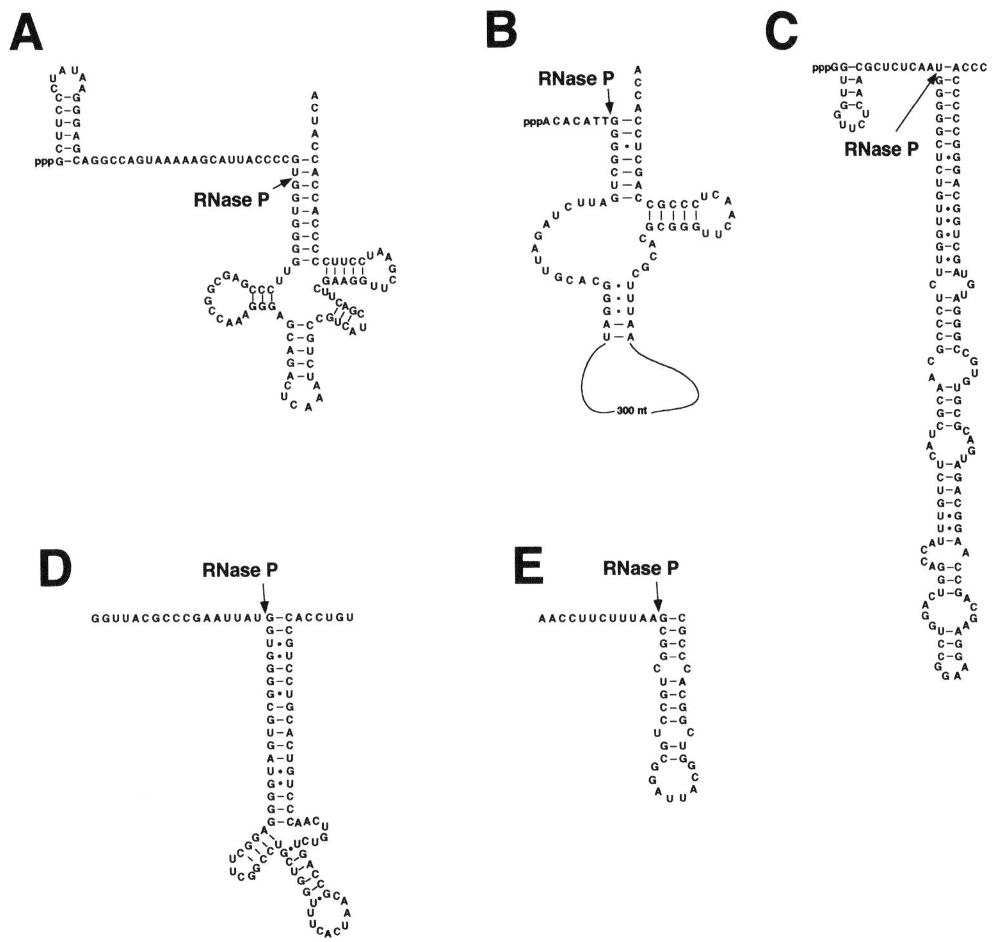

Figure 5. Natural substrates for RNase P from Bacteria. Secondary structures of several substrates are shown. (A) Precursor to $tRNA^{Tyr}$. (B) Precursor to tmRNA. (C) Precursor to 4.5S RNA. (D) Precursor to phages P1 and P7 antisense C4 RNA. (E) *his* operon mRNA.

How can RNase P recognize so many different substrates and cleave them efficiently? A related question concerns the ability of the enzyme to identify the correct cleavage site in so many different substrates with different sequences. This question is of crucial importance because cleavage at the incorrect position in a pre-tRNA would result in an inactive tRNA. In pre-tRNAs there is no conservation in sequence or structure of the 5′ leader but mature tRNA contains some short conserved sequences, such as GTψC. However, although the RNase P RNA contains sequences complementary to this tetranucleotide conserved in the substrates, they are not involved in substrate recognition through base pairing. At present, it is believed that RNase P (with the exception of the 3′-terminal CCA sequence) does not recognize specific nucleotides sequences in its substrates. Rather, it recognizes the overall structure of substrates through multiple contacts.

9.3.1 Role of the 3′-Terminal Sequence in Substrate Recognition

The 3′-terminal CCA sequence in the substrate is an important determinant in cleavage efficiency by *E. coli* RNase P RNA. In the presence of the protein subunit, the preference for CCA-containing substrates is lower. The function of the 3′-terminal CCA sequence in enzyme–substrate interactions has been studied by the standard tools of the enzymologist. The kinetic parameters of the reaction are affected by the presence of the CCA in the sense indicated above: a substrate lacking the 3′-terminal CCA has a much higher K_m than a CCA-containing substrate (Table 2). The affinity of different substrates for the RNase P RNA can be studied by analyzing the complexes on non-denaturing acrylamide gels, where RNA–RNA complexes migrate more slowly than free RNA. By quantitating the amount of pre-tRNA present in the complex when different concentrations of RNA are used, it is possible to estimate the RNA–RNA affinity. This kind of experiment has shown that, at least in *E. coli* and *T. thermophilus*, a CCA-lacking substrate has a much lower affinity for the RNase P RNA than a CCA-containing substrate.[34,35]

One of the most conserved features of bacterial RNase P RNA is a GGU sequence in the 3′ part of the loop connecting helix P15 and helix P16 (nucleotides 292 to 294 in the *E. coli* sequence) in type A RNAse P RNA. In the equivalent structure in type B RNase P RNA (the terminal loop of P15) there is also a conserved GGU sequence (nucleotides 258 to 260 in the *B. subtilis* sequence). The 3′-terminal CCA in pre-tRNAs has been shown to interact with this internal loop in RNase P RNA by the technique of chemical footprinting.[53] In this technique, the RNase P RNA is chemically modified in the absence or the presence of pre-tRNA and the bases that are protected from modification when pre-tRNA is bound are identified. The bases protected from modification are potential pre-tRNA contacts sites. The chemical reagents used are DMS, which methylates N7 of guanine, N1 of adenine and N3 of cytosine, and kethoxal, which modifies N1 and N2 of guanine. All the modifications, excepting N7 of guanine, can be detected by primer extension because reverse transcriptase will stop transcription one base before the modified base. Several nucleotides of the P15–P16 internal loop are protected by pre-tRNA that contains 3′-terminal CCA but not by pre-tRNA lacking the 3′-terminal CCA. The protected nucleotides include the conserved G292 and G293 (*E. coli* numbering), as well as A254, and A255. An interesting result of this study is that equivalent protected sites were found in the RNase P RNA from two additional bacteria,

Table 2. Kinetic parameters of the reaction of *E. coli* RNase P with various substrates.

Substrate [a]	*Enzyme*	K_m *(nM)*	k_{cat} *(min^{-1})*	k_{cat}/K_m *(min^{-1}/mM)*
pTyr	RNA alone	33	0.4	12
	Holoenzyme	29	76	260
pTyr-CCA	RNA alone	500	3	6
	Holoenzyme	200	27	140
p4.5S	RNA alone	11,500	1	0.1
	Holoenzyme	150	60	403
p10AT1	RNA alone	1,460	7	4.7
	Holoenzyme	1010	32	32

[a] pTyr (pre-tRNATyr) terminates in ACCAUCA and is a natural substrate (Figure 5). pTyr-CCA is missing the terminal CCAUCA sequence. p4.5S is the precursor of 4.5S RNA (Figure 5). p10AT1 is like the model susbtrate pAT1 (Figure 7) but with a 5′-leader of 10 nucleotides. Kinetic data taken from Altman and Kirsebom.[3]

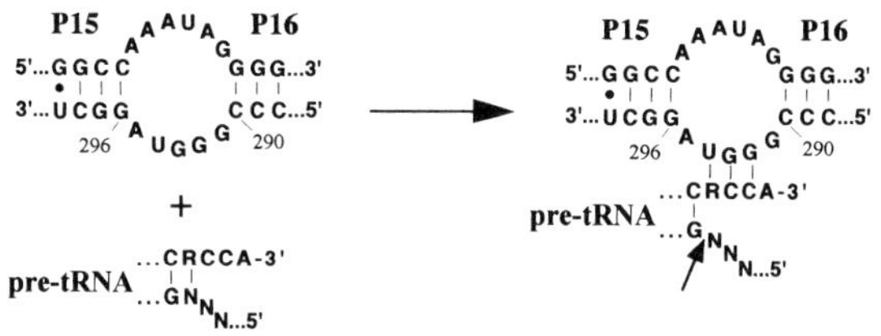

Figure 6. Diagram of the proposed interaction between the 3′-terminal RCCA sequence of pre-tRNAs and RNase P RNA. The RCCA sequence in the substrate base pairs with a highly conserved GGU sequence in the loop connecting helices P15 and P16. This interaction facilitates the partial denaturation of the acceptor stem of the pre-tRNA. Adapted from Kirsebom and Svärd.[43]

Chromatium vinosum and *B. subtilis*. This suggests that the interaction between 3′-terminal CCA and the P15–P16 loop is not an idiosyncratic property of *E. coli* RNase P, but is rather a general property of enzyme–substrate interactions in RNase P. The interaction between 3′-terminal CCA in the substrate and the P15–P16 loop has been further demonstrated by photoaffinity crosslinking.[68] Photoagent-labelled tRNAs with 3′-terminal CCA, CC or C, or lacking all these nucleotides, were mixed with RNase P RNA and irradiated and the sites of crosslinking identified by primer extension. The crosslinking results are fully in agreement with the footprinting experiments described above. Again, similar results were obtained with diverse RNase P RNAs (from *E. coli*, *B. subtilis*, and *Thermotoga maritima*), supporting a general role for CCA interactions with the P15–P16 loop in enzyme–substrate recognition by RNase P.

The interaction between the conserved GGU sequence and the 3′-terminal RCCA sequence (residues that interact with GGU are underlined) in pre-tRNAs is mediated through base pairing between the two Cs of the CCA and the conserved G292 and G293 of RNase P RNA (Figure 6). In addition, the discriminator base (the residue at position +73 in a tRNA precursor, generally a purine) interacts with U294. These interactions would result in partial denaturation of the acceptor stem, facilitating cleavage by RNase P (Figure 6). The base pairing between the conserved GGU sequence in loop P15–P16 and the 3′-terminal RCCA in the substrate has been demonstrated by characterizing the effect of compensatory mutations in both sequences. The use of compensatory mutations is a powerful tool to investigate intra- or intermolecular interactions in nucleic acids. The disruption of a Watson–Crick interaction by the introduction of a mutation in one strand of the helix results in loss or alteration of the activity under study if the interaction is functionally relevant. The loss of activity can also be explained if the identity of the mutated nucleotide itself is important for function, rather than the helical structure in which the wild-type nucleotide participates. However, if a second site mutation in the other strand of the helix ("compensatory mutation") that restores the potential for Watson–Crick interaction also restores the activity under analysis, it is considered strong evidence for functionally relevant base pairing. This approach has been applied to analysis of the interaction between the conserved GGU sequence in loop P15–P16 and the 3′-terminal RCCA.[43,89] When the 3′-terminal RCCA in the substrate was mutated to RGCA, it was miscleaved with high frequency at the −1 position rather than at the correct +1 position by wild-type RNase P RNA. However, if G293 in the RNase P RNA was changed to C, cleavage at the correct position was restored. This result strongly argues for a specific interaction between the first C in the RCCA sequence and G293 in the RNase P RNA. Similar experiments provide support for an interaction between the second C and G292 and between the discriminator purine and U294. What is more, the kinetics of cleavage are altered in a predictable way in the different mutants. When the predicted base pairing is absent, K_m increases and k_{cat} is reduced. When base pairing is restored by compensatory mutations, K_m and k_{cat} return to values close to the originals. The same qualitative results are obtained with RNase P RNA from *Mycoplasma hyopneumoniae* and *Mycobacterium tuberculosis*, arguing again for the generality of this interaction.[89] Substrates lacking the 3′-terminal CCA are still cleaved correctly. This implies that additional interactions are important to dock the substrate with the enzyme and for localization of the cleavage site.

9.3.2 Other Contact Points in Substrates

The T-loop in the pre-tRNA is also a recognition motif used by RNase P RNA. This has been shown by crosslinking, mutagenesis and *in vitro* selection experiments. The distance from the T-loop to the cleavage site is normally 12 bp. It has been suggested that RNase P measures this distance to localize the cleavage site. The enzyme would presumably rely on both the CCA interaction and the T-loop interaction to place the substrate in the correct orientation for cleavage at the +1 position. In support of this mechanism are studies with model substrates in which extra base pairs are added or deleted from the aminoacyl and or the T stem. Addition of 1 or 2 bp in the amino acyl and/or T stem has no effect on the precision of cleavage.[90] When 3 bp are introduced cleavage occurs at positions +1 and +4 if the substrate contains the CCA sequence or exclusively at the +4 position when the substrate lacks the CCA sequence. One must assume that there is a complementary surface or measuring device on M1 RNA to which the important contacts in the substrate bind. The elongated stack formed by helices P5 and P7 appear to be the only candidates for such a surface because it parallels the length of a pre-tRNA contact surface. However, when base pairs have been deleted from these helices, no change in cleavage precision occurs (C. Guerrier-Takada and S. Altman, unpublished).

A systematic analysis of the requirements in the substrate for its recognition by RNase P through mutagenesis, deletions, etc., has provided a wealth of data on the important elements for substrate binding to the enzyme and catalysis. Mutations that disturb the secondary and tertiary structure of the tRNA moiety of the substrate alter the K_m of the reaction. With the technique of photoaffinity labelling and crosslinking, it has been possible to identify interactions between RNase P RNA and pre-tRNA. By using a short-range crosslinking agent, nucleotides adjacent to the cleavage site have been identified.[15] Not surprisingly, those nucleotides are within the catalytic core of the enzyme. The conserved guanosine at the 5′ end of tRNA is adjacent to nucleotides in the short, single-stranded region connecting loops P5 and P15. The nucleotide at position −1 of the pre-tRNA is adjacent to nucleotides in the single-stranded region connecting loops P18 and P2 (Figure 2). Some of the most conserved nucleotides in bacterial RNase P RNA are found in this region. Recent experiments[77] with 4-thiouridine substituted RNase P RNA and pre-tRNAs that have the natural or shortened 3′ termini verify that the internal loop joining P15 and P16 is a critical contact point in M1 RNA and also that multiple binding modes exist for each substrate. Only one of these modes is enhanced by the protein subunit. It is apparent that these multiple binding modes endow RNase P with a versatility that enables it to cleave a variety of substrates.

9.3.3 Model Substrates

In addition to the natural substrates mentioned above (Figure 5), *E. coli* RNase P can process highly reduced model substrates (Figure 7(A)). pAT1[62] is a simple stem–loop analog to the T stem–loop and aminoacyl stem of tRNAs. What is more, there is no need of covalent continuity in the RNA to be cleaved as far as the minimal structural requirements are met (Figure 7(B)). Therefore, the 3′-proximal sequence can be regarded as an "external guide sequence" (EGS) that by base pairing with the 5′-proximal sequence generates a cleavable substrate.[21] This is analogous to the catalytic mechanism of self-cleaving introns, where the phosphodiester bond to be cleaved is placed in the correct position for cleavage by an internal guide sequence. In this case the guide sequence is part of the catalytic intron sequence, while EGS are external. In principle, any RNA could be targeted by a specifically designed EGS for cleavage by RNase P.

The fact that a model substrate that is an analog of the T and aminoacyl stem (pAT1) can be used as substrate by bacterial RNase P, is in agreement with the known requirements for substrate binding, some of which have been described above. The substrate recognition process can be summarized in the following way. RNase P recognizes the conformation of the tRNA domain of pre-tRNA and interacts with nucleotides in the T loop, in the T stem, and in the amino acid acceptor stem. The 3′-terminal RCCA of the precursor establishes a Watson–Crick interaction with the conserved GGU sequence in the P15–P16 internal loop, resulting in partial denaturation of the amino acid acceptor stem, which makes the cleavage site more accessible to the active site. The cleavage site is presumably identified by measuring the distance between the end of the T stem and the cleavage site. Therefore, there have to be multiple determinants for enzyme–substrate binding and cleavage but the role and significance of different determinants might not be the same for all substrates. In addition, although many of the interactions described above have been generalized to several unrelated Bacteria, they might be significant differences in some cases. For instance, although highly conserved, the relevant GGU in RNase P RNA sequence is absent in some bacteria. Most of the RNase P RNAs from cyanobacteria have

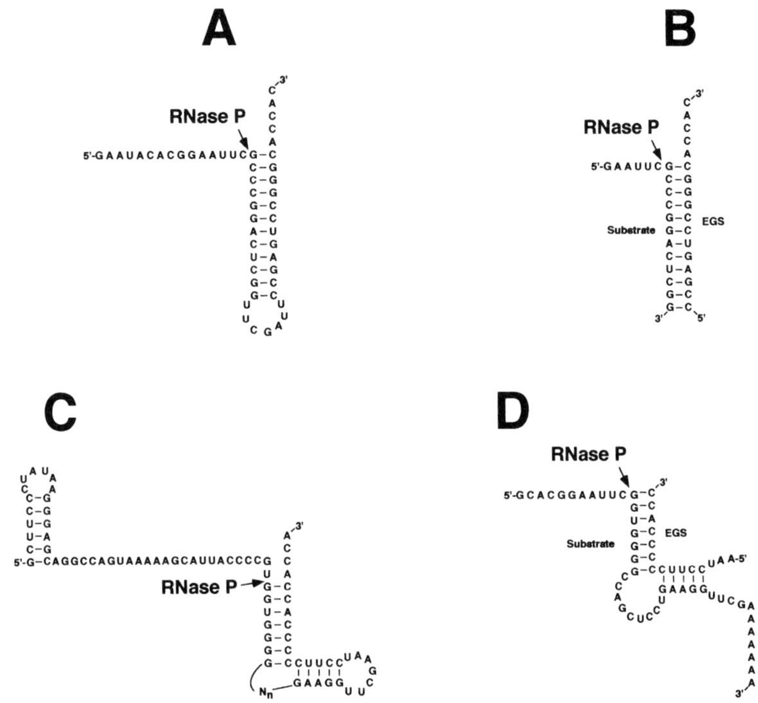

Figure 7. Model substrates for RNase P. (A) pAT1, a small model substrate for *E. coli* RNase P;[62] (B) Target RNA–EGS (external guide sequence) complex for *E. coli* RNase P;[21] (C) Small model substrate for human RNase P;[9,103] (D) Target RNA–EGS complex for human RNase P.[103]

a different structure for this loop and many of them lack the GGU sequence. In some instances there is an helix inserted in the loop.[20,97] It is interesting that tRNA genes in cyanobacteria do not encode the CCA, suggesting an evolutionary correlation between the structure of the P15–P16 internal loop and the primary sequence of the tRNA gene. Therefore, it is possible that RNase P from cyanobacteria interact with their substrate in a way that does not involve the 3′-terminal CCA.[73]

Even more subtle differences can be detected in some bacteria. Nucleotides at position −1 and −2 with respect to the cleavage site are important determinants for cleavage efficiency and accuracy.[44,49,74] *E. coli* RNase P RNA cleaves at the same rate an *E. coli* pretRNA^{Ser} carrying −2A and −2G. However, the *Mycoplasma hyopneumoniae* RNase P RNA (a type B RNA) cleaves at a much reduced rate the −2A precursor.[6] From several of the fully sequenced bacterial genomes it can be deduced that low G + C content Gram-positive bacteria (which contain a type B RNase P RNA) have between 3% and 9% of G at the −2 position in the pre-tRNA genes, while in *E. coli* 28% of the tRNA genes carry a G at the −2 position. This information suggests a correlation between the primary sequence of pre-tRNA genes and RNase P cleavage.

9.3.4 Substrates for Human RNase P

Among eukaryotes, human RNase P has been the best characterized in relation to substrate recognition. Human RNase P has more strict requirements than the bacterial enzyme. For instance, human RNase P cannot cleave minimal model substrates for the bacterial enzyme, such as pAT1 (Figure 7). The minimal substrate requirements for human RNAse P have been determined by systematic deletion of structural elements of the pre-tRNA and by mutagenesis.[9,103] Substrates similar to pAT1 are functional provided that they have a bulge of one to nine nucleotides (Figure 7(C)). The requirement for at least one bulged nucleotide might be important for interaction with the enzyme. A single nucleotide bulge in an RNA helix is a common motif for protein binding in several RNAs. Therefore some of the protein subunits of human RNase P could be involved in substrate recognition through the bulged nucleotide, explaining why the human RNase P is strictly dependent on protein for its function. In contrast with the more strict requirements by the human RNase P, the 3′-terminal CCA, that we have seen is a very relevant recognition element in Bacteria, is not required for substrate recognition by human RNase P. In eukaryotes, the 3′-terminal CCA is not encoded in the gene and is added post-transcriptionally after processing of the tRNA precursor at the 5′ and 3′ ends. Therefore, the RNase P natural substrate lacks CCA.

Similarly to the bacterial system, human RNase P can use an external guide sequence that upon binding to a target RNA, generates a functional substrate, provided that the target RNA–EGS fulfils the minimal requirement of a bulged stem–loop (Figure 7(D)). The promise of using EGS to target an mRNA in human cells for cleavage by RNase P raises the possibility of medical and biotechnological applications of this system.

9.3.5 Chemistry of the Cleavage Reaction

RNase P requires divalent ions for catalysis of which the most efficient is Mg^{2+}. Multiple Mg^{2+} ions are required for optimal activity, and catalysis exhibit a cooperative dependence on Mg^{2+} concentration.[84] Magnesium ion stabilizes the folded RNase P RNA structure, increases enzyme–substrate affinity, and directly aids catalysis through stabilization of the transition state. RNase P RNA can fold and bind substrate in the absence of Mg^{2+} ions with reduced efficiency but there is an absolute requirement of the ion for cleavage. We have mentioned in the introduction that a Mg^{2+} ion is coordinated with a phosphate oxygen at the cleavage site. There are several high affinity cation binding sites in the RNAse P RNA. One of them is in the crucial P15–P16 internal loop.[50] Mg^{2+} ion(s) bound to this site could contribute to catalysis. The pre-tRNA substrate also contributes to binding of Mg^{2+} ions.[67] Mutant pre-tRNAs in the 3′-terminal CCA sequence have weaker affinity for magnesium and reduced rate for the chemical step of cleavage (k_2) correlated with the reduced magnesium affinity. High Mg^{2+} concentrations partially rescued the k_2 defect in the mutants. One can conclude that the 3′-CCA of pre-tRNA participates in binding Mg^{2+} ions that are essential for catalysis. Two or three magnesium ions are required for catalysis and the 2′-OH group of the cleaved phosphodiester bond as well as the *pro*-R_p oxygen at the cleavage site are involved in binding of Mg^{+2} ions.[74,84] The current knowledge of the catalytic mechanism of RNase P can be summarized as follows. The mechanism is, similar to that of other protein RNases, a nucleophilic substitution of an S_N-2 in-line type. In this mechanism there is a trigonal bipyramidal transition state which is stabilized by coordination to Mg^{2+}. The attacking hydroxide is stabilized by interaction with Mg^{2+}. This interaction could be a direct coordination with Mg^{2+} or through hydrogen bonds with hydrated Mg^{2+}.

9.4 THE ROLE OF THE PROTEIN SUBUNIT(S)

In Bacteria, RNase P contains a single protein subunit of about 120 amino acids. A gene (*rnpA*) coding for this protein, which was originally identified in *E. coli*, was recognized thanks to its ability to complement the thermosensitive phenotype of a strain with a mutation in the protein subunit.[33] This thermosensitive strain (A49), originally isolated as a chemically induced mutant with a defect in pre-tRNA processing,[81] carries a point mutation that changes an arginine to an histidine in the protein sequence.[42] Biochemical and genetic studies suggest that the mutant protein is defective in assembly of the holoenzyme, because the thermosensitive phenotype can be suppressed by increasing the intracellular concentration of either the RNA subunit or the (mutant) protein subunit. The *rnpA* gene is located a very short distance downstream from the *rpmH* gene, which codes for ribosomal protein, L34.

The *rnpA* gene from several Bacteria has been identified. In several cases, this was made possible by conservation of genome organization. In many Bacteria the *rpmH–rnpA* operon is close to the well-studied *dnaA* gene. Heterologous complementation of the A49 mutant with expression libraries has also been a useful tool. This approach works because it is generally possible to reconstitute a functional bacterial holoenzyme with subunits from different species.[30,64,71] Finally, the ongoing bacterial genome sequencing projects are also providing a number of RNase P protein sequences. All the *rnpA* genes identified to date are downstream but very close to the *rpmH* and start with a GTG codon.

The sequences of the RNase P protein subunit are poorly conserved among Bacteria. However, they have a number of common properties. First, there is a conserved basic sequence motif essentially shared by all of them. This conserved sequence is called the RNR motif. The consensus RNR motif can be defined as K-X_{4-5}-A-X_2-R-N-X_2-(K,R)-R-X_2-(R,K). Other conserved residues are some aromatic amino acids close to the amino end of the protein.

9.4.1 Crystal Structure of RNase P Protein from *B. subtilis*

The function of the protein and the role of conserved amino acid residues has been studied extensively by mutagenesis and biochemical characterization of the mutant protein.[28] Recently, new light has been shed on the question with the determination of the crystal structure of the RNase P protein from *B. subtilis* at 2.6 Å resolution.[85] The protein contains three α helices and four β strands with a topology $\alpha_A\beta_1\beta_2\beta_3\alpha_B\beta_4\alpha_C$ (Figure 8). Three possible RNA binding regions are identified in the structure. The first RNA binding region comprises the conserved RNR motif that is present in helix B and the preceding loop. Helix B is part of a left-handed βαβ crossover connecting β strand 3 to β strand 4. This is a rather unusual structure, because parallel β strands are usually connected by right-handed connections. As mentioned before, mutagenesis studies,[28] together with its high conservation, support a role for the RNR motif in binding to the RNase P RNA. The second RNA binding region is the central cleft formed by helix A and the central β sheet. In the base of this cleft there are several aromatic residues that are rather conserved in RNase P protein. The spliceosomal protein U1A, whose crystal structure has been determined in a complex with RNA,[69] contains a similar cleft in which exposed aromatic residues bind single-stranded RNA. Therefore, it is possible that the central cleft of RNase P protein binds single-stranded RNA by stacking with the aromatic residues. Actually, photocrosslinking experiments suggest that this cleft is involved in interactions with the single-stranded 5′-leader sequence of one pre-tRNA,[66] but this interaction appears to be nonspecific. Certainly, the 5′-leader sequence is important in mediating binding to M1 RNA alone. The third possible RNA binding region is the loop connecting β strands 2 and 3. This loop has a cluster of negatively charged residues, which is surprising in the context of the highly basic nature of the RNase P protein.

The three-dimensional structure of the RNase P protein reveals an interesting relationship with a domain of two other RNA binding proteins, ribosomal protein S5 and elongation factor G. S5 and EF-G share very little sequence homology with RNase P protein, but both of them contain a left-handed βαβ crossover implicated in RNA binding. Based on this structural homology, it has been speculated[85] that RNase P protein, ribosomal protein S5 and elongation factor G have evolved from a common ancestor early in the evolution of the translation apparatus.

E. coli RNase P protein binds to the RNase P RNA with high affinity (k_D = 0.2–1 nM). The thermodynamic parameters of the interaction and the ionic strength dependence suggest that hydrophobic and stacking interactions play a more significant role in complex formation than do ionic contacts.[29,92] The binding site has been studied by enzymatic and chemical footprinting in *E. coli*[91,98] and *B. subtilis*.[56] The RNase P protein subunit does not recognize a discrete small domain of the RNase P RNA but binds

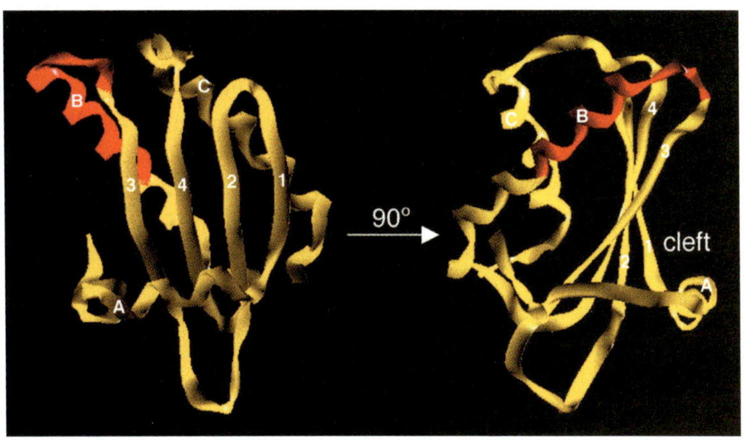

Figure 8. Three-dimensional structure of RNase P protein from *B. subtilis*. Ribbon model drawn with SYBYL according to the crystal structure determined by Stams *et al.*[85] The image on the right has been rotated 90° counterclockwise on the vertical axis to visualize better the central cleft. The conserved RNR motif is highlighted in red. The four β strands are labelled with numbers and the three α helices are labelled with letters.

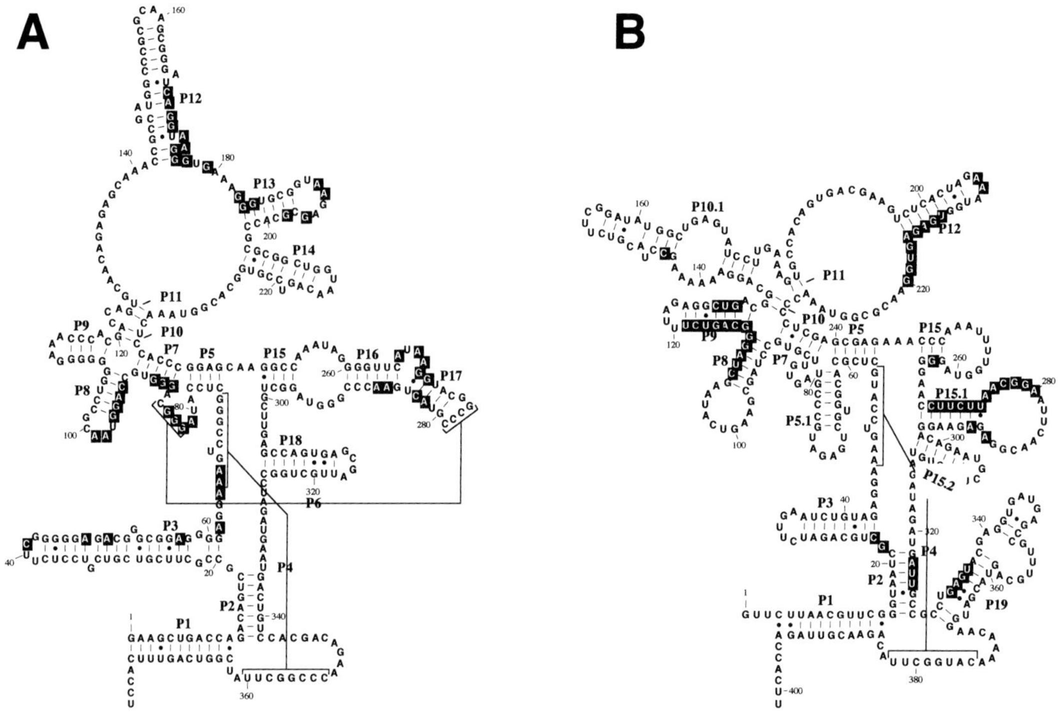

Figure 9. Summary of sites protected in RNase P RNA from *E. coli* (A) or *B. subtilis* (B) by the protein subunit. Sites protected against either RNAse T1[98] or DMS[91] in the *E. coli* RNA or against Fe(II)-EDTA in the *B. subtilis* RNA[56] are shown in black.

to several non-contiguous regions of the RNA (Figure 9), which is different from the mode of binding of several other RNA-binding proteins. It is possible that these potential binding regions are close enough in space to constitute a single binding site or are independently bound by different RNA-binding sites in the protein. We described above three possible binding sites in the crystal structure of the RNase P protein: the protection data, however, are difficult to interpret unambiguously in terms of protein–RNA contacts. Other interpretations are possible, such as conformational changes in the RNA upon binding to the protein. A three-dimensional model of the interaction between the protein and the RNA subunits of bacterial RNase P has yet to be developed.

9.4.2 The Protein Subunit as a Cofactor

In analyzing what the function of the protein subunit in catalysis could be, one must consider first what the kinetic parameters of the reaction affected by the protein are. The reaction catalyzed by the holoenzyme is more efficient (Table 2) than that of the RNA alone, due mainly to effects on k_{cat}. The protein subunit of RNase P alleviates the need for high ionic strength of the RNA-alone reaction. The rate-limiting step in the RNA-alone reaction is product release.[78,93] This is revealed by the fact that under single turnover conditions (excess enzyme) the first round of substrate cleavage is much more rapid than subsequent rounds. In the holoenzyme, there is no difference in the rates of the first and subsequent rounds of cleavage. Also, mature tRNA behaves as an inhibitor of the RNA-alone reaction, but in the presence of the protein inhibition is much reduced. These data resulted in the suggestion[78] that the highly positive protein subunit acts as a localized electrostatic shield facilitating the interaction between the catalytic RNA and the substrate without interfering with rapid product release. In the RNA-alone reaction the high ionic strength required for efficient catalysis provides the shielding effect but this is not localized and product release is slowed.

The function of the protein subunit in bacterial RNase P might be more complex and subtle than just that of a localized electrostatic shield. When the function of the protein is analyzed in a more detailed way, additional interesting features appear. For instance, for some substrates, the protein reduces the K_m several fold.[41] *In vivo* experiments suggest that the protein has a differential effect on the processing of different substrates.[42] Recent detailed studies in the *B. subtilis* system suggest an important role of the protein subunit in catalysis through interaction with the 5′ leader of pre-tRNAs.[16,51,56,66] While yet to be definitively established, these interactions may be mediated by the central cleft in the protein (Figure 8).

The protein subunit of RNase P increases the versatility of the RNA subunit. Mutants in the RNA subunit that have reduced catalytic activity in the absence of protein have wild-type-like behaviour in the presence of the protein.[57] Using *in vitro* selection techniques this idea has been subjected to a direct test.[40] The sequence of the bottom part of the highly conserved helix P4 of *E. coli* RNase P RNA was randomized and active ribozymes selected in the presence or absence of the protein cofactor. In the absence of protein only the wild-type sequence of the RNA was recovered, while in the presence of protein several additional sequences were selected. These mutant RNAs were active only in the presence of the protein. The protein subunit also increases the range of substrates that can be processed. While the RNA alone can efficiently process only pre-tRNAs, the holoenzyme can process a number of additional substrates. This question will be discussed in the next section in the context of evolution of RNase P.

9.4.3 Protein Subunits in Eukaryotes and Archaea

Aside from their identification and sequencing, very little is known of the function of the protein subunits in eukaryotic RNase P.[10,39,87,88] To date no eukaryotic RNase P holoenzyme has been reconstituted from purified RNA and protein subunits. The development of such an *in vitro* system would be an important step forward in the characterization of eukaryotic RNase P.

RNase P from Archaea is intriguing. Several complete archaeal genome sequences are available and a protein homologue to a nuclear RNase P protein has been identified.[32] Chimerical holoenzymes have been reconstituted from some archaeal RNase P RNA and bacterial RNase P protein[65,70] and from *E. coli* RNA and archaeal protein, in one case.[54]

9.5 EVOLUTIONARY CONSIDERATIONS

In spite of the diversity of RNase P holoenzyme structure, all known RNase P RNAs contain a similar core structure. Although only bacterial RNA has catalytic activity in the absence of protein, the non-bacterial RNase P probably requires RNA at the active site and may even be "RNA-based" catalysts[75,94], even though they are inactive without protein. The higher protein content of non-bacterial RNase P can be explained in two ways. First, even if one accepts that catalysis is carried out by RNA, the RNA subunit might be unable to fold into an active conformation in the absence of protein. Co-evolution of the RNA and protein subunits has allowed the protein to take over some of the crucial functions (see below). In this sense, the RNase P RNA from the plastid of *Cyanophora paradoxa* can be considered as an early step of this process. This RNA has a bacterial-like structure but is inactive *in vitro*. However, it must have a conformation that can easily become functional because it can reconstitute a functional holoenzyme with a cyanobacterial RNase P protein subunit.[72] Second, in the more complex environment of non-bacterial systems, proteins might be required for auxiliary functions such as regulation, localization, etc. For instance, two of the human RNase P protein subunits have functional nuclear and nucleolar localization signals.[39]

Another evolutionary issue is the relationship between the protein and the variety of substrates that can be processed. While in general the holoenzyme processes substrates more efficiently than RNA alone (Table 2), the effect of the protein is more pronounced with non-tRNA substrates. From the data in Table 2 it can be concluded that the catalytic efficiency (k_{cat}/K_m) of RNA alone on p4.5S RNA (a non-tRNA substrate) is 100-fold lower than on pre-tRNA, but addition of the protein gives similar catalytic efficiencies for both kinds of substrates. This question has been illustrated also by evolution *in vitro* experiments.[55] In a model substrate, part of the sequence was randomized and the pool of randomized RNA sequences was subjected to several rounds of selection for RNA molecules that were active as substrates of RNase P. Two parallel selection procedures were used, one with RNase P RNA alone and another with holoenzyme. When the selected molecules were analyzed, it was found that selection with RNase P RNA alone produced molecules that were all similar to tRNA molecules. However, selection with the holoenzyme produced a wider diversity of active substrates, some of them similar to 4.5S RNA. This result indicates that the protein allows a greater versatility to the enzyme.

In the presence of the protein, additional substrates like pre-4.5S RNA, can be processed much more efficiently than they are by the RNA alone. As 4.5S RNA functions in protein translocation, therefore, it can be hypothesized that pre-4.5S RNA did not exist before the advent of protein synthesis and could not have been a substrate for RNase P RNA alone.[55] It has been suggested that tRNAs are among the oldest RNA molecules and arose before the origin of protein synthesis.[59] Therefore, primitive pre-tRNAs must have been substrates for RNase P RNA in the absence of proteins. This could explain why pre-tRNA substrates are processed more efficiently by RNase P RNA alone than non-pre-tRNA substrates. One can make the argument, based on experiments with microhelices,[62] that the original RNase P RNA–substrate system comprised much simpler molecules than those we see today. Over time, both the catalytic RNA and its substrates grew in size through accretions of necessary functional domains and this accretion was governed by co-evolution. The advent of proteins contributed to the complexity of this system but also relaxed some of the constraints on the structure of the catalytic RNA.

9.6 ACKNOWLEDGMENTS

Research in the laboratory of A. Vioque is supported by grant PB97-0104 from DGES (Ministry of Education and Culture, Spain), and by the Andalusian Research Plan. Research in the laboratory of S. Altman is supported by NIH grant GM19422. Both laboratories are supported by grants from the Human Frontier Science Organization (RG0291) and the US–Spain Joint Commission on Scientific and Technological Cooperation (98006). We are grateful to Dr. Antonio Díaz for his help with the SYBYL program, to Dr. James Brown for the RNase P database and for unpublished data, and to Dr. David W. Christianson for the coordinates of the *B. subtilis* RNase P protein crystal structure.

9.7 REFERENCES

1. E. Akaboshi, C. Guerrier-Takada and S. Altman, *Biochem. Biophys. Res. Commun.*, 1980, **96**, 831–837.
2. P. Alifano, F. Rivellini, C. Piscitelli, C.M. Arraiano, C.B. Bruni and M.S. Carlomagno, *Genes Dev.*, 1994, **8**, 3021–3031.
3. S. Altman and L.A. Kirsebom, in "The RNA World", eds. R. Gesteland, T. Cech and J. Atkins, Cold Spring Harbor Laboratory Press, Cold Spring Harbor, NY, 2nd ed., 1999, pp. 351–380.
4. M. Baum, A. Cordier and A. Schön, *J. Mol. Biol.*, 1996, **257**, 43–52.
5. A.L. Bothwell, R.L. Garber and S. Altman, *J. Biol. Chem.*, 1976, **251**, 7709–7716.
6. M. Brännvall, J.G. Mattsson, S.G. Svärd and L.A. Kirsebom, *J. Mol. Biol.*, 1998, **283**, 771–783.
7. J.W. Brown, *Nucleic Acids Res.*, 1999, **27**, 314.
8. J.W. Brown, J.M. Nolan, E.S. Haas, M.A.T. Rubio, F. Major and N.R. Pace, *Proc. Natl. Acad. Sci. USA*, 1996, **93**, 3001–3006.
9. G. Carrara, P. Calandra, P. Fruscoloni and G.P. Tocchini-Valentini, *Proc. Natl. Acad. Sci. USA*, 1995, **92**, 2627–2631.
10. J.R. Chamberlain, Y. Lee, W.S. Lane and D.R. Engelke, *Genes Dev.*, 1998, **12**, 1678–1690.
11. J.R. Chamberlain, E. Pagan-Ramos, D.W. Kindelberger and D.R. Engelke, *Nucleic Acids Res.*, 1996, **24**, 3158–3166.
12. J.L. Chen, J.M. Nolan, M.E. Harris and N.R. Pace, *EMBO J.*, 1998, **17**, 1515–1525.
13. J.L. Chen and N.R. Pace, *RNA*, 1997, **3**, 557–560.
14. Y. Chen, X. Li and P. Gegenheimer, *Biochemistry*, 1997, **36**, 2425–2438.
15. E.L. Christian, D.S. McPheeters and M.E. Harris, *Biochemistry*, 1998, **37**, 17618–17628.
16. S.M. Crary, S. Niranjanakumari and C.A. Fierke, *Biochemistry*, 1998, **37**, 9409–9416.
17. S.C. Darr, B. Pace and N.R. Pace, *J. Biol. Chem.*, 1990, **265**, 12927–12932.
18. M.P. Deutscher, in "tRNA: Structure, Biosynthesis and Function", eds. D. Söll and U. RajBhandary, ASM, Washington, 1995, pp. 51–65.
19. T.R. Easterwood and S.C. Harvey, *RNA*, 1997, **3**, 577–585.
20. C. Fingerhut and A. Schön, *FEBS Lett.*, 1998, **428**, 161–164.
21. A.C. Forster and S. Altman, *Science*, 1990, **249**, 783–786.
22. D.N. Frank, A.E. Ellington and N.R. Pace, *RNA*, 1996, **2**, 1179–1188.
23. D.N. Frank, M.E. Harris and N.R. Pace, *Biochemistry*, 1994, **33**, 10800–10808.
24. D.N. Frank and N.R. Pace, *Annu. Rev. Biochem.*, 1998, **67**, 153–180.
25. K. Gardiner and N.R. Pace, *J. Biol. Chem.*, 1980, **255**, 7507–7509.
26. C. Glemarec, J. Kufel, A. Foldesi, T. Maltseva, A. Sandstrom, L.A. Kirsebom and J. Chattopadhyaya, *Nucleic Acids Res.*, 1996, **24**, 2022–2035.
27. B.L. Golden, A.R. Gooding, E.R. Podell and T.R. Cech, *Science*, 1998, **282**, 259–264.
28. V. Gopalan, A.D. Baxevanis, D. Landsman and S. Altman, *J. Mol. Biol.*, 1997, **267**, 818–829.
29. V. Gopalan, S.J. Talbot and S. Altman, in "RNA–Protein Interactions", eds. K. Nagai and I.W. Mattaj, IRL Press, 1994, pp. 103–126.

30. C. Guerrier-Takada, K. Gardiner, T. Marsh, N. Pace and S. Altman, *Cell*, 1983, **35**, 849–857.
31. E.S. Haas, A.B. Banta, J.K. Harris, N.R. Pace and J.W. Brown, *Nucleic Acids Res.*, 1996, **24**, 4775–4782.
32. T.A. Hall and J.W. Brown, *Nucleic Acids Symp., Ser.* 1999, **41**, 92–94.
33. F.G. Hansen, E.B. Hansen and T. Atlung, *Gene*, 1985, **38**, 85–93.
34. W.D. Hardt, J. Schlegl, V.A. Erdmann and R.K. Hartmann, *Nucleic Acids Res.*, 1993, **21**, 3521–3527.
35. W.D. Hardt, J. Schlegl, V.A. Erdmann and R.K. Hartmann, *J. Mol. Biol.*, 1995, **247**, 161–172.
36. M.E. Harris and N.R. Pace, *RNA*, 1995, **1**, 210–218.
37. R.K. Hartmann, J. Heinrich, J. Schlegl and H. Schuster, *Proc. Natl. Acad. Sci. USA*, 1995, **92**, 5822–5826.
38. N. Jarrous, *RNA*, 1999, **5**, 153–157.
39. N. Jarrous, P.S. Eder, C. Guerrier-Takada, C. Hoog and S. Altman, *RNA*, 1998, **4**, 407–417.
40. J.J. Kim, A.F. Kilani, X. Zhan, S. Altman and F. Liu, *RNA*, 1997, **3**, 613–623.
41. L.A. Kirsebom and S. Altman, *J. Mol. Biol.*, 1989, **207**, 837–840.
42. L.A. Kirsebom, M.F. Baer and S. Altman, *J. Mol. Biol.*, 1988, **204**, 879–888.
43. L.A. Kirsebom and S.G. Svärd, *EMBO J.*, 1994, **13**, 4870–4876.
44. R.G. Kleineidam, C. Pitulle, B. Sproat and G. Krupp, *Nucleic Acids Res.*, 1993, **21**, 1097–1101.
45. L. Kline, S. Nishikawa and D. Soll, *J. Biol. Chem.*, 1981, **256**, 5058–5063.
46. R. Kole and S. Altman, *Biochemistry*, 1981, **20**, 1902–1906.
47. Y. Komine, M. Kitabatake, T. Yokogawa, K. Nishikawa and H. Inokuchi, *Proc. Natl. Acad. Sci. USA*, 1994, **91**, 9223–9227.
48. K. Kruger, P.J. Grabowski, A.J. Zaug, J. Sands, D.E. Gottschling and T.R. Cech, *Cell*, 1982, **31**, 147–157.
49. J. Kufel and L.A. Kirsebom, *Proc. Natl. Acad. Sci. USA*, 1996, **93**, 6085–6090.
50. J. Kufel and L.A. Kirsebom, *RNA*, 1998, **4**, 777–788.
51. J.C. Kurz, S. Niranjanakumari and C.A. Fierke, *Biochemistry*, 1998, **37**, 2393–2400.
52. T.E. LaGrandeur, S.C. Darr, E.S. Haas and N.R. Pace, *J. Bacteriol.*, 1993, **175**, 5043–5048.
53. T.E. LaGrandeur, A. Huttenhofer, H.F. Noller and N.R. Pace, *EMBO J.*, 1994, **13**, 3945–3952.
54. N. Lawrence, D. Wesolowski, H. Gold, M. Bartkiewicz, C. Guerrier-Takada, W.H. McClain and S. Altman, *Cold Spring Harb. Symp. Quant. Biol.*, 1987, **52**, 233–238.
55. F. Liu and S. Altman, *Cell*, 1994, **77**, 1093–1100.
56. A. Loria, S. Niranjanakumari, C.A. Fierke and T. Pan, *Biochemistry*, 1998, **37**, 15466–15473.
57. N. Lumelsky and S. Altman, *J. Mol. Biol.*, 1988, **202**, 443–454.
58. Z. Lygerou, P. Mitchell, E. Petfalski, B. Seraphin and D. Tollervey, *Genes Dev.*, 1994, **8**, 1423–1433.
59. N. Maizels and A.M. Weiner, *Proc. Natl. Acad. Sci. USA*, 1994, **91**, 6729–6734.
60. C. Massire, L. Jaeger and E. Westhof, *J. Mol. Biol.*, 1998, **279**, 773–793.
61. C. Massire, L. Jaeger and E. Westhof, *RNA*, 1997, **3**, 553–556.
62. W.H. McClain, C. Guerrier-Takada and S. Altman, *Science*, 1987, **238**, 527–530.
63. M.J. Morales, C.A. Wise, M.J. Hollingsworth and N.C. Martin, *Nucleic Acids Res.*, 1989, **17**, 6865–6881.
64. D.P. Morse and F.J. Schmidt, *Gene*, 1992, **117**, 61–66.
65. D.T. Nieuwlandt, E.S. Haas and C.J. Daniels, *J. Biol. Chem.*, 1991, **266**, 5689–5695.
66. S. Niranjanakumari, T. Stams, S.M. Crary, D.W. Christianson and C.A. Fierke, *Proc. Natl. Acad. Sci. USA*, 1998, **95**, 15212–15217.
67. B.K. Oh, D.N. Frank and N.R. Pace, *Biochemistry*, 1998, **37**, 7277–7283.
68. B.K. Oh and N.R. Pace, *Nucleic Acids Res.*, 1994, **22**, 4087–4094.
69. C. Oubridge, N. Ito, P.R. Evans, C.-H. Teo and K. Nagai, *Nature*, 1994, **372**, 432–438.
70. J.A. Panucci, E.S. Haas, T.A. Hall, J.K. Harris and J.W. Brown, *Proc. Natl. Acad. Sci. USA*, 1999, **96**, 7803–7808.
71. A. Pascual and A. Vioque, *Eur. J. Biochem.*, 1996, **241**, 17–24.
72. A. Pascual and A. Vioque, *FEBS Lett.*, 1999, **442**, 7–10.
73. A. Pascual and A. Vioque, *Proc. Natl. Acad. Sci. USA*, 1999, **96**, 6672–6677.
74. J.P. Perreault and S. Altman, *J. Mol. Biol.*, 1992, **226**, 399–409.
75. T. Pfeiffer, A. Tekos, J.M. Warnecke, D. Drainas, D.R. Engelke, B. Séraphin and R.A. Hartmann, *J. Mol. Biol.* 2000, **298**, 559–565.
76. D.A. Pomeranz-Krummel and S. Altman, *Proc. Natl. Acad. Sci. USA*, 1999, **96**, 11200–11205.
77. D.A. Pomeranz-Krummel and S. Altman, *RNA*, 1999, **5**, 1021–1033.
78. C. Reich, G.J. Olsen, B. Pace and N.R. Pace, *Science*, 1988, **239**, 178–181.
79. M. Reith and J. Munholland, *Plant Mol. Biol. Rep.*, 1995, **13**, 333–335.
80. H. Robertson, S. Altman and J. Smith, *J. Biol. Chem.*, 1972, **247**, 5243–5251.
81. P. Schedl and P. Primakoff, *Proc. Natl. Acad. Sci. USA*, 1973, **70**, 2091–2095.
82. E.L. Shevelev, D.A. Bryant, W. Löffelhardt and H.J. Bohnert, *DNA Res.*, 1995, **2**, 231–234.
83. R.W. Siegel, A.B. Banta, E.S. Haas, J.W. Brown and N.R. Pace, *RNA*, 1996, **2**, 452–462.
84. D. Smith and N.R. Pace, *Biochemistry*, 1993, **32**, 5273–5281.
85. T. Stams, S. Niranjanakumari, C.A. Fierke and D.W. Christianson, *Science*, 1998, **280**, 752–755.
86. B.C. Stark, R. Kole, E.J. Bowman and S. Altman, *Proc. Natl. Acad. Sci. USA*, 1978, **75**, 3717–3721.
87. V. Stolc and S. Altman, *Genes Dev.*, 1997, **11**, 2926–2937.
88. V. Stolc, A. Katz and S. Altman, *Proc. Natl. Acad. Sci. USA*, 1998, **95**, 6716–6721.
89. S.G. Svärd, U. Kagardt and L.A. Kirsebom, *RNA*, 1996, **2**, 463–472.
90. S.G. Svärd and L.A. Kirsebom, *J. Mol. Biol.*, 1992, **227**, 1019–1031.
91. S.J. Talbot and S. Altman, *Biochemistry*, 1994, **33**, 1399–1405.
92. S.J. Talbot and S. Altman, *Biochemistry*, 1994, **33**, 1406–1411.

93. A. Tallsjö and L.A. Kirsebom, *Nucleic Acids Res.*, 1993, **21**, 51–57.
94. B.C. Thomas, J. Chamberlain, D.R. Engelke and P. Gegenheimer, *RNA*, 2000, **6**, 554–562.
95. B.C. Thomas, X. Li and P. Gegenheimer, *RNA*, 2000, **6**, 545–553.
96. M. Turmel, C. Otis and C. Lemieux, *Proc. Natl. Acad. Sci. USA*, 1999, **96**, 10248–10253.
97. A. Vioque, *Nucleic Acids Res.*, 1997, **25**, 3471–3477.
98. A. Vioque, J. Arnez and S. Altman, *J. Mol. Biol.*, 1988, **202**, 835–848.
99. M.J. Wang, N.W. Davis and P. Gegenheimer, *EMBO J.*, 1988, **7**, 1567–1574.
100. D.S. Waugh, C.J. Green and N.R. Pace, *Science*, 1989, **244**, 1569–1571.
101. M. Werner, E. Rosa and S.T. George, *Nucleic Acids Symp. Ser.*, 1997, **36**, 19–21.
102. S.L. Wolin and A.G. Matera, *Genes Dev.*, 1999, **13**, 1–10.
103. Y. Yuan and S. Altman, *EMBO J.*, 1995, **14**, 159–168.
104. A.J. Zaug and T.R. Cech, *Science*, 1986, **231**, 470–475.

10

Ribozyme Selection

ANDREW D. ELLINGTON and MICHAEL P. ROBERTSON

University of Texas at Austin, Austin, TX, USA

10.1 THE HISTORY OF RIBOZYMES AND RIBOZYMES IN HISTORY

While the discoveries that RNase P and the Group I self-splicing intron could catalyze phosphodiester bond rearrangements popularized the notion that nucleic acids could perform functions beyond their traditional roles as informational macromolecules,[1,2] these experiments had been anticipated almost two decades before by speculations that cellular cofactors and RNAs, such as tRNA, were the lineal descendants of ribozymes. In particular, Crick[3] and Orgel[4] were among the first to suggest that a RNA world may have once existed where ribozymes served in place of protein enzymes. The experimental validation of these early musings further emboldened researchers to suggest that the putative RNA world may have been metabolically complex.[5,6]

Further support for the RNA world hypothesis came from studies of both natural and unnatural RNA catalysts. Soon after the discovery of RNA catalysis, several additional natural ribozymes were identified and characterized. Small hairpin and hammerhead ribozymes were found to catalyze the cleavage of specific phosphodiester bonds to yield 5′-hydroxyl and 2′,3′-cyclic phosphate termini (reviewed in Tuschl *et al.*).[7] The sequences and structures of these small ribozymes bore no resemblance to their larger counterparts. Moreover, their mechanisms were vastly dissimilar: RNase P used water as a nucleophile rather than a 2′ hydroxyl to cleave pre-tRNAs, yielding 5′ phosphates and free 3′ hydroxyls. In contrast, the Group I ribozymes used the 3′ hydroxyl of an exogenous guanosine cofactor to initiate a splicing cascade in which the 3′ hydroxyl of one exon was eventually joined to the 5′ phosphate of a second exon and the intervening intron was concomitantly released as a linear RNA. The observed diversities of ribozyme sequences and mechanisms begged the conclusion that either these few remaining ribozymes had descended from a complex bestiary of catalysts, or that ribozyme catalysis was relatively easy to invent. In either instance, the more ribozymes that were found, the more likely it was that a RNA world may have once existed. However, all of the natural ribozymes catalyzed the scission, formation, or rearrangement of phosphodiester bonds. If ribozymes were the progenitors of protein enzymes, then it was likely that they had also once catalyzed a wider variety of reactions, such as those involved in energy-yielding metabolism (e.g., redox reactions) or in the biochemical transformations of simple organic skeletons (e.g., carbon–carbon bond cleavages and formations). These hypothesized ribozymes might now be extinct, but there was potentially a way to generate their doppelgangers: *in vitro* selection.

10.2 *IN VITRO* SELECTION OF RIBOZYMES

In vitro selection has been applied to the study of both natural and unnatural ribozymes.[8] *In vitro* selection is governed by the same principles that are operative during natural selection: first, a pool of heritable diversity is generated. In the case of natural selection, these are the mutations that arise during replication. In the case of *in vitro* selection, the pool can be chemically synthesized with the desired degree of randomness. Second, variants within the pool are selected for function. In the case of natural selections, organismal survival is frequently linked to molecular function; some bacteria can survive in the presence of an antibiotic because they have an enzyme that can hydrolyze that antibiotic. Similarly, in the case of *in vitro* selection molecular survival is made dependent on the ability of ribozymes to catalyze reactions. Finally, the functional variants are preferentially amplified. In the case of natural selection, organismal variants that are more functional are frequently better able to reproduce. It is also possible, of course, that organismal variants can survive based on their reproductive capacities alone, without attendant functionality. In the case of *in vitro* selection, functional ribozymes gain preferential access to reproductive resources (enzymes and nucleotides) and are amplified. In both natural and *in vitro* selection multiple cycles of selection and amplification can optimize the functions of molecules in successive generations. However, in natural selection the variation that exists is episodic, with relatively few new mutations being introduced into each generation, while during *in vitro* selection a much larger range of sequence variation is introduced at the beginning of an experiment and winnowed to a few functional species over multiple cycles. For this reason, it is safe to conclude that *in vitro* selection moves at a much more rapid clip than natural selection and can functionally optimize molecules with much greater efficiency.

In greater detail, *in vitro* selection generally starts with a random sequence nucleic acid population (Figure 1). The nucleic acid population is chemically synthesized as a single-stranded DNA oligonucleotide, initially amplified by the polymerase chain reaction into double-stranded DNA templates, and *in vitro* transcribed into RNA via promoters appended to the PCR primers. There are several types of randomization that can be utilized to examine or generate ribozymes. Partial randomization is used to generate RNA pools that are centered on the wild-type molecule. Depending on the degree of mutagenesis, ribozyme populations that contain anywhere from all possible single to all possible hextuple substitutions can be generated. The level of partial randomization (or "doping") can be varied by varying the composition of the phosphoramidite mix used for chemical synthesis. PCR mutagenesis or other mutational techniques can also be used to generate doped pools, albeit at much lower substitution rates (typically 96–99% wild-type, with an uneven skew of mutations). Such doped pools are useful for delimiting and defining functional sequences and structures. For example, if a given residue at a given position is essential for ribozyme function, then prior to selection the wild-type residue may be present only 70% of the time while following selection the wild-type residue may be present up to 100% of the time. Conversely, if a given residue is neutral for ribozyme function, then there will be little difference between the proportion of the wild-type residue in the pre- and post-selection RNA populations. Similarly, residues that form base-pairs may coordinately change during selection. Alternative pairings will appear as covariations in the primary sequence data, and can be used to map or confirm secondary

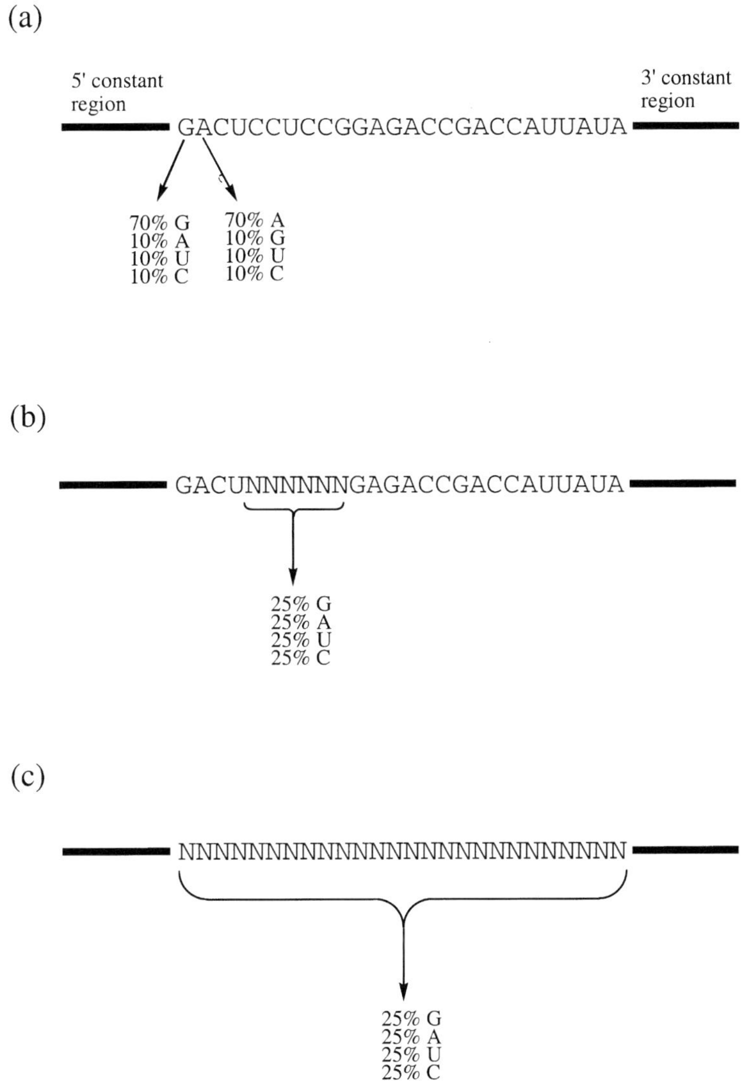

Figure 1. Types of random sequence nucleic acid libraries. (a) Partial randomization. Each position in a constant sequence region is "doped" with mutations. The level of doping can be pre-set by the experimenter. (b) Segmental or "shallow" randomization. A portion of a constant sequence region is completely randomized. The random sequence positions can be grouped (as shown here) or dispersed. Most shallow random pools contain all possible sequences. (c) Complete or "deep" randomization. Most or all of the residues that lie between the constant regions are completely randomized. Most deep random pools do not contain all possible sequences.

structural features and motifs of ribozymes. Covariations are not limited to Watson–Crick pairings, but can also identify or confirm other isosteres such as wobble pairings or purine : purine interactions.

Functional ribozymes can also be selected from completely randomized populations. While the degree of randomization will be absolute (25% of each of the four canonical nucleotides), the length of the random sequence tract can vary. Functional variants of known ribozymes can be selected from RNA pools in which short segments are completely randomized. Partial randomization allows residues important for function to be quickly sieved from those neutral to function over the entire length of a molecule. However, sequence covariations necessary to establish or confirm structural features are relatively rare.

Finally, novel ribozymes with no natural or known counterparts can be selected from "deep random" pools that contain from 30 to over 200 randomized positions. Deep random pools typically span from 10^{13} to 10^{16} different variants, and hence completely cover a sequence space of from 22 ($4^{22} \times 10^{13}$) to 27 ($4^{27} \times 10^{16}$) residues in length (the actual coverage will also be dependent on the length of the random sequence region, a factor not taken into account in this simple calculation; see also Sabetti *et al.*).[9] Many

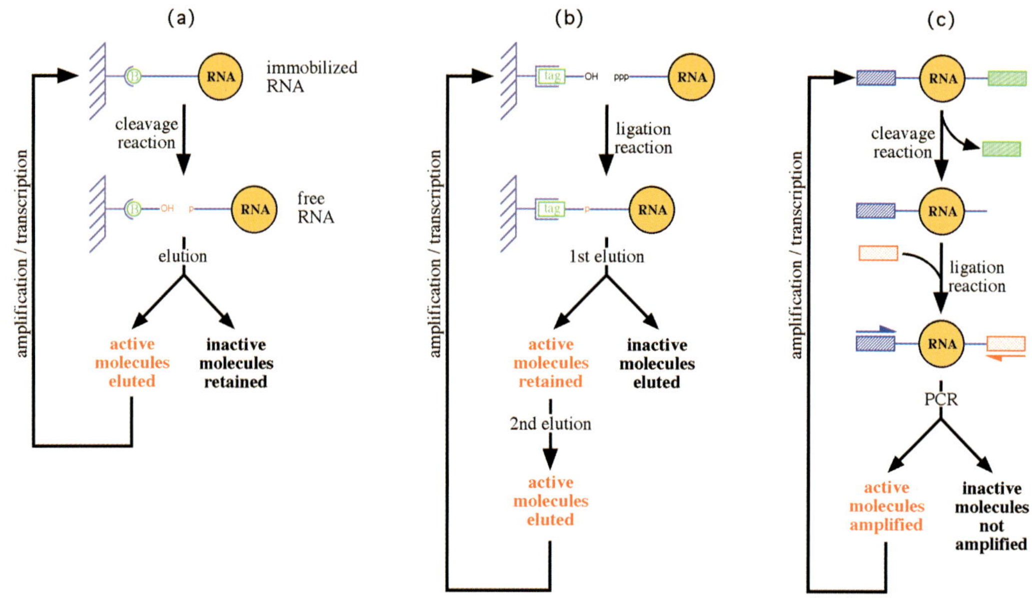

Figure 2. Types of ribozyme selection experiments. These variations on a theme were chosen to demonstrate the versatility of selection protocols. (a) Cleavage selection. A RNA pool is immobilized via a biotin conjugated to its 3′ end. Those RNA molecules that can cleave themselves away from the column are eluted, amplified, and re-derivatized for further cycles of selection. (b) Ligation selection. A "tag" sequence within an oligonucleotide is immobilized via base-pairing to an oligonucleotide affinity column. A random sequence RNA pool is mixed with the immobilized oligonucleotide, and those RNA molecules that can covalently append themselves to the primer are separated from the remainder of the population. The ligated products are eluted from the affinity column, amplified, and further selected for ligation function. (c) Cleavage/ligation selection. Variants in a RNA pool that can cleave away a constant region are isolated and mixed with a different constant region. Those variants that can also perform the reverse reaction, ligation, or otherwise capture the oligonucleotide are then preferentially amplified using primers specific for the second constant region. These reactions all take place in solution, away from a solid support (as could the selections cited in (a) and (b), depending on what isolation and amplification steps are employed).

ribozymes will have more than 30 functional residues, and thus deep random pools may contain only a vanishingly small fraction of all the possible ribozymes that can perform a given function. For example, selections for ribozyme ligases have recovered a variety of novel functional motifs, but at least one of these motifs was predicted to be present at a level of only one in every 2.5×10^{18} different sequences examined.[10]

While there are numerous, different ribozyme activities that can be selected for, the selection experiments themselves fall into one of three classes: cleavage selections, ligation selections (Figure 2), and selections for binding. Cleavage selections remove information (sequence, structural, or chemical) from a ribozyme, while ligation selections add information to a ribozyme. As an example of a cleavage selection, hammerhead ribozyme variants have been tethered to a solid support, and those variants that could chew themselves away from the support were eluted and subsequently amplified.[11] As an example of a ligation selection, Group I self-splicing intron variants have been engineered to carry out a reversal of the first step in the splicing cascade and append an oligonucleotide tag to themselves.[12] The ligation tag was designed to correspond to a primer sequence for the polymerase chain reaction, and the successful variants were directly amplified from the starting population (see also Figure 9). Ligation selections can also encompass the addition of more than just oligonucleotide tags: for example, alkyl transferases that successfully transferred an activated biotin moiety to themselves have been selected from random sequence populations.[13] In some instances, cleavage and ligation selections can be combined. For example, Burke and his co-workers have developed an ingenious method for selecting hairpin ribozyme variants (Figure 2c).[14] In this system the 5′ end of a substrate RNA that contained the cleavage site was covalently attached to the 3′ end of the hairpin ribozyme through a polycytidylate linker, generating a ribozyme that was covalently linked to its substrate. Molecules that were active for both cleavage and ligation were selected by first generating cleaved ribozyme–substrate chimeras that contained a 2′,3′ cyclic phosphate, and then incubating these cleavage products with a new substrate that could participate in the reverse reaction to generate a new 3′ priming site for reverse transcription.

RNA molecules cannot only be directly selected for catalysis, but indirectly selected as well. RNA binding species (aptamers) have been selected from random sequence populations. One of the old saws of bioorganic chemistry is that enzymes catalyze reactions by binding to the transition state of the reaction. Therefore, RNA molecules selected to bind transition state analogues should catalyze the corresponding reaction, just as antibody molecules screened for binding to transition state analogues have proven to be catalytic. Aptamers that could bind to a transition state analogue of an isomerization reaction were selected from a RNA pool that contained 128 randomized positions.[15] One of the aptamers isolated from this selection bound the analogue with a K_i of 7 μM and could catalyze the isomerization of the corresponding biphenyl substrate at a rate 88-fold greater than background. However, such strategies are not uniformly successful: aptamers selected to bind to a Diels–Alder transition state analogue could not catalyze the corresponding Diels–Alder cycloaddition reaction.[16] There are only a few examples of indirect selection for ribozyme catalysis, and in general the catalysts that have been derived seem to be much slower than those derived by direct selection. However, the strength of indirect selection methods is that they are more likely to yield "true" catalysts that can turnover substrates, as opposed to ribozymes that perform a single reaction cycle. While directly selected catalysts have been converted to multiple turnover ribozymes, the separated substrates have tended to be oligonucleotides rather than small molecules such as the biphenyl cited above.

As has been alluded to above, selected ribozyme variants are typically amplified by a combination of reverse transcription, the polymerase chain reaction, and *in vitro* transcription (usually with T7 RNA polymerase). Other amplification procedures have also been used, such as isothermal amplification or 3SR,[17] which mimics the retroviral life cycle. During amplification, mutations arise and can provide fodder for subsequent rounds of selection. Since relatively few of the total number of possible variants are examined in most selection experiments, mutagenic amplification procedures, such as mutagenic PCR, are often incorporated during the amplification step.

10.3 IDENTIFYING FUNCTIONAL SEQUENCES AND STRUCTURES

A variety of natural ribozymes, ranging from the small hairpin and hammerhead ribozymes to the large Group I self-splicing intron and ribonuclease P, have been segmentally randomized and selected for catalytic function. Data from these selection experiments, such as the locations of functionally important residues and sequence covariations, have been used for the development of ribozyme structural models. We will consider several particularly illuminating examples.

The relatively small size (30–50 residues) of the hairpin and hammerhead ribozymes precludes the formation of extensive tertiary structural contacts, and thus their secondary structures are the key to their overall folds. A secondary structure had been proposed for the hairpin ribozyme (Figure 3). However, initial attempts at confirming this structure were limited by the dearth of natural sequences available for comparative analysis (only three hairpin ribozymes were known, and these contained little sequence variation). An *in vitro* cleavage–ligation selection was therefore used to generate an artificial phylogeny of functional ribozymes.[18] Initially, secondary structural features of the hairpin ribozyme were established by selecting functional ribozymes from populations in which each of the helices was randomized in turn.[19,20] Helices 1 and 2 flank the cleavage–ligation junction and may be involved in positioning catalytically important residues. These helices were separately but completely randomized, and active variants were selected. The artificial phylogeny of selected sequences confirmed almost all of the previously proposed pairing interactions for helices 1 and 2 and, in addition, suggested a catalytic role for particular residues. Similarly, selected sequences supported the pairing scheme for helix 3. The existence of helix 4 had not previously been conclusively established, and secondary structural models ranged from those that contained no helix to those that contained helices of variable length. Analysis of selected ribozymes revealed that three predicted base-pairs contributed to catalytic activity.

The participation of formally single-stranded joining regions in secondary structural interactions has also been probed.[18] A four-nucleotide region that spanned the cleavage–ligation junction was completely randomized, and functional variants were again selected. A guanosine at a position immediately 3′ of the cleavage–ligation junction was identified as essential for activity.

Further analyses of the hairpin ribozyme focused on locating tertiary structural interactions.[19] It was thought that tertiary structural interactions would most likely reside in the single-stranded joining regions of the molecule. A series of pools was constructed in which the internal loop sequences J1/2, J2/1,

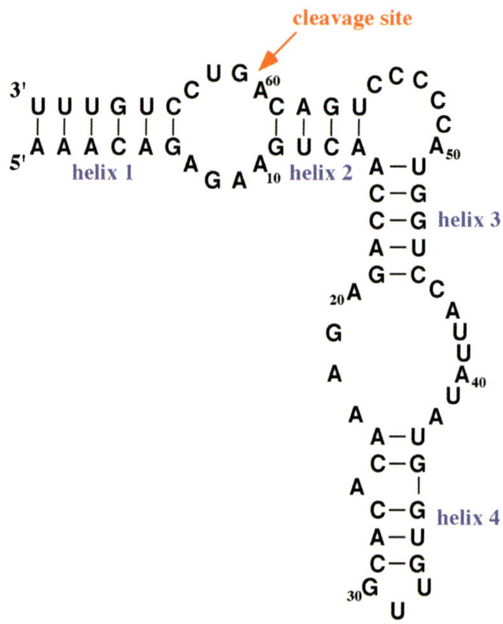

Figure 3. Hairpin ribozyme. The secondary structure of the hairpin ribozyme used by Burke and his coworkers for cleavage/ligation selection experiments is shown. The helices are labeled; single-stranded regions are named according to which helices they fall between (e.g., J1/2 falls between helices 1 and 2, reading in the 5′ direction. J2/1 also falls between helices 1 and 2, but in the opposite orientation). The cleavage junction is indicated by an arrow.

J3/4, and J4/3 were segmentally randomized, either alone or pairwise. Sequence analysis of selected variants revealed that the same results were obtained regardless of whether single-stranded regions were randomized independently or pairwise.

Finally, in order to identify functional residues and interactions that may have been dispersed throughout the hairpin ribozyme, a doped pool was constructed that contained, on average, five substitutions per molecule. After selection for active variants, a double substitution (G21U, U39C) was identified that had higher activity than one of the single substitutions (G21U) in isolation. This result could be interpreted to mean that positions 21 and 39 were involved in a tertiary structural contact. To better interpret this result, a series of site-specific mutations were constructed at positions 21 and 39. The activities of the substituted ribozymes were inconsistent with a tertiary structural interaction, but were consistent with the hypothesis that U39C was a nonspecific suppressor of multiple, different substitutions of G21.

The hammerhead ribozyme (Figure 4) has also been examined using an *in vitro* cleavage selection. In an attempt to optimize utilization of a RNA substrate containing AUA at the cleavage site, Nakamaye

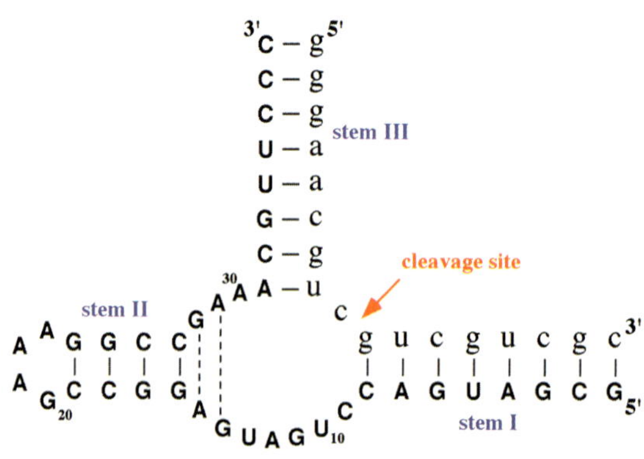

Figure 4. Hammerhead ribozyme. The secondary structure of a hammerhead ribozyme similar to those used in the selection experiments that have been cited is shown. The helices are labeled. Sequential, non-Watson–Crick G : A pairings are represented by dashed lines. The cleavage junction is indicated by an arrow.

and Eckstein[21] completely randomized three residues in the core of a hammerhead ribozyme found in barley yellow dwarf virus. After three rounds of selection and amplification, two isolates showed slight improvements in the efficiency of cleavage of the AUA-containing substrate. However, it proved difficult to identify variants with much higher cleavage efficiencies when the entire core of the hammerhead ribozyme was randomized. Similarly, when ten residues within the core of a UCA-cleaving hammerhead ribozyme were randomized and active variants were selected, only two sequence changes in the consensus core were noted: A9U and A7U.[22] Interestingly, both of these changes actually decreased the speed and efficiency of the hammerhead ribozyme but were selected because they could be more readily amplified than the wild-type hammerhead consensus. Tang and Breaker[23] also randomized 14 residues in the core of the hammerhead ribozyme, including all residues known to be critical for function, and selected variants that could cleave themselves. However, in contrast to previous selection experiments, Tang and Breaker inhibited cleavage reactions that occurred during transcription by designing an "allosteric trigger" into the ribozyme. An anti-ATP aptamer was appended to stem II of the hammerhead ribozyme in such a way that cleavage was inhibited in the presence of ATP and encouraged in its absence. After six rounds of selection and amplification, the 10^{13} molecules in the initial pool had been winnowed to one major variant, the consensus hammerhead sequence (a U7C substitution was also observed, but had previously been shown to occur in natural ribozyme sequences as well). In addition to the hammerhead ribozyme itself, two other minor sequence classes were observed that did not appear to be mere variants of the hammerhead ribozyme, but were in fact completely different sequence classes. Both of these classes of ribozymes showed much less dependence on the engineered allosteric trigger, and could readily cleave themselves even in the presence of ATP. Both classes of ribozymes were roughly as active as the hammerhead ribozyme in 20 mM $MgCl_2$. However, under lower, more physiologically relevant magnesium concentrations both classes showed enhanced cleavage activity relative to the selected hammerhead ribozymes. Consistent with these findings, additional rounds of selection in either the presence of ATP or in the presence of low concentrations of magnesium (2 mM) allowed the novel Class I ribozymes (Figure 5) to predominate over the consensus hammerhead in the selected population. Finally, [24] randomized not only the core residues of the hammerhead but the residues corresponding to stem II as well (22 residues total). In addition, these authors also altered the cleavage site from GUC to AUG, a sequence not normally recognized by the natural hammerhead. After 13 cycles of cleavage selection and amplification several different sequence classes were observed. One of the most populous and active classes was predicted to form a three-helix junction separated by short single-stranded joining regions. This ribozyme was not only hammerhead-like in terms of its secondary structure, but the core sequences were similar to those found in the hammerhead. The selected ribozyme could efficiently ($k_{cat} \times 0.5$ min^{-1}) cleave the phosphodiester bond following the unnatural AUG and other non-canonical triplets.

In vitro ligation selections with the Group I ribozyme (Figure 6) have also proven useful in establishing or confirming secondary structural features. In initial experiments, residues that abutted a pseudoknot structure in the catalytic core were probed.[12] Nine nucleotides within the J3/4 and J7/3 joining regions and the P3 stem were completely randomized, and active ribozyme variants were selected in three cycles. Following selection, only wild-type residues were found at positions J3/4-1 and J3/4-2, suggesting a critical structural or catalytic role for these residues. Other residues were found to covary in a manner that was consistent with the existence of direct interactions.

Long-range tertiary structural interactions, such as contacts between the coaxial stacks of the Group I ribozyme, have also been verified by *in vitro* selection experiments.[25] Sequence covariations observed in the natural Group I intron phylogeny had previously prompted Michel *et al.*[26] to propose that triple base-pairs formed between P4 and J6/7, and between P6 and J3/4. The existence of some of these triple base-pairs was confirmed by site-directed mutational analysis, but proof for others remained elusive. To determine the rules for base-triple formation in more detail and to potentially gain insights into the multiple interactions within this region an *in vitro* selection experiment was designed based on the Group I self-splicing intron sunY. The initial pool contained 14 randomized residues scattered through P4, P6, J3/4, and J6/7 and spanned 2.7×10^8 different variants. After three rounds of selection, the ligation ability of the selected population was actually 1.5-fold better than that of the wild-type ribozyme. Sequences from the most active clones in the final population identified four invariant residues: J3/4-1 and J3/4-2 (as in the previous selection), and the base-pair P6-2. Sequence covariations were also recovered that could be modeled as identical or isosteric to base triples that had previously been predicted to form between J6/7-1 and P4-1, and J6/7-2 and P4-2.

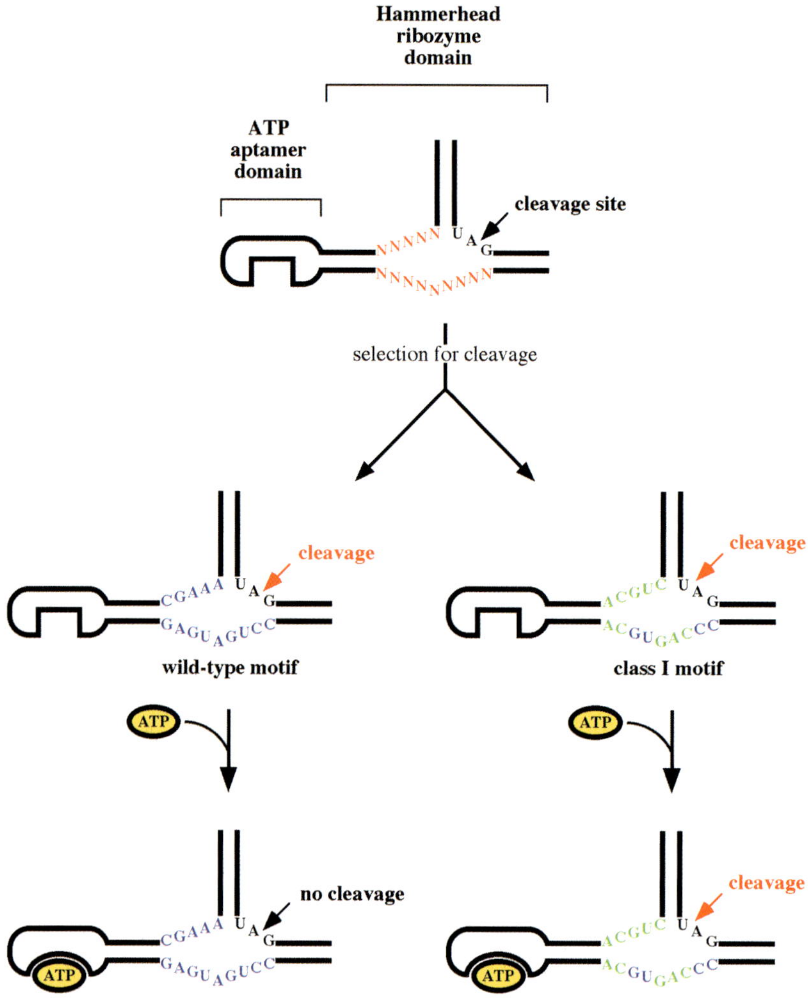

Figure 5. Transmogrification of the hammerhead ribozyme. Tang and Breaker[23] selected functional ribozymes from a randomized population based on the hammerhead ribozyme. Some of the selected variants (left, blue) were similar in sequence to the wild-type hammerhead and were responsive to the "allosteric trigger" built into the selection. Other variants (right) showed limited similarity to the hammerhead ribozyme (blue) but appeared to represent novel sequence motifs (green). Note that while the hammerhead cleaved following A that the novel motif cleaved following the previous residue, U. The novel ribozymes also were not responsive to the allosteric trigger. Orientation of the hammerhead and hammerhead-like ribozymes is as in Fig. 4.

The aforementioned selection experiments were carried out with the hope of defining the Group I structure. Additional experiments have sought to use the Group I ribozyme as a model for better understanding general principles of RNA folding and structure. Michel and his co-workers had previously used phylogenetic and biochemical analyses to identify a common interaction between GNRA tetraloops and common sequence motifs in Group I and Group II introns.[27–29] Tetraloops with the sequence GAAA interacted preferentially with a 11-nucleotide motif CCUAAG... UAUGG that formed an internal loop secondary structure, while tetraloops with the sequences GURA interacted with two contiguous base-pairs in a continuous helix: GUAA interacted with C:G, C:G pairs, while GUGA interacted with C:G, U:A pairs (Figure 7). The GAAA docking site has been resolved in atomic detail,[30] while the GURA docking sites have been most easily rationalized by a model in which the third and fourth residues of the tetraloop formed minor groove base-triples with the corresponding base-pairs, with the third residue of the tetraloop discriminating between the second base-pair.

To better determine whether the rules for tetraloop recognition that have been gleaned from phylogenetic analyses of natural ribozyme variants were complete or merely represented a subset of possible rules, Costa and Michel[31] generated an artificial phylogeny of tetraloop recognition sequences. As a model for the tetraloop: receptor interface, Costa and Michel chose an interaction between L2 and P8 that had been proposed to occur in a variety of Group I ribozymes. As a starting point for selection

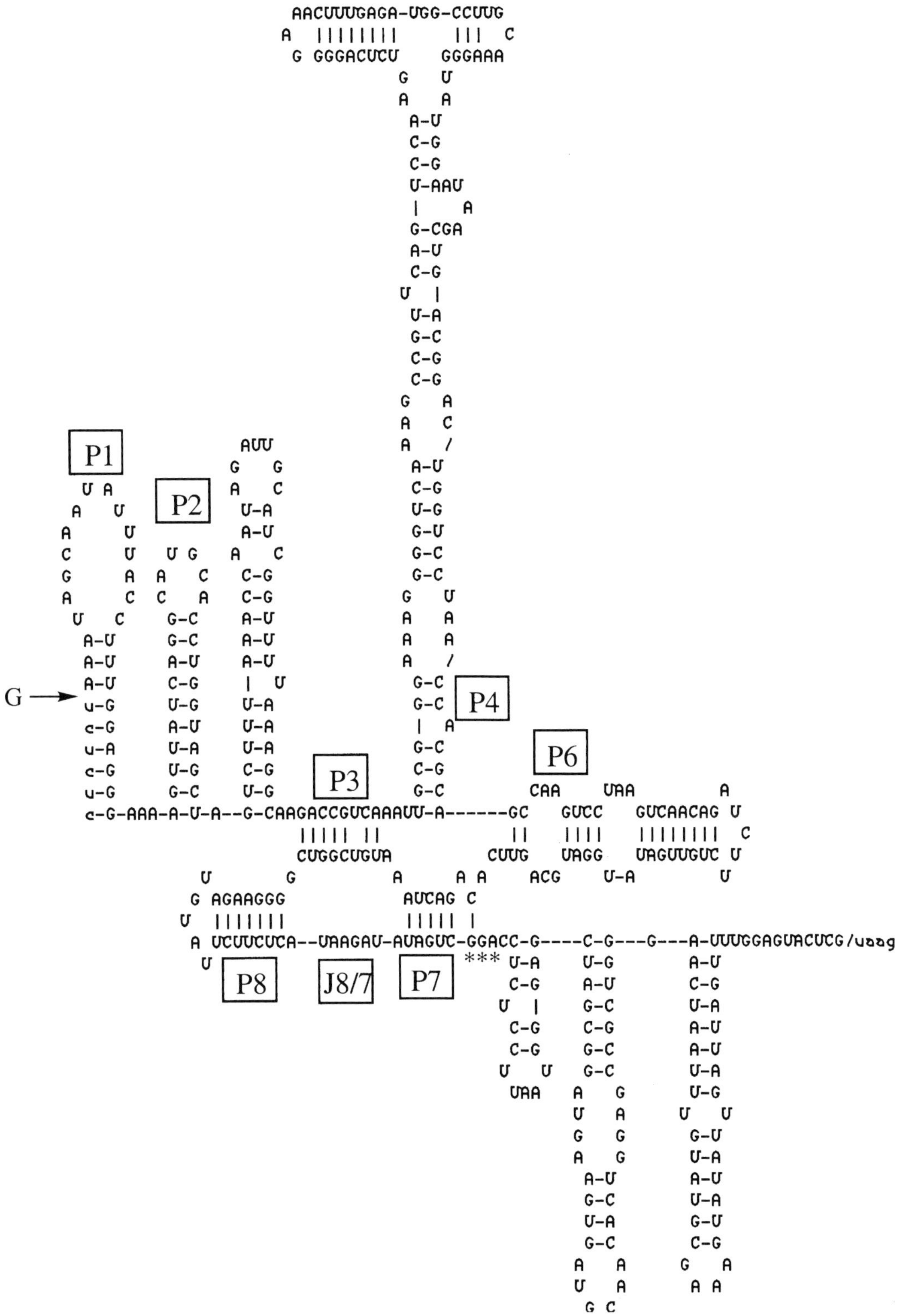

Figure 6. *Tetrahymena* Group I self-splicing intron. The sequence and secondary structure of the ribozyme are shown. Intron sequences are in capital letters, while exon sequences are in lower case. Paired stems or helices are denoted in order as P1, P2, and so forth. Single-stranded joining regions lie between some of the paired stems and are denoted as Jn/n, where "n" indicates the stems 5′ and 3′ of the joining region, respectively (e.g., J8/7 is labeled). Individual residues can be labeled according to their positions within paired stems (e.g., P3-8, the eighth residue within P3) or within joining regions (e.g., J6/7-3, the third residue within J6/7). The site of attack by an exogenous guanosine on P1 that initiates the splicing cascade is indicated by an arrow. The approximate positions of residues 312, 313, and 314, residues that mutated in response to a selection for utilization of DNA as a substrate, are shown by "***".

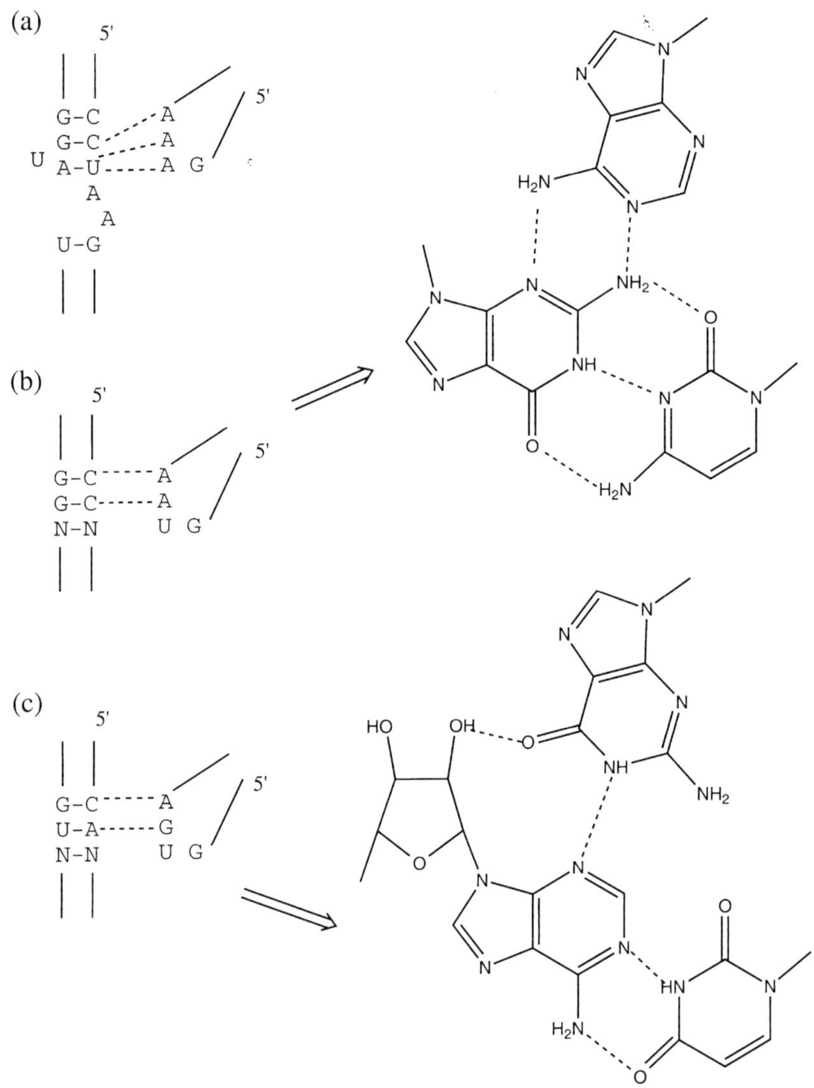

Figure 7. Tetraloop receptors. (a) Interactions between the 11-nucleotide motif 5′ CCUAAG... UAUGG and a GAAA tetraloop, redrawn from Cate *et al.*[30] The A–U base-pair in the helix is a reverse Hoogsteen pairing. Hydrogen bonds are indicated by dashed lines. (b) Proposed interactions between a helical tetraloop receptor and a GUAA tetraloop, as proposed by Jaeger *et al.* and Costa and Michel.[28,31] (c) Proposed interactions between a helical tetraloop receptor and a GUGA tetraloop.

experiments, the bacteriophage T4 td self-splicing intron was divided into two pieces: a P1–P2 substrate that contained the tetraloop, and a catalytic core into which the substrate could dock. The P8 stem of the catalytic core was replaced with 21 random sequence residues, and variants that could catalyze the attack of the 3′ terminal guanosine on the 5′ intron–exon junction of the substrate (the reverse of the second step in splicing) were selected. Since the L2–P8 interaction had been shown to be essential for efficient catalysis, all and only those variants that could repair this interaction should have been selected. Two different constant sequence substrates were used: one containing a GAAA tetraloop, and one containing a GUGA tetraloop. Following selection, the predominant class of ribozymes that could interact with the GAAA tetraloop contained the 11-nucleotide internal loop motif previously identified by phylogenetic analysis. Similarly, the predominant class of ribozymes that could interact with the GUGA tetraloop contained consecutive C : G and U : A base-pairs in a continuous stem. These results supported the notion that natural selection is an efficient search algorithm for optimal interactions. However, additional, unnatural binding motifs were also identified. For example, a variant of the 11-nucleotide internal loop motif efficiently bound GAAA tetraloops. Similarly, ribozymes containing helices with G:C, C:G stacks followed by a short internal loop were found to bind GUGA tetraloops and, unlike their natural counterparts, could discriminate strongly against GCGA tetraloops. While some of these novel tetraloop receptors were found to have natural counterparts, many of the selected sequences had no correspondents even though they were as kinetically competent as the natural receptors.

10.4 IMPROVING AND CHANGING CATALYSIS

In vitro selection can be used to alter the chemical and structural properties of natural ribozymes from their norms. In the most complete set of experiments to date, Joyce and co-workers have subjected several aspects of the Group I ribozyme mechanism to artificial selection pressure. To evolutionarily modify the self-splicing intron, the Scripps researchers developed a ligation selection scheme in which a primer binds to the 5′ end of the ribozyme (mimicking the 5′ intron–exon junction) and is attacked by the 3′ terminal guanosine. This engineered reaction is essentially the reverse of the second step in the splicing cascade.

Initially, the chemical specificity of the Group I ribozyme was changed. The wild-type self-splicing intron cleaves specific phosphodiester bonds within RNA substrates by using at least one, and probably two, magnesium ions to polarize the 3′ oxygens of endogenous and exogenous guanosine substrates.[32] Starting from a small pool of deletion variants, Robertson and Joyce[33] were able to identify a ribozyme that could better utilize a DNA primer as a substrate. However, the rate enhancement that was achieved was quite small, and appeared to be due to a general improvement in catalysis rather than to a specific improvement in DNA cleavage ability. To explore whether additional sequence changes might further improve catalysis with DNA substrates, Beaudry and Joyce[34] partially randomized 140 positions in the catalytic core of the *Tetrahymena* ribozyme at a rate of 5% and prepared a starting population that completely spanned all four-error substitutions. After ten cycles of selection and amplification the selected molecules had an aggregate activity that was 30-fold greater than the wild-type parent toward DNA; the most active individual clones could cleave DNA substrates up to 100-fold more efficiently.

While there were some common functional themes amongst the evolved, DNA-cleaving Group I variants, different ribozymes chose different evolutionary strategies. Catalytic improvements could be attributed to changes in both K_m for the DNA substrate and k_{cat} for the reaction. Further, most of the catalytic improvement was due to changes in only four residues, although substitutions at a number of other positions also contributed to the altered activities. The most active clones have a double mutation, GA to UG at positions 313 and 314, that disrupts a pairing in the wild-type ribozyme that would normally position the terminal guanosine adjacent to the 3′ intron–exon junction. These ribozymes not only cleave DNA more readily, but also decrease the rate of hydrolysis at the 3′ intron–exon junction. In contrast, other ribozyme variants had sacrificed fidelity for activity and could splice themselves into different positions on the DNA primer.

Because the DNA cleavage activity of the ribozyme population looked as though it was still improving by generation 10, Tsang and Joyce[35] continued the selection for an additional 17 generations. To increase the stringency of the selection, the concentration of the DNA substrate was decreased by 50-fold; this modification of the protocol should have favored those catalysts that could bind substrate most tightly and/or those catalysts that could more efficiently react following substrate binding. In fact, an examination of the kinetic and substrate binding properties of individual variants and the evolving pool, respectively, showed that by generation 18 K_m had dropped about 50-fold relative to generation 9, while K_d had decreased about 10-fold. At this stage the variants were saturated with substrate, and further improvements in catalysis were encouraged by reducing the time allowed for the splicing reaction to occur by 12-fold. Again, the population responded, this time by increasing k_{cat} about 50-fold between generations 18 and 27. Overall, the catalytic efficiency (k_{cat}/K_m) of individual generation 27 ribozymes had improved by 1000-fold relative to generation 9 and by 100,000-fold relative to the wild-type parent.

However, while the evolved ribozymes had drastically improved DNA-cleaving abilities they also retained the ability to cleave RNA. In fact, the generation 27 ribozymes still cleaved RNA approximately 10-fold better than DNA. To further improve the DNA-cleaving abilities of these ribozymes and to reverse their substrate preference, additional rounds of selection were performed in which the DNA substrate was mixed with a RNA oligonucleotide corresponding to the product of the DNA cleavage reaction.[36] Those ribozymes that could preferentially bind the DNA substrate relative to the RNA product analogue could also potentially cleave the DNA substrate and capture the sequence tag necessary for amplification. In contrast, those ribozymes that could not discriminate between DNA and RNA oligonucleotides would likely be inhibited by binding of the RNA product analogue. After 36 additional generations of *in vitro* selection, the ribozymes accumulated additional mutations that allowed them to cleave the DNA oligonucleotide 50-fold better than the generation 27 ribozymes (k_{cat}/K_m), and to utilize the DNA substrate roughly 100-fold better than the corresponding RNA substrate.

Unexpectedly, the generation 27 ribozymes could also carry out the cleavage of linkages other than phosphodiester bonds.[37] As with the original, engineered ribozyme, substrates that could pair with the internal guide sequence were positioned adjacent to the activated terminal guanosine of the ribozyme. However, the evolved ribozyme had apparently either increased the nucleophilicity of the hydroxyl moieties on the guanosine and/or reduced the specificity of the attack, since a phosphodiester bond between deoxyribose and arabinose could be efficiently cleaved. Moreover, when an amide bond was introduced between a 3′ terminal amine and a 5′ terminal carboxylate it could also be cleaved by the ribozyme, which formed a transient ester linkage to the cleavage product. The amide linkage could be introduced between two modified DNA molecules, or between DNA and an amino acid or peptide. While the rate of amide hydrolysis was only ca. 10^{-5} to 10^{-6} per min this is still roughly 1000-to 10,000-fold faster than the background reaction.

10.5 INVENTING CATALYSIS: THE ROAD TO THE RIBOSOME

It has been hypothesized that ribozymes preceded protein enzymes during evolution. If so, then it is likely that ribozymes invented translation, and that ribozymes should be able to carry out many of the reactions that occur during translation. This hypothesis has recently received support from *in vitro* selection experiments that have successfully recapitulated the activities of some of the individual steps in translation (Figure 8; for a review see Hager *et al.*).[38] First, while contemporary ribozymes have proven adept at phosphodiester bond transfers, ancient ribozymes in any nascent metabolism would have also had to build and manipulate high-energy phosphoanhydride bonds. For example, the synthesis and utilization of ATP would likely have been critical to almost any aspect of metabolism. Lorsch and Szostak[39] have selected ribozymes that can carry out reactions similar to cellular kinases, which typically transfer the gamma phosphate of ATP to a substrate. An ingenious selection scheme was devised in which a RNA pool was first mixed with an ATP analogue (ATP-γS) that contained a sulfur in place of one of the terminal, non-bridging oxygen atoms. Those RNA molecules that could self-phosphorothiolate themselves were captured on a thiol affinity column. Several major classes of kinases were identified that utilized either the 5′ hydroxyl or internal 2′ hydroxyl residues as nucleophiles. The rate acceleration of one of the fastest ribozymes was estimated to be an astounding 10^9-fold. While several of the kinases could also utilize ATP as a substrate, they preferred the thiol analogue by roughly two orders of magnitude. Further kinetic analysis of one of the classes indicates that the ribozyme mechanism is relatively simple: the ATP cofactor binds adjacent to a 5′ hydroxyl and the enzyme promotes the formation of a metaphosphate-like transition state.[40] Similarly, Huang and Yarus[41] have selected RNA molecules that catalyze the conjugation of phosphate-containing ligands to a 5′ alpha phosphate, as opposed to a 5′ hydroxyl. These ribozymes were isolated by first selecting pool RNAs that could hydrolyze their initiating 5′ triphosphates to monophosphates. The monophosphorylated species were subsequently circularized, separated from linear species by gel electrophoresis, and preferentially amplified. The nascent pyrophosphatase activity derived from this selection proved to be relatively nonspecific, and nucleophiles other than water, such as pyrophosphate, could also attack the 5′ triphosphate and displace the 5′ beta and gamma phosphates. The pyrophosphate exchange reaction was further enhanced by selecting for RNA molecules that could immobilize themselves on a UTP column (forming a tetraphosphate linkage). Immobilized species were amplified *in situ*, purified, and re-selected. After a total of 20 cycles of the first selection method and four cycles of the second selection method a predominant family of catalysts was identified; this family enhanced the rate of pyrophosphate exchange by at least 500,000-fold compared to unselected RNAs. Moreover, when assayed with a variety of substrates that contained terminal phosphates, the pyrophosphate-exchanging ribozyme could catalyze its conjugation to virtually all of them, from nucleoside monophosphates (forming a 5′–5′ pyrophosphate bond) to nucleoside triphosphates (forming a tetraphosphate linkage) to $N\varepsilon$-phosphate arginine (forming a pyrophosphoramidate bond).[42] The remarkable versatility of this ribozyme indicates that it is possible that primordial ribo-organisms could have generated and manipulated a wide variety of metabolic intermediates containing high-energy phosphoanhydride bonds.

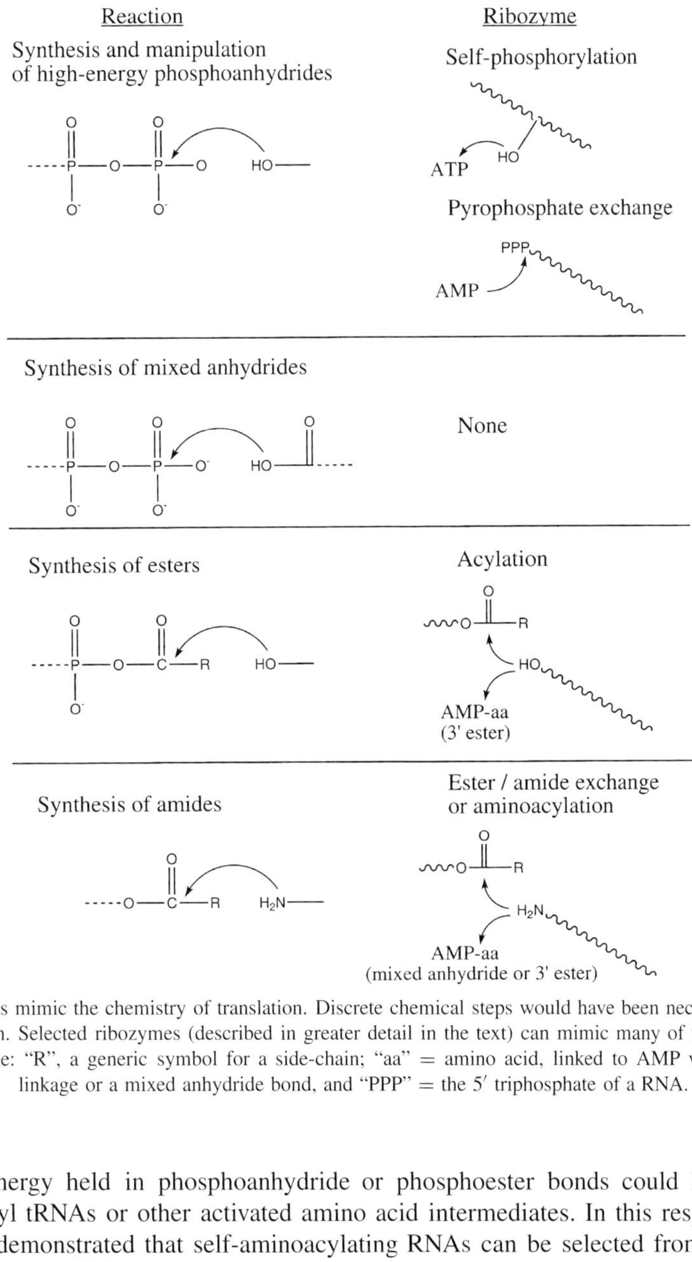

Figure 8. Ribozymes mimic the chemistry of translation. Discrete chemical steps would have been necessary for the metabolic invention of translation. Selected ribozymes (described in greater detail in the text) can mimic many of these. The abbreviations in this diagram include: "R", a generic symbol for a side-chain; "aa" = amino acid, linked to AMP via either a (2′) 3′ ester linkage or a mixed anhydride bond, and "PPP" = the 5′ triphosphate of a RNA.

Second, the energy held in phosphoanhydride or phosphoester bonds could have been used to generate aminoacyl tRNAs or other activated amino acid intermediates. In this respect, Yarus and his co-workers have demonstrated that self-aminoacylating RNAs can be selected from random sequence pools.[43] An oligonucleotide pool that contained 50 random sequence positions was incubated with phenylalanyl-AMP, the same mixed anhydride intermediate that occurs during the biosynthesis of the ester bond of phenylalanyl tRNA. RNA molecules that attached a phenylalanyl residue to themselves were further derivatized with a naphthoxyacetyl moiety, isolated on a reversed-phase column based on their increased hydrophobicity, and subsequently amplified and re-selected. After 11 cycles of selection and amplification individual clones were analyzed and assayed for their ability to acylate themselves. One of the clones was found to be highly reactive, and was estimated to accelerate the rate of ester formation by roughly 250,000-fold (however, see Illangasekare *et al.*[44] for corrections). Interestingly, the clone is aminoacylated on its 2′ (3′) terminus, just as tRNA is. Further examination of the self-aminoacylating species revealed that multiple different species were catalytic, and that the catalytic core of the most active species can be pared down to a mere 43 residue RNA of which only 17 residues were in the initial random sequence population.[44] While the ribozyme was specific for the AMP leaving group, it differed from cellular tRNA synthetases in that it showed little preference for its cognate amino acid phenylalanine and could also conjugate itself to serine or alanine.[45] However, a different isolate (RNA 77) from the same selection does show a marked preference for phenylalanine (or tyrosine) over other amino acids.[46] Nonetheless, the plethora and simplicity of the selected catalysts bolster the contention that aminoacylation activity should have been accessible even to primordial RNA molecules.

Third, activated amino acids could have undergone additional chemical transformations, such as the conversion of the ester linkages of aminoacyl tRNAs to the amide linkages of peptide backbones. A number of unnatural ribozymes have been isolated that can perform such reactions. For example, Lohse and Szostak[47] have selected a ribozyme with acyl transferase activity. An oligonucleotide that was esterified at its 2′ (3′) hydroxyl with a methionine residue that was in turn biotinylated via its alpha amino group, was annealed to a random sequence population. Those RNA species that catalyzed the transfer of the methionine to themselves in turn captured the biotin moiety and were immobilized on a streptavidin column. The selected RNA species were then amplified and re-selected for acyl transferase activity. After 11 cycles of selection and amplification the activity of the pool had increased 10,000-fold, and the predominant class of selected ribozymes catalyzed the self-esterification of their 5′ hydroxyls. Just as isolated ribosomes have been shown to transfer activated amino acids to either hydroxyl or amine groups, the selected ribozyme can also use a 5′ amine as a nucleophile to generate an amide bond. More recently, Zhang and Cech[48] have selected ribozymes that can also synthesize an amide (peptide) bond; one ribozyme may partially resemble the portion of 23S rRNA implicated in peptidyl-transferase activity.[49] While Lohse and Szostak relied on an endogenous hydroxyl to act as a nucleophile, Zhang and Cech appended an amine moiety to their pool by tethering the amino acid phenylalanine to a 5′ guanosine monophosphorothioate (see Figure 14). The derivatized pool was again mixed with a suitably activated amino acid, a biotinylated methionine that formed an ester with the (2′) 3′ hydroxyl of adenosine monophosphate. Ribozymes that could form a peptide bond and displace the AMP leaving group were immobilized, amplified, re-derivatized, and again reacted with the activated methionine. After 19 cycles of selection and amplification the RNA pool exhibited substantial activity, and individual ribozymes were cloned and sequenced. In contrast to the results of Lohse and Szostak, Zhang and Cech observed several different classes of catalysts. One of the most populous and active catalysts had a k_{cat} of 0.05 min^{-1}, and could accelerate peptide bond formation by a million-fold relative to the uncatalyzed reaction. In addition, this ribozyme has been engineered to act in *trans* on two small substrates.[49] The fact that amide bond formation can be carried out by a relatively short, unmodified RNA provides experimental support for the hypothesis that ribosomal RNA may have arisen in a RNA world.

10.6 ASSESSING THE BREADTH AND DEPTH OF SELECTED CATALYSTS: THE LIGASES

While the examples cited above are remarkable both in terms of the range of reactions that are catalyzed and in terms of catalytic and kinetic properties of the ribozymes themselves, it is difficult to draw generalized conclusions from the individual experiments. The failure to develop a more general understanding of "ribozymology" is in part a result of the fact that until recently true comparative enzymology has been all but impossible. There were only a few natural ribozymes, and the unnatural ribozymes were selected from different pools, under different buffer conditions and stringencies, to catalyze quite different reactions. However, as more results have accumulated it has become possible to compare the catalytic and kinetic properties of at least one class of selected nucleic acid enzymes, the ribozyme and deoxyribozyme ligases. Bartel and Szostak[50] originally selected ribozyme ligases from a long and complex random sequence population. The initial pool spanned 220 random sequence positions and contained over 10^{15} different species. The selection procedure was conceptually simple but technically challenging: RNA sequences that could append a constant sequence "tag" to their 5′ ends were preferentially removed from the population by affinity purification and preferentially amplified by reverse transcription and PCR (Figure 9). Following selection, the putative catalysts could be re-generated by nested PCR, and the cycle could be carried out iteratively. After four cycles of selection and amplification, ligase activity could be detected in the selected population. An additional three cycles of selection were carried out in which additional mutations were incorporated into the population via error-prone PCR, and a final three rounds of selection were carried out under increasingly stringent, low magnesium conditions. The population continued to improve in each round, so that by Round 10 the population could catalyze the ligation of RNA strands over 7 million-fold faster than the uncatalyzed reaction. The number of selected ribozymes and corresponding complexity of the selected

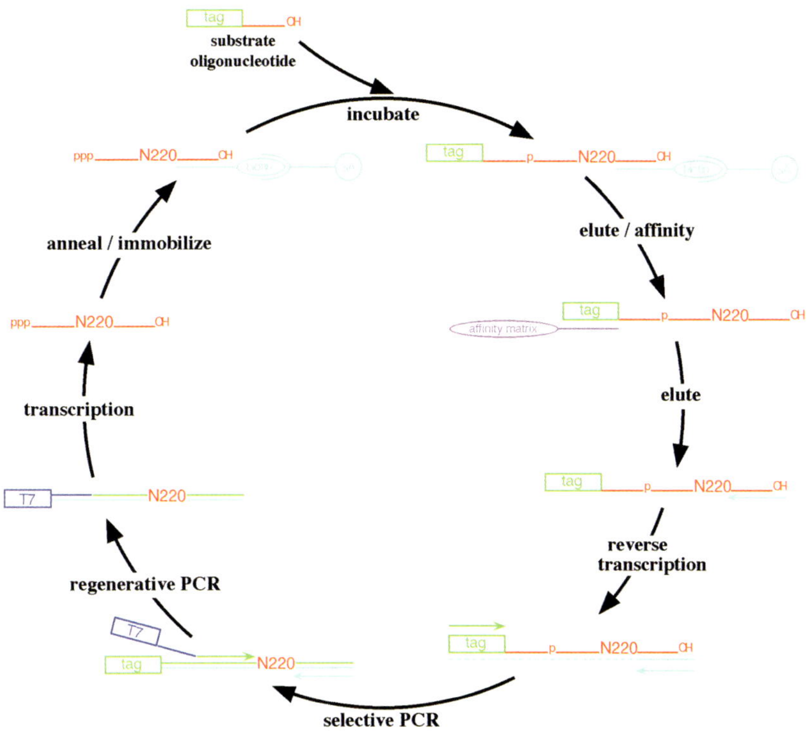

Figure 9. Selection of ribozyme ligases. This scheme was employed by Bartel and Szostak[50] to isolate ribozyme ligases from a RNA pool that spanned 220 random sequence positions. A biotinylated oligonucleotide complementary to the 3′ end of the pool was immobilized on a streptavidin column and was used in turn to immobilize pool RNAs. Immobilization prevented aggregation in the presence of high concentrations of magnesium. Those species that could ligate a substrate ribooligonucleotide to themselves were captured by affinity chromatography and preferentially amplified using primers specific for a "tag" sequence embedded in the substrate. A second round of PCR amplification was required to append the T7 RNA polymerase promoter to the double-stranded DNA templates. Transcription of the PCR products gave rise to RNA molecules that could be introduced into subsequent rounds of selection.

population was roughly assessed by looking for restriction polymorphisms. Interestingly, variants that predominated at early rounds of the selection were often displaced in later rounds, suggesting either that the programmed or spontaneous introduction of mutations led to the production of faster ribozyme variants during the course of the selection.

Individual ribozymes were cloned from the Round 10 population, and their sequences and kinetic properties were determined.[10,51] Of 66 clones that were sequenced, 45 were from a single predominant, homologous family, while the remaining 21 could be divided into 6 additional, evolutionarily related families. Several families showed primary and secondary structural similarities, and could in turn be grouped into three diverse classes of ligases. There was a single representative of Class I, three other families formed Class II, and the two remaining families were independent examples of Class III (Figure 10). Both Class II and Class III ligases formed 2′–5′ phosphodiester bonds. The Class I and Class II ligases could be engineered to act on substrates provided in *trans*, but their predicted secondary structures suggested that this was only because they could bind an alternative conformation of the hairpin stem originally provided in *cis*. This hypothesis was further bolstered by the fact that the Class II ligases were rate-limited by product release.

The selected ligases were surprisingly complex, both in terms of their functional primary sequences and their predicted secondary structures. The Class I ligase was partially randomized and re-selected for ligation function.[10,51] As discussed in Section 10.2, some residues were partially or completely conserved following selection and were presumed to be important for function; at least 93 residues are present in the core catalytic domain. Other residues covaried with one another following selection, verifying a predicted secondary structure that included several apparently stacked helical junctions and two pseudoknots (Figure 10).

Because of its speed and complexity, the Class I ligase has been used as a starting point for additional experimentation. For example, Wright and Joyce[52] have developed a scheme for the continuous evolution

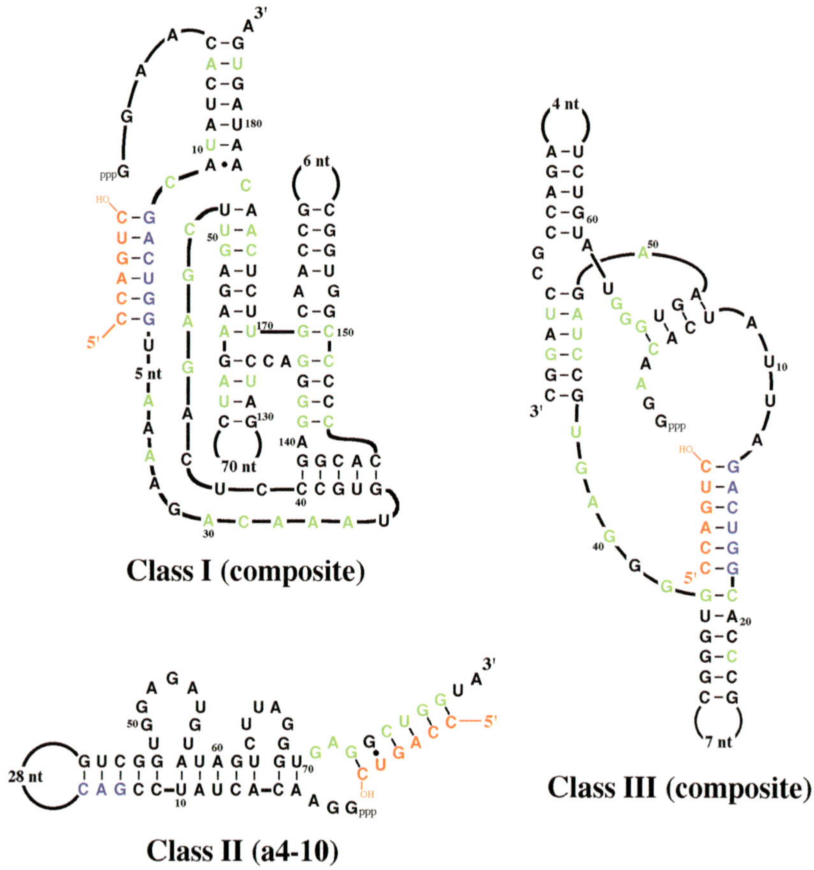

Figure 10. Selected ligases. Bartel and Szostak[50] discovered three major classes of ribozyme ligases by *in vitro* selection. The predicted secondary structures of these classes are shown. The Class I and Class III secondary structures are composites based on deletion and mutational analyses, while a single deletion variant from the Class II ligases, a4–10, is shown. Residues in red are the substrates for ligation, while residues in blue are the substrate binding sites originally programmed into the pools. Residues in green are highly conserved between functional mutational variants or independent isolates.

of the Class I ribozyme (Figure 11). In this scheme, the ligation junction of the Class I ligase has been altered so that it spans a T7 RNA polymerase promoter site. In the absence of ligation, the RNA cannot be amplified by isothermal amplification, because no promoter exists to generate new RNA. In contrast, variants that ligate effectively supply their own promoter and are preferentially amplified. Because isothermal amplification is a continuous process, any new RNA molecules that are generated by transcription of a ligated template, including any mutational variants, are simultaneously and iteratively subjected to this stringent requirement for survival. The continuously evolving ribozymes could be propagated by simple serial dilution of the population into new "food" (substrates, enzymes, and nucleotides), just as with a growing population of bacteria, rather than by the sequential isolation of ligated RNA, amplified DNA, and amplified RNA intermediates. Thus, over time it is expected that the fastest and/or most fecund species should predominate. After 100 cycles of continuous evolution, the catalytic efficiency of at least one member of the population had improved by over 10,000-fold, and its exponential growth rate had improved by over nine orders of magnitude. The continuously evolved ribozyme ligase was found to be much faster than Bartel and Szostak's[50] original ribozyme, and even faster than Ekland *et al.*'s[10] optimized ligase (albeit under slightly different reaction conditions).

It should be noted that the use of the Class I ligase was essential to the success of these experiments. Previous selections for ligation activity had been carried out using a Group II ribozyme and isothermal amplification, yet these did not yield a continuously evolving quasispecies.[53] The speed of the natural ribozyme was slow relative to the speed of the amplification scheme; so slow, in fact, that most ribozymes may have been converted by reverse transcription into double-stranded (and hence inactive) molecules before they could catalyze primer addition. The relatively slow and inefficient amplification of the natural ribozyme abetted the evolution of an interesting parasite, "RNA Z", which could more efficiently feed

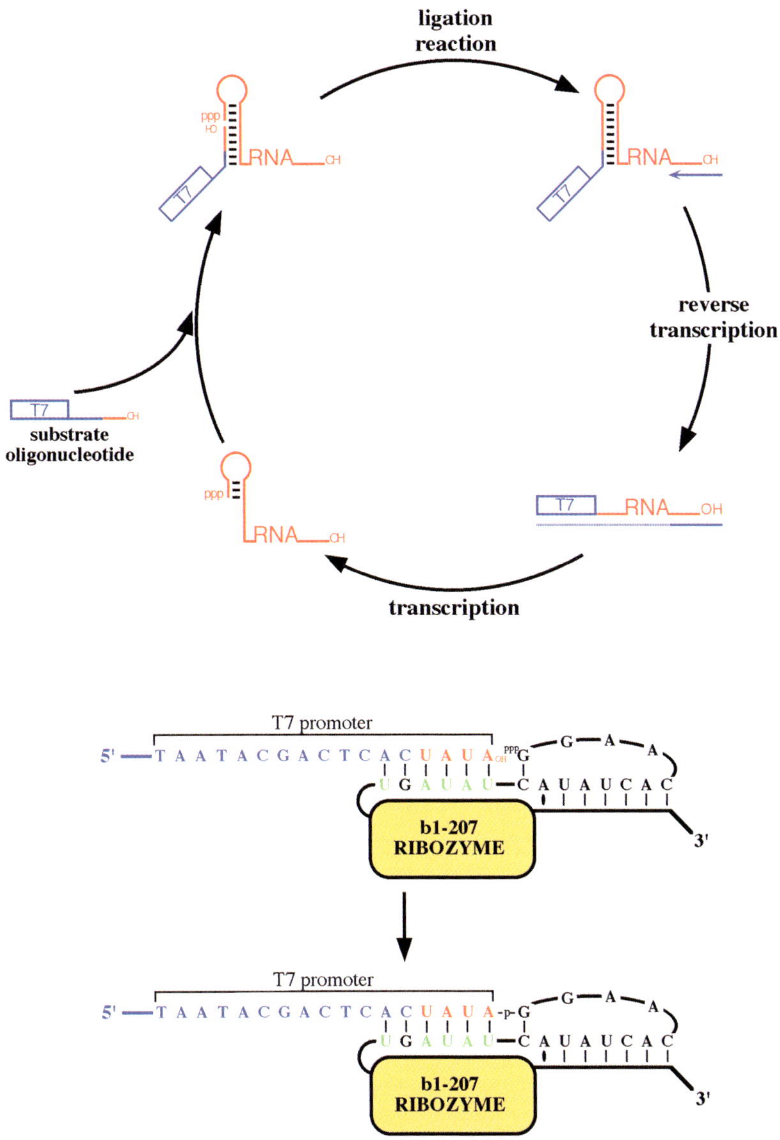

Figure 11. Continuous evolution of ligase activity. The scheme for continuous evolution is similar to the original scheme for the selection of ligase activity (Fig. 9). However, the key difference between these schemes is that the ligation junction has been altered to correspond to the T7 RNA polymerase promoter. Residues in blue are DNA in the ligation substrate, while residues in red are RNA. Residues in green have been mutated in the "core" ribozyme to accommodate the new ligation substrate. Upon ligation and reverse transcription, a hybrid molecule is created in which a double-stranded DNA that includes the T7 RNA polymerase promoter abuts a RNA : DNA duplex. This hybrid can be used as a template for transcription, and thus the entire cycle of selection and amplification that is normally carried out in discrete steps can be carried out continuously.

off of the isothermal amplification scheme.[53] This fate could potentially also have befallen the Class I ligase: the variant that Wright and Joyce started with was a factor of 10,000 slower than its parent, and was also originally too slow to participate in the continuous evolution scheme. Therefore, a population of sequence variants centered on the Class I ligase was generated, and more efficient ligators were identified by 15 rounds of conventional selection. The resultant population was roughly 100-fold more active than the parent ribozyme, but variants in this population were still not active enough to undergo continuous evolution. Ligation activity was further increased by carrying out 100 cycles of a simplified selection protocol, in which ribozymes were allowed to react for five minutes and were then transferred to the amplification reaction mixture. The quasispecies generated by this "manual" evolution scheme was finally adequate to kick start continuous evolution. What is particularly remarkable is that the "capacity" of the Bartel ligase for functional evolution was great enough to withstand adaptation to a completely new ligation junction and a 10,000-fold declination in activity.[54]

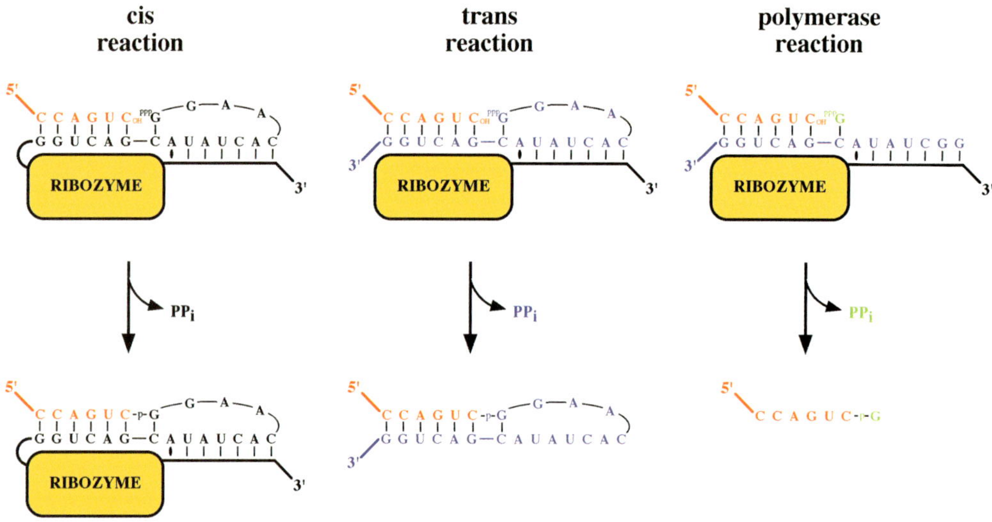

Figure 12. Converting a ribozyme ligase to a ribozyme polymerase. The Class I ligase selected by Bartel and Szostak[50] can be engineered to carry out a limited, template-directed polymerization reaction.[55] The original ribozyme (left, "*cis* reaction") ligates an oligonucleotide substrate to itself with concomitant displacement of pyrophosphate. The ribozyme's template can be split away from the remainder of the ribozyme (middle, "*trans* reaction"). The core ribozyme can bind the template in *trans* and can catalyze the ligation of the substrate and template to form a hairpin stem. Finally, the ligation junction can be sheared away and the core ribozyme will still catalyze the template-directed addition of nucleoside triphosphates. If additional residues are included in the template between the substrate : template and template : ribozyme pairings then the sequential, template-directed addition of mononucleotides can be observed.

Finally, Ekland and Bartel[55] were able to convert the Class I ligase into a template-directed polymerase by separating the enzyme into three domains. The "template" domain could pair with the "catalytic" domain, and the "substrate" domain could in turn pair with the "template" (Figure 12). Following this division, it was found that a single nucleotide could be added to the substrate in a template-directed manner at a position corresponding to the original ligation junction. Amazingly, when additional, unpaired residues were inserted between the substrate–template and template–catalyst pairings the substrate could nonetheless be serially and faithfully extended in a template-directed manner.

While the various Bartel ligases are perhaps the best-studied of the selected ribozymes, other selection experiments have yielded additional ligases with novel sequences and mechanisms (Figure 13). Chapman and Szostak[56] selected a ribozyme ligase that formed a 5′, 5′ P^1, P^4-tetraphosphate linkage. The original pool spanned 90 random sequence positions, and the ligation substrate was a short, biotinylated oligophosphoramidate with an imidazole leaving group. Members of the original pool that could append themselves to the ligation substrate were sieved from the remainder of the population using a streptavidin column, and were subsequently amplified and re-selected. After eight cycles of selection and amplification a major class of ligases was identified. Similarly, Hager and Szostak[57] selected a ribozyme ligase from a pool that spanned 210 random sequence positions and that initiated with a 5′–5′ pyrophosphate linkage (AppG). The ligation substrate could pair with the 5′ constant region of the pool to form a hairpin stem that juxtaposed the substrate's 3′ hydroxyl with the pool's 5′ pyrophosphate. This ligation junction was designed to mimic the reaction intermediate normally formed by T4 DNA ligase, and the oligonucleotide pair was in fact an efficient substrate for the protein ligase. The similarity to the T4 ligase reaction mechanism was further emphasized by the inclusion of an ATP binding site (anti-adenosine aptamer) in the core of the pool. After only four cycles of selection and amplification the population was dominated by a single clone. Six additional rounds of selection were carried out using a mutagenic amplification regime. In parallel, a minor variant from round 4 was separated from the majority population and further selected for ligase activity. These secondary selections yielded ribozymes related to those originally observed in the round 4 populations, but with roughly 10- to 100-fold increases in ligase activity over the majority clone from round 4. All of the final variants that were tested formed 3′–5′ phosphodiester bonds and were specific for the 5′–5′ pyrophosphate linkage but not for the AMP leaving group. Finally, DNA as well as RNA can act as a catalyst (as we will see in greater detail in the next section), and a deoxyribozyme has been selected that can ligate DNA to itself.[58] A constant sequence substrate was activated with a phophorimidazolide and mixed

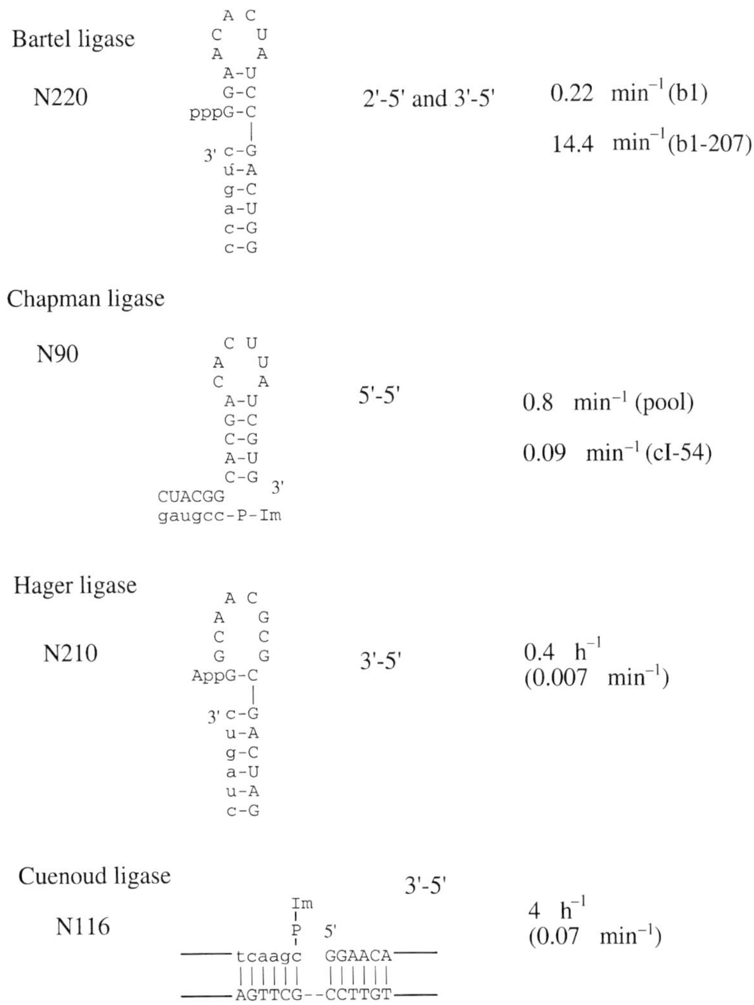

Figure 13. Chemistries of selected ligases. The programmed ligation junctions for the four different ligases described in this review are shown. None of these ligation junctions were actually used by the selected nucleic acid enzymes. In the first three instances, a 3′ hydroxyl present at the termini of a RNA pool was meant to displace a leaving group from the 5′ end of a phosphorylated substrate; the Chapman ligase departed from this scheme and catalyzed the formation of a 5′–5′ tetraphosphate linkage. For the Cuenoud ligase, a 5′ hydroxyl present at the termini of a DNA pool was meant to displace a leaving group from the 3′ end of a phosphorylated substrate. Other characteristics of the selected ligases, including the lengths of the starting random sequence pools, the types of linkages generated by ligation, and rates, are shown. The values reported for the rates of catalysis may only apply to a few members of a selected population.

with a random sequence DNA pool, and those species that acquired the constant sequence tag were preferentially amplified. After nine cycles of selection and amplification the ligation activity of the pool had been substantially enhanced. The deoxyribozyme was relatively slow (0.07 min^{-1}, k_{cat}) but was still 3400-fold faster than the templated background reaction.

10.7 METALLORIBOZYMES

Although it is apparent that the catalytic parameters of known ribozymes can be substantially altered by mutation and that novel ribozymes can be selected from random sequence populations, there are limits to what the chemistries inherent to the four canonical nucleotides can accomplish. Ribozymes composed only of G, A, U, and C lack acids or bases with pK_a values near 7 and have few strong nucleophiles. In consequence, many of the natural RNA catalysts have been shown to be metalloenzymes, possibly because metals are much better nucleophiles than most of the chemical moieties found on RNA. Since the choice and positioning of metals for catalysis is strongly dependent on RNA sequence and structure, these properties have also proven to be fodder for selection experiments. The ability of the hammerhead ribozyme to utilize magnesium in catalysis has been optimized.[59] A small random sequence library that contained from one to four random sequence positions in stem II was selected for its ability to cleave a target RNA in the presence of sequentially smaller concentrations of magnesium. Active complexes were separated from inactive by gel mobility shift, and at the conclusion of the selection variants were identified that could cleave the target 20-fold faster than the initial pool. The Group I self-splicing intron has been used again as an excellent model system for evolutionary manipulations, this time to alter the metal specificity of ribozyme catalysis. Although the *Tetrahymena* ribozyme cannot normally utilize calcium for RNA cleavage, Lehman and Joyce[60] started from the same doped sequence population used to isolate DNA-cleaving ribozymes and selected variants that could function in the presence of calcium and only trace amounts of magnesium. After eight cycles of selection and amplification, the selected population was 170-fold more active in the presence of calcium than the wild-type; individual clones were up to 300-fold more active. However, the selected ribozymes were still 1000 to 10,000-fold less active in calcium than the wild-type ribozyme was in magnesium. As was the case with ribozymes selected for DNA cleavage, the activity of the selected variants was due largely to specific changes at relatively few (six) positions. When the selection for calcium utilization was carried out an additional four generations[61] the population improved an additional threefold in a standard assay. The calcium-dependent ribozymes could now cleave RNA to the same extent as the wild-type enzyme with magnesium, although they were still kinetically slower. Similarly, Frank and Pace[62] selected RNase P variants from a partially randomized population that could cleave an appended tRNA substrate from themselves in the presence of calcium rather than the normative magnesium. Amazingly, a single point mutation in the highly conserved P4 pseudoknot structure conferred a 90-fold alteration in the metal specificity of the ribozyme. These results provide strong supporting evidence for the existence of specific binding sites for catalytic metals in ribozymes.

Not only can the metal binding properties of natural ribozymes be modified, but artificial metalloribozymes can be generated as well. Pan and Uhlenbeck[63] set out to map the residues and structures involved in a lead-catalyzed tRNA self-cleavage event that had previously been observed. The tRNA molecule was segmentally randomized, circularized, and those variants that could linearize themselves were selected and amplified. After six cycles of selection and amplification catalysts were identified that performed lead-dependent cleavage reactions at a variety of sites. One variant could be pared to an active stem–internal loop–stem structure that was quite different from that of the parental tRNA. The cleavage reaction was highly specific for lead, and appeared to use lead hydroxide as a base.[64] As expected, the mechanism of cleavage involved the formation of a 2′,3′ cyclic phosphodiester bond.[65] Subsequently, however, the lead : RNA complex also catalyzed the resolution of the cyclic phosphate into a 3′ monophosphate ester, just as ribonuclease A does.

The importance of metals for nucleic acid catalysis is even more apparent when we examine the selection of DNA enzymes (deoxyribozymes or "dinozymes"). Although DNA is best known as an informational macromolecule, rather than as a progenitor of shape and catalytic functionality, the secondary and tertiary structural interactions that DNA is capable of should be almost as complex as those already observed in structured RNAs. To this end, several groups have selected "deoxyribozymes" or "dinozymes" from random sequence populations (reviewed in Burgstaller and Famulok, Bashkin, Breaker).[66–68] Initially, Breaker and Joyce[69] selected a DNA enzyme that could cleave a RNA substrate. A DNA pool was synthesized that contained a single ribotide at the junction between the constant and random sequence regions. The pool was immobilized via a terminal biotin moiety, the cleavage reaction was initiated by the addition of lead cation (Pb^{2+}), and those variants that could cleave themselves from the column were eluted, amplified, and re-selected. After only five cycles of selection and amplification the population was significantly enriched in self-cleaving variants. A major variant was identified, and as expected performed self-scission at the labile ribose linkage. The major variant could be engineered to act in *trans* as a mere 38-nucleotide deoxyribozyme. Despite the fact that the catalyst was both short and composed entirely of DNA, it exhibited a rate enhancement of roughly 100,000-fold compared to the uncatalyzed reaction (in the presence of lead). Mechanistic studies of the lead-dependent deoxyribozyme in which lanthanides were substituted for lead indicated that a single metal ion was involved in catalysis and that binding to the metal occurred almost exclusively via the 2′ hydroxyl of the unique ribotide.[70] Using the same selection scheme, Breaker and Joyce[71] selected deoxyribozymes that could cleave the phosphodiester linkage adjacent to the ribose in the presence of either magnesium, manganese, or zinc. The deoxyribozymes selected in the presence of manganese and zinc exhibited at least some activity in the presence of non-cognate metals, with the exception of magnesium. Clones from these selected populations frequently showed sequence similarity to the deoxyribozyme originally selected in the presence of lead. In contrast, the deoxyribozyme selected in the presence of magnesium was uniquely active with magnesium and was dominated by a single species whose sequence and structure were unlike those of the lead cleavage motif. Although the nascent rate of magnesium-dependent cleavage of the phosphodiester linkage adjacent to ribose was three orders of magnitude slower than the rate of lead-dependent cleavage, the magnesium cleavage motif gave a catalytic improvement (k_{cat}) of 100,000-fold, similar to the improvement seen with the lead cleavage motif. Faulhammer and Famulok[72,73] also selected magnesium-dependent deoxyribozymes that could cleave a phosphodiester bond adjacent to a single ribose.

Building on these results, Carmi *et al.*[74] selected deoxyribozymes that could cleave an all-DNA substrate. While the selection scheme was similar to that previously employed by Breaker and Joyce,[69] the reaction was initiated with copper (Cu^{2+}) rather than lead cations. After seven cycles of selection and amplification, two major sequence classes of catalysts were identified. Both classes attacked sequences in their constant regions, but appeared to utilize different cleavage chemistries. Both ribozymes appeared to rely on radicals to initiate cleavage, but the first class (CA1) required ascorbate (included during the original selection) for radical generation, while the second class (CA3) could cleave in the presence of copper alone. Both classes of deoxyribozymes generated a range of cleavage products, consistent with localized radical generation and diffusion from an enzyme active site. Although the maximal rates of the selected catalysts were roughly 10- to 100-fold slower than the rates observed with the deoxyribozymes that cleaved a RNA–DNA linkage (above), the rate enhancement was greater (10^6-fold above background) because of the greater stability of the all-DNA substrate. The second class of ribozymes was further optimized by additional rounds of selection and amplification, and was engineered to act as a restriction endonuclease for single-stranded DNA.[75] The hybridizing arms of the optimized catalyst could be altered to recognize and form duplexes with different DNA substrates. Surprisingly, some cross-reaction between different substrates was observed, due in part to the fact that the deoxyribozyme recognizes its substrates not only via duplex formation, but also via the presentation of a triplex strand.

Finally, Santoro and Joyce[76] selected a deoxyribozyme that can quickly and efficiently cleave RNA of almost any sequence. Instead of including a single ribonucleotide in the constant region, a RNA dodecamer was incorporated into the 5′ end of a DNA pool as a potential cleavage target, and the reaction was initiated by the addition of magnesium. After ten cycles of selection and amplification a wide variety of sequences were found that could cleave the RNA oligonucleotide. Two deoxyribozymes were identified that could form obvious base-pairs with their RNA substrates; outside of the hybridizing arms, these deoxyribozymes contained only 13 and 15 residues. As expected, these deoxyribozymes were dependent on magnesium to facilitate RNA cleavage. The deoxyribozyme that contained a catalytic core of 15 deoxyribonucleotides (10–23) could cleave its RNA substrate in *trans*, could be rationally altered to target other RNA sequences, and had a second-order rate constant (k_{cat}/K_m) in excess of 10^8 under multiple turnover conditions, the same order of magnitude as a protein enzyme such as ribonuclease A.

The metalloribozymes that have so far been considered in this section have catalyzed phosphodiester bond rearrangements, primarily cleavage reactions. However, metals can also aid in other types of catalysis and can even themselves be substrates for catalysis. Conn *et al.*[77] selected a RNA molecule from a random sequence population that could non-covalently bind to a porphyrin, *N*-methylmesoporphyrin IX (NMM). This compound is a distorted porphyrin analogue and is thought to conformationally resemble the transition state for porphyrin metallation. Antibodies that bind NMM can catalyze porphyrin metallation, and it was hypothesized that anti-NMM aptamers might also catalyze porphyrin metallation. In fact, a small (35-nucleotide) aptamer could catalyze the copper metallation of porphyrins. The k_{cat}/K_m values for the artificial catalyst rivaled those of protein ferrochelatases. Li and Sen[78] selected anti-NMM aptamers from DNA libraries and these also proved to be deoxyribozyme chelatases. The rate enhancement was approximately 1700-fold over the background reaction (k_{cat}) and was only 40-fold less efficient than the corresponding catalytic antibody (k_{cat}/K_m). The deoxyribozyme could catalyze the metallation of meso-, deutero-, and protoporphyrin IX, while the catalytic antibody was specific for mesoporphyrin alone. Further optimization of the deoxyribozyme[79] reduced its size to only 24-nucleotides while optimization of buffer conditions resulted in an improvement in catalysis to 3700-fold above background (k_{cat}/K_m), a value comparable to that of catalytic antibodies and natural ferrochelatases. Based on its high guanosine content and dependence on potassium, the deoxyribozyme was thought to form some manner of G-quartet structure.

10.8 MODIFIED CATALYSTS

While the folded structure of a ribozyme may create an environment that augments the chemistries of the individual bases, sugars, or phosphate backbone, a quick comparison between the building blocks of ribozymes and the building blocks of protein enzymes makes obvious why most of the RNA world was supplanted roughly 3.5 billion years ago. To the extent that metal ions enhance ribozyme catalysis, it might be expected that other chemical moieties would also yield improvements in either the versatility or speed of ribozymes. Chemical modifications can be introduced into random sequence pools in two ways: incorporation at the termini of RNA or DNA, or incorporation internally via modified nucleotides. Both methods have yielded functional catalysts.

We have previously seen that an amino group incorporated at the termini of a RNA led to the selection of an amide synthase (Figure 14).[48] A similar strategy was employed by Wiegand *et al.*[80] for the selection of amide synthases. These authors conjugated an amine to the beginning of RNA molecules by ligating a modified DNA oligonucleotide to the RNA pool. In addition, though, the RNA pool contained a modified uridine residue, 5-imidazole uridine, in place of uridine (Figure 14). The modified pool was mixed with AMP-biotin, and biotinylated RNAs were separated on streptavidin, amplified, and re-selected. After

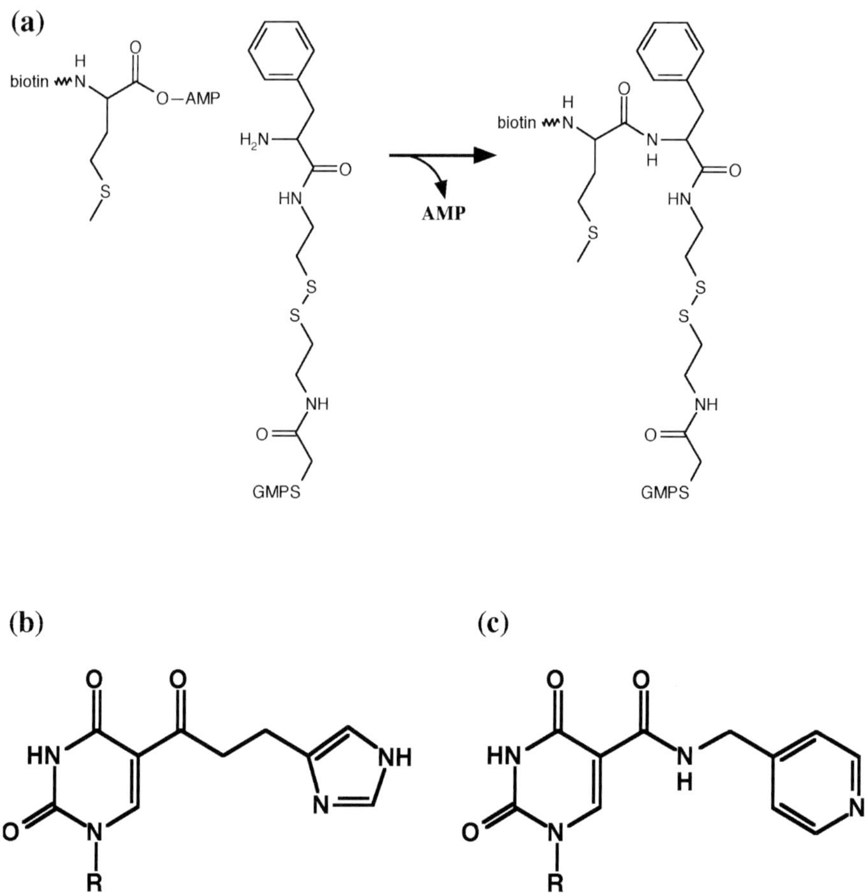

Figure 14. Ribozyme modifications. (a) Derivatization of the 5′ end of a ribozyme. Zhang and Cech[48] initiated transcripts with guanosine monophosphorothioate (GMPS), and alkylated the single sulfur moiety on these transcripts with *N*-bromoacetyl-*N*′-phenylalanyl cystamine. The final products thus contained an amino acid (phenylalanine) joined to RNA via a reversible disulfide bond. The derivatized pool was mixed with biotinylated methionine esterified to the (2′) 3′ hydroxyl of AMP and ribozymes that could exchange the external ester for an internal amide were selected. (b) The modified base 5-imidazole-UTP used by Wiegand *et al.*[80] for the selection of ribozyme amide synthases. (c) The modified base 5-pyridylmethylcarboxamide-UTP used by Tarasow *et al.*[81] for the selection of ribozyme Diels–Alder synthases. R = ribose 5′ triphosphate.

sixteen cycles of selection and amplification, numerous catalysts remained in the population but, as was the case for the sulfur alkylases described above, many of these catalysts contained a common, short-sequence motif. The rate enhancements of the selected catalysts were on the order of 10^4- to 10^5-fold, similar to the 10^6-fold rate enhancement claimed by Zhang and Cech[48] for their amide ligases. The ribozyme recognized the biotin moiety that it attached to itself, but did not appear to recognize the leaving group, as UMP, AMP, and ribose monophosphate derivatives were all effective donors. These experiments not only reinforced the notion that amide synthases could have existed in a putative RNA world, but also showed that chemical augmentation could significantly influence RNA catalysis. At least one of the selected catalysts was absolutely dependent on the modified nucleotide for activity, although the precise role of the modification in catalysis was unknown. The characterized ribozyme could function in the absence of divalent metals, but its reaction rate was enhanced by divalent ions such as magnesium and calcium and was most significantly increased by the addition of Cu^{2+}. Interestingly, the copper ions appeared to contribute to substrate binding rather than to chemical catalysis.

Terminal modifications have also allowed ribozymes to take advantage of sulfur chemistry. Wecker *et al.*[82] synthesized a RNA pool that initiated with a guanosine monophosphorothioate. When this pool was incubated with an activated variant of the peptide bradykinin (*N*-bromoacetyl-bradykinin) individual RNA species were alkylated at their 5′ ends and could be separated from the remainder of the population on [(β-acryloylamino)phenyl]mercuric chloride polyacrylamide gels. In this gel system, RNA molecules with unmodified thiolates are retarded relative to RNA species with blocked thiolates, such as those that have appended bradykinin to their 5′ ends. After twelve cycles of selection via gels and affinity chromatography on thiol columns the RNA population showed roughly 200-fold improvement in alkylation activity. Similarly, Unrau and Bartel[83] have used thiol-trapping gels to identify thiouridine synthetases. A RNA pool that terminated in a phosphoribosyl pyrophosphate moiety was incubated with 4-thiouracil, and those species that could form glycosidic bonds were selected. After eleven cycles of selection and amplification the small nucleobase substrate was joined to the selected pool over ten million times faster (k_{cat}/K_m) than to the unselected pool. The resultant ribozymes showed great specificity for 4-thiouracil, and reacted slowly or not at all with other thiol-bearing pyrimidine bases and uracil itself.

Eaton and his co-workers have also used modified nucleotides to move ribozymes into new catalytic realms. Tarasow *et al.*[81] report the selection of ribozymes that can catalyze carbon–carbon bond formation, enhancing the rate of a Diels–Alder reaction. One "half" of the product, the diene, was conjugated to the end of a RNA pool via ligation. The pool again contained a modified uridine, 5-pyridylmethylcarboxamid-uridine (Figure 14), in place of the natural nucleotide. A dienophile conjugated to biotin was mixed with the pool, and biotinylated species were selected over twelve cycles. The selected sequences fell into several families, but the families were again related by a core motif. One of the selected species enhanced the rate of the Diels–Alder cycloaddition by 800-fold. As was observed with the modified amide synthases, catalytic activity was dependent on the presence of the modified nucleotide in transcripts.

10.9 CONCLUSIONS

10.9.1 Natural Ribozymes and Selection: Structure Determination

The benefits of artificial phylogenetic analyses for nucleic acid structure prediction become apparent when RNA and protein catalysts are compared. The primary sequence of both types of biopolymers determines their global architecture, which in turn determines their catalytic functions. However, protein folds are determined partially by the formation of secondary structural elements, and largely by the establishment of tertiary packing interactions in a hydrophobic core. The secondary structures of proteins can sometimes be predicted from primary sequence, but the further identification of tertiary structural interactions is difficult. In consequence, predicting the three-dimensional structure of a protein from primary sequence remains a daunting problem.

In contrast, some of the features that make nucleic acid catalysts less structurally diverse than proteins make their structures far easier to predict. Whereas protein catalysts are derived from a repertoire of 20 amino acids with a wide array of chemical properties, ribozymes are built from a cast of four chemically similar nucleotides: adenosine, guanosine, cytidine, and uridine. Whereas the interactions between amino acids in protein cores are idiosyncratic and highly dependent on context, the interactions between nucleotides in functional RNAs tend to occur along a single plane and are much more independent of context. In short, a G:C base-pair in one helix is structurally similar to a G:C base-pair in another helix. Moreover, the global folds of ribozymes are determined largely by secondary structural elements, and only partially by tertiary contacts. Therefore, establishing a solid secondary structure is a major step towards determining the overall three-dimensional structure of a ribozyme.

We have seen that selection was useful for confirming the proposed secondary structure of the hairpin ribozyme. It might also have proven useful for determining or confirming the secondary structure of the Group I ribozyme, except for the fact that abundant natural phylogenetic data already existed. In consequence, artificial phylogenetic analyses have been most useful in determining the structure of new ribozymes. The Class I Bartel ligase is as large and easily as complex as large, natural ribozymes such as the Group I ribozyme. The secondary structure of the Class I ligase was established almost solely based on sequence covariations that were observed following partial randomization and re-selection for function.

10.9.2 Natural Ribozymes and Selection: Nature's Search Engine

An interesting question that can be uniquely explored by *in vitro* selection experiments is to what extent are natural ribozymes functionally optimal. Or, in other words, are there other ribozymes with different sequences that are functionally superior to natural ribozymes? Admittedly, a precise answer to this question is unobtainable, since both function and survival are highly dependent on the environment in which a catalyst is observed. It would be impossible in a test tube setting to recapitulate the millions to billions of years of selection pressures that have yielded Group I ribozymes. Conversely, few ribozymes selected in test tubes have been optimized for function in organisms, and it is unlikely that unnatural ribozymes will be released into the wild solely to observe how they fare. Nonetheless, as long as we are cognizant of the caveat that the natural and unnatural worlds have different criteria for survival, then we can compare the results of natural and unnatural selection experiments.

From the vantage of the experimentalist, Nature's search engine appears to be an astoundingly good one. Even when all fourteen residues in the core of the hammerhead ribozyme are randomized and the population is selected for cleavage function the selected ribozymes "relax" back to the wild-type sequence. The eleven nucleotide tetraloop receptor motif in the Group I ribozyme can be recovered by *in vitro* selection. Similarly, Frank *et al.*[84] randomized 22 residues in and around the P4 (pseudoknotted) stem of ribonuclease P. After ten cycles of selection for cleavage away from a tethered tRNA molecule, every position within the randomized region had reverted to wild-type!

Of course, there is also ample evidence that there are many apparently unnatural catalysts that can function as well or better than their natural counterparts. For example, Costa and Michel[31] recovered both wild-type and non-wild-type tetraloop receptors in the same experiment. Tang and Breaker[23] found non-hammerhead motifs in the same cleavage selection that yielded hammerhead ribozymes. Cleavage selections in which metals other than magnesium were included in the reaction have yielded unnatural motifs.

How can Nature's apparent optimality be reconciled with the apparently limited scope of Nature's catalysts? By assuming that Nature can identify local rather than global optima. While the hammerhead may in fact be a globally optimum solution to the problem of how to cleave double-stranded RNA that contains a UH (H = not G) target it is interesting to note that the selection experiments that did not return the hammerhead ribozyme either had an altered stem II[23] or a randomized stem II[24] in addition to a randomized hammerhead core. Similarly, large RNA molecules are not functionally constrained to their natural sequences. Tuschl *et al.*[85] synthesized random sequence libraries that contained U2 and U6 spliceosomal RNAs at their core. The libraries were selected for their ability to catalyze the ligation of an oligonucleotide substrate to themselves. It was hoped that this selection would yield spliceosomal-like catalysts that were independent of proteins, and that analysis of these catalysts might yield clues to the origin of the spliceosome. However, the selected catalysts bore little or no resemblance to the spliceosome either in terms of sequence or reaction mechanism.

These results can all be reconciled by assuming that ribozymes are sparsely represented in any given sequence space. During the course of evolution Nature may have chanced upon a given functional peak, and climbed to the top of that peak by point mutation and selection. However, precisely because Nature optimizes catalysts largely by point mutation it may have proven difficult to visit other, widely separated peaks that had equivalent function. At a guess, Nature may have visited at least all possible 15-mers in sequence space (e.g., the hammerhead), during the course of evolution but not all possible 30-mers or greater.

10.9.3 Unnatural Ribozymes and Selection: The Role of Metals

Many, but not all, of the ribozymes and deoxyribozymes we have considered are metalloenzymes. Since selection experiments to some extent examine the universe of possible catalysts (however, see Section 9.5), then by reviewing whether and how metals are adopted by selected catalysts the general importance of metals in nucleic acid catalysis can be assessed. Two questions are fundamental: first, are metals important for catalysis, and second, which metals are important for catalysis.

It is frequently assumed that divalent cations are critical for ribozyme catalysis.[86] Such an assumption at first glance seems reasonable, given the limited chemistries available to nucleic acids via the functional groups of the nucleotides. However, ribozymes may have other catalytic mechanisms available to them that obviate the role of metals in catalysis, such as the conformational stabilization of transition states or activated intermediates. Thus, while it is surprising, it is nonetheless reasonable that recent experiments have revealed that natural ribozymes are not necessarily metalloenzymes. Murray *et al.*[87] have assayed the hammerhead, hairpin and VS ribozymes in buffers devoid of divalent cations, and found that these ribozymes can catalyze self-cleavage in the presence of 4 M LiCl at rates that are identical to those in the presence of magnesium.

Unnatural ribozymes also do not seem to absolutely require divalent metal ions for catalysis. Faulhammer and Famulok[73] identified deoxyribozymes that could cleave a phosphodiester bond adjacent to a ribose and that required no divalent metal ion for catalysis. The rates of cleavage for the metal-independent deoxyribozymes were similar to those of the metal-dependent deoxyribozymes (ca. 0.005 min^{-1}). Deoxyribozymes selected by Geyer and Sen[70] were also metal-independent and also cleaved the targeted bond at a similar rate (ca. 0.01 min^{-1}). These values are from 1 to 3 orders of magnitude slower than those observed for natural ribozymes such as the hammerhead, and for other selected deoxyribozymes.[70] In contrast, Jayasena and Gold[88] have selected self-cleaving ribozymes from a random sequence population using a selection method similar to that used to select lead-dependent, tRNA-like ribozymes.[63] Thus, with the proviso that monovalent salts or general acids and bases can to some extent "repair" deficits in catalysis resulting from a lack of divalents, divalent metal ions may be worth factors of 100 to 1000 in the overall rates of nucleic acid catalysis.

Selection experiments can also reveal whether some metals are more amenable to certain types of catalysis than others. For example, several selections for catalysts have been carried out using a mixture of metals, yet the selected catalysts have a decided preference for one or more of the individual metals. The deoxyribozyme ligase[58] was isolated in the presence of both magnesium and zinc, yet required only zinc for activity. The capping ribozyme[41] was isolated in the presence of magnesium, calcium, zinc, and manganese, yet required only calcium for activity. The novel reactions catalyzed by the Diels–Alder synthase[81] and the amide synthase[80] required copper for activity, although a number of other transition metals were originally present in the selection buffer. While it is interesting that selected catalysts show preferences for individual metals, the preferences are not always absolute. For example, the "leadzyme" motif selected by Breaker and Joyce[53,71] could utilize a variety of other metals for catalysis, while the deoxyribozyme ligase could utilize copper in place of zinc.

10.9.4 Unnatural Ribozymes and Selection: The Role of Modifications

One of the newest innovations in ribozyme chemistry is the inclusion of modified bases in selection reactions. While the modified nucleotides that were incorporated into the selections for the Diels–Alder synthase and amide synthases were clearly essential for the catalytic activity of the ribozymes, it is unclear whether similar ribozymes could have been selected in the absence of the modified nucleotides and what the kinetic parameters of such unmodified ribozymes might be. It has been claimed that since previous selections for a Diels–Alder synthase with all-RNA pools had failed,[16] that it was the presence of a modified nucleotide that led to the later, successful selection of catalysts.[81] However, another more critical distinction between these experiments was that the all-RNA selection was an indirect selection for transition-state affinity, while the modified RNA selection was a direct selection for catalysis. In fact, Seelig and Jaschke[89] later selected an all-RNA ribozyme that could catalyze a Diels–Alder reaction.

A different, more tenable comparison between all RNA and modified RNA catalysts might be the comparison between an all-RNA amide synthase[48] and a modified RNA amide synthase.[80] Although the amine conjugates, selection procedures, and metal specificities of the selected ribozymes also differed, the substrates used for selection were similar and the two catalysts have surprisingly similar k_{cat}/K_m values. Further doubt regarding the superiority of chemically modified nucleic acid catalysts comes from a selection that attempted to produce an amide synthase but instead yielded an ester synthase. Jenne and Famulok[90] synthesized a RNA pool that was similar to those previously used by Zhang and Cech[48] and Wiegand *et al.*[80] for the selection of amide synthases: the pool containing a dipeptide, citrulline–cysteine, conjugated to its 5′ end. The delta amino group of citrulline would have been expected to have a nucleophilicity similar to the alpha amino group of the phenylalanine at the termini of the pool generated by Zhang and Cech[48] and the alkyl amino group of the pool generated by Wiegand *et al.*[80] However, after thirteen cycles of selection using a (2′) 3′ AMP ester of biotinylated phenylalanine as a substrate at least one of the selected ribozymes catalyzed the transfer of the amino acid ester to an internal 2′ hydroxyl rather than to the presumably much more nucleophilic terminal amine. Despite the ambiguity of these results, it is intuitively resonant, if not proven, that augmenting an evolving biopolymer with more or better chemical groups should yield faster, more specific catalysts.

The chemical augmentation of ribozymes need not occur via covalent modification or the incorporation of modified bases. Like protein enzymes, ribozymes could presumably non-covalently bind cofactors that would assist in catalysis. However, it has proven difficult to select for such catalysts. Faulhammer and Famulok[72,73] attempted to select for histidine-dependent deoxyribozyme cleavage of a phosphodiester bond adjacent to ribose, but identified only metal-dependent or metal-independent catalysts. In contrast, Roth and Breaker[91] have recently reported the selection of histidine-dependent deoxyribozymes using almost exactly the same selection procedure. The reaction is very specific for histidine and accelerates the rate of phosphodiester bond cleavage by roughly 10-million fold.

10.9.5 Unnatural Ribozymes and Selection: You Get What You Select

One caveat to the interpretation of the results of selection experiments is that while these experiments are carefully designed to produce catalysts with a desired activity, the catalysts that are eventually generated must in fact adhere to only one imperative: survival. In general, attempts to direct the reactivities of selected ribozymes are successful. For example, experiments and pools designed to select for ligase activity typically yielded ligases, not phosphatases. However, it is interesting to note that design features that are not implicitly required for survival are often blatantly ignored by selected ribozymes. For example, the Bartel ligases were selected from a pool that could form a hairpin stem at the ligation junction, just as the Group I ribozyme does (Figures 10 and 13). However, none of the selected catalysts used the designed stem in the original pool, and instead chose to generate new ligation junctions. The Class I ligase breaks up the designed stem by forming a pseudoknot, and only one residue holds the 3′ portion of the ligation junction in place; the Class II ligase does not use Watson–Crick pairing at the junction, instead relying on a series of purine residues that may form non-canonical pairings; and the Class III ligase uses Watson–Crick pairing for the 5′ portion of the ligation junction but not the 3′ portion. The Class I and III ligases use their original constant regions to bind the substrate for ligation, but the Class II ligase generates an internal template for the substrate as well. The Chapman, Hager, and Cuenoud ligases also generated their own internal substrate-binding sites from random sequence despite the fact that constant sequence ligation junctions were in each instance designed into the selections.

The tendency for selected ribozymes to develop their own unique catalytic mechanisms rather than acceding to the wishes of their designers is also apparent when non-oligonucleotide substrate binding sites are considered. Although the Lorsch kinases utilized ATP as a substrate, they did not coherently utilize the anti-adenosine aptamer included within the original random sequence pool. Multiple mutations accumulated in the anti-adenosine aptamers programmed into the original pool, and the K_m of the ribozymes for ATP was much higher than the original K_d of the aptamer. One of the best characterized aptamers had a K_m of 3 mM for ATP, while the original anti-adenosine aptamer had a K_d

of 1 μM for ATP.[40] Similarly, although the Hager ligases utilized AMP as a leaving group they appear to have ignored the anti-adenosine aptamer originally provided them. GMP is also efficiently used as a leaving group, and the final, selected ligases contained multiple mutations in their internal anti-adenosine aptamers.

Other steps in the selection procedure besides catalysis can also influence the nature of the survivors. Selections that involve immobilization run the risk of selection for binding species rather than catalysts. While this has not been reported, Jenne and Famulok[90] had hoped to select for amide synthases by first capturing an activated biotin then eluting catalysts from a streptavidin column by reversing the disulfide conjugation of a 5′ terminal amine. However, they noted that as the selection progressed increasingly higher concentrations of disulfide (up to 2 M) was required to elute captured species. Upon closer examination, it became apparent that the selected ribozymes were *trans*acylases and that their elution was likely due to disruption of biotin : streptavidin interactions rather than reduction of a (non-essential) disulfide bond.

Finally, there are multiple ways in which the selection process can be skewed, diverted, or thwarted during the amplification procedure. For example, replication "parasites" can arise during selections for catalysis, such as the RNA Z species that aborted the original attempts to develop a continuous selection scheme. While such parasites are an extreme example of how amplification can skew the course of a selection, the amplification protocol may influence the nature of selected catalysts as much as the selection protocol. In the selections for the Bartel and Hager ligases, the ligation junction was also a template for reverse transcription. In contrast, the selection that produced the Chapman ligases did not require that the ligation junction serve as a template for reverse transcription. Thus, in the selections for Bartel and Hager ligases only replicable (3′–5′ and the surprising 2′–5′) ligation junctions were identified, while in the selection for the Chapman ligases non-replicable (5′–5′) ligation junctions were identified.

10.9.6 The RNA World: Ribozymes Versus Protein Enzymes

Table 1 shows the rates and relative rate enhancements for many of the ribozymes and deoxyribozymes discussed in this review. What is surprising is the relative uniformity of nucleic acid catalysis over a variety of catalysts, selection procedures, pools, metal ions, and chemistries. When all is said and done, ribozymes and deoxyribozymes are not very fast compared to their protein counterparts. Overall, most nucleic acid catalysts have k_{cat} values of 0.01 min^{-1} to 1.0 min^{-1} and have k_{cat}/K_m values of roughly 100–1000, compared with values that are typically three orders of magnitude or more higher for proteins. The apparent exceptions to this rule, such as the Lohse *trans*acylase, have "artificially" high k_{cat}/K_m values that are driven by the extremely low K_m values for oligonucleotide substrates, or else catalyze reactions that were extremely facile to begin with, such as metallation of porphyrins.

What are the reasons behind this unflattering comparison between ribozymes and proteins? One might suggest that it is just that proteins have had so much longer to evolve, that billions of years in the wild will of necessity produce catalysts that are far more optimized and elegant than those that are produced within weeks or months in a laboratory. However, it should be recognized that the selective constraints present in any *in vitro* selection experiment are likely much, much more stringent than the constraints on molecular evolution that are present in the wild. There can be little or no neutral evolution during the initial rounds of a selection experiment: there is purely a highly streamlined selection for function. Moreover, as we have previously seen, the number of molecules that are queried for function during an *in vitro* selection experiment dwarfs the number of variants that may be present in the wild. Thus, even though it is likely that Nature has carried out more generations of selection than experimentalists it is unlikely that the catalysts that have been generated by *in vitro* selection are any less optimal.

A more optimistic caveat may be that the ribozymes of the ancient or modern RNA world will receive enormous assistance from compounds that contain functional groups with better pK_as and nucleophilicity, such as Eaton's modified bases or Breaker's histidine cofactor. It will be interesting during the next several years to observe whether or not the speed and specificity of selected catalysts improves along with improvements in nucleotide chemistry.

A final constraint on the speed of selected catalysts may be the way in which the selection experiments are carried out. While it is true that the selections are stringent, they are, with few exceptions, direct selections, selections for single-turnover reactions. Since the nucleic acid catalysts are not selected for

Table 1 Characteristics of selected nucleic acid enzymes.

Ribozyme	*Rate* [a]	*Acceleration* [b]	*Metal*	*Ref.*
Natural ribozymes				
Group I DNA cleavage	3×10^7 (m.M)$^{-1}$ (k/K)	100,000 rt wt	Mg	36
Group I Ca utilization		300 rt wt	Ca or Mg	58,59
Unnatural ribozymes				
Self-kinase	6000 (m.M)$^{-1}$ (k/K)	10^9 rt hydrolysis	Mg	40
Pyrophosphate exchange	0.03 min^{-1} (assay)	500,000 rt Rd0	Ca	42,43
Aminocylation	70 (m.M)$^{-1}$ (2nd order)	250,000 rt Rd0	Mg/Ca	44,46
Transacylation	4×10^6 (m.M)$^{-1}$ (k/K)	10,000 rt Rd0	Mg	47
	0.19 min^{-1} (k)			
	350 (m.M)$^{-1}$ (k/K)		Mg	89
Amide bond formation	260 (m.M)$^{-1}$ (k/K)	10^6 rt uncatalyzed	Mg	48
	290 (m.M)$^{-1}$ (k/K)	100,000 rt uncatalyzed	Mg/Ca Cu	81
Ligases				
Bartel Class I	100 min^{-1} (mt)	10^9 rt templated, uncatalyzed	Mg	9
	1×10^7 (k/K)			
Chapman	0.1 min^{-1} (k)	1000 rt uncatalyzed	Mg	54
Hager	0.007 min^{-1} (st)	500,000 rt templated, uncatalyzed	Mg	54
Cuenoud	0.07 min^{-1} (mt)	3400 rt templated	Zn or Cu	55
		100,000 rt untemplated		
Nucleases				
RNA cleavage by DNA	0.2 min^{-1} (st)	100,000 rt uncatalyzed	Pb	52,68
	0.02 min^{-1} (st)		Mg	70
	0.006 min^{-1} (st)		Mg or Ca	72
	3.4 min^{-1} (mt)	10^9 rt uncatalyzed	Mg	75
	0.01 min^{-1} (st)		–	69
DNA cleavage by DNA	0.04 min^{-1} (st)	10^6 rt uncatalyzed	Cu	73
Chelatases				
Ribozyme	125,000 (m.M)$^{-1}$ (k/K)	460 rt uncatalyzed	Cu	76
Deoxyribozyme	40,000 (m.M)$^{-1}$ (k/K)	3700 rt uncatalyzed	Cu or Zn	78
Modified ribozymes				
Diels–Alder synthase	240 (m.M)$^{-1}$ (k/K)	800 rt uncatalyzed	Mg/Ca or Cu	82

[a] Abbreviations include: $k = k_{cat}$; $k/K = k_{cat}/K_m$; m.M = minutes.molar; mt = multiple turnover; Rd0 = Round 0; rt = relative to; st = single-turnover.
[b] The rates and rate enhancements that are reported are generally taken directly from the primary literature. Usually only one of several rates or an average rate is reported. No attempt is made to reconcile apparently contradictory values (e.g., the differences between estimates for rate acceleration by Conn *et al.*[77] and Li and Sen[78] for similar reactions) or to judge the relevance of the uncatalyzed reaction (Round 0 pools vs. templated reactions vs. untemplated reactions).

more than a single turnover, it is possible that this structure in turn imposes an unseen limitation of the rate of catalysis that can be obtained.

10.9.7 The RNA World: Is Bigger Better?

The natural ribozymes range in size from the tiny hammerhead (ca. 20 nucleotides in length) to the giant Group I and Group II self-splicing introns (>200 nucleotides in length). Selected ribozymes span a similar size range: the Yarus aminoacylase had only 17 selected nucleotides in a core of 43 nucleotides, while the Bartel Class I ligase could not be reduced to less than 112 nucleotides in length. It might be *a priori* expected that the functional complexity of a catalyst should be related to its informational or structural complexity; that is, an enzyme that performs a difficult chemical reaction might be expected to require more residues arrayed in a more complex architecture than an enzyme that performs a simple chemical reaction.

Unfortunately, it is difficult to correlate the sizes and functional complexities of selected nucleic acid catalysts, since the reactions that are selected for are frequently quite different and often contrived. The hypothetical metric that relates size and complexity might be expected to be most easily observed within a given class of ribozymes: for example, a small ribozyme performing a chemical reaction might be expected to be less efficient than a much larger ribozyme performing the same reaction. This hypothesis has been bolstered by the careful analysis of the complexities of selected ribozyme ligases carried out by Bartel and his co-workers. Upon initial selection, the Class III ribozyme ligase contained at least 19 functionally important residues and could catalyze self-ligation at a rate of ca. 0.005 min^{-1}. The Class I ribozyme ligase contained upwards of 93 structurally or functionally important residues and could catalyze self-ligation at a rate of ca. 0.22 min^{-1}. More importantly, partial randomization and re-selection of the Class III ligase yielded a 38-fold functional improvement, while optimization of the Class I ligase yielded a 700-fold functional improvement (David Bartel, personal communication). Based on these results, it can be tentatively concluded that the size of the Class I ligase relative to the Class III ligase led not only to an initial selective advantage, but also to a much more robust evolutionary potential.

Interestingly, the Class I ribozyme ligase should not even exist. Based on the number of conserved and semi-conserved positions that were found following partial randomization and re-selection, the probability of the Class I ribozyme ligase existing in a pool of 220 random sequence nucleotides would have been ca. 4×10^{-19}. Since the initial pool contained only ca. 1.4×10^{15} sequences, the Class I ligase should only have been found once in every 2000 experiments. While it is possible that Bartel and Szostak were merely extremely lucky, the more likely explanation is also more intriguing: catalysts as complex as, but different in sequence than, the Class I ligase must be relatively abundant, at least in the universe of 220-mers.

To numerically address the relationship between pool size and catalytic activity, Sabetti *et al.*[9] appended long (148 residue) random sequence pools to the Class II and Class III Bartel ligases and assayed the catalytic activities of the resultant constructs. Depending on the placement of the pool relative to the ribozyme, the effects on the activities of the ribozymes ranged from zero to 18-fold inhibition. When individual clones were analyzed, the median effect was 5-fold inhibition. The primary conclusion of this study is that longer sequences should not suppress the selection of shorter catalysts that may be included within. Given the previous observation that the selection of the Class I ribozyme ligase was an improbable event, it is safe to conclude that long pools are inherently better for the selection of complex catalysts.

However, this conclusion is not the same as proving that higher order sequence spaces of necessity contain numerous catalysts whose information content and complexity would rival that of the Class I ligase. This hypothesis strongly predicts that if the selection that produced the Class I ligase were to be carried out again using a different pool of the same length and complexity, then an equally functional, equally complex ribozyme with a completely different sequence and structure should be selected. While to our knowledge this precise experiment has not been carried out, Robertson and Ellington[92] have selected ribozyme ligases from a random sequence population that spanned 90 positions and that also contained ca. 10^{15} different sequence variants. Like the Bartel ligases, the Robertson ligases displace pyrophosphate from the 5′ end of transcripts. The optimized, selected Robertson ligase had a self-ligation rate of only 0.2 min^{-1} and had roughly 25 functionally important residues within its catalytic core. Making the tenuous assumption that these experiments were carried out in a similar fashion, either Bartel and Szostak were lucky, other researchers were unlucky, or the length of the pool does in fact have a profound effect on the complexity of selected catalysts. Arguing against this analysis, the Hager ligases were selected from a random sequence population that spanned 210 positions and that contained ca. 10^{15} different sequence variants. The Hager ligases displace a nucleoside monophosphate rather than a pyrophosphate from their 5′ termini but again the reactions are chemically similar. The selected Hager ligases could not be significantly foreshortened by deletion analysis, suggesting that they might be internally complex. However, in contrast to the Bartel ligases the Hager ligases were extremely slow, with the most active carrying out self-ligation at a rate of only ca. 0.4 h^{-1} (0.0067 min^{-1}).

10.9.8 The RNA World: What Was It Like?

As was pointed out in the introduction, comparative biochemistry can be used to make a strong case for the existence of a metabolically complex RNA world. This case has now been amply supported by the discovery of ribozymes that can catalyze many different reactions that may have occurred in the RNA world. However, at least two intellectual hurdles remain in trying to fully conceptualize what occurred during the early evolution of life. First, it is unclear how the RNA world came to be. Reactions that mimic prebiotic chemistry at best yield short oligonucleotides, not long random sequence populations. Thus, it is unclear where the fodder for the earliest "*in vivo*" selection experiments may have come from. One scenario is that short, self-replicating oligonucleotides (or oligonucleotide-like molecules, such as peptide nucleic acids, PNAs) grew and diversified until they could acquire catalytic function. The nascent evidence that suggests a correlation between ribozyme and ribozyme complexity bolsters this hypothesis. It is possible that beyond some "critical length" many different catalytic functions became accessible to otherwise simplistic replicators. Second, it is unclear how long the RNA world may have existed. While it can be argued that the ribosome and tRNA are remnants of the reign of RNA, microfossils that are strikingly similar to modern organisms existed more than 3 billion years ago. Assuming that organismal morphology is a reliable marker of identity (a tenuous assumption), then cells substantially similar to those that exist today, and that included a full complement of proteinaceous catalysts, may have existed as much as 3.5 billion years ago. Since the Earth's environment would not have been clement for abiogenesis until roughly 4 billion years ago this leaves a narrow window (relatively speaking) for the evolution of replication, cellularization, and complex metabolism. The results summarized in this review may provide a basis for reconciling our view of the RNA world with these caveats. Despite the best efforts of experimentalists, RNA catalysts are slow and relatively non-specific, and seem to operate best on their own kind. If the metabolism of ancient organisms were populated with the best and brightest ribozymes from the present, then these organisms would have replicated slowly, trapped chemical energy grudgingly, and incorporated carbon inefficiently. Overall, the RNA world would have been short and metabolically brutish. However, the very features of RNA catalysts that make them unacceptable as organismal mainstays may have fortuitously rendered them into optimal interlopers between origins and translation. We have seen that even short RNA molecules can catalyze reactions, albeit slowly, and that such short RNA molecules can work with a variety of substrates. As a particularly good example, the pyrophosphate exchange enzyme selected by Huang and Yarus[42] can in effect explore the combinatorial chemistry of 5′ end modification. The amide synthase of Wiegand *et al.*[80] can carry out a similar range of derivatization reactions given the presence of a terminal amine. Thus, the one advantage of an early RNA world would have been that many, many different molecular species would have been invented, most inadvertently. Irrespective of the error rate of replication, the error rate of metabolism, the synthesis of unintended or unwanted compounds, would likely have been huge. Thus, the RNA world would have inexorably slouched towards the invention of molecules with greater catalytic abilities, proteins.

10.10 REFERENCES

1. K. Kruger, P.J. Grabowski, A.J. Zaug, J. Sands, D.E. Gottschling and T.R. Cech, *Cell*, 1982, **31**, 147–157.
2. C. Guerrier-Takada, K. Gardiner, T. Marsh, N. Pace and S. Altman, *Cell*, 1983, **35**, 849–857.
3. F.H. Crick, *J. Mol. Biol.*, 1968, **38**, 367–379.
4. L.E. Orgel, *J. Mol. Biol.*, 1968, **38**, 381–393.
5. S.A. Benner, A.D. Ellington and A. Tauer, *Proc. Natl. Acad. Sci. USA*, 1989, **86**, 7054–7058.
6. S.A. Benner, M.A. Cohen, G.H. Gonnet, D.B. Berkowitz and K.P. Johnsson, in "The RNA World", eds. R.F. Gesteland and J.F. Atkins, Cold Spring Harbor Press, Cold Spring Harbor, New York, 1993, pp. 27–70.
7. T. Tuschl, J.B. Thomson and F. Eckstein, *Curr. Opin. Struct. Biol.*, 1995, **5**, 296–302.
8. S. Baskerville, D. Frank and A.D. Ellington, in "RNA Structure and Function", eds. R.W. Simons and M. Grunberg-Manago, CSH Laboratory Press, Cold Spring Harbor, NY, 1998, pp. 203–251.
9. P.C. Sabetti, P.J. Unrau and D.P. Bartel, *Chem. Biol.*, 1997, **4**, 767–774.
10. E.H. Ekland, J.W. Szostak and D.P. Bartel, *Science*, 1995, **269**, 364–370.
11. M. Ishizaka, T. Oshima and T. Tani, *Biophys. Res. Comm.*, 1995, **214**, 403–409.
12. R. Green, A.D. Ellington and J.W. Szostak, *Nature*, 1990, **347**, 406–408.
13. C. Wilson and J.W. Szostak, *Nature*, 1995, **374**, 777–782.
14. B. Sargueil and J.M. Burke, *Meth. Mol. Biol.*, 1997, **74**, 289–300.
15. J.R. Prudent, T. Uno and P.G. Schultz, *Science*, 1994, **264**, 1924–1927.
16. K.N. Morris, T.M. Tarasow, C.M. Julin, S.L. Simons, D. Hilvert and L. Gold, *Proc. Natl. Acad. Sci. USA*, 1994, **91**, 13028–13032.
17. J.C. Guatelli, K.M. Whitfield, D.Y. Kwoh, K.J. Barringer, D.D. Richman and T.R. Gingeras, *Proc. Natl. Acad. Sci. USA*, 1990, **87**, 1874–1878.
18. A. Berzal-Herranz, S. Joseph and J.M. Burke, *Genes Dev.*, 1992, **6**, 129–134.
19. A. Berzal-Herranz, S. Joseph, B.M. Chowira, S.E. Butcher and J.M. Burke, *EMBO J.*, 1993, **12**, 2567–2573.
20. S. Joseph, A. Berzal-Herranz, B.M. Chowira, S.E. Butcher and J.M. Burke, *Genes Dev.*, 1993, **7**, 130–138.
21. K.L. Nakamaye and F. Eckstein, *Biochemistry*, 1994, **33**, 1271–1277.
22. N.K. Vaish, P.A. Heaton and F. Eckstein, *Biochemistry*, 1997, **36**, 6495–6501.

23. J. Tang and R.R. Breaker, *RNA*, 1997, **3**, 914–925.
24. N.K. Vaish, P.A. Heaton, O. Federova and F. Eckstein, *Proc. Natl. Acad. Sci. USA*, 1998, **95**, 2158–2162.
25. R. Green and J.W. Szostak, *J. Mol. Biol.*, 1994, **235**, 140–155.
26. F. Michel, A.D. Ellington, S. Couture and J.W. Szostak, *Nature*, 1990, **347**, 578–580.
27. F. Michel and E. Westhof, *J. Mol. Biol.*, 1990, **216**, 585–610.
28. L. Jaeger, F. Michel and E. Westhof, *J. Mol. Biol.*, 1994, **236**, 1271–1276.
29. M. Costa and F. Michel, *EMBO J.*, 1995, **14**, 1276–1285.
30. J.H. Cate, A.R. Gooding, E. Podell, K. Zhou, B.L. Golden, A.A. Szewczak, C.E. Kundrot, T.R. Cech and J.A. Doudna, *Science*, 1996, **273**, 1696–1699.
31. M. Costa and F. Michel, *EMBO J.*, 1997, **16**, 3289–3302.
32. L.B. Weinstein, B.C. Jones, R. Cosstick and T.R. Cech, *Nature*, 1997, **388**, 805–808.
33. D.L. Robertson and G.F. Joyce, *Nature*, 1990, **344**, 467–468.
34. A.A. Beaudry and G.F. Joyce, *Science*, 1992, **257**, 635–641.
35. J. Tsang and G.F. Joyce, *Biochemistry*, 1994, **33**, 5966–5973.
36. J. Tsang and G.F. Joyce, *J. Mol. Biol.*, 1996, **262**, 31–42.
37. X. Dai, A.D. Mesmaeker and G.F. Joyce, *Science*, 1995, **267**, 237–240.
38. A.J. Hager, J.D. Pollard and J.W. Szostak, *Chem. Biol.*, 1996, **3**, 717–725.
39. J.R. Lorsch and J.W. Szostak, *Nature*, 1994, **371**, 31–36.
40. J.R. Lorsch and J.W. Szostak, *Biochemistry*, 1995, **34**, 15315–15327.
41. F. Huang and M. Yarus, *Biochemistry*, 1997, **36**, 6557–6563.
42. F. Huang and M. Yarus, *Proc. Natl. Acad. Sci. USA*, 1997, **94**, 8965–8969.
43. M. Illangasekare, G. Sanchez, T. Nickles and M. Yarus, *Science*, 1995, **267**, 643–647.
44. M. Illangasekare, O. Kovalchuke and M. Yarus, *J. Mol. Biol.*, 1997, **274**, 519–529.
45. M. Illangasekare and M. Yarus, *J. Mol. Biol.*, 1997, **268**, 631–639.
46. M. Illangasekare and M. Yarus, *Proc. Natl. Acad. Sci. USA*, 1999, **96**, 5470–5475.
47. P.A. Lohse and J.W. Szostak, *Nature*, 1996, **381**, 442–444.
48. B. Zhang and T.R. Cech, *Nature*, 1997, **390**, 96–100.
49. B. Zhang and T.R. Cech, *Chem. Biol.*, 1998, **5**, 539–553.
50. D.P. Bartel and J.W. Szostak, *Science*, 1993, **261**, 1411–1418.
51. E.H. Ekland and D.P. Bartel, *Nucleic Acids Res.*, 1995, **23**, 3231–3238.
52. M.C. Wright and G.F. Joyce, *Science*, 1997, **276**, 614–617.
53. R.R. Breaker and G.F. Joyce, *Proc. Natl. Acad. Sci. USA*, 1994, **91**, 6093–6097.
54. M.P. Robertson and A.D. Ellington, *Curr. Biol.*, 1997, **7**, R376–R379.
55. E.H. Ekland and D.P. Bartel, *Nature*, 1996, **382**, 373–376.
56. K.B. Chapman and J.W. Szostak, *Chem. Biol.*, 1995, **2**, 325–333.
57. A.J. Hager and J.W. Szostak, *Chem. Biol.*, 1997, **4**, 607–617.
58. B. Cuenoud and J.W. Szostak, *Nature*, 1995, **375**, 611–614.
59. M. Zillmann, S.E. Limauro and J. Goodchild, *RNA*, 1997, **3**, 734–747.
60. N. Lehman and G.F. Joyce, *Nature*, 1993, **361**, 182–185.
61. N. Lehman and G.F. Joyce, *Curr. Biol.*, 1993, **3**, 723–734.
62. D.N. Frank and N.R. Pace, *Proc. Natl. Acad. Sci. USA*, 1997, **94**, 14355–14360.
63. T. Pan and O.C. Uhlenbeck, *Biochemistry*, 1992, **31**, 3887–3895.
64. T. Pan, B. Dichtl and O.C. Uhlenbeck, *Biochemistry*, 1994, **33**, 9561–9565.
65. T. Pan and O.C. Uhlenbeck, *Nature*, 1992, **358**, 560–563.
66. P. Burgstaller and M. Famulok, *Angew. Chem. Int. Ed. Engl.*, 1995, **34**, 1189–1192.
67. J.K. Bashkin, *Curr. Biol.*, 1997, **7**, R286–R288.
68. R.R. Breaker, *Nat. Biotechnol.*, 1997, **15**, 427–431.
69. R.R. Breaker and G.F. Joyce, *Chem. Biol.*, 1994, **1**, 223–229.
70. C.R. Geyer and D. Sen, *J. Mol. Biol.*, 1998, **275**, 483–489.
71. R.R. Breaker and G.F. Joyce, *Chem. Biol.*, 1995, **2**, 655–660.
72. D. Faulhammer and M. Famulok, *Angew. Chem. Int. Ed. Engl.*, 1996, **35**, 2837–2841.
73. D. Faulhammer and M. Famulok, *J. Mol. Biol.*, 1997, **269**, 188–202.
74. N. Carmi, L.A. Shultz and R.R. Breaker, *Chem. Biol.*, 1996, **3**, 1039–1046.
75. N. Carmi, S.R. Balkhi and R.R. Breaker, *Proc. Natl. Acad. Sci. USA*, 1998, **95**, 2233–2237.
76. S.W. Santoro and G.F. Joyce, *Proc. Natl. Acad. Sci. USA*, 1997, **94**, 4262–4266.
77. M.M. Conn, J.R. Prudent and P.G. Schultz, *J. Am. Chem. Soc.*, 1996, **118**, 7012–7013.
78. Y. Li and D. Sen, *Nat. Struct. Biol.*, 1996, **3**, 743–747.
79. Y. Li and D. Sen, *Biochemistry*, 1997, **36**, 5589–5599.
80. T.W. Wiegand, R.C. Janssen and B.E. Eaton, *Chem. Biol.*, 1997, **4**, 675–683.
81. T.M. Tarasow, S.L. Tarasow and B.E. Eaton, *Nature*, 1997, **389**, 54–57.
82. M. Wecker, D. Smith and L. Gold, *RNA*, 1996, **2**, 982–994.
83. P.J. Unrau and D.P. Bartel, *Nature*, 1998, **395**, 260–263.
84. D.N. Frank, A.D. Ellington and N.R. Pace, *RNA*, 1996, **2**, 1179–1188.
85. T. Tuschl, P.A. Sharp and D.B. Bartel, *EMBO J.*, 1998, **17**, 2637–2650.
86. M. Yarus, *FASEB J.*, 1993, **7**, 31–39.
87. J.B. Murray, A.A. Seyhan, N.G. Walter, J.M. Burke and W.G. Scott, *Chem. Biol.*, 1998, **5**, 587–595.

88. V.K. Jayasena and L. Gold, *Proc. Natl. Acad. Sci. USA*, 1997, **94**, 10612–10617.
89. B. Seelig and A. Jaschke, *Chem. Biol.*, 1999, **6**, 167–176.
90. A. Jenne and A. Famulok, *Chem. Biol.*, 1998, **5**, 23–34.
91. A. Roth and R.R. Breaker, *Proc. Natl. Acad. Sci. USA*, 1998, **95**, 6027–6031.
92. M.P. Robertson and A.D. Ellington, *Nat. Biotechnol.*, 1999, **17**, 62–66.

11

Ribozyme Enzymology

JULIANE K. STRAUSS-SOUKUP and SCOTT A. STROBEL
Yale University, New Haven, CT, USA

11.1 INTRODUCTION

The identification of RNA molecules that serve as catalytic machines was a monumental discovery. In the early 1980s, it was demonstrated that an intron within the pre-rRNA of *Tetrahymena thermophila* had the ability to splice itself out of the transcript, independent of proteins.[1] Soon afterward, the RNA subunit of ribonuclease P (RNase P) was identified as the catalytic portion of this ribonucleoprotein complex.[2] These discoveries led to the theory that biology may have evolved from an "RNA world" where RNA could take on the dual roles of information storage and biocatalyst.

This chapter presents an overview of the reaction mechanisms used by RNA enzymes. Most biocatalysts are proteins and they perform this role efficiently due to the chemistry available from their 20 amino acid side chains. RNA lacks this functional group diversity, and therefore it was originally not thought to be a likely candidate for a catalytic molecule. The authors discuss the ways in which RNA may achieve efficient catalysis despite the lack of diversity in its nucleoside residues. The next sections of the chapter describe the secondary and tertiary structural elements of RNA enzymes, as well as the role metal ions play in RNA catalysis. Finally, a thermodynamic and kinetic profile of two well-characterized RNA enzymes, the *Tetrahymena* group I intron and the hammerhead ribozyme, are reviewed.

11.2 TYPES OF RNA CATALYSIS AND REACTION MECHANISMS

11.2.1 Group I/Group II Nucleophilic Attack

RNA enzymes use different reaction mechanisms for catalysis (Figure 1). Group I and group II splicing involves a two-step transesterification reaction, where an exchange of phosphate diesters occurs with no net change in the number of ester linkages.[3] Group I introns use an external nucleophile and group II introns use an internal nucleophile in the first step of splicing (Figure 1(a), (b)).[3] In the first step of group I splicing, the 3′-OH of an exogenous nucleophile attacks the 5′ splice site. The comparable step in group II splicing is carried out by the 2′-OH of an internal bulged A nucleotide. Both reactions generate a 3′-OH group at the end of the 5′ exon, which acts as a nucleophile for the second transesterification reaction, resulting in exon ligation. The group I and group II introns are single-turnover enzymes, and therefore they do not act as true catalysts *in vivo*.[3] The *Tetrahymena*

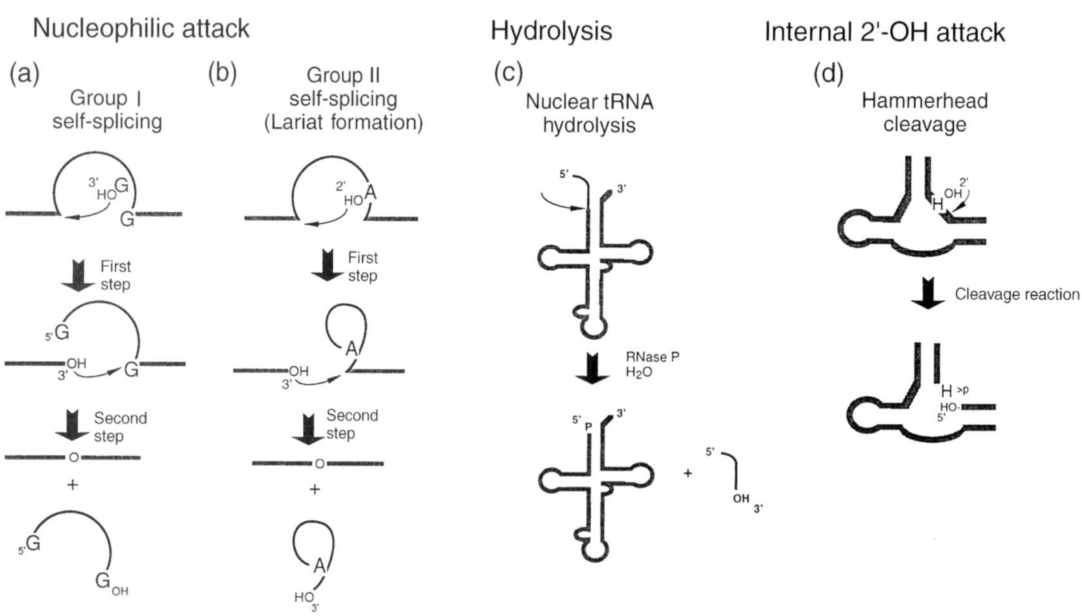

Figure 1 Splicing mechanisms of precursor RNAs. (a–b) Nucleophilic attack by an external (a) or internal (b) nucleotide. Thin lines indicate introns. Thick lines indicate flanking exons. (c) Hydrolysis. Thick line indicates mature tRNA. Thin line denotes pre-tRNA that is removed. (d) Internal 2′-OH attack. Thick line indicates ribozyme (after Cech[3]).

group I intron has a rate constant for the chemical step of ~350 min^{-1}, which corresponds to a rate enhancement of ~10^{11}-fold relative to the solution reaction.[4] The self-splicing ai5γ group II intron has a slower chemical step (~0.03 min^{-1}), and it provides ~10^7-fold rate enhancement over solution hydrolysis.[5]

11.2.2 Hydrolysis by External H_2O or OH^-

The second major ribozyme reaction mechanism involves hydrolysis by an external H_2O or OH^- (Figure 1(c)). RNase P and in some cases group I and group II introns use H_2O or OH^- as a nucleophile. RNase P catalyzes the removal of a block of nucleotides at the 5′ end of a pre-tRNA molecule to produce the mature 5′ terminus (reviewed in reference 3). RNase P cleaves to generate a 5′-phosphate and 3′-OH termini.[3] RNase P is the only known catalytic RNA that can act as a true enzyme, with multiple turnover kinetics *in vivo*.[2,6–8] RNase P is made up of both protein and RNA components.[9] The RNA component has been identified as the catalytic subunit, but the protein is needed for activity *in vivo*.[3] RNase P processes the precursors to all the cellular tRNAs (~20 molecules), which indicates that it can recognize related structural elements. The RNase P RNA has a rate constant of ~180 min^{-1}, which corresponds to a rate enhancement of ~10^{11}-fold relative to the solution reaction.[10]

11.2.3 Internal 2′-OH Attack

The third class of reaction mechanisms involves attack by an internal 2′-OH (Figure 1(d)). The small ribozymes which include the hammerhead, hairpin, hepatitis delta virus, and *Neurospora* Varkud satellite (VS) RNAs, all use an internal 2′-OH adjacent to the cleavage site as the nucleophile (reviewed in reference 11). The RNA enzymes have different secondary and tertiary structures, but they react by the same mechanism. They cleave to generate 2′,3′-cyclic phosphates and 5′-OH termini. Small ribozymes are responsible for processing plant pathogenic RNA or human virus RNA during rolling circle replication in order to generate monomer length RNAs.[12] The hammerhead ribozyme has a rate constant for the chemical step of ~1 min^{-1}, which corresponds to a rate enhancement of ~10^8-fold relative to the solution reaction.[13]

11.3 SECONDARY AND TERTIARY STRUCTURES OF RNA ENZYMES

The global structure of a ribozyme plays a major role in its ability to perform catalysis. RNA enzymes fold into compact structures that involve secondary and tertiary structural contacts. Two-dimensional models of the group I, group II, RNase P, hammerhead, and hairpin ribozymes are shown in Figure 2. These models have been determined by comparative sequence analysis. The formation of an active site for the chemical reaction is dependent on tertiary interactions that can form the catalytic core. Helix packing allows regions of the RNA enzymes to fit closely together in order to correctly position those groups involved in catalysis.

11.3.1 Secondary Structure

11.3.1.1 Group I introns

In the case of group I introns, the tertiary structure of the RNA is important for binding of the substrate and nucleotide cofactor, and for the cleavage reaction. A secondary structure of the *Tetrahymena* intron was proposed by Michel and Davies (Figure 2(a)).[14–16] The intron forms nine paired regions, P1–P9, with joining regions between the helices.[14,16] The ribozyme has two main components, the P4–P6 domain and the P3–P7–P8 domain, which requires P4–P6 for folding.[17,18] These two domains form the active site of the enzyme, into which the P1 helix docks. Paired region P1 includes the 5′ splice site and is formed by base pairing of the 3′ end of the 5′ exon with the internal guide sequence (IGS), contained within the intron.[19] The interactions between domains P4–P6 and P3–P7–P8 and the P1 helix

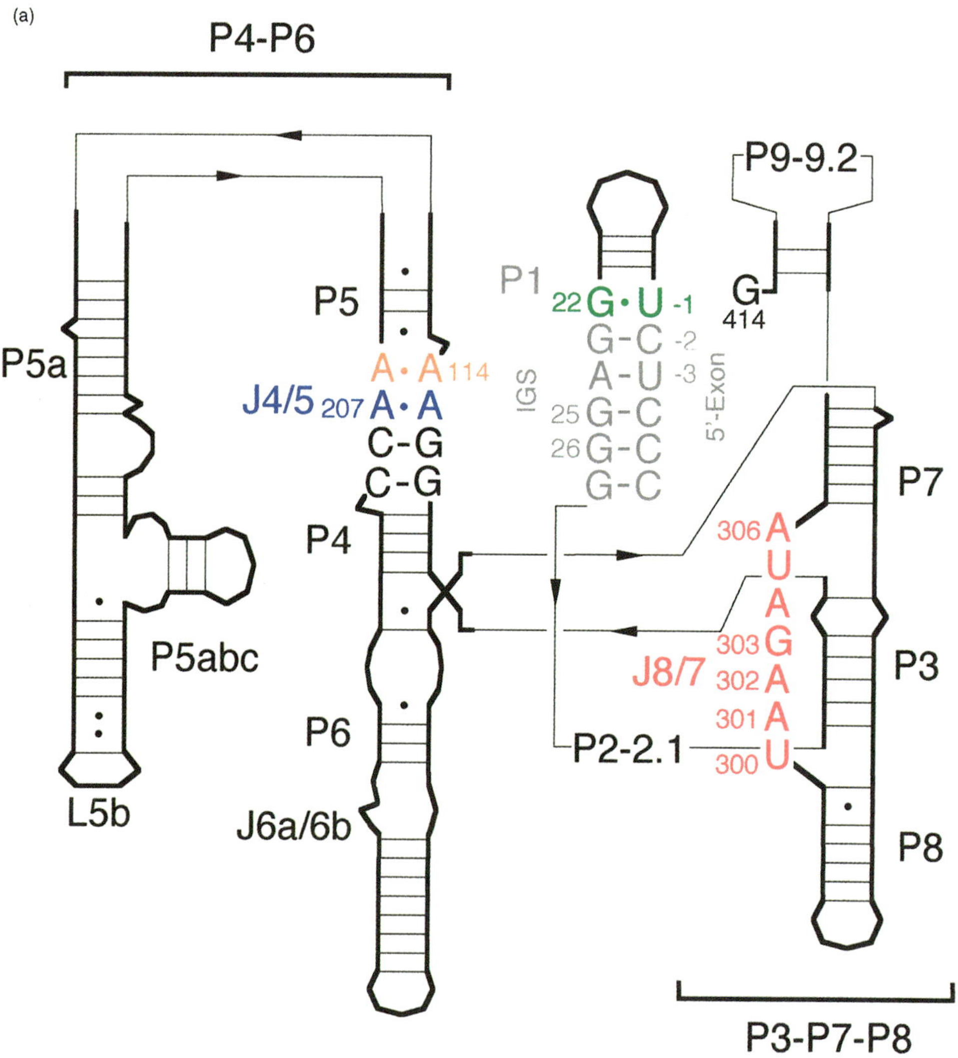

Figure 2 Secondary structure of (a) *Tetrahymena* group I intron, P1 helix (gray), G22 • (-1) base pair (green), J8/7 strand (red), sheared A • A pairs in J4/5 (blue and yellow).

create the correct catalytic core for RNA splicing.[20–22] Group I introns also contain a guanosine binding site in the major groove of the P7 helix,[23] a terminal guanosine that defines the 3′-splice site,[23–25] and joiner regions J4/5 and J8/7 that are important for alignment of the P1 helix into the active site for nucleophilic attack by guanosine.[26–30] Tertiary hydrogen bonding, stacking interactions, and metal ions allow for close packing of the RNA into the active site of this highly compact ribozyme. The P2, P5, and P9 helices are considered peripheral domains located outside of the ribozyme core.

The global structure of the *Tetrahymena* intron was probed using Fe^{II}-EDTA to distinguish the inside from the outside of the RNA.[20,31] In the absence of Mg^{2+}, cleavage occurred throughout the intron, indicating that the RNA backbone was unfolded and accessible.[31] After adding Mg^{2+}, areas of the RNA backbone were protected, which indicated formation of tertiary structure. These areas included regions within P4, P6, P3, P7, and P8 that are phylogenetically conserved.

11.3.1.2 Group II introns

Group II introns are organized into a set of six helical domains that emanate from a central wheel[14] (Figure 2(b)). The secondary structure was formed using phylogenetic data and mutational and chemical modification studies.[32] Domains I and V are essential for catalytic function and the branched adenosine within domain VI contains the nucleophile for the transesterification reaction.[33] Domain I contains two

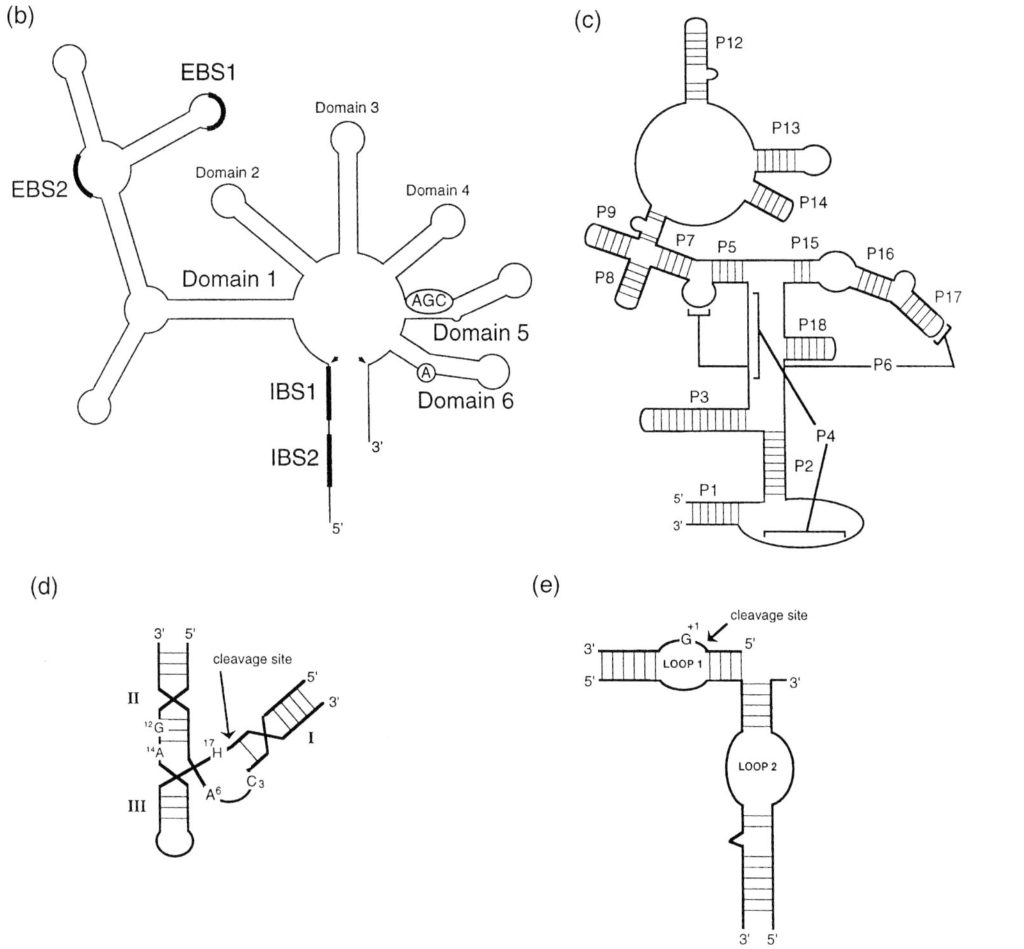

Figure 2 (continued) Secondary structure of (b) group II intron; (c) RNase P RNA; (d) hammerhead ribozyme, H = A, U or C nucleotides; and (e) hairpin ribozyme.

sequence segments, exon-binding sites 1 and 2 (EBS1 and EBS2), that form tertiary contacts to the intron-binding sites 1 and 2 (IBS1 and IBS2), which are found in the 5′ exon.[34,35] The two EBS–IBS interactions are essential for 5′ splice site selection, lariat formation, and stability. Domain V is the "catalytic core" of the intron[36,37] and includes the chemically essential and phylogenetically conserved AGC triad.[38] Domains II, III, and IV are peripheral domains that enhance catalysis and contribute to the overall structure. High salt concentrations are needed *in vitro* for catalytic activity, therefore it is thought that a protein cofactor is utilized *in vivo*.[33]

11.3.1.3 RNase P

The holoenzyme of RNase P from *E. coli* is made up of a 377 nucleotide RNA (M1 RNA) and a 13.8 kDa protein (the *C5* protein).[9] C5 is important *in vivo* for maximal efficiency of cleavage at physiological salt concentrations.[9] The secondary structure of the RNA was determined by comparative sequence analysis with other eubacterial M1-like RNAs (Figure 2(c)).[39,40] Some of the helices have been identified by phylogenetic comparisons. The RNA has 18 paired regions. Two of these helices, P4 and P6, are formed between nonadjacent single-stranded regions.[41] Helix P4 and the immediately adjacent nucleotides are the most highly conserved structural elements in the RNA.[42] This region is thought to be the catalytic center of the RNA enzyme. Tetraloop–receptor interactions and pseudoknots are two of the tertiary interactions identified in the global structure of RNase P RNA.[43] A global structure model was proposed that included three multihelix domains surrounding the universally conserved P4 helix.[41]

11.3.1.4 Hammerhead

The hammerhead ribozyme is one of the smallest catalytic RNAs, comprising just 30 nucleotides. The secondary structure (Figure 2(d)) of the hammerhead involves three Watson–Crick base-paired helices (numbered I–III).[13,44,45] The helical regions are joined by conserved single-stranded segments. Mutagenesis experiments have demonstrated that two of these nonhelical segments are required for catalysis (the single-stranded regions, C_3–A_9 and G_{12}–A_{14}).[44,46] In addition, the nucleotide at the cleavage site must be unpaired.

11.3.1.5 Hairpin

The hairpin ribozyme is composed of two pairs of helix–loop–helix segments (Figure 2(e)).[47] The smaller symmetrical loop contains the reactive phosphodiester.[48] The secondary structure has been determined by data from mutagenesis, phylogenetic comparisons, *in vitro* selections, cross-linking, and chemical protection experiments.[47,49–51] The ribozyme interacts with its substrate through two short helices on either side of loop 1.[51] Active molecules all contain G at the +1 position.[51] The working model for catalysis by this ribozyme is that the two looped regions assemble to form an active site. Although no tertiary interactions have been identified between the two loops, several have been proposed.[52,53]

11.3.1.6 Hepatitis delta virus

The smallest hepatitis delta virus (HDV) ribozyme found to have catalytic activity *in vitro* is ~85 nucleotides long.[54] The secondary structure involves four paired regions, P1–P4. Tertiary interactions between P1 and P2 form a type of pseudoknot.[25] Support for this model comes from mutagenesis studies and the secondary structure generated from active *trans*-acting ribozymes with novel sequences.[55] The cleavage site resides in P1, and therefore, this helix is involved in cleavage site selection.[55,56] P1 and P3 have length requirements, whereas the lengths of P2 and P4 can be varied since they only play structural roles.[55] The crystal structure of the HDV ribozyme has verified many of the proposed secondary interactions and identified an additional short helix within the structure. This results in a double-nested pseudoknot in the L3 loop with the joining region J4/2.[57]

11.3.1.7 Neurospora **VS RNA**

Neurospora contain VS RNA, which is an 881 nucleotide single-stranded circular molecule.[58] A secondary structure was proposed for the self-cleaving region of the VS RNA, nucleotides 617–881, based on site-directed mutagenesis and chemical modification structure probing.[59] This catalytic RNA segment contains six paired regions with intervening single-stranded loops and bulges. In helices II–VI, site-directed base substitutions decreased or abolished activity, while compensatory substitutions restored activity.[59] The site of cleavage is in the single-stranded region of helix I. Mutations at certain positions in helix I inactivated the ribozyme and compensatory mutations did not restore activity. This suggested a conformational change during formation of the active structure that involved these nucleotides.[59] Helix VI contains a GNRA tetraloop, which could be involved in a long-range tertiary interaction. Damage selection and modification experiments have also identified nucleotides within the junction regions that are important for structure and function.[59,60]

11.3.2 Tertiary Structure

11.3.2.1 Tetrahymenagroup I intron

The three-dimensional structure of a ribozyme is essential for its ability to act catalytically. Requirements such as substrate and cofactor binding, positioning of the scissile phosphate, and the actual catalytic reaction, are not possible without an intricate tertiary structure. While three-dimensional models have been proposed for many of the RNA enzymes, crystal structures of only two ribozymes and one ribozyme domain are available. The three-dimensional structures of both the hammerhead and HDV ribozymes and the P4–P6 domain of the *Tetrahymena* group I intron have been determined by X-ray crystallography.

A crystal structure of the P4–P6 domain of the *Tetrahymena* intron has provided detailed information about the tertiary structure of a tightly packed RNA fold.[61] The P4–P6 domain is made up of two arrays of stacked helices that are packed against one another at two distant regions of the structure.[61] These helices are held together by a GNRA tetraloop in one helix that interacts through hydrogen bonds and base stacking with an 11-nt tetraloop receptor in another helix (Figure 2(a)).[61] At three points within the P4–P6 domain, adenosine platforms were seen. In the platform, the adenosines are side by side and form a pseudo base pair.[61,62] Platforms allow the minor groove to widen and become more accessible for base pairing and stacking.

An A-rich bulge within P5a bridges two parallel helical stacks (Figure 2(a)).[61] The backbone of the A-rich bulge makes a corkscrew turn. Its bases are flipped out and they interact with residues of the P4 helix on one side and the three-helix junction within P5abc on the other face.[61] Formation of ribose zippers and metal ion binding allows close interactions between RNA helices.[61] Pairs of riboses interact through hydrogen bonds to form ribose zippers in the A-rich bulge and in the GAAA tetraloop. The close packing of phosphates in nearby helices are also mediated by hydrated magnesium ions.[63,64]

The J4/5 region was proposed to interact with the P1 substrate helix within the catalytic core, and the G • U wobble pair (positions 22 and -1, respectively, within P1) has been shown to cross-link to adenosines within this region (Figures 2(a) and 3).[22,28–30] In the crystal, adjacent adenosines in J4/5 forms sheared homopurine base pairs.[15,21,22,61] This configuration forces the functional groups of the adenosines into the minor groove, where they are accessible to interact with the P1 helix.

The structure of the P4–P6 domain served as a starting point for the modeling of other elements within the catalytic core based upon biochemical tertiary constraints. Nucleotide analogue interference mapping (NAIM) has been used to identify important functional groups within the *Tetrahymena* intron.[65] These studies identified an RNA helix packing motif between the G · U wobble pair at the cleavage site in P1 and the two sheared A–A pairs in J4/5, termed the wobble pair receptor (Figure 3).[28] This interaction is important for 5′ splice site selection.

Using the NAIM results, nucleotide analogue interference suppression (NAIS) experiments were designed. NAIS aids in the identification of specific hydrogen bonding partners within an RNA tertiary structure.[28,65] Using this approach, if a tertiary interaction is disrupted by deletion or alteration of one functional group, then no energetic penalty will result from deletion or alteration of the second functional group.[28] This approach confirmed the hydrogen bonding partners between the G • U wobble pair and the sheared A • A pairs in J4/5.[28] Additional experiments using NAIS demonstrated that the P1 helix and the J8/7 single-stranded region form an extended minor groove triple helix within the catalytic core of the ribozyme.[66] These data revealed the importance of the J8/7 strand in organizing several helices within the active site (Figure 4).

11.3.2.2 Hammerhead

Three crystal structures have been solved for the hammerhead ribozyme.[67–70] In one structure, a DNA strand containing the unpaired nucleotide at the cleavage site was base-paired to an RNA strand to form helices I and III.[67] The DNA strand acted as an inhibitor since it lacked the 2′ OH at the cleavage site. The first all-RNA crystal structure contained an RNA enzyme bound to a substrate that had a 2′-methoxy group at the cleavage site.[69,70] The second all-RNA structure used a substrate with a *tallo*-5′-C-methyl ribose modification at the cleavage site that created a kinetic bottleneck at the final step of the cleavage reaction.[68] This system resulted in a crystal structure of the conformation during self-cleavage. In the crystals, the four conserved nucleotides, 5′-$C_3U_4G_5A_6$-3′, form a sharp turn that is almost identical to the conformation of the uridine turn in tRNA (Figure 2(d)).[71,72] The rest of the catalytic core forms a non-Watson–Crick duplex with a divalent ion binding site. In the first two structures solved, the cleavage site phosphate was in a standard helical conformation. This does not allow for in-line attack of the 2′-OH, which is the mechanism used by the ribozyme.[67,69,70] The *tallo*-5′-C-methyl ribose substrate resulted in a conformation in the crystal structure where the intermediate is correctly oriented for in-line attack.[68]

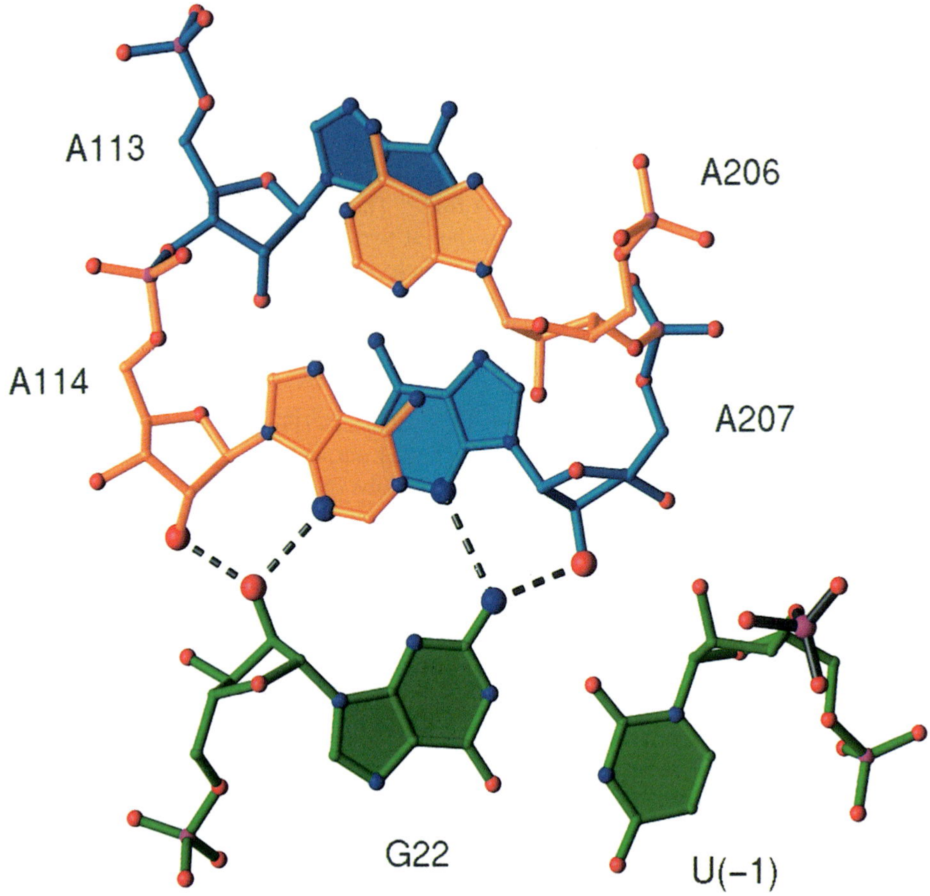

Figure 3 P1 helix docking into the active site of the *Tetrahymena* ribozyme. Tertiary hydrogen bonds between G22 and the sheared A • A pairs are shown; G22 • (-1) base pair (green), sheared A • A pairs in J4/5 (blue and yellow).

11.4 THE ROLE OF METALS IN RNA CATALYSIS

Most ribozymes are considered metalloenzymes because divalent metal ions are important for activity. Cationic metals shield the negative charge of the RNA phosphate backbone, which allows them to fold into complex structures.[73,74] In addition, metal ions can bind specific sites within an RNA and play a direct role in catalysis. Therefore, metal ions play a key role in forming the intricate shape of an active ribozyme.

Many ribozymes use divalent ions, usually Mg^{2+}, for RNA folding and catalysis. Mg^{2+} binds six ligands, possesses octahedral geometry, and has a high preference for binding to oxygen relative to Mn^{2+}, another divalent ion.[75] RNA molecules have a number of sites that can interact with metal ions in solution.[76] These sites include oxygen atoms on the ribose sugar, including the 2′-OH and 5′- and 3′-oxygens. Metal sites also include phosphate oxygens and base carbonyls.

11.4.1 Metal Ions and RNA Folding

11.4.1.1 Site-specific metal ion binding

The identification of metal binding sites within large RNAs is a challenging biochemical problem, though the development of two techniques, metal rescue and metal ion cleavage, has assisted in this endeavor. Interference assays are performed in the presence of different metal ions to determine the sites of ion coordination. RNA molecules are transcribed with a small percentage of ribonucleotide analogues that contain phosphorothioate diester linkages and the active RNAs are selected from the

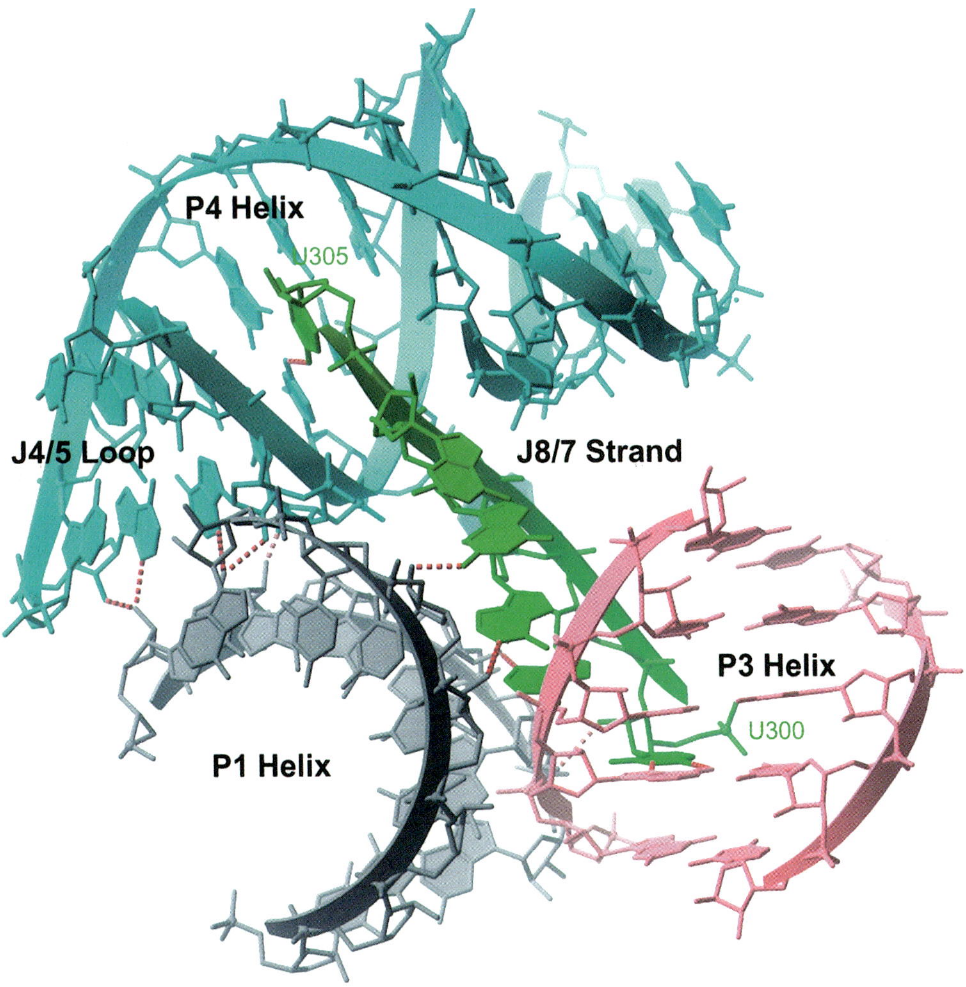

Figure 4 Model of the *Tetrahymena* intron catalytic core, composed of helices P1 (gray), P3 (pink), P4 (blue), and joining regions J4/5 (blue) and J8/7 (green). Tertiary hydrogen bonds are indicated by dotted lines.

overall population by their ability to splice in the presence of Mg^{2+}.[77] The phosphorothioate linkages are cleaved with iodine and the cleavage products are resolved by electrophoresis.[78] The active fractions will exhibit radioactive bands at all the positions where a phosphorothioate linkage was unimportant for activity. Phosphorothioate interference corresponds to sites of RNA–metal or RNA–RNA interactions.[64,77,79,80] The role of phosphates can be distinguished by repeating the experiment in the presence of Mn^{2+}, which binds sulfur with an affinity similar to oxygen.[81] Phosphate oxygens important for catalysis that coordinate metals are tolerant of the phosphorothioate linkage when Mn^{2+} is present. Those positions that can be "rescued" by Mn^{2+} suggest that phosphates are directly coordinated to a metal ion. This type of experiment was performed on the *Tetrahymena* group I intron and several sites within the catalytic core were identified as potential metal binding sites.[77] A similar approach has been used to map metal sites within other RNAs.[42,77]

Metal ion cleavage experiment can also identify sites where metal ions may be present. At high pH, Mg^{2+} can cleave RNA because the main species of Mg^{2+} is $Mg(OH)_2$.[82] At neutral pH, $Pb(OH)_2$ can perform a similar reaction.[83,84] The metal hydroxides act as bases to remove protons from the 2′-OHs, which causes the remaining oxygen to nucleophylically attack the phosphodiester linkage and produce a 2′,3′-cyclic phosphate and 5′-OH termini. Although this technique cannot identify specific sites of metal ion binding, it can define a vicinity within which metal ions are likely to be found. For example, the site of Pb^{2+} cleavage in tRNA was similar to the site of metal binding in the crystal.[83,85,86] Metal ion cleavage performed on the *Tetrahymena* intron identified a group of potential metal sites within the catalytic core.[87] In addition, the potential metal binding sites of RNase P were determined by this method.[82,88]

Metal binding sites within RNA have also been mapped using Fenton chemistry since Fe^{2+} can occupy Mg^{2+} ion binding sites.[89] Incubation of RNAs with Fe^{2+}, ascorbate, and hydrogen peroxide yields hydroxyl radicals that cleave the RNA near sites of metal binding. In contrast to Fe^{2+}/EDTA footprinting, which maps the solvent-accessible surface of an RNA, this technique identifies metal sites on the inside of an RNA.[89]

In addition to nonspecific binding, metal ions can also bind to specific sites within an RNA molecule and aid in the formation of tertiary contacts. Metal ions can inner-sphere coordinate to specific functional groups or are fully hydrated and bind the RNA by outer-sphere coordination. It was first shown that metals may bind specific sites when crystals of the $tRNA^{Phe}$ revealed sites of Mg^{2+} binding.[90–92] The crystal structure of the P4–P6 domain from *Tetrahymena* includes three types of metal binding sites.[61,63,64,93] First, hydrated magnesium ions bind to the major groove of G • U pairs by outer-sphere coordination. Second, Mg^{2+} ions bind to RNA through direct inner-sphere coordination. In these examples, Mg^{2+} coordinates to specific functional groups of the bases, such as phosphate oxygens and the O-6 carbonyls. These interactions help form a metal ion core that may drive folding.[63,64] Finally, three sites of monovalent ion binding have been identified by X-ray crystallography and NAIM.[93] These ions make inner-sphere contacts to functional groups below the A–A platforms.

In the hammerhead structure a metal ion is bound to a *pro*-R_p phosphoryl oxygen and to a ring nitrogen.[67–70] When a phosphorothioate was incorporated at the *pro*-R_p oxygen there was a decrease in catalytic activity in the presence of Mg^{2+}.[44,94] This result can be explained if the Mg^{2+} plays a role in structure or catalysis.

11.4.2 Role of Metal Ions in Catalysis by *Tetrahymena* Group I Intron

Although metal ions are required for RNA folding, they also play an essential role in the actual catalytic reactions of most ribozymes. A two-metal ion mechanism for catalytic RNAs was modeled after the catalytic mechanism of protein enzymes, such as *E. coli* DNA polymerase, that employ metals for activity.[95] Steitz and Steitz hypothesized that there are specific binding sites for two metals at ribozyme active sites, and that these metals may play a major role in stabilizing the transition state of the reaction. Large ribozymes catalyze attack on the scissile phosphate by either an exogenous guanosine (group I), water (RNase P and sometimes group II), or an internal adenosine (group II). In the first step of splicing of group I and II introns, the R-OH group is activated for nucleophilic attack by a Mg^{2+} ion (metal A). A second Mg^{2+} ion (metal B) is proposed to act as a Lewis acid to stabilize the leaving oxyanion and the transition state intermediate. During the reaction, the phosphorus at the active site goes through a trigonal bipyrimidal coordination state. The reactions result in a substantial negative charge on the leaving group, which is neutralized by direct metal ion coordination. In the second step of group I splicing, the Mg^{2+} ions reverse roles. Mg^{2+} ion (A) facilitates the 3′-oxyanion leaving group of the terminal G, and Mg^{2+} ion (B) activates the attacking hydroxyl, the 3′-OH of the 5′-exon. Unlike the small ribozymes, large ribozymes do not use an adjacent 2′-OH for catalysis. Therefore, the size of the large ribozymes may be required for the formation of a complex structure that can bind nucleophiles and the essential metal ions.

The *Tetrahymena* ribozyme undergoes splicing in the presence of Mg^{2+}.[96–98] The L-21 ScaI ribozyme, was constructed that can perform a reaction analogous to the first step of splicing and act as a multiple turnover enzyme (see below).[4,99,100] A 3′-phosphorothiolate linkage was placed at the scissile phosphate in order to determine the role of metal coordination to the 3′ oxygen leaving group.[101] Cleavage of this substrate in the presence of Mg^{2+} was undetectable. Catalytic activity was partially restored when the reaction was carried out in Mn^{2+}, which is a thiophilic metal. This result identified a metal ion binding site in the catalytic core, and suggested that this metal ion directly coordinates to the 3′ leaving oxyanion in the transition state (Figure 5(a), metal #1).

A second metal ion was shown to be involved in catalysis by the *Tetrahymena* intron using a ribozyme that could perform the reverse of the first step of splicing with the substrate IspU, 3′-(thioinosylyl)-(3′-5′)-uridine.[102] This substrate, with a 3′-bridging sulfur substitution, was active only in the presence of thiophilic metals (Cd^{2+} or Mn^{2+}) due to the special linkage. This result indicated that again stabilization of the leaving group was important. The principle of microscopic reversibility argued that in the forward reaction this metal site must be activating the 3′-OH of the nucleophile (Figure 5(a), metal #2). This supported the idea that metal ions at two different sites were involved in catalysis.

A third possible metal ion was postulated to interact with the 2′-OH of the exogenous guanosine

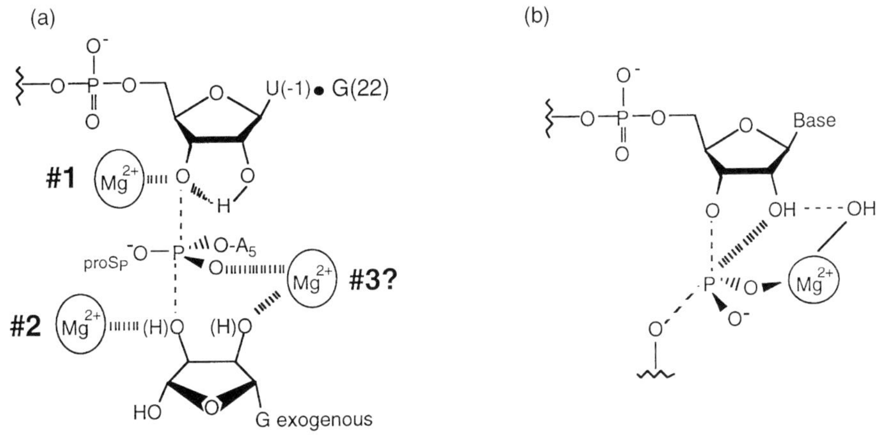

Figure 5 Proposed catalytic metal ions in the *Tetrahymena* group I intron (a) and the hammerhead ribozyme (b).

(Figure 5(a), metal #3).[103] This metal ion may play a role in catalysis, since it has an effect on the rate of transesterification. However, there may be some overlap in that one metal ion may be able to coordinate to both the 2′- and 3′-OHs of the exogenous guanosine.

11.4.3 Role of Metal Ions in Catalysis by Hammerhead Ribozyme

The small ribozymes catalyze intramolecular attack of an adjacent 2′-OH group on the phosphodiester to produce a 2′,3′-cyclic phosphates and 5′-OH termini. Metal ions have been found to be important for the catalytic reactions of most members of this class. A metal-coordinated hydroxyl group may act as a general base to deprotonate the 2′-OH (Figure 5(b)).[83] Metal ions may also be directly coordinated to the phosphoryl oxygen (Figure 5(b)).[104] This could stabilize the increased negative charge of the oxyanion in the transition state or it may cause the phosphorus to be more susceptible to attack. The metal ion may also help stabilize the developing negative charge on the leaving group.

Metal ions are important for both structure and catalysis in the hammerhead ribozyme. NMR and nuclease protection experiments have determined that divalent ions are not required for folding, and that monovalent or higher valent polycations can compensate for the absence of Mg^{2+}.[105,106] Divalent ions are, however, absolutely required for catalytic activity. Experiments were performed to measure the pH-dependent catalytic cleavage rate as a function of metal ion identity.[107] For all the metals tested, the rate increased with pH, which indicated that general base catalysis was occurring. Therefore, it was proposed that a metal hydroxide in the catalytic core deprotonates the 2′-OH at the cleavage site, promoting nucleophilic attack on the scissile phosphate (Figure 5(b)). Metal rescue experiments indicated that the phosphoryl oxygen at the cleavage site was directly coordinated to a metal ion in the transition state of the reaction.[108] This indicated two distinct roles for metal ions in hammerhead catalysis.[109]

In the crystal structures of the hammerhead, sites of metal binding have been identified.[67–70] All three structures show divalent ion binding to sites important for structure. Two of the structures[68–70] also show divalent metal ions bound to sites important for catalysis. In one structure, all of the Mn^{2+} sites identified reappeared as Mg^{2+} sites.[70] In addition, another Mg^{2+} bound directly to the *pro*-R phosphate oxygen adjacent to the scissile bond. This additional Mg^{2+} sites was only found in the conformational intermediate structure, indicating its relevance to the cleavage mechanism. This structure is adopted before the conformation amenable to in-line attack forms. The mechanism of cleavage was then thought to involve either: (i) two Mg^{2+} ions that initiate cleavage, one bound to the *pro*-R phosphate oxygen and the other involved in a metal hydroxide that attacks the cleavage site 2′-OH; or (ii) a single Mg^{2+} ion that binds to the *pro*-R phosphate oxygen and also provides the hydroxide that initiates cleavage.

Work by Lott *et al.* favors the two-metal ion mechanism.[109] Using a hammerhead ribozyme construct, the rate of cleavage of an RNA substrate was studied in the presence of a constant amount of Mg^{2+}, with La^{3+} as a competing metal ion. k_{obs} was measured as a function of La^{3+} concentration. Bound Mg^{2+} was displaced by La^{3+} as its concentration was increased. Because a kinetic parameter was monitored, the resulting activity titration as a function of La^{3+} ion concentration defined the relative amounts of the metal ions in the transition state of the rate-limiting step. The two metals had very different effects at the two binding sites. It was postulated that the metal ion at the first binding site coordinates the attacking 2′-oxygen and lowers the pK_a of the attached proton. The metal ion at the second binding site helps catalyze the reaction by absorbing the negative charge that accumulates on the leaving group 5′-oxygen in the transition state.

11.4.4 Metal Ions Play a Passive Role in the Hairpin Ribozyme

Somewhat surprisingly, data has shown that metal ions play a passive role in the cleavage reaction of the hairpin ribozyme.[110,111] These data are the exception to the rule that metal ions are required for catalysis. In this instance it is thought that metal ions are merely present for structural purposes. It has been shown that cobalt hexammine can substitute for Mg^{2+} in cleavage reactions of the hairpin ribozyme without a change in catalytic activity.[110,111] This result indicated that inner-sphere coordination was not essential to the hairpin ribozyme. Substrates were used that had an equal mixture of R_p- and S_p-phosphorothioate linkages at the scissile phosphate.[110] In the presence of Mg^{2+}, these substrates had activity similar to that of the wild-type substrate, which indicated that Mg^{2+} is not directly bound to a phosphorus center and that it is not absolutely required for catalysis.[110] In another set of experiments, two substrates were synthesized, one had an R_p-phosphorothioate at the scissile phosphate and a second had an S_p-phosphorothioate, and no thio effects were detected.[111] Furthermore, addition of monovalent cations enhanced activity in reactions with cobalt hexammine, but inhibited activity in reactions with magnesium.[110] Since cobalt hexammine is a poor hydrogen bond donor, the addition of monovalent ions may allow specific sites to be occupied by hydrated cations required for catalysis. Activity may be inhibited with magnesium and monovalent ions present because of competition between the two for the Mg^{2+} binding sites.

This section has concentrated on describing ribozymes as metalloenzymes. Several experiments have demonstrated that metal ions, especially divalent ions, are required for RNA molecules to fold into their catalytically active conformations. Metal ions shield the negative charge on RNA and bind to specific sites within the molecule. In addition to their structural role, metal ions also contribute to the catalytic reactions of ribozymes. Because RNA molecules lack functionalities with a pK_a near neutrality, metal ions may help in stabilizing the developing negative charge on the leaving oxyanions and the attacking hydroxyls.

11.5 KINETICS OF THE *TETRAHYMENA* AND HAMMERHEAD RIBOZYME REACTIONS

11.5.1 *Tetrahymena* Group I Intron

It was first reported in 1981 that RNA molecules were capable of self-splicing; in other words that RNA could act as an enzyme.[1] The first such catalytic RNA was the group I intron of *Tetrahymena* rRNA. Since then a vast amount of data on the kinetics of the self-splicing reactions have been collected on this intron. This includes kinetic analysis of both steps of splicing and the development of intron constructs capable of multiple turnover catalysis.

11.5.1.1 Reaction components

Since the discovery of catalytic RNA, numerous experiments have been performed to decipher the necessary components and chemical mechanism of catalysis. It was initially determined that divalent cations and guanosine were essential for splicing.[1,112,113] ATP, CTP, and UTP could not substitute for GTP, but guanosine, 5′-GMP, and 5′-GDP could be utilized.[1] This indicated that GTP did not provide an energy source since the number of phosphates was unimportant. The apparent lack of an energy requirement for splicing suggested a phosphoester transfer mechanism in which each cleavage step was coupled to a ligation step.[1] It was also determined that guanosine analogues lacking a 2′- or 3′-OH could not initiate splicing and that guanosine was added to the 5′ end of the intervening sequence (IVS) during its excision.[1]

11.5.1.2 Self-splicing mechanism

The splicing reaction proceeds by inversion of configuration at the active site phosphorus, using an S_N2 reaction or "in-line attack" mechanism.[114,115] This type of pathway is used by protein phosphotransferases and phosphatases.[116,117] S_N2 reactions proceed through a trigonal bipyramidal configuration in the transition state.[116] The stereochemical course of RNA catalysis was determined using a stereo specific phosphorothioate substitution at the cleavage site and snake venom phosphodiesterase (which has a 1700-fold greater activity on R_p isomers).[114,115] It was determined that the ribozyme reaction proceeded with inversion of configuration at the phosphorus. In reactions with substrates that contained a phosphorothioate at the R_p oxygen, the rate of the reaction was decreased 1000-fold relative to unsubstituted substrates. The presence of sulfur may have interfered with coordination of Mg^{2+} to the phosphate oxygen in the chemical transition state.

The phosphorothioate substrates were also used to identify the rate-limiting step of splicing. Sulfur was substituted for the *pro*-R_p (nonbridging) phosphoryl oxygen at the scissile phosphate.[118] Although there was no effect on the equilibrium binding constant of the substrate, there was a 2.3-fold decrease for the reaction with guanosine and a seven-fold decrease for hydrolysis.[118] These "thio-effects" indicated that chemistry was the rate-limiting step because of the inherent ability of phosphorothioate linkages to be cleaved. No rescue of activity was seen in the presence of Mn^{2+}, indicating that the *pro*-R_p oxygen is not directly coordinated to Mg^{2+}. Substitution of the *pro*-S_p oxygen with sulfur gave a thio effect of $\sim 10^3$-fold, indicating that there probably exists a direct contact between the *pro*-S_p oxygen of the substrate and a metal ion or a functional group in the ribozyme active site.

11.5.1.3 Kinetics of Tetrahymena *splicing — first step of splicing*

After self-splicing of the *Tetrahymena* rRNA precursor, the intervening sequence undergoes cyclization and site-specific hydrolysis to produce an intron lacking the first 19 nucleotides, termed the L-19 ribozyme.[99,112,119] The cyclization reaction proceeds after the L1 loop of P1 base pairs to the IGS.[120] The terminal G then nucleophilically attacks the 5′ end of the intron to release the first 19 nucleotides.[120] To investigate the first step of splicing, a form of the intron lacking the first 21 nucleotides and the terminal G at the 3′-end was designed, termed the L-21 ScaI ribozyme.[121] This RNA cannot undergo cyclization or the second step of splicing, but it can perform the first step of splicing with multiple turnover kinetics in the presence of a substrate oligonucleotide and exogenous G.[4,122] The oligonucleotide binds to the complementary IGS to form the P1 helix which is docked into the catalytic core of the enzyme.[4,122]

The L-21 ScaI ribozyme was used to determine a number of kinetic parameters for the first step of splicing.[4,122] Pulse-chase experiments were performed using various amounts of enzyme (RNA), substrate, and guanosine cofactor in order to determine individual rates and binding constants. Initially, binding of guanosine cofactor and substrate to the enzyme was shown to be independent and random.[4] More careful studies demonstrated that there is a modest level of cooperativity between guanosine and substrate binding and anticooperativity between product and guanosine binding to the enzyme.[123,124] In reactions where the concentration of guanosine was saturating, and the concentration of substrate was subsaturating, the rate-limiting step was substrate binding (k_{on}^{S}).[4] Under these $(k_{cat} \cdot K_m^{-1})^S$ conditions, every RNA substrate that bound was cleaved because the energy barrier for substrate release was higher than the energy required to proceed with the chemical step. Under k_{cat} conditions, where the concentrations of both substrate and guanosine were saturating, the release of product from the active site was rate limiting.[4,118] Finally, under single turnover $(k_{cat} \cdot K_m^{-1})^G$ conditions, where all of the substrate was bound to the enzyme and G was also saturating the chemical step (k_c) is rate limiting at lower pH. At higher pH, substrate binding is rate limiting. Under multiple turnover conditions, the rate-limiting step becomes release of the product.[4] Mutations within the ribozyme or substrate weaken E • S complex binding, thereby increasing the multiple turnover rate.[25]

There is a pH dependence for the single turnover reaction, $E \cdot S + G \rightarrow$ products, with a pK_{app} of 6.9.[126] The titratable groups of RNA have pK_a values $<\sim 4$ and $>\sim 9$. When the 2′ position of the uridine at the cleavage site in the substrate was modified to -H or -F, the substrate reacted at very different rates at low pH, but achieved the same limiting rate at high pH.[126] Substitution of the *pro*-R_p oxygen by sulfur resulted in a "thio effect" only at pH below 6.9. This suggested that the chemical cleavage step is rate limiting at low pH and that a conformational step is rate limiting at higher pH. The pH dependence also indicated that a proton may be lost from the E • S • G ternary complex prior to chemical cleavage.

11.5.1.4 Kinetics of **Tetrahymena** *splicing — second step of splicing*

The L-21 G414 ribozyme construct was used to study the second step of splicing.[127] This intron is missing the first 21 nucleotides, but it has the 3′ terminal guanosine G414. Reactions with this ribozyme were investigated in the forward and reverse directions, where G414 acts as a leaving group in the forward reaction and as a nucleophile in the reverse reaction.[127] The rate constant of the chemical step for the reaction with the L-21 G414 construct was identical to that obtained for the reaction of the L-21 ScaI ribozyme with exogenous G.[4,127,128] This result supported the idea that there is a single G binding site and that the orientation of the bound G is the same for both steps of splicing. The results also proposed that the L-21 G414 ribozyme exists primarily with the terminal G docked into the G-binding site. This docking was destabilized ~100-fold when an electron-withdrawing group was attached to G414.

The chemical step of the exon ligation reaction reached an internal equilibrium near 1, which suggested that the bound substrate and product are thermodynamically matched and that there is energetic symmetry within the active site.[127] Slow dissociation of the 5′ exon analogue relative to a ligated exon analogue suggested a kinetic mechanism to ensure efficient ligation of exons.[127]

11.5.1.5 P1 helix docking

As mentioned above, docking of the P1 helix into the catalytic core is essential for catalysis. Binding of substrate RNA to the ribozyme has been shown to occur by a two-step process. The P1 helix docks into tertiary interactions in different registers after binding of the substrate.[129,130] This indicated that the ribozyme could act processively. The ability of the P1 helix to change registers without dissociation of the substrate led to a two-step model for substrate binding. In this model, an open complex is formed first through base-pairing interactions within P1, then a closed complex forms involving tertiary interactions.[129] The same conclusions were drawn from fluorescence-detected stopped-flow experiments using a pyrene-labeled RNA substrate. Pyrene experienced three different microenvironments during binding, indicating conformational changes as the P1 helix was formed.[131,132] These correspond to the open and closed complexes, while the third conformation is that of the cleavage step.

The G22 • U(−1) wobble pair at the cleavage site is one of the only universally conserved nucleotide pairs outside the catalytic core.[26,133] This pair contributes significantly to P1 helix docking and reaction specificity, which is essential for catalysis. The importance of G • U base pair conservation was shown by *in vitro* mutagenesis.[26] Single base changes in either the IGS or the 5′-exon (excluding mutations in the G • U pair) disrupted splicing, but the compensatory mutations restored activity. Only the A • C wobble pair retained a significant fraction of the wild-type activity.[29,133,134] The secondary and tertiary structural elements associated with the G • U wobble pair were found to be essential for splice site selection since mutation to a G–C pair reduced catalytic activity and this substrate was bound ~10–20-fold weaker than one with a G • U wobble pair.[133,135,136]

Unnatural nucleotide analogues were used to probe the contributions made by the G • U pair. It was demonstrated that the 2′-OH and the N-2 exocyclic amine of G were important for hydrogen bonding to the catalytic core.[29,134] The amine is the only sequence-specific contact within the helix that contributes to 5′ splice site selection and transition state stabilization. The amine makes its contribution (2.5 kcal mol^{-1}) in the context of all three wobble pairs tested, but does not contribute (<0.8 kcal mol^{-1}) when presented with a Watson–Crick geometry.[29] The amine also makes a small contribution to chemical transition state stability (1.0 kcal mol^{-1} relative to an I–U pair) that is largely independent of the stability given by the 2′-OH of U(−1).[134] This indicates that tertiary contacts between the exocyclic amine of guanine and its hydrogen bonding partner in the active site are improved during the chemical transition. Results from NAIM indicate that the 2′-OH and N-2 exocyclic amine of the G22 hydrogen bond into the minor groove of two consecutively stacked sheared A • A pairs in J4/5.[28] This wobble–wobble receptor interaction may be the only sequence-specific tertiary contact between P1 and the catalytic core.[26]

Six sequence nonspecific 2′-OH groups participate in P1 helix docking (Figure 2(a)).[30,137–139] C(-2) and U(-3) are found in the substrate strand and G22, G23, G25, and G26 are in the IGS. The P1 helix conforms to A-form geometry, therefore all the 2′-OH groups lie within the minor groove on the same face of the helix.[76] This region of the helix is about a quarter turn away from the exocyclic amine of the G22 • U(−1) pair. The 2′-OHs at positions G22 and G25 played a critical role in docking P1 into the catalytic core, contributing 2.6 and 2.1 kcal mol^{-1}, respectively.[30] Docking was moderately stabilized (~1 kcal mol^{-1}) by G23 and G26 within the IGS (see positions in Figure 2(a)).[30,127]

By contrast, the 2′-OH at the (-1) position had very little effect on ground state binding, but had a large effect on the chemical step.[137] The 2′-OH at the cleavage site of the substrate was replaced with other substituents to investigate their effect on the chemical step.[128] The total transition state stability by the ribozyme was 4.8 kcal mol^{-1} greater when a ribose was present at the cleavage site compared to a deoxyribose.[128] This effect is specific to the chemical transition state because a series of 2′ substituents followed a linear free energy relationship between the rate of the chemical step and the pK_a of the leaving group in this order: 2′-F_2 > -F > -H.[137] It was assumed that electron withdrawing groups helped accelerate the rate. Stability from the 2′-OH at U(−1) is proposed to result from donation of an intramolecular hydrogen bond to the incipient 3′-oxyanion in the transition state.[128] This is specific to the transition state because it is not until the bond between the 3′ oxygen atom and the phosphorus atom begins to break that there is an accumulation of negative charge on the 3′-oxygen.[128]

The hydrogen bonding partners for many tertiary interactions within the catalytic core have been identified. A tertiary interaction with the U(-3) position was mapped to a phylogenetically conserved adenine (A302) in J8/7 (Figure 4).[15,140] A hydrogen bond between the 2′-OH of U(-3) and the N-1 of A302 was demonstrated by DMS footprinting and mutagenesis.[27] NAIS demonstrated that the 2′-OH of G25 and the 2′-OH of A301 were energetically coupled, probably through a direct hydrogen bond (Figure 4).[66] The C(-2) and G26 2′-OHs also contribute moderately to P1 docking (~1 kcal mol^{-1}),[30,139,141] and based upon NAIS experiments they form hydrogen bonds with the N-2 amine of G303 and the 2′-OH of U300, respectively (Figure 4).[66]

These biochemical data, in conjunction with the crystal structure of the P4–P6 domain, have allowed modeling of most of the catalytic core of the *Tetrahymena* intron.[66] In the model, the J8/7 single strand forms a 4 bp triple helix in the minor groove of P1 and it interacts with both strands of the P1 helix (Figure 4). Each base triple involves at least one 2′-OH, so DNA molecules cannot form this complex. In addition, mutational analysis has identified a base triple between U305 and a major groove face of P4.[142,143] Inclusion of this interaction into the J8/7 triplex placed U305 in a non-A form geometry. A304 within J8/7 bridged the interactions between P1 and P4.[66] The J8/7 region has also been shown by NAIM and mutational analysis to interact with P3.[15,80] Residues U300 and A301 may make hydrogen bonding contacts to P1 and P3, placing J8/7 in the middle of a group of tightly packed helices (Figure 4). J8/7 is also the joining region between P7 and P8, so it is actually in close contact with at least five helices within the catalytic core of the group 1 intron.

11.5.1.6 Guanosine binding

A guanosine cofactor is required for 5′ splice site cleavage and exon ligation, and it binds to a specific site within the ribozyme with a K_m of ~60–90 μM.[113,124] Other guanosine analogues exhibited higher K_m values. Based on the free energy of binding values determined, it seemed as though the intron could discriminate between different cofactors.[113] Indeed, deoxyguanosine and dideoxyguanosine are competitive inhibitors of the splicing reaction, indicating the importance of the 2′-OH.[98]

The guanosine binding site was identified using guanosine analogues and mutant ribozymes.[23] The highly conserved G264•C311 base pair within P7 was specifically mutated to A–U.[23] The A–U mutation greatly reduced the rate of 5′ and 3′ cleavage, but there was only a slight further decrease when the 3′ terminal G was changed to an A.[23] The A–U mutant preferred 2-amino purine (2-AP) as a cofactor over G. This showed that G was bound to the G264 • C311 base pair, since 2-AP bound to an A–U pair is isosteric to G bound to a G–C pair. These results indicated there is a single G binding site since the same mutation (A–U) affected the G residue requirement for both 5′ and 3′ splicing.[23]

The guanosine cofactor binds in a nonplanar (or axial) configuration with respect to the G264 • C311 base pair.[144] The activity of mutant ribozymes with substrates which had hydrogen bond donors and acceptors switched provided proof of an interaction between guanosine and A265 (right below the binding site G264 • C311 base pair). This additional interaction was only possible with axial binding of G.

Identification of the guanosine binding site made it possible to demonstrate that both splicing reactions were catalyzed within the same active site.[25] Mutation of the G264 • C311 base pair to C264 • G311 resulted in preference for adenosine as the substrate for cleavage at the 5′ splice site.[25] The exon ligation step could not be performed with these mutations. However, the second step of splicing could occur if the G at the 3′ splice site was changed to an A.[25] This result indicated that a single determinant specified nucleoside binding for both steps of splicing. In addition, the rate of the chemical step for reactions of the L-21 ScaI intron with bound exogenous guanosine and the L-21 G414 intron with its 3′ terminal guanosine were approximately the same. This supported a single G-binding site model and led to the proposal that the overall active site architecture and orientation remained the same in both steps. Subsequent experiments identified a positive entropy change for guanosine binding and for the chemical step of the ribozyme reaction.[145] The increase in disorder in the ground state could be due to a conformational change or to the release of bound water. The positive entropy for reaching the transition state of the chemical step was probably due to the release of bound water from a catalytic Mg^{2+} ion(s).

11.5.2 Hammerhead Ribozyme

11.5.2.1 Hammerhead constructs

The *in cis* hammerhead ribozyme was dissected into a catalytic domain and its complementary substrate to generate an *in trans* cleaving ribozyme that was used to determine rate constants for the reaction.[146–148] The hammerhead constructs studied contained a large ribozyme RNA, with most of the conserved nucleotides and a small substrate RNA with the cleavage site.[148] The constructs had the same essential unpaired nucleotides at the junctions of the helices and the same helix II (see Figure 2(d)). The constructs differed in the lengths and sequences of helices I and III.[148]

11.5.2.2 Kinetics of hammerhead cleavage

Under multiple turnover conditions, ribozyme constructs with fewer than six base pairs in each of the substrate binding helices I and III exhibited substrate binding and dissociation rates that were fast relative to k_{cat}.[147,148] Product release was also fast and therefore k_{cat} represented the rate of phosphodiester bond cleavage, which was ~1–2 min^{-1}. This corresponded to a rate enhancement of ~10^8 over the solution reaction.[13,147,148] Lengthening of helices I and III increased stability and in turn drastically reduced the dissociation rates of substrate and product. k_{cat} decreased such that a hammerhead ribozyme with five base pairs in both helices I and III had a k_{cat} of ~1.4 min^{-1}, but a construct with eight base pairs in each helix had a k_{cat} of only ~0.008 min^{-1}.[147,148] Substrate binding strength had a modest effect on the rate constant of the reaction. Increases in helix stability resulted primarily in decreased dissociation rates with little effect on association rates. Mutations to the ribozyme or substrate that eliminated catalysis increased the stability of the hammerhead. Therefore, substrate destabilization may play a role in hammerhead catalysis. The equilibrium and kinetic properties of the ribozyme–product complexes resembled those expected for a simple RNA duplex.[148] Substrate binding rate constants fell within the range expected for the formation of simple RNA duplexes. Product release was faster than the rate-limiting step of chemistry for a hammerhead construct that contained five base pairs between the product and ribozyme. For constructs where products bound by seven base pairs to the ribozyme, the rate of product dissociation was rate limiting.[148] With these small RNAs, dissociation rate constants were directly proportional to the stability of the helix.

One specific hammerhead construct, the HH16 ribozyme, was studied in particular detail.[147] This ribozyme had eight potential base pairs in each of the substrate recognition helices. This construct was designed to be stable enough to determine product binding and to investigate reversal of the cleavage step. HH16 cleaved bound substrate with a rate constant of ~1 min^{-1}, similar to what was found in the previous experiments.[147] The rate of ligation of the 5′ product and 3′ product to form substrate was only ~0.008 min^{-1}. Therefore, HH16 preferred to maintain products on the ribozyme. Product and substrate association rates were between 10^7 and 10^8 M^{-1} min^{-1}, which are values similar to those for simple helices.[147] The stabilities of ribozyme–product complexes did not indicate any additional tertiary interactions. Finally, the dissociation constant for the binding of substrate to the ribozyme is $\sim 10^{-17}$ M, indicating tight binding between these two molecules.

In order to determine which phosphates are important for catalytic activity, phosphorothioate linkages were incorporated. These substitutions identified four phosphates in the conserved central core that play a role in the self-cleavage reaction.[94] In addition, substitution of sulfur for a nonbridging oxygen at the cleavage site reduced the affinity of an essential Mg^{2+} ion involved in efficient cleavage.[94] The reaction proceeded normally in the presence of Mn^{2+}. These results suggested that a divalent ion is coordinated directly to the phosphate at the cleavage site.

A 1998 crystal structure of the hammerhead captured a conformation that is in agreement with the geometry needed for in-line attack, the mechanism thought to be used by the hammerhead ribozyme for catalysis.[68] Previous crystal structures have identified structures in the ground state which did not contain the proper positioning for an S_N2 reaction.[69,70] This result suggested that a conformational change must take place prior to cleavage, allowing the in-line geometry to form.[68] In addition, the presence of metal ions in the newest structure is also in agreement with the kinetic pathway thought to be utilized. This structure lends support to the biochemical assays that have studied the kinetics of hammerhead self-cleavage.

11.6 OTHER CATALYTIC RNAS

Using *in vitro* selection, a number of ribozymes have been identified for their ability to catalyze reactions at phosphodiester bonds. For example, a leadzyme was identified from a yeast $tRNA^{Phe}$ selection.[83,149] The best Pb^{2+}-dependent ribozyme exhibited a cleavage rate 20-fold higher than that of the original yeast $tRNA^{Phe}$ molecule and had a substantially different geometry than the original tRNA. RNA ligases have also been isolated by selection from a random library.[150] These ribozymes catalyze the ligation of a $5'$ triphosphate with a $3'$-OH (a $5' \rightarrow 3'$ ligase).[151,152] The rate enhancement ($\sim 10^{10}$-fold over background) exhibited by this ribozyme is the highest for a selected catalytic RNA.[151] Another ribozyme was selected for its catalytic activity as an RNA polynucleotide kinase.[153] The reaction involved transfer of a γ-thiophosphate from ATP-γS to either a $5'$-OH or an internal $2'$-OH of an RNA molecule. The best ribokinase had a K_m for ATP-γS of $\sim$40 nM.[153] This value is close to that determined for the protein enzyme, T4 polynucleotide kinase. In addition, the k_{cat} was only $\sim 10^4$-fold slower than that of the protein enzyme.

In vitro selection has also identified ribozymes that are able to catalyze reactions at carbon centers. The unmodified *Tetrahymena* ribozyme weakly catalyzed the hydrolysis of an aminoacyl ester bond at a rate about 5–15 times greater than the uncatalyzed rate.[154] The carboxylate ester was targeted to the catalytic core of the ribozyme by covalently attaching an amino acid to an oligonucleotide that bound to the internal guide sequence. Another RNA was selected for its ability to aminoacylate its own $2'$ ($3'$) terminus when phenylalanyl-monophosphate was present.[155] This reaction is similar to the aminoacyl group transfer performed by protein aminoacyl-transfer RNA synthetases. Further selections have identified RNAs capable of binding to non-nucleotide substrates and performing self-alkylation reactions.[156] In addition, RNAs were evolved that had acyl transferase activity and could transfer an RNA-linked amino acid to their own $5'$-amino-modified terminus, analogous to peptidyl transfer on the ribosome.[157] In 1997, a ribozyme was selected for its ability to perform the same peptidyl transferase reaction as the ribosome.[158] These RNAs could join amino acids by a peptide bond when given two substrates, one that had an amino acid (*N*-blocked methionine) esterified to the $2'$ ($3'$)-oxygen of adenosine and the other which was an acceptor amino acid (phenylalanine) with a free amino group.

11.7 CONCLUSIONS

This description of ribozyme enzymology has demonstrated the importance of structure and metal ions for catalytic activity. While the group I and hammerhead ribozymes are the most fully characterized examples of catalytic RNAs, the principles learned from these examples are likely to be applicable to other structured RNAs including the other examples of catalytic RNAs.[125]

ADDENDUM

A 5.0 Å crystal structure of the P3–P9 regions of the *Tetrahymena* group I intron has recently been solved. There appear to be binding pockets for both the 5′-splice site helix and the guanosine cofactor. Therefore, the ribozyme seems to contain a preorganized active site.[159]

11.8 REFERENCES

1. T.R. Cech, A.J. Zaug and P.J. Grabowski, *Cell*, 1981, **27**, 487.
2. C. Guerrier-Takada, K. Gardiner, T. Marsh, N. Pace and S. Altman,, *Cell*, 1983, **35**, 849.
3. T.R. Cech, in "The RNA World", eds. R.F. Gesteland and J.F. Atkins, Cold Spring Harbor Laboratory Press, Cold Spring Harbor, NY, 1993,
4. D. Herschlag and T.R. Cech, *Biochemistry*, 1990, **29**, 10–159.
5. W.J. Michels and A.M. Pyle, *Biochemistry*, 1995, **34**, 2965.
6. C. Guerrier-Takada and S. Altman, *Science*, 1984, **223**, 285.
7. T. Ikemura, Y. Shimura, H. Sakano and H. Ozeki, *J. Mol. Biol.*, 1975, **96**, 69.
8. P. Schedl and P. Primakoff, *Proc. Natl. Acad. Sci. USA*, 1973, **70**, 2091.
9. S. Altman, *Adv. Enzymol. Rel. Areas Mol. Biol.*, 1989, **62**, 1.
10. J.A. Beebe and C.A. Fierke, *Biochemistry*, 1994, **33**, 10–294.
11. T. Pan, D.M. Long and O.C. Uhlenbeck, in "The RNA World", eds. R.F. Gesteland and J.F. Atkins, Cold Spring Harbor Laboratory Press, Cold Spring Harbor, NY, 1993,
12. G.A. Prody, J.T. Bakos, J.M. Buzayan, I.R. Schneider and G. Bruening, *Science*, 1986, **231**, 1577.
13. O.C. Uhlenbeck, *Nature*, 1987, **328**, 596.
14. F. Michel, A. Jacquier and B. Dujon, *Biochimie*, 1982, **64**, 867.
15. F. Michel and E. Westhof, *J. Mol. Biol*, 1990, **216**, 585.
16. R.W. Davies, R.B. Waring, J.A. Ray, T.A. Brown and C. Scazzocchio, *Nature*, 1982, **300**, 719.
17. D.W. Celander and T.R. Cech, *Science*, 1991, **251**, 401.
18. E.A. Doherty and J.A. Doudna, *Biochemistry*, 1997, **36**, 3159.
19. R.B. Waring, P. Towner, S.J. Minter and R.W. Davies, *Nature*, 1986, **321**, 133.
20. D.W. Celander and T.R. Cech, *Biochemistry*, 1990, **29**, 1355.
21. J.-F. Wang and T.R. Cech, *Science*, 1992, **256**, 526.
22. J.-F. Wang, W.D. Downs and T.R. Cech, *Science*, 1993, **260**, 504.
23. F. Michel, M. Hanna, R. Green, D.P. Bartel and J.W. Szostak, *Nature*, 1989, **342**, 391.
24. T. Inoue, F.X. Sullivan and T.R. Cech, *J. Mol. Biol*, 1986, **189**, 143.
25. M.D. Been and A.T. Perrotta, *Science*, 1991, **252**, 434.
26. F.L. Murphy and T.R. Cech, *Proc. Natl. Acad. Sci. USA*, 1989, **86**, 9218.
27. A.M. Pyle, F.L. Murphy and T.R. Cech, *Nature*, 1992, **358**, 123.
28. S.A. Stobel, L. Ortoleva-Donnelly, S.P. Ryder, J.H. Cate and E. Moncoeur, *Nat. Struc. Biol*, 1998, **5**, 60.
29. S.A. Strobel and T.R. Cech, *Science*, 1995, **267**, 675.
30. S.A. Strobel and T.R. Cech, *Biochemistry*, 1993, **32**, 13–593.
31. J.A. Latham and T.R. Cech, *Science*, 1989, **245**, 276.
32. G. Chanfreau and A. Jacquier, *Science*, 1994, **266**, 1383.
33. A.M. Pyle, in "Catalytic RNA", eds. F. Eckstein and D.M.J. Lilley, Springer, New York, 1996,
34. A. Jacquier and M. Rosbash, *Science*, 1986, **234**, 1099.
35. A. Jacquier and F. Michel, *Cell*, 1987, **50**, 17.
36. K.A. Jarrell, R.C. Dietrich and P.S. Perlman, *Mol. Cell. Biol*, 1988, **8**, 2361.
37. F. Michel, K. Umensono and H. Ozeki, *Gene*, 1989, **82**, 5.
38. C.L. Peebles, M. Zhang, P.S. Perlman and J.S. Franzen, *Proc. Natl. Acad. Sci. USA*, 1995, **92**, 4422.
39. B.D. James, G.J. Olsen, J. Liu and N. Pace, *Cell*, 1988, **52**, 19.
40. E.S. Haas, A.B. Banta, J.K. Harris, N.R. Pace and J.W. Brown, *Nucleic Acids Res*, 1996, **24**, 4775.
41. M.E. Harris, A.V. Kazantsev, J.-L. Chen and N.R. Pace, *RNA*, 1997, **3**, 561.
42. M.E. Harris and N.R. Pace, *RNA*, 1995, **1**, 210.
43. C. Massire, L. Jaeger and E. Westhof, *RNA*, 1997, **3**, 553.
44. D.E. Ruffner, G.D. Stormo and O.C. Uhlenbeck, *Biochemistry*, 1990, **29**, 10–695.
45. R.H. Symons, *Annu. Rev. Biochem*, 1992, **61**, 641.
46. A.C. Forster and R.H. Symons, *Cell*, 1987, **49**, 211.
47. A. Hampel and R. Tritz, *Biochemistry*, 1989, **28**, 4929.
48. L.A. Hegg and M.J. Fedor, *Biochemistry*, 1995, **34**, 15–813.
49. A. Hampel, R. Tritz, M. Hicks and P. Cruz, *Nucleic Acids Res*, 1990, **18**, 299.
50. L. Rubino, M.E. Tousignant, G. Steger and J.M. Kaper, *J. Gen. Virol*, 1990, **71**, 1897.
51. A. Berzal-Herranz, S. Joseph and J.M. Burke, *Genes Dev.*, 1992, **6**, 129.
52. P.A. Feldstein and G. Bruening, *Nucleic Acids Res*, 1993, **21**, 1991.
53. Y. Komatsu, M. Koizumi, N. Sekiguchi and E. Ohtsuka, *Nucleic Acids Res*, 1993, **21**, 185.

54. M.D. Been, *Trends Bio. Sci*, 1994, **19**, 251.
55. M.D. Been, A.T. Perrotta and S.P. Rosenstein, *Biochemistry*, 1992, **31**, 11–843.
56. A.T. Perrotta and M.D. Been, *Nucleic Acids Res*, 1990, **18**, 6821.
57. A. Ferre-D'Amare, K.H. Zhou and J.A. Doudna, *Nature*, 1998, **395**, 567.
58. B.J. Saville and R.A. Collins, *Cell*, 1990, **61**, 685.
59. T.L. Beattie, J.E. Olive and R.A. Collins, *Proc. Natl. Acad. Sci. USA*, 1995, **92**, 4686.
60. T.L. Beattie and R.A. Collins, *J. Mol. Biol*, 1997, **267**, 830.
61. J.H. Cate, A.R. Gooding, E. Podell, K. Zhou, B.L. Golden, C.E. Kundrot, T.R. Cech and J.A. Doudna, *Science*, 1996, **273**, 1678.
62. J.H. Cate, A.R. Gooding, E. Podell, K. Zhou, B.L. Golden, A.A. Szewczak, C.E. Kundrot, T.R. Cech and J.A. Doudna, *Science*, 1996, **273**, 1696.
63. J.H. Cate and J.A. Doudna, *Structure*, 1996, **4**, 1221.
64. J.H. Cate, R.L. Hanna and J.A. Doudna, *Nat. Struc. Biol*, 1997, **4**, 553.
65. S.A. Strobel and K. Shetty, *Proc. Natl. Acad. Sci. USA*, 1997, **94**, 2903.
66. A. Szewczak, L. Ortoleva-Donnelly, S. Ryder, E. Moncoeur and S. Strobel, *Nat. Struct. Biol., 1998,* , 1998, **5**, 1037.
67. H.W. Pley, K.M. Flaherty and D.B. McKay, *Nature*, 1994, **372**, 68.
68. J.B. Murray, D.P. Terwey, L. Maloney, A. Karpiesky, N. Usman, L. Beigelman and W.G. Scott, *Cell*, 1998, **92**, 665.
69. W.G. Scott, J.T. Finch and A. Klug, *Cell*, 1995, **81**, 991.
70. W.G. Scott, J.B. Murray, J.R.P. Arnold, B.L. Stoddard and A. Klug, *Science*, 1996, **274**, 2065.
71. G.J. Quigley and A. Rich, *Science*, 1976, **194**, 796.
72. D.B. McKay, *RNA*, 1996, **2**, 395.
73. M. Record and T. Lohman, *Biopolymers*, 1978, **17**, 159.
74. G. Manning, *Q. Rev. Biophys.*, 1978, **11**, 179.
75. J. Huheey, "Inorganic Chemistry", Harper & Row, New York, 1985.
76. W. Saenger, "Principles of Nucleic Acid Structure", Springer-Verlag, New York, 1984.
77. E.L. Christian and M. Yarus, *J. Mol. Biol.*, 1992, **228**, 743.
78. G. Gish and F. Eckstein, *Science*, 1988, **240**, 1520.
79. E.L. Christian and M. Yarus, *Biochemistry*, 1993, **32**, 4475.
80. L. Ortoleva-Donnelly, A.A. Szewczak, R.R. Gutell and S.A. Strobel, *RNA*, 1998, **4**, 498.
81. V.L. Pecoraro, J.D. Hermes and W.W. Cleland, *Biochemistry*, 1984, **23**, 5262.
82. S. Kazakov and S. Altman, *Proc. Natl. Acad. Sci. USA*, 1991, **88**, 9193.
83. R.S. Brown, J.C. Dewan and A. Klug, *Biochemistry*, 1985, **24**, 4785.
84. L.S. Behlen, J.R. Sampson, A.B. DiRenzo and O.C. Uhlenbeck, *Biochemistry*, 1990, **29**, 25.
85. J.R. Rubin and M. Sundaralingam, *J. Biomol. Struct. Dyn*, 1983, **1**, 639.
86. G. Dirheimer, J.P. Ebel, J. Bonnet, J. Gangloff, G. Keith, B. Krebs, B. Kuntzel, A. Roy, J. Weissenbach and C. Werner, *Biochimie*, 1972, **54**, 127.
87. B. Streicher, U. von Ahsen and R. Schroeder, *Nucleic Acids Res*, 1993, **21**, 311.
88. J. Ciesiolka, W. Hardt, J. Schlegl, V.A. Erdmann and R.R. Hartmann, *Eur. J. Biochem*, 1994, **219**, 49.
89. C. Berens, B. Streicher, R. Schroeder and W. Hillen, *Chem. Biol*, 1998, **5**, 163.
90. A. Jack, J.E. Ladner, D. Rhodes, R.S. Brown and A. Klug, *J. Mol. Biol*, 1977, **111**, 315.
91. S.R. Holbrook, J.L. Sussman, R.W. Warrant, G.M. Church and S.-H. Kim, *Nucleic Acids Res*, 1977, **4**, 2811.
92. G.J. Quigley, M.M. Teeter and A. Rich, *Proc. Natl. Acad. Sci. USA*, 1978, **75**, 64.
93. S. Basu, R.P. Rambo, J. Strauss-Soukop, J.H. Cate, A.R. Ferre-D'Amare, S.A. Strobel and J.A. Doudna, *Nat. Struct. Biol.,*, 1998, **5**, 986.
94. D.E. Ruffner and O.C. Uhlenbeck, *Nucleic Acids Res*, 1990, **18**, 6025.
95. T.A. Steitz and J.A. Steitz, *Proc. Natl. Acad. Sci. USA*, 1993, **90**, 6498.
96. K. Kruger, P.J. Grabowski, A.J. Zaug, J. Sands, D.E. Gottschling and T.R. Cech, *Cell*, 1982, **31**, 147.
97. C.A. Grosshans and T.R. Cech, *Biochemistry*, 1989, **28**, 6888.
98. B.L. Bass and T.R. Cech, *Biochemistry*, 1986, **25**, 4473.
99. A.J. Zaug and T.R. Cech, *Science*, 1986, **231**, 470.
100. A.J. Zaug and T.R. Cech, *Biochemistry*, 1986, **25**, 4478.
101. J.A. Piccirilli, J.S. Vyle, M.H. Caruthers and T.R. Cech, *Nature*, 1993, **361**, 85.
102. L.B. Weinstein, B.C.N.M. Jones, R. Cosstick and T.R. Cech, *Nature*, 1997, **388**, 805.
103. A.S. Sjogren, E. Pettersson, B.M. Sjoberg and R. Stromberg, *Nucleic Acids Res.*, 1997, **25**, 648.
104. A.M. Pyle, *Science*, 1993, **261**, 709.
105. H.A. Heus and A. Pardi, *J. Mol. Biol.*, 1991, **217**, 113.
106. R.A.J. Hodgson, N.J. Shirley and R.H. Symons, *Nucleic Acids Res*, 1994, **22**, 1620.
107. S.C. Dahm, W.B. Derrick and O.C. Uhlenbeck, *Biochemistry*, 1993, **32**, 13–040.
108. S.C. Dahm and O.C. Ohlenbeck, *Biochemistry*, 1991, **30**, 9464.
109. W. Lott, B. Pontius and P. von Hippel, *Proc. Natl. Acad. Sci. USA*, 1998, **95**, 542.
110. A. Hampel and J.A. Cowan, *Chem. Biol*, 1997, **4**, 513.
111. K.J. Young, F. Gill and J.A. Grasby, *Nucleic Acids Res*, 1997, **25**, 3760.
112. A.J. Zaug, P.J. Grabowski and T.R. Cech, *Nature*, 1983, **301**, 578.
113. B.L. Bass and T.R. Cech, *Nature*, 1984, **308**, 820.
114. J.A. McSwiggen and T.R. Cech, *Science*, 1989, **244**, 679.
115. J. Rajagopal, J.A. Doudna and J.W. Szostak, *Science*, 1989, **244**, 692.

116. J.R. Knowles, *Annu Rev. Biochem*, 1980, **49**, 877.
117. F. Eckstein, *Annu Rev. Biochem*, 1985, **54**, 367.
118. D. Herschlag, J.A. Piccirilli and T.R. Cech, *Biochemistry*, 1991, **30**, 4844.
119. A.J. Zaug, J.R. Kent and T.R. Cech, *Science*, 1984, **224**, 574.
120. M.D. Been and T.R. Cech, *Cell*, 1986, **47**, 207.
121. A.J. Zaug, C.A. Grosshans and T.R. Cech, *Biochemistry*, 1988, **27**, 8924.
122. D. Herschlag and T.R. Cech, *Biochemistry*, 1990, **29**, 10–172.
123. P.C. Bevilacqua, K.A. Johnson and D.H. Turner, *Proc. Natl. Acad. Sci. USA*, 1993, **90**, 8357.
124. T.S. McConnell, T.R. Cech and D. Herschlag, *Proc. Natl. Acad. Sci. USA*, 1993, **90**, 8362.
125. B. Young, D. Hershlag and T.R. Cech, *Cell*, 1991, **67**, 1007.
126. D. Hershlag and M. Khosla, *Biochemistry*, 1994, **33**, 5291.
127. R. Mei and D. Herschlag, *Biochemistry*, 1996, **35**, 5796.
128. D. Herschlag, F. Eckstein and T.R. Cech, *Biochemistry*, 1993, **32**, 8312.
129. D. Herschlag, *Biochemistry*, 1992, **31**, 1386.
130. S.A. Strobel and T.R. Cech, *Nat. Struc. Biol.*, 1994, **1**, 13.
131. P.C. Bevilacqua, R. Kierzek, K.A. Johnson and D.H. Turner, *Science*, 1992, **258**, 1355.
132. P.C. Bevilacqua, Y. Li and D.H. Turner, *Biochemistry*, 1994, **33**, 11–340.
133. J.A. Douna, B.P. Cormack and J.W. Szostak, *Proc. Natl. Acad. Sci. USA*, 1989, **86**, 7402.
134. S.A. Strobel and T.R. Cech, *Biochemistry*, 1996, **35**, 1201.
135. A.M. Pyle, S. Moran, S.A. Strobel, T. Chapman, D.H. Turner and T.R. Cech, *Biochemistry*, 1994, **33**, 13–856.
136. D. Knitt, G. Narlikar and D. Herschlag, *Biochemistry*, 1994, **33**, 13–864.
137. D. Herschlag, F. Eckstein and T.R. Cech, *Biochemistry*, 1993, **32**, 8299.
138. P.C. Bevilacqua and D.H. Turner, *Biochemistry*, 1991, **30**, 10–632.
139. A.M. Pyle and T.R. Cech, *Nature*, 1991, **350**, 628.
140. T.R. Cech, S.H. Damberger and R.R. Gutell, *Nat. Struct. Biol.*, 1994, **1**, 273.
141. G.J. Narlikar, M. Khosla, N. Usman and D. Herschlag, *Biochemistry*, 1997, **36**, 2465.
142. M.A. Tanner, E.M. Anderson, R.R. Gutell and T.R. Cech, *RNA*, 1997, **3**, 1037.
143. M.A. Tanner and T.R. Cech, *Science*, 1997, **275**, 847.
144. M. Yarus, M. Illangesekare and E. Christian, *J. Mol. Biol.*, 1991, **222**, 995.
145. T.S. McConnell and T.R. Cech, *Biochemistry*, 1995, **34**, 4056.
146. J. Haseloff and W. Gerlach, *Nature*, 1988, **334**, 585.
147. K. Hertel, D. Herschlag and O. Uhlenbeck, *Biochemistry*, 1994, **33**, 3374.
148. M. Fedor and O.C. Uhlenbeck, *Biochemistry*, 1992, **31**, 12–042.
149. T. Pan and O.C. Uhlenbeck, *Biochemistry*, 1992, **31**, 3887.
150. D. Bartel and J. Szostak, *Science*, 1993, **261**, 1411.
151. E. Ekland, J. Szostak and D. Bartel, *Science*, 1995, **269**, 364.
152. E. Ekland and D. Bartel, *Nucleic Acids Res*, 1995, **23**, 3231.
153. J. Lorsch and J. Szostak, *Nature*, 1994, **371**, 31.
154. J.A. Piccirilli, T.S. McConnell, A.J. Zaug, H.F. Noller and T.R. Cech, *Science*, 1992, **256**, 1420.
155. M. Illangasekare, G. Sanchez, T. Nickles and M. Yarus, *Science*, 1995, **267**, 643.
156. C. Wilson and J.W. Szostak, *Nature*, 1995, **374**, 777.
157. P. Lohse and J. Szostak, *Nature*, 1996, **381**, 442.
158. B. Zhang and T.R. Cech, *Nature*, 1997, **390**, 96.
159. B.L. Golden, A.R. Gooding, E.R. Podell and T.R. Cech, *Science,*, 1998, **282**, 259.

12

Viroids

ROBERT H. SYMONS
University of Adelaide, Glen Osmond, SA, Australia

12.1 INTRODUCTION

12.1.1 Nature of Viroids

Viroids are the smallest pathogenic agents yet described. They are single-stranded circular RNA molecules which vary in length from 246 to 463 nucleotides and are found only in plants. Of the 27 viroids characterized so far (Table 1; Figure 1), 25 infect dicotyledonous plants and the other two infect monocotyledonous plants. The diseases caused by some of these viroids are of considerable agricultural importance. The coconut cadang cadang viroid found in the Philippines, which consists of only 246 nucleotides kills most of the palms it infects. For all other viroids, death of infected plants is unusual, and phenotypic effects vary from essentially no symptoms to various degrees of debilitation and dwarfing.

All evidence indicates that viroids do not code for any proteins. Consequently they must rely completely on normal plant processes for their replication and spread throughout the plant. The host processes involved in their replication are slowly being characterized but essentially nothing is known about how they cause disease. Because of their small size and intriguing biological properties, they have attracted considerable interest, yet there are only a few groups worldwide who are characterizing them at the molecular level.

The reader will find references 1–12 a rich source of historical and background information on viroids. The book by Diener[1] is the classic text and remains a valuable reference work. References 2–4 are multiauthor volumes, while references 5–12 are review articles that cover the more molecular aspects of viroid structure, function, and evolution.

12.1.2 An Historical Perspective

Viroid-caused diseases are a phenomenon of the twentieth century in the sense that these diseases were recognized and described in agricultural crops, and later shown to be of viroid aetiology, only in the twentieth century. In 1967, Diener and Raymer[13] concluded that PSTV is a free, low molecular weight RNA on the basis of its low sedimentation rate and sensitivity to ribonuclease, a result confirmed by Semancik and Weathers[14] for CEV. Accumulating evidence further confirmed the small, protein-free, RNA nature of these infectious agents and the ability of RNA eluted from a specific RNA band on a polyacrylamide gel to be infectious when inoculated on a susceptible host plant.[15,16] The term viroid was proposed as a generic term for PSTV in 1971[15] and is now widely accepted. The first sequence of a viroid was published[17] in 1978 (PSTV) and the sequences of other viroids appeared progressively thereafter.

12.2 CLASSIFICATION OF VIROIDS

Table 1 lists the 27 viroids together with an abbreviation used for each one and their size in number of nucleotides. It is becoming increasingly common to add a small "d" at the end of the abbreviation to indicate a viroid; the original abbreviations are used in this review. In many viroids, the length can vary from one to several nucleotides, as indicated in Table 1, while in two viroids, CCCV and CEV, terminal repeats of nucleotides can substantially increase the size (see Section 12.5.2).

On the basis of comparative sequence analysis between the different viroids as well as other features, the 27 viroids can be divided into two main groups, the ASBV-group with three members and the PSTV-group for the remainder. There is a further subdivision of the PSTV-group into three subgroups (Figure 1). Many aspects of their properties relevant to this classification are considered throughout this review.

12.3 ISOLATION, PURIFICATION, AND SEQUENCING OF VIROIDS

All the viroids listed in Table 1 were originally identified in field samples taken from plants showing symptoms of diseases, especially of agricultural crops. Most, but not all, viroids can be readily maintained in the greenhouse on herbaceous or woody host plants. For example, tomato is a very useful

Table 1 Classification of viroids.

Viroid-group	*Viroid-subgroup*	*Viroid*	*Abbreviation*	*Length (nucleotides)*
A. ASBV-group	ASBV-subgroup	Avocado sunblotch viroid	ASBV	246–250
		Chrysanthemum chlorotic mottle viroid	CChMV	398–399
		Peach latent mosaic viroid	PLMV	337
B. PSTV-group	B1. PSTV-subgroup	Chrysanthemum stunt viroid	CSV	354–356
		Citrus exocortis viroid	CEV	370–463
		Citrus viroid IV	CV IV	284
		Coconut cadang cadang viroid	CCCV	246–346
		Coconut tinangaja viroid	CTiV	254
		Columnea latent viroid	CLV	370–373
		Hop latent viroid	HLV	256
		Hop stunt viroid	HSV	297–303
		Iresine viroid	IrV	370
		Mexican papita viroid	MPV	360
		Potato spindle tuber viroid	PSTV	359–360
		Tomato apical stunt viroid	TASV	360–363
		Tomato planta macho viroid	TPMV	360
	B2. ASSV-subgroup	Apple scar skin viroid	ASSV	329–330
		Apple dimple fruit viroid	ADFV	306
		Australian grapevine viroid	AGV	369
		Citrus bent leaf viroid	CBLV	318
		Citrus viroid III	CV III	294–297
		Grapevine yellow speckle viroid-1	GYSV-1	366–368
		Grapevine yellow speckle viroid-2	GYSV-2	363
		Pear blister canker viroid	PBCV	315–316
	B3. CbV-subgroup	Coleus blumei viroid 1	CbV 1	248–251
		Coleus blumei viroid 2	CbV 2	301
		Coleus blumei viroid 3	CbV 3	361–364

host for PSTV, CEV, and TPMV and chrysanthemum for CSV and CChMV. Some viroids of woody species have a very limited host range: CCCV and CTiV only infect members of the palm family, GYSV-1 and GYSV-2 are only found in grape vines, and ASBV is limited to avocado and related species. In contrast, CEV, which was first isolated from citrus as its name implies, has a host range that includes species in several different families of plants, and PSTV can infect many species in at least 12 different families. Appearance of symptoms on inoculated plants usually occurs within a month for herbaceous hosts. However, in coconut seedlings inoculated with CCCV and avocado seedlings infected with ASBV symptoms may not appear for one to two years. In all cases, it is preferable to maintain day temperatures of inoculated plants at 25–30 °C since higher than usual greenhouse temperatures favor viroid replication.

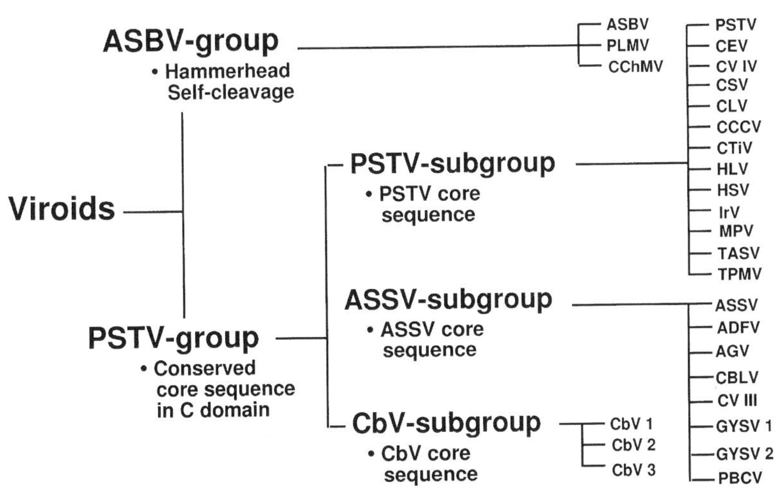

Figure 1 Classification of viroids. The key features used for the division of viroids into two groups and of the PSTV-group into three subgroups are indicated. Viroid abbreviations are defined in Table 1.

Infection of plants for experimental purposes is most commonly done by rubbing crude extracts of viroid-infected plants, partly purified extracts, or purified viroid on leaves of uninfected host plants, or, in the case of more woody species, by slashing through a drop of extract placed on a stem of a host plant. Infectious cDNA clones of viroids have proven of great benefit for the characterization of viroid replication and pathogenicity since they can also be used for inoculation; rubbing purified cDNA clones on the leaves of susceptible plants usually suffices. The great advantage here is that a single sequence variant can be used for inoculation, in contrast to RNA extracts of infected plants, which may contain many sequence variants. Obviously, also, cDNA clones can be used for mutagenesis experiments to investigate structure/function relationships in viroids.

Many viroids are present in infected plants at levels which allow their purification by gel electrophoresis of crude or partly purified RNA extracts. In the case of viroids that are present at low concentrations, for example GYSV-1 and GYSV-2, larger amounts of starting material and extra concentration steps are needed. The fractionation of RNAs by chromatography on CF11 cellulose is commonly used here.

Viroids were originally sequenced by partial RNase digestion, and gel electrophoresis purification of individual fragments followed by enzymatic sequencing using several nucleases of known specificity.[18,19] The method is reasonably rapid but problems may arise when different sequence variants exist as a population in an infected plant (see Section 12.5). The preparation of cDNA clones of a viroid is now obviously the method of choice but some initial sequence information is necessary to define appropriate oligonucleotide primers for the synthesis of first-strand cDNA.

12.4 DOMAIN MODEL FOR PSTV GROUP OF VIROIDS

From an examination of sequence homology among members of the PSTV-subgroup of viroids that had been sequenced prior to 1985, a domain structural model[20] for viroids was developed (Figure 2) which has stood the test of time. The domain model was based on the rod-like secondary structure of viroids as predicted from sequence comparisons and was consistent with the early electron microscope images of PSTV and CCCV.[21–23] The three members of the ASBV-group of viroids do not conform to this model. However, as more members of this group are described, it is possible that an appropriate domain model will be developed.

The boundaries between neighboring domains of the PSTV-group of viroids were determined by comparative sequence analysis between pairs of viroids. In many, but not all, such comparisons, the boundaries between domains were defined by sharp changes in sequence homology, from high to low or vice versa. Different pairwise viroid comparisons were consistent in defining the exact position of the boundaries. The pathogenic domain P was so called because it is associated with symptom expression, at least for the PSTV-subgroup members at the time when the model was developed.[7,18–20] Likewise, the variable domain V was so named because it showed the greatest sequence variability between closely related viroids. The C or central conserved domain is the most highly conserved domain in pairwise

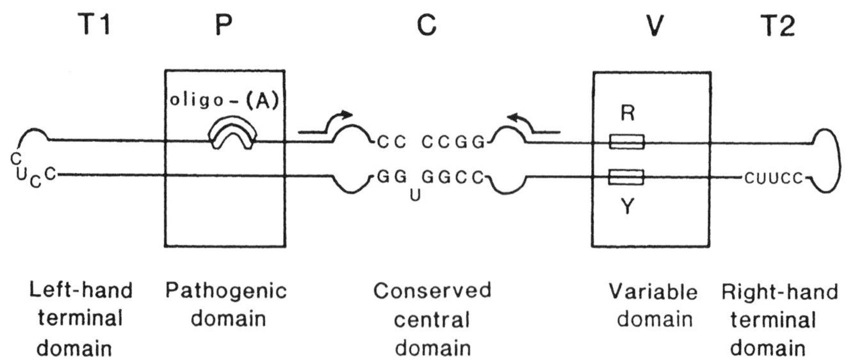

Figure 2 Domain model for the PSTV-group of viroids.[18,20]. This 1985 model was based on sequences of seven viroids known at that time in the PSTV-subgroup. Boundaries between the domains were determined by marked changes in sequence homology on pairwise comparisons between viroids. Conserved nucleotides are indicated. R and Y indicate a short oligopurine-oligopyrimidine helix. Arrows depict an inverted repeat sequence with the potential to form a hairpin stem. Later models also contained a similar inverted repeat sequence in the bottom strand of the C domain (Figures 3–5).

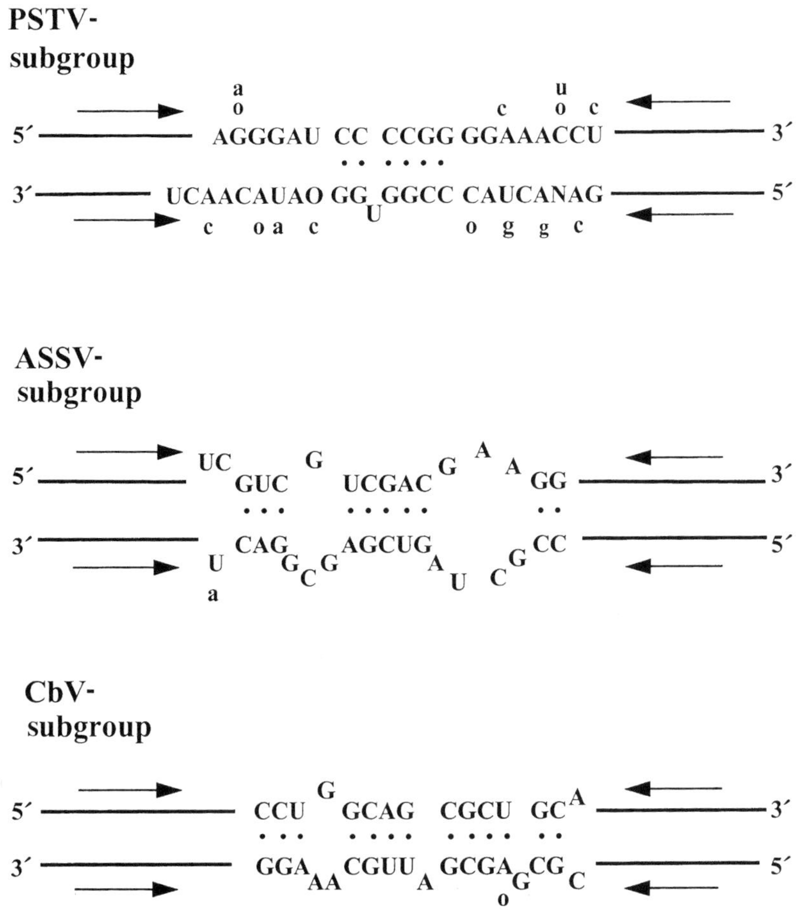

Figure 3 CCRs within the C domain of the viroids and within the three subgroups of the PSTV-group of viroids (see Figures 1 and 2). The sequences given for the PSTV- and ASSV-subgroups are those of PSTV and ASSV, respectively. Variations of sequence in one or more of the subgroup members are given in lower case above or below the relevant residue. O indicates a single base deletion.

comparisons between viroids. The U-bulged helix and the inverted repeat sequence, indicated by the arrows in Figure 2, are highly conserved in the PSTV-subgroup members.

The publication of sequences for members of the ASSV-subgroup and CbV-subgroup viroids confirmed that the domain model developed for the PSTV-subgroup viroids is directly applicable to them also and this allowed a further elaboration of the viroid classification scheme[24] with the result shown in Figure 1.

The C domains of three PSTV-subgroups are compared in Figure 3. They each show inverted repeat sequences, indicated by the arrows, and a central core sequence specific to each subgroup; this core sequence is often referred to as the central conserved region (CCR) within the C domain. Another conserved feature of the C domain is the potential ability of the conserved sequences in the top strand of the C domain to form palindromic structures[25,26] (Figure 4) in tandem monomeric repeats. Likewise, the presence of the inverted repeat sequences allows the potential formation of two hairpin-loop structures, as indicated schematically in Figure 5. Whether such structures are biologically significant is not known; this is discussed further in Section 12.8.

12.5 SEQUENCE PATTERNS AND VARIATION IN VIROID SEQUENCES

12.5.1 Sequence Variants within Viroid Populations

The ability to produce full-length cDNA clones of purified viroids followed by the sequencing of many individual clones from a single viroid isolate has provided us with important information on the

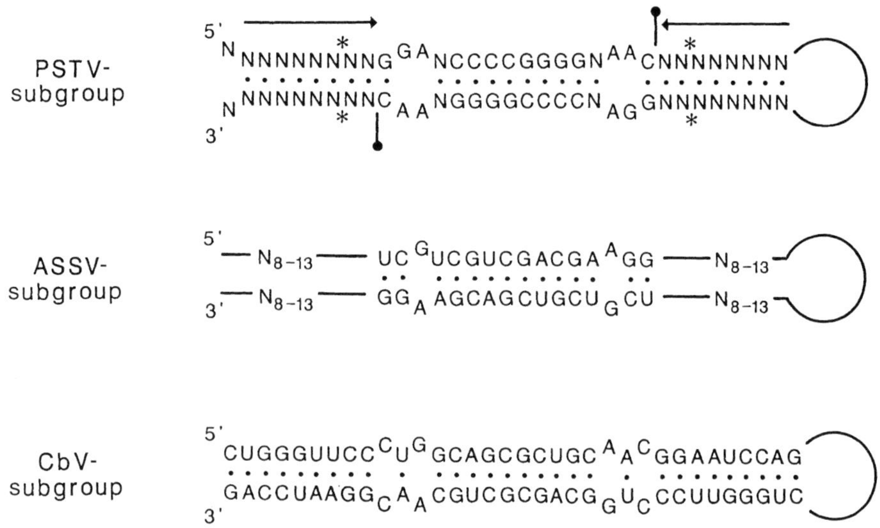

Figure 4 Palindromic structures that are possible in tandem monomeric repeats of members of each subgroup of the PSTV-group of viroids.[25,26] Sequences are from the top strand of the C domain while the specific sequences are from the CCR within the C domain except for CbV where the CCR sequence is yet to be determined since CbV-1, CbV-2, and CbV-3 have high sequence homology in the C domain.[27] Reverse arrows on the top strand of PSTV-subgroup indicate location of inverted repeat sequences for the three subgroups of viroids. N is a nonconserved nucleotide. In HSV, a U is inserted at the positions marked by a vertical bar. Asterisks denote the sites of insertion of two unpaired nucleotides in HLV. (Reproduced by permission of CRC Press from "Viroids: Pathogens at the Frontiers of Life," 1991.)

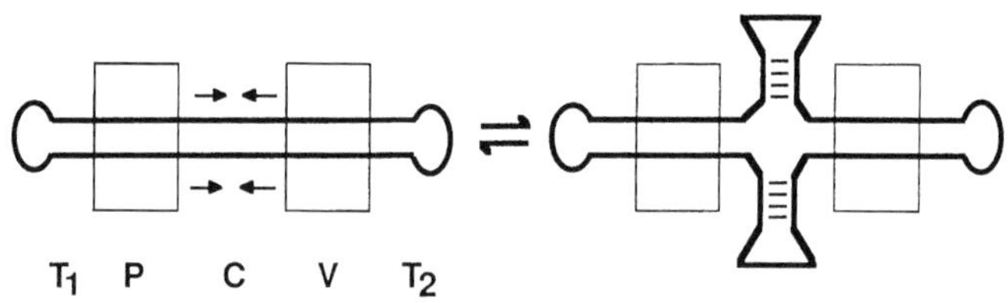

Figure 5 Schematic diagram of potential alternative hairpin structures within the C domain of members in the PSTV-group of viroids. Left-hand side: domain model of Figure 2. with two pairs of inverted repeat sequences indicated by arrows. Right-hand side: potential stem-loop structures formed in the C domain involving the inverted repeat sequences. (Reproduced by permission of Oxford University Press from *Nucleic Acids Res.*, 1997, **25**, 2683.)

dynamics of viroid populations. Viroids, being RNA pathogens, are likely to mutate during replication over a period of time and on transfer from one host to another. Not all mutants so produced will be viable but it would be expected that some will be and that these will replicate to a level detectable in a viroid preparation. Hence, it can be predicted that a purified viroid sample will contain more than one sequence variant where a sequence variant is defined as an individual viroid molecule of defined sequence.

These predictions have been amply verified by comprehensive sequence analysis of isolates of several viroids. Perhaps the most extensive analyses have been done with CEV.[7,29,30] Some isolates seem to contain only one sequence variant, others contain two, whereas another CEV isolate contained nine different sequence variants in a total of 20 cDNA clones sequenced. The latter result indicates a much greater number of sequence variants than the nine sequenced in that CEV population, which was isolated from a single field orange tree. Relevant here is that, if a particular sequence variant is present at a level of 5% of the total variants, then at least 20 cDNA clones would have to be sequenced to provide a reasonable possibility of detecting it. The sequence variations found in 15 Australian sequence variants of CEV as well as two Californian variants were mostly located in the P and V domains and were concentrated into two smaller regions within the P and V domains.

In an attempt to define the region of the viroid molecule which determines the severity of symptom expression, cDNA clones were prepared from two isolates of CEV, one which gave severe symptoms on tomato and the other which gave mild symptoms.[31] The left-hand part of a severe isolate was joined with the right-hand part of a mild isolate through the C domain, and vice versa. Infectivity of such constructs on tomato seedlings indicated that severe symptoms were determined by the P domain.

There is no information about how viroids produce symptoms in infected plants (Section 12.9) and hence no molecular explanation for the results of domain-swapping experiments. In the case of CEV and CCCV, viroid replication appears to occur in the nucleus where the viroids accumulate[32] so that symptom expression is probably a consequence of some interference with normal nuclear function.

Sequence variants have been reported in populations of other viroids of the PSTV-group, e.g., PSTV, HSV, CCCV, and CSV, and hence this appears to be a common feature of viroid infection. In the case of the ASBV-group of viroids, 16 sequence variants were found in isolates prepared from three different Australian avocado trees.[33] Sequence length varied from 246 to 251 nt with sequence variation occurring mainly in the region of the left and right terminal loops.

12.5.2 Terminal Repeat Sequences in at least Two Viroids

In 1982, the sequencing of CCCV provided the first evidence for sequence duplications in the T2 domain.[34] It was already known that longer forms of CCCV accumulated in infected coconut palms as disease symptoms progressively become worse.[35] The 246/247 nt CCCV present early in infection is gradually replaced by one or more larger forms in which the T2 region and part of the V domain are duplicated (Figure 6).[7,34,35] Four forms have been identified with duplications of 41, 50, 55, and 100 (2 × 50) nucleotides. The mechanism by which such duplications are produced is not known but it is possible that a jumping RNA polymerase switching from one template strand to another during RNA

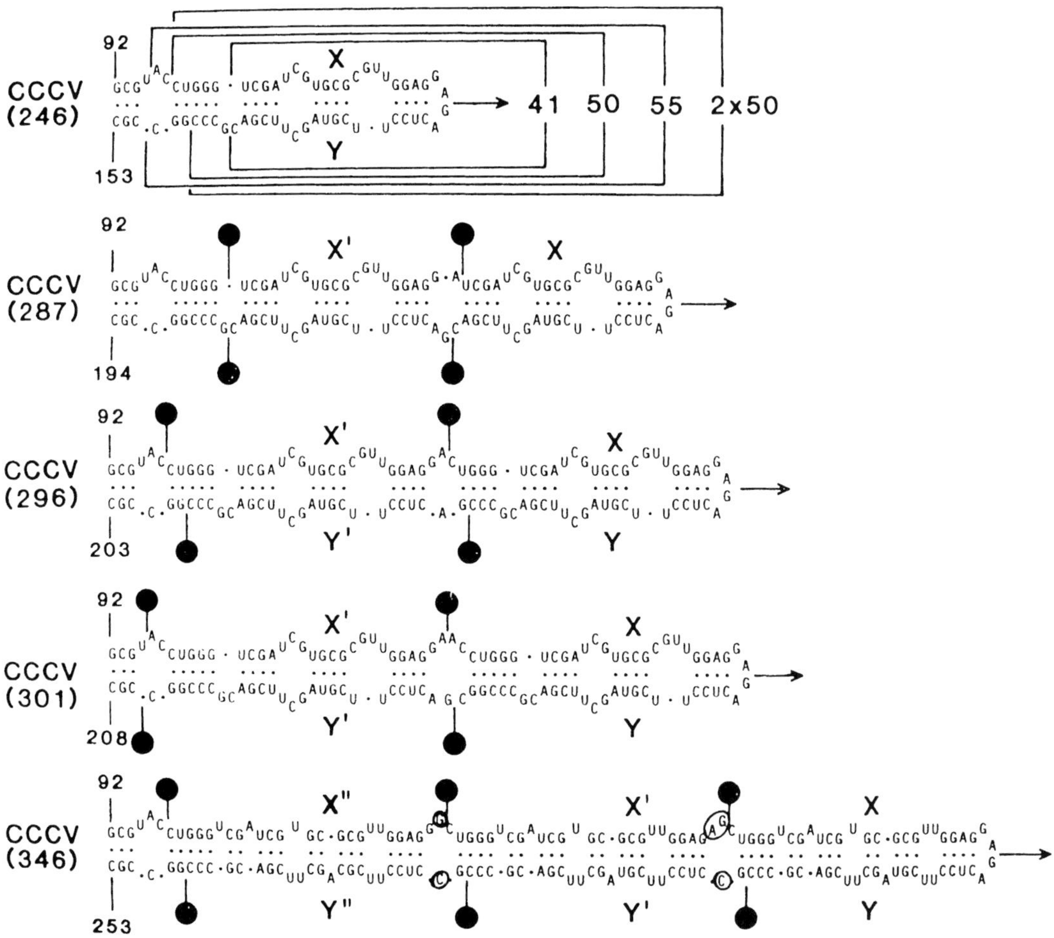

Figure 6 Partial sequence duplications of the T2 domain of CCCV.[7] The two sequences X and Y in CCCV (246) with a total of 41, 50, 55, and 2 × 50 nucleotides are duplicated as indicated. The arrows pointing to the right mark the boundaries of the X and Y sequences, while the filled circles mark the boundaries of the duplicated sequences. Circled nucleotides are sites of mutation in sequence variants. (Reproduced by permission of CRC Press from "Viroids: Pathogens at the Frontiers of Life," 1991.)

synthesis may synthesize specific repeat sequences. Obviously, such sequence variants are viable but it is feasible that many other variants are produced which are not viable or are only replicated poorly and thus accumulate at below detectable levels.

An unusual variant of CEV, CEV D-92, was detected when an inoculum maintained on *Gynura aurantiacum* was used to infect a hybrid tomato (*Lycopericon esculentum* × L. *peruvianum*).[30] This new variant, of length 463 nt, contains an extra 92 nucleotides made up essentially of two 46 nt repeats which begin within the V domain at a site comparable to that found for the repeat sequences of CCCV. Hence, very similar duplications can occur in viroids which infect monocot and dicot hosts. For the CEV 463 nt variant, there was a dramatic reduction in symptom expression as compared with the 371 nt CEV parent when it was inoculated back onto *Gynura*.

12.5.3 RNA Rearrangements and their Potential Role in Viroid Evolution

The domain model described suggests that viroids evolve by the rearrangement of domains between two or more viroids infecting the same cell, followed by further mutation. For example, the T1 and T2 domains of CLV show high sequence homology to the corresponding domains in PSTV and TASV, respectively, and the boundaries are sharply defined by pairwise comparisons between members of the PSTV-subgroup[36] (Figure 7). Domain exchange at precise boundaries provides a nice model for recombination events but subdomain lengths also appear capable of exchanging as indicated by the presence in CLV of subdomain lengths of TPMV sequences in the P domain and of HSV and PSTV sequences in the C domain (Figure 7).

Examples of even more scrambled mixtures of sequences are found in the ASSV-subgroup of viroids.[37] For example, AGV contains sequences which appear to have been derived from CEV, PSTV, ASSV, and GYSV. These segments have sequence homologies which vary from about 50 to 100% of those in the putative parent viroids. Not only does AGV provide another example of RNA rearrangements within domains but it was also the first example in which rearrangements appear to have taken place between viroids belonging to two separate PSTV-subgroups (Table 1, Figure 1).

12.5.4 Poly(pur) and Poly(pyr) Bias in Viroid Sequences

All evidence indicates that viroids can replicate in plant cells without the assistance of any helper virus; this is in contrast to viral single-stranded circular and linear satellite RNAs of roughly the same size range which are dependent on a helper. We have little understanding of why these two groups of RNAs differ this way. However, one aspect which may be relevant is the pattern of purine and pyrimidine tracts, which can make up substantial portions of the viroid genomes and are much less frequent in helper virus-dependent RNAs.[38] Intriguingly, a single-stranded circular human RNA of 1700 nt, hepatitis delta virus RNA with some viroid-like features, also has much greater than average distribution of these purine and pyrimidine tracts.[38]

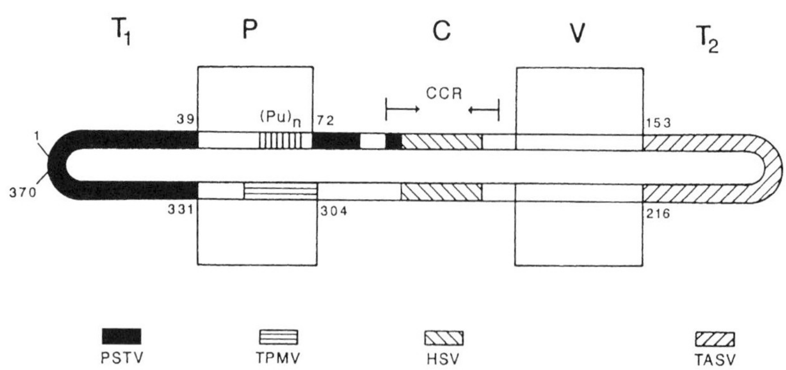

Figure 7 Schematic diagram of CLV showing regions of high sequence homology to other viroids. Residue numbers at boundaries of domains are given for this 370 nt viroid. (Reproduced by permission of CRC Press from "Viroids: Pathogens at the Frontiers of Life," 1991.)

The biological significance of this difference in the extent of distribution of these poly(pur) and poly(pyr) tracts is not known but it has been suggested that it may be related to differences in the replication strategies of the two groups.[38] For example, all evidence points to a role of the host nuclear DNA dependent RNA polymerase II in the replication of the PSTV-group of viroids and the hepatitis delta virus RNA, whereas the helper dependent satellite RNAs are replicated by RNA polymerases coded for by the helper viral RNAs.[39]

12.6 REPLICATION OF VIROIDS

12.6.1 Rolling Circle Mechanism of Replication

There seems to be universal agreement that viroids are replicated by a rolling circle mechanism, as first proposed in 1984[40] (Figure 8), and the finer aspects of their replication are slowly being unraveled. In one of the two variations of the rolling circle mechanism (Figure 8(a)), the infectious plus RNA is copied continuously by an RNA polymerase to produce a long minus RNA strand. Specific cleavage of the (−) strand, either enzymatically or RNA catalyzed, produces monomeric (−) strands which are circularized by a host RNA ligase. Continuous copying of this (−) RNA produces a continuous (+) strand which is cleaved at monomeric lengths that are then circularized to provide the circular (+) RNAs which accumulate in the infected cell.

The three members of the ASBV-group are the only viroids which follow this pathway, as they have the ability to self-catalyze specific cleavage of both the (+) and (−) strands and multimeric (−) RNA does not accumulate in infected plants, at least for ASBV.[41–43]

In the other variation of the rolling circle model (Figure 8(b)), which is predicted to be the pathway for all members of the PSTV-group of viroids, the long linear (−) strand is not cleaved and can be detected in infected plants.[44] This (−) strand is then copied to provide the linear (+) strand which is specifically cleaved to monomers that are then circularized to give the progeny (+) viroid. Support for this model is indicated by the presence of high molecular weight (−) RNA in infected plants and the absence of unit length (−) species.

12.6.2 Host Enzymes involved in Viroid RNA Synthesis

There is no evidence that any viroids can code for any functional polypeptides. Only short potential open reading frames are found in these RNAs, which indicates that viroid replication must be mediated by host RNA polymerases. Inhibition of viroid replication, at least for PSTV and CEV, by alpha-amanitin points to a role for the nuclear host-encoded DNA polymerase RNA II in viroid replication.[11,39,45] There are no detailed studies as to how an enzyme that usually copies DNA is induced to copy an invading viroid RNA. Eukaryotic RNA polymerases are complex, multicomponent enzymes and no-one has yet accepted the challenge of characterizing their viroid RNA copying properties *in vitro*. One significant aspect of this problem would be trying to replicate *in vitro* the structural forms of the (+) and (−) viroid

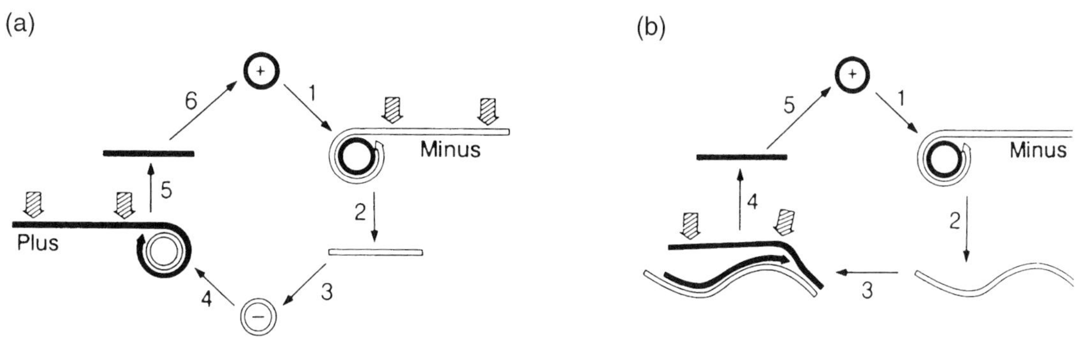

Figure 8 Two rolling circle models for the replication of viroids.40. (a) Pathway followed by three members of the ASBV-group where both minus and plus RNAs self-cleave by the hammerhead structure. (b) Pathway followed by members of the PSTV-group of viroids where only the plus RNA is processed.

RNAs *in vivo* where they are likely to be quite different to the strongly folded RNAs isolated from infected plants.

Of the three viroids identified in the ASBV-group, ASBV is found in chloroplasts (see Section 12.6.3) so that its replication is predicted to be by the more prokaryote-like chloroplast DNA dependent RNA polymerase. The location of PLMV and CChMV in infected plants has yet to be determined but it seems reasonable to expect that this will also be in the chloroplasts.

12.6.3 Where does Viroid Replication occur within the Cell?

Unraveling the mechanism of viroid replication obviously requires knowledge of the site of replication within the cell. *In situ* hybridization approaches with analysis at both the electron microscope and confocal levels can identify sites of accumulation which may or may not also be the sites of synthesis. Sorting out the fine details of synthesis and accumulation possibilities will be a difficult task, and has yet to be done for any viroid.

Earlier studies have shown the inhibition of replication of PSTV and CEV by alpha-amanitin,[45] thereby implicating the host nuclear DNA-dependent RNA polymerase II in viroid synthesis. The first definitive *in situ* hybridization studies were carried out by Harders *et al.*[46] using nuclei purified from PSTV-infected tomato plants and biotinylated RNA probes followed by fluorescence detection by confocal laser scanning microscopy. Viroids were detected in 6–18% of infected nuclei and were homogeneously distributed throughout the nucleoli with minor distribution in the nucleoplasm.

The first study to localize viroids by *in situ* hybridization at the electron microscope level made use of biotinylated RNA probes for CCCV and ASBV, and detection using gold-labeled monoclonal antibiotin antibodies.[47] CCCV was located in the nucleus and nucleoplasm of cells of infected leaves of oil palm (*Elaeis guineensis*). Both PSTV and CCCV are in the same PSTV-subgroup of viroids so that the viroid distribution in nuclei is reasonably consistent between monocot and dicot plants. However, ASBV, the type member of the ASBV-group, was located within the chloroplasts, mostly on the thylakoid membranes of cells from infected leaves of avocado (*Persea americana*) (Figure 9), indicating a fundamental difference between the two group of viroids.[47] An independent study also localized ASBV in chloroplasts.[48]

In a more detailed study, the tissue and intracellular distribution of CCCV and CEV, both members of the PSTV-subgroup, was determined by *in situ* hybridization and both confocal laser scanning and

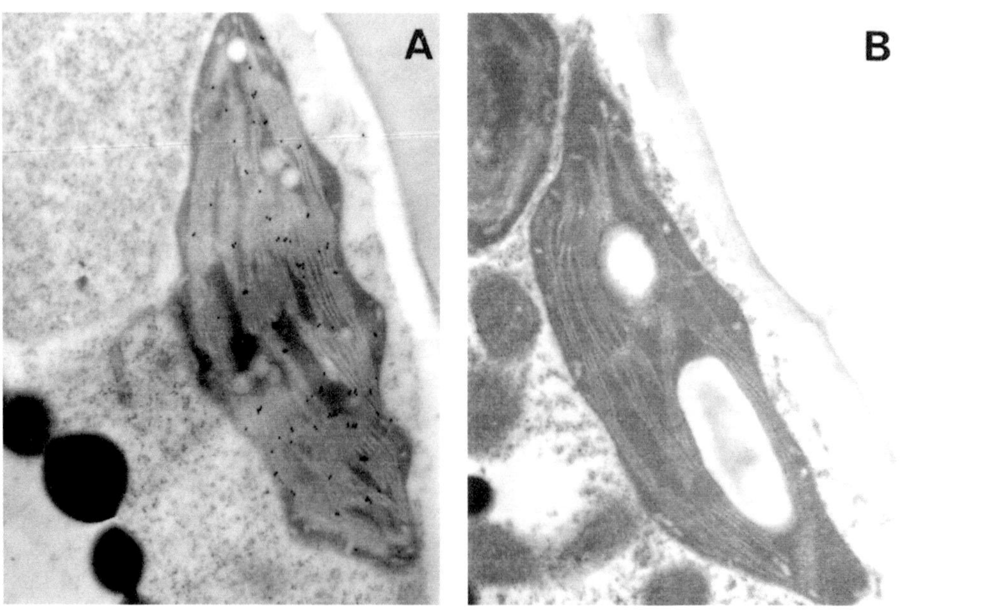

Figure 9 Localization of ASBV on thylakoid membranes of avocado chloroplasts by *in situ* hybridization with biotinylated RNA probes.[47] Hybridized probes detected by direct immunogold localization using 15 nm diameter gold-labeled antibiotin monoclonal antibodies. Section of mature mesophyll leaf cell from ASBV-infected avocado. (Reproduced in part by permission of BIOS Scientific Publishers from *Plant J.*, 1994, **6**, 99.)

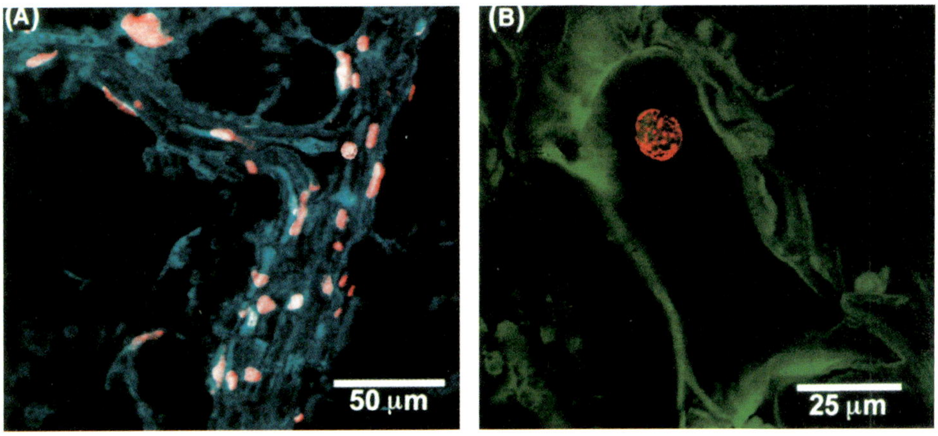

Figure 10 Localization by *in situ* hybridization of CEV in infected tomato by confocal microscopy. (a) Branched vascular bundle (blue) from CEV-infected tomato leaf tissue containing abundant CEV signal (red/orange). (b) Single mesophyll cell from CEV-infected tomato leaf showing cell nucleus with viroid signal (red/orange) and cell structure by autofluorescence (green). (Reproduced in part by permission of BIOS Scientific Publishers from *Plant J.*, 1996, **9**, 457.)

transmission electron microscopy.[49] Both viroids were found in the vascular tissues as well as in the nuclei of mesophyll cells of infected oil palm and tomato, respectively. At the subnuclear level, however, CEV was distributed across the entire nucleus (Figure 10) in contrast to CCCV which was mostly concentrated in the nucleolus with the remainder distributed throughout the nucleoplasm.

The differences between the reports of the two laboratories on the distribution within the nucleus of the closely related viroids CEV and PSTV need to be considered. The whole nuclear distribution of CEV in sections of infected tomato tissue[47] contrasts with the nucleolar localization of PSTV, also in tomato.[46] It is possible that the purification of nuclei prior to *in situ* hybridization may have caused a redistribution of the PSTV into the nucleolus, especially since there is no fundamental reason why PSTV and CEV should show different nuclear distributions. A more comprehensive side-by-side repetition of the two experiments is needed to resolve these differences.

Obviously, we have a long way to go to gain a better appreciation of what is happening in the nucleus for the PSTV-group of viroids. Studies of nuclear distribution as a function of time after infection would be a good start as well as the inclusion of another couple of viroids from the PSTV-group which also infect tomato.

12.7 PROCESSING REACTION *IN VITRO* IN THREE VIROIDS VIA THE HAMMERHEAD SELF-CLEAVAGE REACTION

12.7.1 Hammerhead Self-cleavage Structure

The hammerhead self-cleavage structure was first identified in both the (+) and (−) forms of ASBV[41] and so named and further defined during the characterization of the self-cleavage of the 324 nt virusoid or viroid-like satellite RNA of lucerne transient streak virus (vLTSV)[50–52] and the (+) form of the 359 nt satellite RNA of tobacco ringspot virus.[53] Hammerhead self-cleavage has also been identifed in two other members of the ASBV-group of viroids, PLMV and CChMV, where both the (+) and (−) RNAs self-cleave.[42,43] The history and experimental approaches involved in the discovery and characterization of the hammerhead self-cleavage reaction is covered in several reviews.[4–6,54,55]

Two-dimensional hammerhead structures for the plus strand of ASBV as well as of the RNA transcript of the newt satellite II DNA, one of only two nonpathogenic RNAs so far identified that self-cleaves, are shown in Figures 11(a) and 11(b). The boxed nucleotides are highly conserved in the hammerhead self-cleaving viroid and satellite RNAs identified so far. The two base pair stem III of (+) ASBV is inherently unstable and that of the newt hammerhead structure, which has only a two base hairpin loop, even more so. However, the (−) ASBV hammerhead stem III, which has a three base pair stem III and a three base hairpin loop, is more stable.

For the two other viroids which contain (+) and (−) hammerhead structures, PLMV and CChMV,

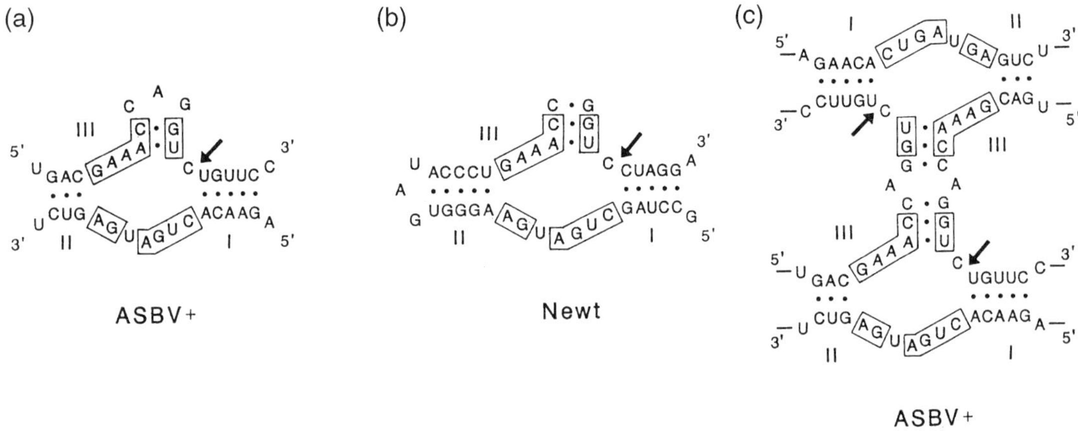

Figure 11 Single (a) and (b) and double (c) hammerhead structures. The newt double-hammerhead structure has the same arrangements as for plus ASBV. (Reproduced by permission of Elsevier Science from *Trends Biochem. Sci.*, 1989, **14**, 445.)

the stem III contains from six to eight base pairs and hence should be quite stable.[42,43] The potential instability of stem III in the (+) ASBV and newt hammerhead structures led to the prediction that a double hammerhead structure might be involved in self-cleavage (Figure 11(c)). This was verified for (+) ASBV by mutational studies *in vitro*,[56] described in Section 12.7.2. In the case of the newt hammerhead structure, support for a double-hammerhead structure *in vitro* was indicated by a concentration dependence on the rate of self-cleavage.[56]

12.7.2 Double-hammerhead Structure in the Processing of the Plus Sequence of ASBV

Experimental proof of the double-hammerhead structure for ASBV was provided[56] using single base mutations of cDNA prepared from a tandem dimer repeat of the 247 nt ASBV in a SP6 RNA polymerase transcription vector (Figure 12). Either or both of the conserved GAAAC sequences (GAAAC-(A) or (B)) were mutated to GAAC in the dimeric cDNA. Mutation of both GAAACs, as predicted, abolished self-cleavage at both sites, SC-1 and SC-2. When only GAAC-(A) was mutated, self-cleavage at site SC-1 was unaffected, whereas self-cleavage at site SC-2 was abolished. Likewise, the mutation of GAAAC-(B) to GAAC abolished self-cleavage at SC-1 and not at SC-2. Both results are completely consistent with the double-hammerhead models shown in Figures 11(c) and 12(c).

The differences between the two dimeric structures and the interpretation of the mutation data can be best visualized in the diagrams of Figure 12(b), containing two single hammerheads in the dimeric transcript, and the double-hammerhead structure in Figure 12(c). Note, for example, that the GAAAC-(A) sequence in Figure 12(b) (thick black section, labeled A) is juxtapositioned next to SC-1 in the single-hammerhead structure, whereas it is juxtapositioned next to SC-2 in the double-hammerhead structure of Figure 12(c). Hence, the GAAC-(A) mutation in Figure 12(b) directly inhibits self-cleavage at site SC-1 whereas, in the double-hammerhead structure of Figure 12(c), it inhibits self-cleavage at site SC-2.44

An interesting variation of the result just described was observed[57] when the same kinds of transcriptions were carried out with dimeric cDNA clones of (−) ASBV, where the three base pair stem III and three base loop (not shown) should provide greater stability to stem III than the two base pair stem III of (+) ASBV. However, transcription of the minus cDNA clones showed that self-cleavage was still by the double-hammerhead structure. During these transcription reactions, the RNA would be expected to fold progressively into tertiary structures as it became separated from the RNA polymerase.

When gel-purified full-length dimeric transcripts of the (+) and (−) dimeric RNA were incubated under self-cleavage conditions, the (+) RNA still cleaved via a double-hammerhead structure. However, the (−) ASBV now self-cleaved via a single-hammerhead structure, indicating that stem III of the (−) ASBV hammerhead is, as predicted, more stable than that of the (+) ASBV RNA.[57]

In the case of the newt RNA hammerhead, with an even less stable stem III, a short (40-mer) RNA transcript containing the newt hammerhead sequence was incubated over a range of concentrations under

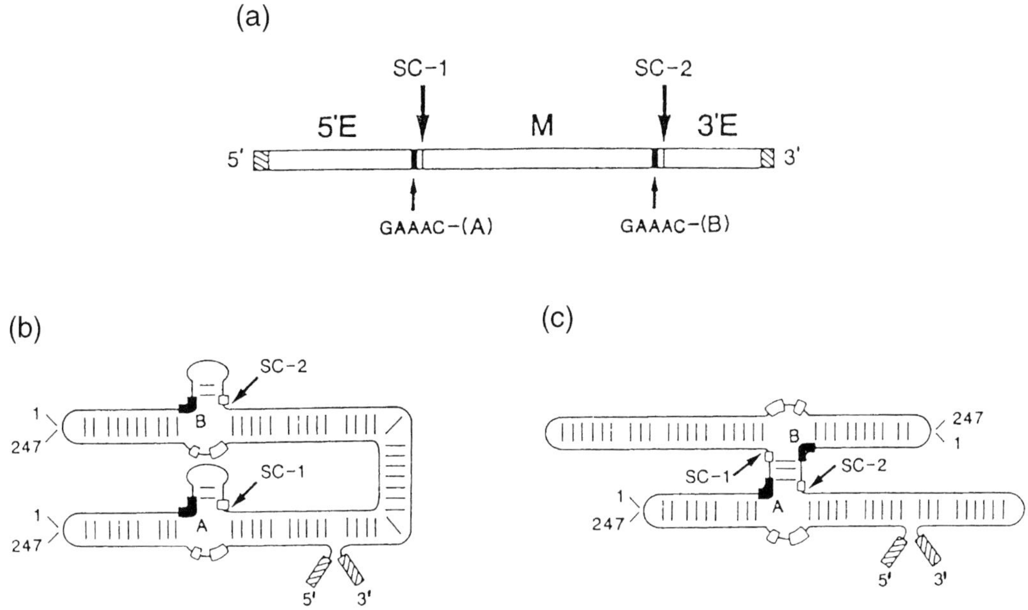

Figure 12 Experimental approach to demonstrate the presence of a double-hammerhead structure in RNA transcripts of a dimeric cDNA clone of ASBV.[56] (a) Schematic diagram of RNA transcript with self-cleavage sites SC-1 and SC-2. The conserved GAAAC in the hammerhead structure was mutated to GAAC in position A and/or B. (b) and (c) Single- and double-hammerhead arrangements, respectively, of dimeric RNA transcripts of plus ASBV. Self-cleavage occurs between residues 55 and 56 in plus ASBV and between residues 69 and 70 in minus ASBV.[44] (Reproduced by permission of CRC Press from *Crit. Rev. Plant Sci.*, 1991, **10**, 189.)

self-cleavage conditions. The extent of self-cleavage was dependent on RNA concentration and showed roughly second-order kinetics, consistent with a bimolecular, double-hammerhead reaction.[56,58]

12.7.3 Crystallization of the Hammerhead

The simplistic two-dimensional structure of the hammerhead[41,50] has served its purpose well for the design of many successful experiments to characterize the self-cleavage reaction. The challenge of determining its three-dimensional structure was accepted by the crystallographers and nine years after the initial publications of the self-cleavage reaction, the first analysis of a crystal structure was published.[59] For this work, a 34-mer ribozyme and a 13-mer deoxynucleotide substrate were hybridized together and crystallized, and the X-ray crystallographic structure determined at 2.6 Å. This was followed soon after by the structure of an all-RNA ribozyme where self-cleavage was blocked by a 2′-methyl group on the 2′-hydroxyl at the cleavage site.[60] Finally, the crystal structure of an unmodified hammerhead RNA in the absence of divalent metal ions was solved and it was shown that this ribozyme can cleave itself in the crystal when divalent ions are added.[61]

The three-dimensional structures developed from the three approaches are essentially the same, and a diagrammatic representation of the structure is given in Figure 13. Stem III forms the stem of a Y-shaped structure where the arms are formed by stems I and II. The reader is referred to the original papers[59–61] for details.

12.7.4 Self-cleavage Hammerhead Reaction *in Vivo*

Because ASBV cleaves itself *in vitro*, it seemed likely that this self-cleavage reaction will have a central role in the rolling circle replication of ASBV *in vivo*. Providing proof of this is much more difficult. The host range of ASBV is very narrow and essentially limited to avocados and related woody species which do not lend themselves to simple experiments using infectious cDNA clones. In experiments done many years ago to show the infectivity of purified ASBV on young avocado seedlings, it took a year for symptoms to appear. Likewise, PLMV offers no better prospects because of its very

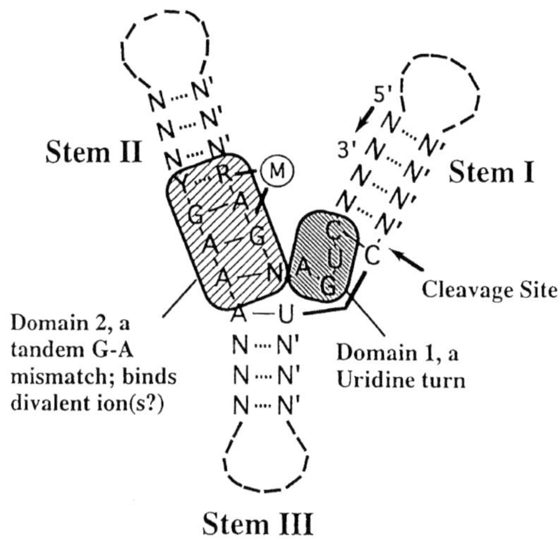

Figure 13 Schematic drawing of one hammerhead molecule as derived from its crystal three-dimensional structure.[58–61] Non-Watson–Crick base pairs are shown as thin lines. M, divalent cation. (Reproduced by permission of Spring-Verlag from *Nucleic Acids Mol. Biol.*, 1996, **10**, 161.)

limited host range, again with a difficult woody species. However, in 1997, the characterization of CChMV with its herbaceous chrysanthemum[43] host raised the prospect of repeating the experiment described below where a viroid-like satellite RNA or virusoid of vLTSV was used to provide evidence for the involvement of the hammerhead self-cleavage reaction *in vivo*.

The approach used in this experiment[63] was to mutate an infectious cDNA clone of vLTSV such that the (−) RNA self-cleavage was eliminated and that of the (+) RNA left intact. In the wild-type situation where both the (+) and (−) RNAs self-cleave, only monomeric and short polymeric species of (+) and (−) RNAs are found in infected plants in the presence of the specific helper virus.[44] When the (−) RNA self-cleavage is eliminated by appropriate mutation, it can be predicted that high molecular weight polymeric (−) species will replace the monomeric and short polymeric (−) sequences of the wild type. Such high polymeric (−) species are found in plants infected with three virusoids where the (+) RNAs self-cleave *in vitro* via the hammerhead structure but the (−) RNAs do not self-cleave[64] and have no known self-cleavage activity.[41] Hence, three cDNA clones of vLTSV were mutated at three sites to eliminate self-cleavage *in vitro* in the (−) RNA.[63] When the excised inserts from these clones were co-inoculated with the helper virus LTSV on an appropriate herbaceous host plant, high molecular weight (−) vLTSV RNA was detected as expected while the control nonmutated vLTSV gave only the wild-type lower molecular weight species.

It was a surprise, however, to find that the mutated virusoids also produced some monomeric (−) vLTSV progeny in infected plants.[63] Sequence analysis of these progeny showed that 8–20% contained reversions and pseudoreversions of the introduced mutations which allowed self-cleavage in these progeny. Overall, the results provide convincing evidence for a role of hammerhead self-cleavage in the replication of at least the (−) RNA, but, as in many biological experiments, the experimental data are never as precise as the models on which we design our experiments.

In a less direct approach, the structure of a series of RNAs extracted from ASBV-infected avocado were characterized.[65] Sequence analysis of different (+) and (−) forms of ASBV were consistent with the symmetric rolling circle mechanism (Figure 8) and the predicted role of the hammerhead self-cleavage reaction.

12.7.5 Hammerhead Self-cleavage *in trans* — Application to Cleavage of Target RNAs

In the original single-hammerhead structure of (+) and (−) ASBV, the sequences making up the two halves of each structure originate from opposite strands in the central region of the molecule,[41] and self-cleavage *in trans* was demonstrated using two large fragments of the (−) ASBV molecule.[52] It was Uhlenbeck[66] who first recognized the potential for hammerhead self-cleavage *in trans* and who used a variety of short oligonucleotides to study the reaction and to use part of the hammerhead domain as a nuclease to cleave sequences embedded in different RNAs. The simplest construct of 19 nt RNA could cause rapid and specific cleavage of a 24 nt substrate and it had all the properties of an RNA enzyme or ribozyme, especially as it catalyzed multiple turnover of substrate.

Haseloff and Gerlach[67] extended this approach by the design of small ribozymes based on the hammerhead structure for the specific cleavage at three sites of an 835 nt RNA transcript of the bacterial chloramphenicol acetyl transferase gene *in vitro*. The approach was to find GUC sequences in the target RNA to define the cleavage site 3′ to the C residue and then to design the ribozyme around the core sequence of the (+) hammerhead domain of the 359 nt satellite RNA of tobacco ringspot virus which had been characterized by Bruening and his colleagues.[53,54] When the three ribozymes were in molar excess of the substrate, almost complete cleavage of the substrate could be obtained after 60 min at 50 °C. When the RNA substrate was in excess, each of the ribozyme constructs participated in greater than 10 cleavage events at each of the three sites in 75 min.[67]

The publication in 1988 of this work stimulated considerable effort by others to develop similar ribozyme constructs for their own particular system *in vitro* and to extend the approach to *in vivo* situations. There has been an explosion of effort in this area and the reader is referred to the three chapters on ribozymes in this volume for further details (Chapters 10, 11, and 16).

12.8 WHAT IS THE PROCESSING REACTION DURING ROLLING CIRCLE REPLICATION IN THE PSTV GROUP OF VIROIDS?

12.8.1 The Processing Reaction is likely to be RNA Catalyzed

In contrast to the well-characterized hammerhead self-cleavage reaction in the (+) and (−) RNAs of the three members of the ASBV-group of viroids, as well as evidence indicating a role for such reactions in rolling circle replication, very little is known about the processing event in the PSTV-group of viroids. In the few cases where such viroids have been investigated, the presence of low levels of high molecular (−) RNA species and essentially only monomeric (+) species strongly indicates that a processing reaction is only involved in the (+) species.[44,68,69] It can be reasonably safely predicted that this will be the situation for all members of the PSTV-group of viroids because of the conserved features of this viroid group.

Mainly because of the absence of any type of RNA catalyzed processing reaction in members of the PSTV-group of viroids, there has been a tendency to assume that processing of the multimeric (+) species is enzyme catalyzed although no such potential enzyme has been identified.[70,71] There has been one exception in a preliminary report[72] showing a 1–5% conversion of a dimeric RNA transcript of PSTV to monomeric linear PSTV plus two end fragments when the dimeric RNA was incubated under conditions used for Group I splicing reactions. The cleavage site was placed between residues 250–270 which is in the bottom strand of the C domain of PSTV and in a highly prospective region for a processing site. No further information has been published on this initial observation.

In view of the hammerhead self-cleavage reactions in the three members of the ASBV-group of viroids and in a number of circular single-stranded satellite RNAs associated with plant viruses,[11,43] it seems reasonable to predict that the processing reaction in the rolling circle replication of members of the PSTV-group of viroids will be RNA catalyzed also. The most likely cause of lack of success in providing clear evidence of such a reaction is the ability of single-stranded RNA to fold into multiple conformations *in vitro*, none of which are the active form required for self-cleavage. The author's own experience with the 324 nt viroid-like satellite RNA or virusoid of vLTSV provides a good example.[50–52] Various length RNA transcripts only self-cleaved after heat denaturation and snap cooling followed by assembly of the reaction mixture plus Mg^{2+} on ice. No self-cleavage occurred when the heated and snap cooled RNA was allowed to warm up prior to the addition of Mg^{2+}. In other words, conditions had to be empirically determined to form an active conformation capable of self-cleavage. Hence, it was considered likely that success in demonstrating a specific processing reaction in members of the PSTV-group of viroids would be crucially dependent on developing folding procedures where at least a small fraction of the population of RNA molecules would be in an active conformation in the presence of a divalent or multivalent cation.

12.8.2 Initial Indications of a Specific Processing Site

The CCR within the C domain of viroids, is a likely site for any type of specific processing reaction. The author's initial experiments exploring this region employed mutagenesis and infectivity studies on longer than unit length transcripts of cDNA clones of CEV. A potential *in vivo* processing site was identified in the upper strand of the CCR which corresponds to G96–G98 of the PSTV sequence.[25] However, this site was not unique in such experiments as would be necessary for a specific self-cleavage site, since mutation in other parts of the CEV molecules showed that the basic requirements for infectivity appeared to be the ability of RNA transcripts to form a short double-stranded region of viroid and vector sequences at the junction of the two termini.[33] Presumably, some processing reaction *in vivo* allowed the elimination of the vector sequences within the double-stranded region and restoration of the circular CEV.

12.8.3 A Common Mechanism in the CCR for all PSTV-group Viroids?

Given that there will be a common processing mechanism for all PSTV-group viroids, then it is highly likely to occur in a region conserved between its three subgroups (Figure 3). The most conspicuous feature they have in common is the potential for their inverted repeat sequences to form the two hairpin structures depicted in Figure 5. The sequences of the CCR in each subgroup vary (Figure 3) but the general two-dimensional structure of their hairpin loops is similar. That there is something unusual about these sequences, and hence about the structure of the CCR, is indicated by the specific cross-linking in PSTV of G98 to U260 that occurs on irradiation of the purified viroid with UV-light (Figure 14).[73] Obviously, there must be a juxtaposition of the two bases, presumably in a coplanar structure, for such cross-linking to occur. The specific UV-cross-linking also occurred[74] when a 55 nt RNA transcript constructed from sequences within the C domain was used; this RNA contained the sequence G88 to U114 coupled to the sequence G245 to A272 (see Figure 14). Since neither hairpin loop of Figure 14 is possible in the construct, a rod-like structure appears to be required for cross-linking to occur.

Obviously, we have too little structural information to explore possible conserved three-dimensional structures within the C domains of the three PSTV-subgroups by model building. At some point it may be possible for the RNA crystallographers to turn their attention to the sequences within the C domain of the PSTV-group viroids.

12.8.4 Is the Self-cleavage Site in the Bottom Strand of the CCR of the PSTV-subgroup of Viroids?

The author's group have searched for a specific self-cleavage site in the RNA sequences which comprise the C domain and have used CCCV for this work because it is the smallest member of the PSTV-subgroup. The approach was to prepare cDNA clones of the sequences in the C domain; thus,

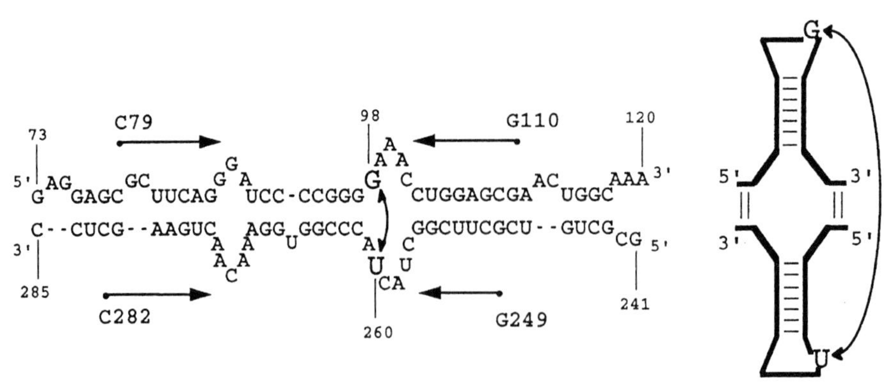

Figure 14 Sequence of the C domain of PSTV in the rod-like structure (left) and the potential hairpin stem-loop structure that can be formed from it (right). The four arrows represent the inverted repeat sequences.[73,74] the base residues of which are numbered. The UV-induced cross-linking between G98 and U260 is shown by a double-headed arrow. (Reproduced by permission of Oxford University Press from *Nucleic Acids Res.*, 1997, 25, 2683.)

one clone contained the top strand of the C domain connected via a small loop to the bottom strand and the other contained the bottom strand connected to the top strand. A wide range of denaturation and reannealing approaches were explored using RNA transcripts prepared from these cDNA clones as well as various divalent and polyvalent cations, to see if self-cleavage could be detected by polyacrylamide gel electrophoresis.

One set of specific conditions identified a self-cleavage site (about 5% self-cleavage) on the left-hand side of the bottom hairpin loop[75] in Figure 5. The site is certainly in a conserved region within the members of the PSTV-subgroup and must be considered as a potential *in vivo* processing site. Proving that it actually is such a site is much more difficult and an important approach here would be mutagenesis analysis of infectious cDNA viroid clones. A different viroid would have to be used for this work since CCCV is quite impractical as its host range is restricted to the coconut and oil palms and related species in the palm family. Infectious cDNA clones of CEV are available with tomato seedlings as the experimental host plant. Initial experiments would involve single base mutations on either side of, or near to, the putative processing site and determination of their effect on infectivity. Two-dimensional structure modeling around this site-would permit a more rational design of mutations.

12.9 HOW DO VIROIDS EXERT THEIR PATHOGENIC EFFECTS?

12.9.1 What do we know about Pathogenicity?

Viroids do not code for any proteins or polypepetides which implies that pathogenic effects must be mediated in some way by interaction with one or more cellular constituents. Beyond this there are essentially no data on the molecular basis of viroid-induced pathogenic responses in plants. Speculation has centered around base-pair interactions with small nuclear and cytoplasmic RNAs, activation of protein kinases as a triggering event in viroid pathogenesis, the inhibitory binding of viroids to specific proteins, etc.[12,76] Obtaining concrete experimental evidence is likely to remain very difficult, especially as it will be necessary to distinguish between primary and secondary events.

Symptom expression can vary widely, depending on the particular viroid and the host plant. CCCV is essentially lethal to coconut and oil palms in the Philippines, with death coming several years after the appearance of the first symptoms. The smallest viroid certainly packs the biggest punch. Whether this highly virulent effect is related to the very limited monocot host range of CCCV is not known. ASBV also has a very limited host range within the dicot family Lauraceae where it causes symptoms ranging from severe to so mild that some trees are asymptomatic. Symptoms can also vary widely in the same tree from year to year.

We also have no molecular explanation as to how small sequence variations in a viroid can convert a severe PSTV variant into a mild one, except to suggest that sequence variation may modulate interactions with particular host components. It has, however, been possible to map pathogenic effects to particular domains within members of the PSTV-subgroup of viroids. Sequence analysis of sequence variants of PSTV, CEV, and HSV showed that essentially all symptom-affecting sequence differences for each viroid are located in the P and V domains.[8] In the case of CEV, sequence analysis of 17 sequence variants showed that sequence variation occurred within regions within the P and V domain,[7] further narrowing the part of CEV molecule likely to influence pathogenicity.

Unraveling the molecular basis of pathogenicity remains a most difficult task. Naturally occurring isolates of viroids, defined as viroid isolated from a single plant, usually contain a mixture of more than one sequence variant. It is therefore essential for such work to use infectious cDNA clones of the viroid under study to ensure that plants are inoculated with only one sequence variant. This approach does not remove the problem, it only delays it, since mutants of the inoculated viroid appear in progeny viroids. The site of replication of each viroid in the cell is obviously important in the details of pathogenicity but, as considered earlier (Section 12.6.3), the site of accumulation may not be the site of synthesis.

12.9.2 What Determines the Host Range of Viroids?

As for pathogenicity, we really have no explanations at the molecular level as to what determines host range, mainly because we know so little about the specificity of the interactions between the invading viroid and host processes. The three viroids in the ASBV-group all have very limited host ranges: ASBV to some members of the family Lauraceae, PLMV to a few species in the family Rosaceae, and CChMV to only the chrysanthemum in the very large Compositae family. A common feature of these three viroids is that both the plus and minus RNAs of each viroid self-cleave via the hammerhead structure. There is no information as to whether there is any functional correlation between limited host range and hammerhead self-cleavage. ASBV accumulates on the thylakoid membranes of avocado chloroplast;[47] it will obviously be of considerable interest to determine if both PLMV and CChMV also accumulate on the thylakoid membranes of their host plants.

Within the PSTV-subgroup of viroids, the host range varies from a very limited range for CCCV, as considered in Section 12.9.1, to quite wide ranges for PSTV, CEV, and HSV. Members in the ASSV- and CbV-subgroups seem to have restricted host ranges, but this could change as more members of these groups are discovered and characterized.

12.10 REFERENCES

1. T.O. Diener, "Viroids and Viroid Diseases", Wiley, New York, 1979.
2. T.O. Diener, "The Viroids", Plenum Press, New York, 1987.
3. J.S. Semancik, "Viroids and Viroid-like Pathogens", Academic Press, New York, 1987.
4. R.H. Symons, "Seminars in Virology", W. B. Saunders, Philadelphia, PA, 1990.
5. R.H. Symons, *Crit. Rev. Plant Sci.*, 1991, **10**, 189.
6. R.H. Symons, *Annu. Rev. Biochem.*, 1992, **61**, 641.
7. P. Keese, J.E. Visvader and R.H. Symons, in "RNA Genetics", eds. E. Domingo, J. Hollard and P. Ahlquist, CRC Press, Boca Raton, FL, 1988, 71
8. R.H. Symons, *Mol. Plant Microbe Interact.*, 1991, **4**, 111.
9. D. Riesner, *Mol. Plant Microbe Interact.*, 1991, **4**, 122.
10. T.O. Diener, *Virus Genes*, 1996, **11**, 47.
11. R.H. Symons, *Nucleic Acids Res.*, 1997, **25**, 2683.
12. R. Flores, F.D. Serio and C. Hernandez, *Semin. Virol.*, 1997, **8**, 65.
13. T.O. Diener and W.B. Raymer, *Science*, 1967, **158**, 378.
14. J.S. Semancik and L.G. Weathers, *Phytopathology*, 1968, **58**, 1067.
15. T.O. Diener, *Virology*, 1971, **45**, 411.
16. J.S. Semancik and L.G. Weathers, *Nature New Biology*, 1972, **237**, 242.
17. H.J. Gross, H. Domdey, C. Lossow, P. Jank, M. Raba, H. Alberty and H.L. Sanger, *Nature*, 1978, **273**, 203.
18. P. Keese and R.H. Symons, in "Viroids and Viroid-like Pathogens", ed. J.S. Semancik, Academic Press, New York, 1987, 1
19. P. Keese and R.H. Symons, in "The Viroids", ed. T.O. Diener, Plenum Press, New York, 1987, 37
20. P. Keese and R.H. Symons, *Proc. Natl. Acad. Sci. USA*, 1985, **82**, 4582.
21. H.L. Sanger, G. Klotz, D. Riesner, H.J. Gross and A.K. Kleinschmidt, *Proc. Natl. Acad. Sci. USA*, 1976, **73**, 3852.
22. J.M. Sogo, T.H. Koller and T.O. Diener, *Virology*, 1973, **55**, 70.
23. J.W. Randles and T. Hatta, *Virology*, 1979, **96**, 47.
24. A.M. Koltunow and M.A. Rezaian, *Intervirology*, 1989, **30**, 194.
25. J.E. Visvader, A.C. Forster and R.H. Symons, *Nucleic Acids Res.*, 1985, **13**, 5843.
26. T.O. Diener, *Proc. Natl. Acad. Sci. USA*, 1986, **83**, 58.
27. R.L. Spieker, *J. Gen. Virol.*, 1996, **77**, 2839.
28. J.L. McInnes and R.H. Symons, in "Viroids: Pathogens at the Frontiers of Life", ed. K. Maramoroch, CRC Press, Boca Raton, FL, 1991, 21
29. J.E. Visvader and R.H. Symons, *Nucleic Acids Res.*, 1985, **13**, 2907.
30. J.S. Semancik, J.A. Szychowski, A.G. Rakowski and R.H. Symons, *J. Gen. Virol.*, 1994, **75**, 727.
31. J.E. Visvader and R.H. Symons, *EMBO J.*, 1986, **5**, 2051.
32. R.G. Bonfiglioli, D.R. Webb and R.H. Symons, *Plant J.*, 1996, **9**, 457.
33. A.G. Rakowski and R.H. Symons, *Virology*, 1989, **173**, 352.
34. J. Haseloff, N.A. Mohamed and R.H. Symons, *Nature*, 1982, **299**, 316.
35. J.S. Imperial, J.B. Rodriguez and J.W. Randles, *J. Gen. Virol.*, 1981, **56**, 77.
36. R. Hammond, D.R. Smith and T.O. Diener, *Nucleic Acids Res.*, 1989, **17**, 10–083.
37. M.A. Rezaian, *Nucleic Acids Res.*, 1990, **18**, 1813.
38. A.D. Branch, S.E. Lee, O.D. Neel and H.D. Robertson, *Nucleic Acids Res.*, 1993, **21**, 3529.
39. I.M. Schindler and H.P. Muhlbach, *Plant Sci.*, 1992, **84**, 221.
40. A.D. Branch and H.D. Robertson, *Science*, 1984, **223**, 450.
41. C.J. Hutchins, P.D. Rathjen, A.C. Forster and R.H. Symons, *Nucleic Acids Res.*, 1986, **14**, 3627.
42. C. Hernandez and R. Flores, *Proc. Natl. Acad. Sci. USA*, 1992, **89**, 3711.
43. B. Navarro and R. Flores, *Proc. Natl. Acad. Sci. USA*, 1997, **94**, 11–262.
44. C.J. Hutchins, P. Keese, J.E. Visvader, P.D. Rathjen, J.L. McInnes and R.H. Symons, *Plant Mol. Biol*, 1985, **4**, 293.
45. R. Flores and J.S. Semancik, *Proc. Natl. Acad. Sci. USA*, 1982, **79**, 6285.

46. J. Harders, N. Lukacs, M. Robert-Nicoud, T.M. Jovin and D. Riesner, *EMBO J.*, 1989, **8**, 3941.
47. R.G. Bonfiglioli, G.I. McFadden and R.H. Symons, *Plant J.*, 1994, **6**, 99.
48. M.I. Lima, M.E.N. Fonseca, R. Flores and E.W. Kitajima, *Arch. Virol.*, 1994, **138**, 385.
49. R.G. Bonfiglioli, D.R. Webb and R.H. Symons, *Plant J.*, 1996, **9**, 457.
50. A.C. Forster and R.H. Symons, *Cell*, 1987, **49**, 211.
51. A.C. Forster and R.H. Symons, *Cell*, 1987, **50**, 9.
52. A.C. Forster, A.C. Jeffries, C.C. Sheldon and R.H. Symons, *Cold Spring Harbor Symp. Quant. Biol.*, 1987, **52**, 249.
53. G.A. Prody, J.T. Bakos, J.M. Buzayan, I.R. Schneider and G. Bruening, *Science*, 1986, **231**, 1577.
54. G. Bruening, *Semin. Virol.*, 1990, **1**, 127.
55. K.R. Birikh, P.A. Heaton and F. Eckstein, *Eur. J. Biochem.*, 1997, **245**, 1.
56. R.H. Symons, *Trends Biochem. Sci.*, 1989, **14**, 445.
57. A.C. Forster, C. Davies, C.C. Sheldon, A.C. Jeffries and R.H. Symons, *Nature*, 1988, **334**, 265.
58. C. Davies, C.C. Sheldon and R.H. Symons, *Nucleic Acids Res.*, 1991, **19**, 1893.
59. H.W. Pley, K.M. Flaherty and D.B. McKay, *Nature*, 1994, **372**, 68.
60. W.G. Scott, J.T. Finch and A. Klug, *Cell*, 1995, **81**, 991.
61. W.G. Scott, J.B. Murray, J.R.P. Arnold, B.L. Stoddard and A. Klug, *Science*, 1996, **274**, 2065.
62. D.G. McKay, *Nucleic Acids Mol. Biol.*, 1996, **10**, 161.
63. C. Davies, C.C. Sheldon and R.H. Symons, *Nucleic Acids Res.*, 1991, **19**, 1893.
64. C.C. Sheldon, A.C. Jeffries, C. Davies and R.H. Symons, *Nucleic Acids Mol. Biol.*, 1990, **4**, 227.
65. J.A. Daros, J.F. Marcos, C. Hernandez and R. Flores, *Proc. Natl. Acad. Sci. USA*, 1994, **91**, 12–813.
66. O.C. Uhlenbeck, *Nature*, 1987, **328**, 596.
67. J. Haseloff and W.L. Gerlach, *Nature*, 1988, **334**, 585.
68. A.D. Branch, H.D. Robertson and E. Dickson, *Proc. Natl. Acad. Sci. USA*, 1981, **78**, 6381.
69. R.A. Owens and T.O. Diener, *Proc. Natl. Acad. Sci. USA*, 1982, **79**, 113.
70. G. Steger, T. Baumstark, M. Morchen, M. Tabler, M. Tsagris, H.L. Sanger and D. Riesner, *J. Mol. Biol.*, 1992, **227**, 719.
71. T. Baumstark and D. Riesner, *Nucleic Acids Res.*, 1995, **23**, 4246.
72. H.D. Robertson, D.L. Rosen and A.D. Branch, *Virology*, 1985, **142**, 441.
73. A.D. Branch, B.J. Benenfeld and H.D. Robertson, *Proc. Natl. Acad. Sci. USA*, 1985, **82**, 6590.
74. C.P. Paul, B.J. Levine, H.D. Robertson and A.D. Branch, *FEBS Lett.*, 1992, **305**, 9.
75. Y. Liu and R.H. Symons, *RNA*, 1998, **4**, 418.
76. T.O. Diener, R.W. Hammond, T. Black and M.G. Katze, *Biochimie*, 1993, **75**, 533.

13

Structural Elements of Ribosomal RNA

STEVEN T. GREGORY, MICHAEL O'CONNOR and ALBERT E. DAHLBERG
Brown University, Providence, RI, USA

13.1 INTRODUCTION

Ribosomal RNAs (rRNAs) represent an immense challenge to structural biochemists and molecular biologists who attempt to define their role as the structural and functional core of the translational machinery. All ribosomes consist of two subunits of unequal mass, the smaller containing a 16S–18S rRNA molecule, the larger containing a 23S–28S rRNA molecule in addition to, in most organisms, a small 5S rRNA molecule. Complexed with numerous ribosomal proteins, these massive rRNAs must fold into compact ribonucleoprotein particles, interact with factors for initiation, elongation and termination, bind and differentiate among numerous tRNA molecules based on their interaction with mRNA, and move the tRNA–tRNA–mRNA complex through the ribosome in a precise and concerted fashion. Deciphering the three-dimensional structure of the large rRNAs promises to reveal not only many new underlying principles that govern RNA higher order structure and folding, but also how large RNA molecules change conformation to achieve the coordinated movement of multiple substrates.

While the three-dimensional structures of either 16S or 23S rRNA have yet to be determined at high resolution, the substantial effort directed toward this objective has nevertheless proven fruitful. There is great confidence in the highly detailed secondary structure models derived by the method of comparative sequence analysis and a number of tertiary interactions are now being detected by this approach.[1] Comparative sequence analysis has also identified secondary structure motifs, such as tetraloops and E loops, which are likely to play key roles in establishing and maintaining the tertiary structure of the ribosome. The role of tetraloops and their interaction with specific docking sites (receptors) in the formation of tertiary structure of smaller RNAs has been examined in detail,[2] and their participation in rRNA architecture is expected to be equally important. In addition, high-resolution structure determination of other secondary structure building blocks and functionally important structural elements will continue to reveal new modes of tertiary interaction.

The accumulated data from cross-linking studies and footprinting studies places a sufficient number of constraints on the three-dimensional folding of 16S rRNA for the construction of several models.[3–5] Remarkable advances are also being made using X-ray crystallography and cryoelectron microscopy which, it is anticipated, will culminate in no less than an atomic resolution structure of the ribosome.[6] In this chapter, the authors review several aspects of what is currently known about rRNA secondary and tertiary structure, and describe in some detail the structure of several important functional sites that have been characterized at high resolution.

13.2 SECONDARY STRUCTURE ELEMENTS OF rRNAS

13.2.1 Secondary Structure Models Derived by Comparative Sequence Analysis

Ribosomal RNA secondary structure models have been derived primarily via comparative sequence analysis. The principle of the comparative approach, first applied to prediction of the secondary structures of tRNAs, has more recently been applied to other RNAs including rRNAs.[1] The underlying principle of this method is the presumption that homologous RNA molecules, having experienced similar selective pressures throughout their evolution, will adopt similar three-dimensional structures, despite having very different nucleotide sequences. Thus, while the nucleotide sequence of any individual RNA molecule is compatible with any of a vast number of alternative secondary structures, only one of these structures will be compatible with the nucleotide sequences of all homologous RNAs. In practice, two bases are inferred to engage in hydrogen-bonded base pairing if they covary in a concerted fashion among a large number of RNA molecules. The greater the number of phylogenetically independent exchanges, the greater the certainty of the base pairing interaction. Examples of nonconcerted changes are taken as evidence against base pairing. Formally, mechanisms other than direct base–base interaction can be invoked to explain covariation, although base pairing is the most intuitively satisfying for a series of consecutive covariations that follow Watson–Crick pairing rules. In addition to the standard Watson–Crick exchanges (i.e., A–U to G–C, G–C to C–G, etc.), noncanonical exchanges (such as U–U to C–C, U–G to C–A, A–G to G–A, etc.) are also detectable by covariation analysis. Thus, covariation is not restricted *a priori* by any specific set of pairing rules based on the hydrogen bonding arrangements of the pairing bases. Nevertheless, the vast majority of helices in rRNAs are constructed of antiparallel, canonical Watson–Crick or G–U wobble pairs (roughly 90%[1]) with occasional noncanonical pairing interactions embedded within standard helices. In some cases these noncanonical pairings comprise parts of conserved motifs that may eventually be found to participate in RNA–RNA, RNA–protein, or RNA–ligand interactions (Figure 1).

The secondary structures of all rRNAs have been divided (somewhat arbitrarily) into domains; three for 16S rRNAs and six for 23S rRNAs. It is unknown how structurally or functionally independent these divisions are, although the 3′ minor domain of 16S has been assembled *in vitro* with ribosomal proteins into a discrete particle,[7] while individual domains of 23S rRNA have been found to vary in their ability to stimulate peptide bond formation.[8]

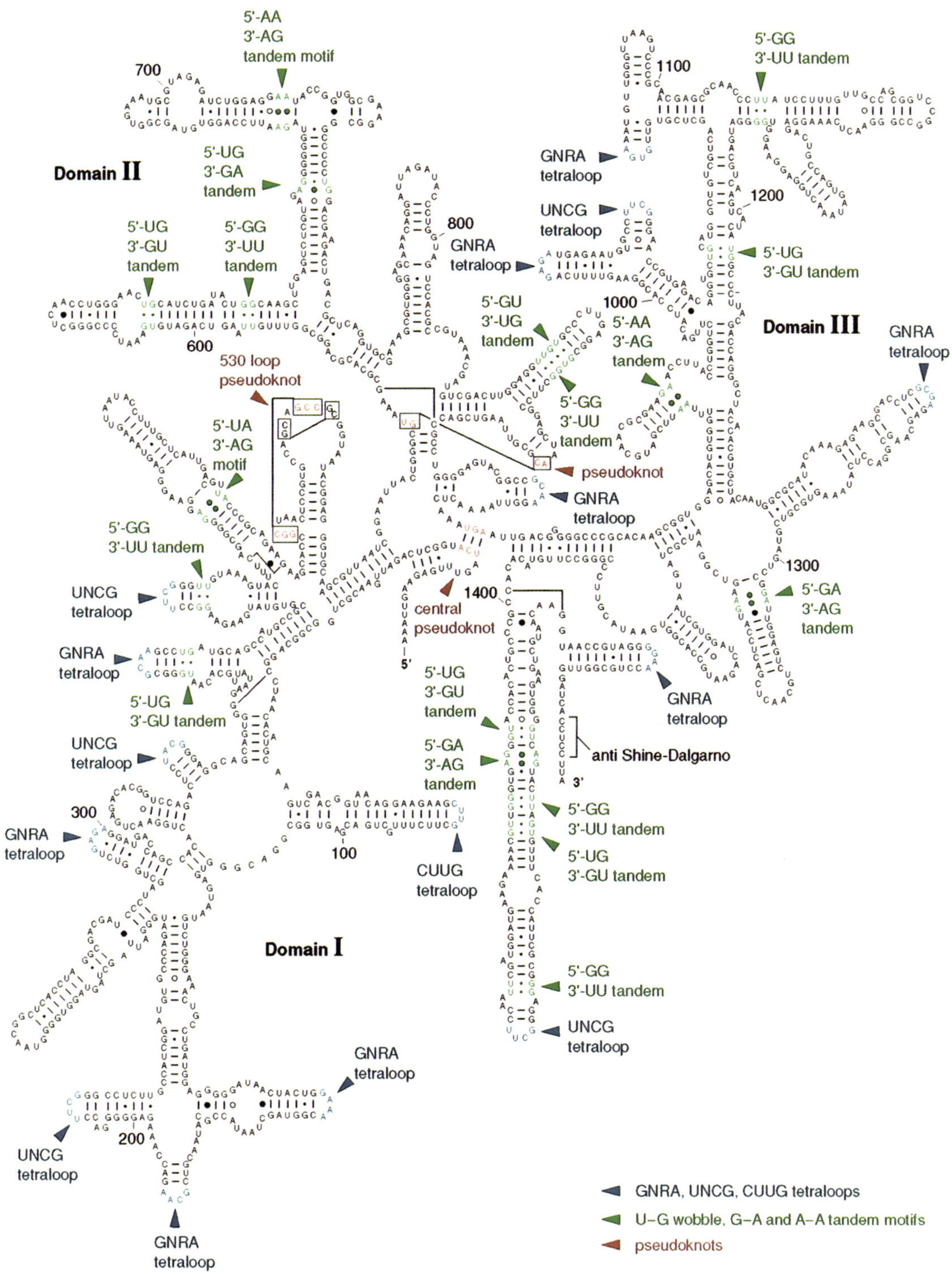

Figure 1 Secondary structure model of *E. coli* 16S rRNA derived by comparative sequence analysis.[1] Indicated are GNRA, UNCG, and CUUG tetraloops (blue), G–U and purine–purine base pairing motifs (green), and pseudoknots (red). The structure has been modified to include G–U and purine–purine base pairing interactions described in references 9 and 13.

The universal sequence conservation of certain regions of rRNAs makes them resistant to structural modeling via comparative sequence analysis. Unfortunately, it is the structures of these regions that are of the greatest interest as their conservation implies a critical association with ribosome functions. Establishing the structures of such sites will require high-resolution biophysical techniques.

13.2.2 Secondary Structure Elements and Motifs

The secondary structures of rRNAs are composed largely of short Watson–Crick helices that are either connected to one another by internal loops and multihelix junctions, or capped by terminal loops (see Figure 1). The three-dimensional structures of multihelix junctions have not been solved, whereas the structures of a number of internal loops have been examined by NMR spectroscopy or inferred from related structures in other RNA molecules. Internal loops can also be thought of as a series of consecutive mismatches (paired or unpaired), and can be symmetric (an equal number of residues on each strand) or asymmetric (an unequal number on each strand), the latter possibly resulting in bulged residues.

Several internal loop or mismatch motifs have been identified by comparative sequence analysis. A high frequency of purine–purine mismatches exist at certain locations in rRNAs.[9] These mismatches frequently occur in tandems and in nonrandom arrangements. Thus, while 5′-GA/3′-AG, 5′-AA/3′-AG, and 5′-AA/3′-GG occur at high frequency, 5′-AG/3′-GA is extremely rare. These patterns are rationalized as resulting from the formation of a type II, or "sheared," base pairing configuration. The type II sheared GA pair consists of hydrogen bonds from N2G to N7A and N6A to N3G. This configuration positions the G toward the major groove and the A toward the minor groove, with an underwinding of the helix. In contrast with a single sheared G-A pair, two consecutive G-A pairs facing one another in the 5′-GA/3′-AG arrangement can be embedded within a Watson–Crick helix without substantially distorting the neighboring backbone conformation. The structure of this motif has been observed directly in model duplexes,[10] the hammerhead ribozyme,[11] and the *Tetrahymena* ribozyme as the related 5′-AA/3′-AA motif.[2] A highly conserved 5′-GA/3′-AG tandem occurs in 16S rRNA at positions 1305–1306/1332–1331, flanked by two mismatches creating the structure 5′-GGAU/3′-AAGU (Figure 1[9]). A related motif, 5′-UA/3′-AG, consists of a U–A reverse Hoogsteen pair followed by a type II sheared A–G pair.[9,12] This motif is highly conserved at positions 486–487/448–447 of 16S rRNA (Figure 1) and at several locations in 23S rRNA.[9]

Secondary structure motifs involving G-U pairings have also been identified by comparative sequence analysis.[13] These include 5′-UG/3′-GU, 5′-GG/3′-UU, and 5′-UG/3′-GA tandems. It is expected that such motifs most often involve G–U pairs in the wobble configuration. In the case of 5-UG/3′-GU, opposing wobble pairs may "compensate" for deficiencies in stacking caused by the wobble configuration. In the case of 5′-UG/3′-GA, the G moved toward the minor groove by the wobble configuration may enhance stacking of the minor groove adenosine in the sheared configuration. Consecutive G–U pairs in the wobble configuration project successive N-2 amino groups into the minor groove. Such an arrangement might serve as a protein recognition site or an anchoring site for RNA–RNA interaction. 5′-GA/3′-AG and 5′-UG/3′-GU motifs occur in a nonrandom fashion and are likely to play an important, albeit as yet undefined, role in rRNA structure and function.

Several consecutive noncanonical base pairing interactions combine to form an asymmetric internal loop found in 5S and 23S rRNAs. Called the E loop motif after the E loop of eukaryotic 5S rRNA,[12] this structure consists of a type II sheared A–G pair, a U–A reverse Hoogsteen pair (N3U–N7A, O2U–N6A), a bulged G residue, and a symmetric (N-6–N-7) A–A pair. While one backbone strand conforms closely to A form geometry, the other strand displays a striking local reversal of direction to accommodate the bulged G residue and the symmetric A–A pair (both of which are *anti*). It has been proposed that the eukaryotic E loop belongs to a larger family of loops, the S loops, which might serve as docking modules for long-range RNA–RNA interactions.[14] Interestingly, the E loop of *E. coli* 5S rRNA[15] differs substantially from that of eukaryotic 5S rRNA. In addition to a sheared A–G pair and a U–A reverse Hoogsteen pair, the loop also contains a G–G (N-1, N-2–O-6) pair, a head-to-head type I A–G (N-6A–O-6G) pair, and a nonwobble G–U (N-1, N-2G–O-4U) pair. The bacterial loop also lacks the bulged G residue and backbone reversal characteristic of the eukaryotic loop.

The most frequently found single class of terminal loops found in rRNA are the tetranucleotide loops or tetraloops[16] (see Figure 1). While originally identified in rRNAs, they are also found in many other RNAs and are likely to be one of the most common secondary structure motifs in rRNAs. Three main classes of tetraloops, GNRA, UNCG, and CUUG, comprise half the terminal loops in rRNAs. These loops have in common a highly ordered and compact structure with extensive hydrogen bonding. In particular, UNCG tetraloops are highly stable thermodynamically and may function as nucleation sites for RNA folding. GNRA tetraloops, on the other hand, are important tertiary structure elements, docking into the minor groove of RNA helices at highly specific tetraloop receptor sequences.[2,17]

Interestingly, the GNRA fold has also been found in larger loops,[18] suggesting that it may exist hidden in a variety of terminal and internal loops and at multihelix junctions. The GAAA tetraloop–receptor complex can also withstand substantial variation in both the loop and receptor structures, suggesting that a wide range of equivalent or similar docking interactions may exist.[19] This variability in the structure of such complexes will make them more difficult to recognize by sequence analysis, but may prove to be a fundamental building block for RNA architecture.

13.3 HIGH-RESOLUTION STRUCTURES

Recent advances in methodologies for the synthesis of RNA molecules and their analysis by NMR spectroscopy have led to the structure solutions for several rRNA analogues. As described in previous chapters, UNCG,[20] GNRA,[21] and CUUG[22] classes of tetraloops comprise half the terminal loops in rRNAs. More recently, the structures of loops and helices unique to specific functional sites of rRNAs have also been solved. These studies reveal some of the remarkable variety of conformations that RNAs can adopt, while at the same time providing insight into some of the common themes of rRNA secondary structure.

The practical limitations of the methodologies used to derive high-resolution structures prevent their accounting for tertiary interactions or interactions with ribosomal proteins. Thus, it is possible, and perhaps likely, that the conformation of particular elements differ in some details in the context of an intact functioning ribosome. In addition, the structure of an oligonucleotide analogue may represent only one of several possible alternate conformations adopted in the ribosome during the translation cycle. Its relevance to ribosome structure can, to some degree, be measured by its compatibility with results from mutagenesis, structure probing, and other biochemical methodologies using intact ribosomes. Disagreements between structural data derived with analogues and data from intact ribosomes may be indicative of the existence of tertiary interactions or conformational dynamics *in situ*.

Each of the structures described below was analyzed with the goal of revealing features important to a particular aspect of ribosome function. While some of these structures, such as the decoding center and P loop analogues, probably do not represent common secondary structure motifs used in rRNA architecture, they do illustrate the potential for RNA to adopt unusual conformations to form sites of interaction with specific ligands. The GUAAUA hexaloop may be found to be representative of a class of six base loops, while the dimethyl A stem loop is a common GNRA tetraloop with a dramatically different conformation due to base methylations.

13.3.1 The Decoding Region (16S rRNA)

The decoding region consists of a segment of domain III of 16S rRNA that is intimately associated with codon–anticodon interaction (Figure 2(a)). The anticodon of tRNA in the P site can be cross-linked to the highly conserved C1400,[23] as can mRNA.[24] Chemical footprinting experiments with tRNA define a subset of nucleotides potentially interacting with the codon–anticodon complex.[25] Aminoglycoside antibiotics, which promote mistranslation, protect a similar set of nucleotides,[26] and resistance to these antibiotics occurs via mutations or base methylations in this region (reviewed in reference 27).

Aminoglycoside antibiotics bind specifically to oligonucleotide analogues of the decoding region.[28,29] The structure of a 27 nucleotide analogue corresponding to C1404–C1412 and G1488–G1497 of the decoding region complexed with the aminoglycoside antibiotic paromomycin, has been examined by NMR spectroscopy.[30] This RNA consists of a helical stem interrupted by an asymmetric internal loop (Figure 2(a)). The loop contains a closing U1406–U1495 base pair followed by a Watson–Crick C1407–G1494 pair. A1408 engages in a noncanonical pair via its N-6 by hydrogen bonding to the N7 of A1493. A1492 is unpaired and bulged. These noncanonical base pairing interactions and the conformation of A1408, A1492, and A1493 are essential for formation of the aminoglycoside binding site. A Watson–Crick base pair between C1409 and G1491 is also essential for aminoglycoside binding, and mutations disrupting this base pair confer aminoglycoside resistance (reviewed in reference 27). Mutations at this base pair also affect translational accuracy.[31] The ability of this RNA to bind drugs supports the contention that it represents the native structure of the corresponding region of 16S rRNA in the ribosome.

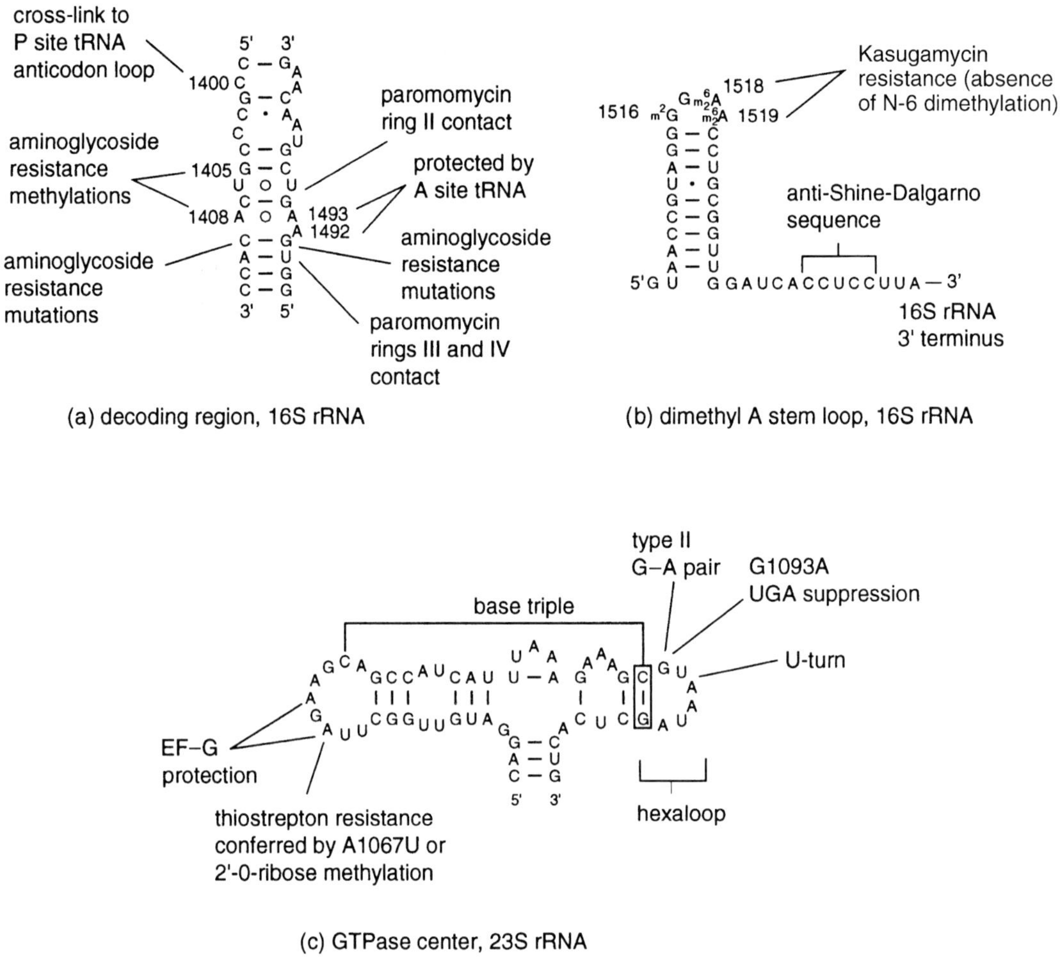

Figure 2 Secondary structures of functional sites characterized at high resolution by NMR spectroscopy. (a) The decoding region of 16S rRNA[30], indicating sites of mutations, protections by tRNA, and sites of contact with the aminoglycoside antibiotic paromomycin detected by NMR spectroscopy. (b) The dimethyl A stem loop[39] and the adjacent 3′ terminus of 16S rRNA. The anti-Shine–Dalgarno sequence which base pairs with mRNA during translation initiation is indicated. (c) The GTPase center of 23S rRNA. The two loops which interact with ribosomal protein L11 and the antibiotic thiostrepton are brought into close proximity by a base triple between the G1092-A1099 base pair and C1072.[46] The structure of the 5′-GUAAUA-3′ hexaloop has been solved by NMR.[47]

Binding of paromomycin occurs in the major groove within the internal loop. Ring II of paromomycin spans the U1406–U1495 and C1407–G1496 base pairs; amino groups in ring II make hydrogen bonds to the O-4 of U1495 and N-7 of G1494. The A1408–A1493 pair, together with the bulged A1492, form a binding pocket for ring I of paromomycin. All eukaryal small subunit rRNAs have a G at position 1408 and bind aminoglycosides much less effectively than bacterial ribosomes. The C1409–G1491 base pair forms the "floor" of the drug-binding pocket.

Aminoglycoside-producing organisms typically avoid the inhibitory effects of the antibiotic through modification of specific rRNA residues. Methylation of A1408 prevents formation of the A1408–A1493 pair and prevents binding of a range of aminoglycosides. Resistance to kanamycin and gentamycin in the producing organisms is achieved through methylation of G1405; NMR studies have placed ring III of gentamicin close to G1405, and consequently, methylation may sterically hinder binding of these antibiotics. Analysis of RNA interactions with the two-, three-, and four-ring aminoglycosides (neamine, ribostamycin, and neomycin, respectively) indicate that rings I and II (found in all these aminoglycosides) interact with RNA in much the same way as paromomycin, while rings III and IV of the neomycin class of antibiotics make sequence-independent interactions with the phosphate backbone of the RNA, near the A1410–U1490 base pair. Comparison of the free and paromomycin-bound forms of the RNA indicates that in the absence of the antibiotic, A1492 and A1493 in the internal loop assume a dynamic, flexible conformation.[32] Binding of paromomycin in the major groove results in a shift of these residues towards the minor groove and a stabilization of their conformation. Antibiotic binding also altered the pattern of hydrogen bonding between A1408 and A1493; one hydrogen bond was found in the free form of the RNA while two occurred in the drug-bound form. Aminoglycoside antibiotics promote misreading of the genetic code and prevent translocation by increasing the nonspecific (codon-independent) affinity of tRNAs for the A site. Three bases implicated in aminoglycoside–ribosome interactions, A1408, A1492, and A1493 are also protected from DMS modification by mRNA-dependent, A site tRNA binding.[25] These considerations have led to the suggestion that the universally conserved A1492 and A1493 may be involved in recognition of the correctly-formed codon–anticodon complex, and stabilization of their conformation that occurs upon drug binding may induce a high-affinity binding site for the codon–anticodon complex, with a concomitant loss in the specificity of decoding.[30,32] This study is notable as the first characterization at atomic resolution of the mode of interaction of an antibiotic and rRNA. Such studies are potentially of great importance for the rational design of novel antibiotic inhibitors of ribosome function.

13.3.2 The Dimethyl A Loop (16S rRNA)

One of the most conserved structural features of 16S-like rRNAs is the 3′ terminal dimethyl A loop (Figure 2(b)). This tetranucleotide loop has been cross-linked to domain IV of 23S rRNA[33] and the helix containing this loop has been cross-linked to initiation factor 3 (IF3),[34] suggesting it plays a role in subunit association during the initiation phase of translation. In prokaryotic ribosomes the loop has the sequence GGAA (positions 1516–1519, *E. coli* numbering), while its eukaryotic counterpart is UGAA. This loop is heavily modified, with 2-methylguanosine at the first position in prokaryotic 16S rRNAs, and *N*-6-dimethyladenosine at the third and fourth positions in all 16S-like rRNAs examined. Bacterial ribosomes lacking the *N*-6-dimethylations are resistant to the action of the antibiotic kasugamycin, an inhibitor of translation initiation,[35] and are less accurate.[36] While little or no growth defect in prokaryotes is associated with deficiency in this modification, and only mild effects on ribosome function are observed *in vitro*,[37] its absence is lethal in yeast.[38]

The structure of the fully modified loop was recently solved by NMR spectroscopy.[39] Despite having a canonical GNRA tetraloop consensus sequence, the dimethyl A loop differs substantially in its conformation from that of the corresponding unmodified loop.[21] While m^2G1516 is positioned as expected for a GNRA loop, *N*-6 dimethylation of A1519 prevents formation of a closing type II sheared G1516–A1519 pair characteristic of the GNRA family. Rather than being coplanar with m^2G1516, $m_2{}^6$A1519 is tilted upward toward the loop apex. The adjacent $m_2{}^6$A1518 stacks on $m_2{}^6$A1519. These methylations prohibit the extensive hydrogen bonding interactions found in GNRA loops, but only partially destabilize the loop.[39,40] The unusual conformation of the dimethyladenosines also changes the backbone conformation such that the turning phosphate is after A1519 rather than after G1517 (which also has a C2′ *endo* ribose pucker) and has the effect of positioning the bases in the major groove rather than in the minor groove as in GNRA tetraloops. While the NMR structure of the unmodified eukaryotic UGAA tetraloop has been solved,[41] it remains to be seen if the *N*-6-dimethylated UGAA loop adopts a conformation similar to that of the prokaryotic GGAA loop. Comparison of the methylated and unmodified structures illustrates the potential for modified nucleotides to affect a specific loop conformation.

13.3.3 The 1093–1098 5′-GUAAUA-3′ Hexaloop (23S rRNA)

The 1093–1098 5′-GUAAUA-3′ hexaloop, located in a three helix junction structure in domain II of 23S rRNA, comprises part of the binding sites for elongation factors G (EF-G) and Tu (EF-Tu) and ribosomal protein L11 (Figure 2(c)). Based on chemical modification experiments with ribosomal protein L11 and the antibiotic thiostrepton,[42,43] and footprinting and hydroxyl radical probing with EF-G,[44,45] the junction structure is thought to fold back on itself such that the two terminal loops, U1065 to A1073 and G1093 to A1098, are in close proximity. This is mediated in part by a base triple between the C1092–G1099 base pair and C1072.[46] Mutations and post-transcriptional modifications affect binding of the antibiotic thiostrepton, ribosomal protein L11, and EF-G.[42]

The structure of the G1093 to A1098 5′-GUAAUA-3′ hexaloop has been determined by NMR spectroscopy.[47] The loop contains a closing type II sheared G–A pair such as that seen in GNRA tetraloops. Interestingly, a G1093A mutation which confers a nonsense suppressor phenotype[31,48] could potentially form an analogous type II A–A pair. In addition, a U turn or uridine turn[49] occurs at residues U1094, A1095, and A1096, with a hydrogen bond between the O-2′ of U1094 and the N-7 of A1096 and a hydrogen bond between the N-3 of U1094 and the phosphate O-2 of U1097. Uridine turns are found in the anticodon and T loops of tRNA and in GNRA tetraloops.[50] The 5′-GUAAUA-3′ hexaloop provides an additional example of the existence of this secondary structure motif in terminal loops. U1097 is extended into solution, perhaps to engage in tertiary or RNA–protein interactions. Whether or not this structure is representative of a class of hexaloops remains to be determined.

13.3.4 The P Loop (23S rRNA)

The P loop, residues G2250 to C2254, located in domain V of 23S rRNA, forms a Watson–Crick base pair with the 3′ terminal CCA of P site tRNA,[51] and is an essential component of the peptidyl transferase center. Mutations in the loop affect translational fidelity[52] and cause severe defects in peptidyl transferase activity.[51,53]

The structure of an oligonucleotide analogue of the P loop was solved by NMR spectroscopy[54] (Figure 3). The most striking feature of the loop is the splayed out position of the loop bases, making them accessible for base pairing with the CCA end of tRNA. This arrangement is very different from that of other loops such as tetraloops and the anticodon loops of tRNAs, in which loop bases extend the stacking arrangement of the stem helix. G2252, which forms the canonical base pair with C74 of tRNA, and G2251 stack on one another facing the major groove of the helix. A reverse Watson–Crick base pair between G2251 and C75 of tRNA has been proposed.[54] Not surprisingly, mutations at G2251 or G2252 produce dominant lethal phenotypes and cause increased frameshifting and readthrough of nonsense codons. The G2253A and G2253U mutations have little or no detectable phenotype, in contrast to the G2253C mutation which produces a slow-growth, misreading phenotype.[52]

Interestingly, G2253, while protected from kethoxal modification by tRNA,[55,56] is located distant from G2251 and G2252 on the minor groove side of the stem. Protection of G2253 may result from a dynamic tertiary interaction induced by binding of tRNA in the P site.[54] G2250 and C2254 form an unusual noncanonical pair, with the N-4 of C2254 hydrogen bonding to the N-3 of G2250 and the N-2 of G2250 hydrogen bonding to the N-3 of C2254. However, C2254 is reactive to DMS in intact ribosomes but becomes unreactive upon binding of A site tRNA,[55] suggesting that the oligonucleotide analogue mimics the structure of the 2250 loop in ribosomes with a filled A site, but that these bases are unpaired with an empty A site. It seems likely, therefore, that the P loop is dynamic in structure, and that such conformational changes may be important for ribosome function.

13.3.5 The Sarcin–Ricin Loop (23S rRNA)

The NMR-derived structure of the eukaryotic sarcin–ricin loop is remarkable for the number of noncanonical pairing interactions forming a large internal loop[57] (Figure 4). The entire structure is essentially a composite of two secondary structure motifs observed elsewhere in ribosomal RNAs, a GNRA tetraloop and the E loop of eukaryotic 5S rRNA. The internal "E" loop consists of a type II sheared G–A pair, a U–A reverse Hoogsteen pair (N3U–N7A, O2U–N6A), a bulged G residue positioned in the major groove, and a symmetric A–A pair (N-6–N-7). The latter positions are an A–C juxtaposition in *E. coli* 23S rRNA. The backbone reverses direction at the symmetric A–A pair and is kinked at the following bulged G residue. A cross-strand stack occurs between the sheared G–A pair and the U–A reverse Hoogsteen pair. This internal loop is analogous in conformation to loop E of eukaryotic 5S rRNA.[12] The terminal GAGA tetraloop is separated from the internal loop by a Watson–Crick C–G pair.

The extensive interest in the sarcin–ricin loop stems from its interaction with elongation factors G and Tu,[44,58] and from the effects on protein synthesis of the cytotoxins α-sarcin and ricin.[59,60] In chemical footprinting experiments, EF–G protects the bulged G2655 and the tetraloop nucleotides A2660 and G2661, while EF–Tu protects G2655 and G2661 and to a weaker degree A2660 and A2665 (of the U–A reverse Hoogsteen pair).[44] Mutations in the sarcin–ricin loop affect the binding of the EF–Tu–tRNA–GTP ternary complex to the ribosome and perturb the accuracy of translation.[61–63] Hydroxyl radicals

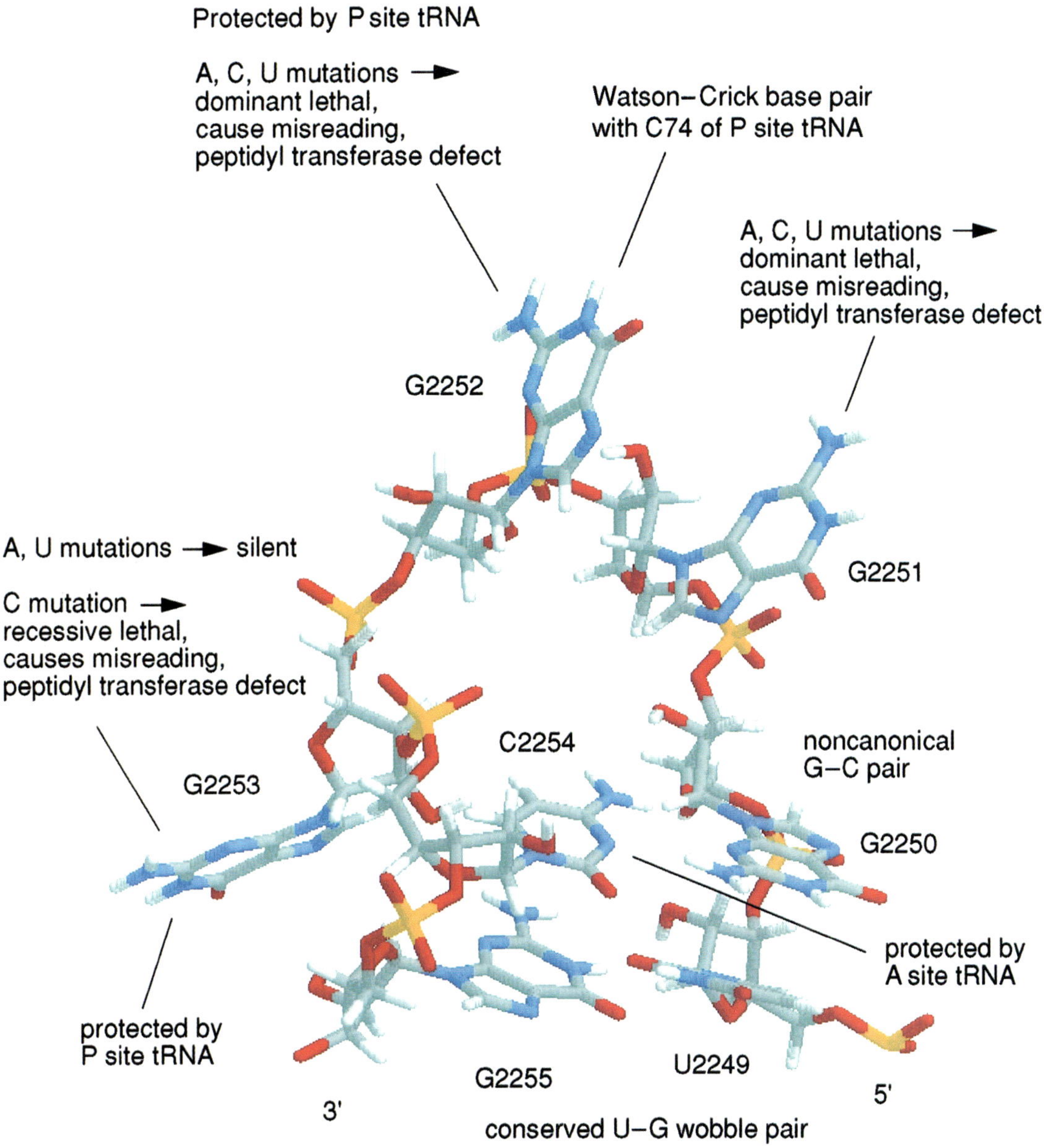

Figure 3 NMR-derived structure of the P loop of 23S rRNA.[54] Sites of interaction with the CCA terminus of P site tRNA are indicated, as are the phenotypes of mutations in the loop (pdb number 1VOP).

generated by Fe^{II}–EDTA tethered to amino acid 650 of EF–G produce cleavages in the sarcin–ricin loop in the GAGA tetraloop and in the 1093–1098 hexaloop in domain II,[45] placing these two structural elements in close proximity to EF–G and to one another.

13.4 TERTIARY STRUCTURE OF rRNAS

Comparative sequence analysis continues to be the most effective method for revealing tertiary interactions such as pseudoknots, lone pairs, and base triples. While some of these interactions, such as pseudoknots, tend to be composed primarily of Watson–Crick pairings, others, such as base triples, are more difficult to recognize as they do not covary in a concerted fashion. Presumably other interactions have escaped detection for this same reason. Nevertheless, the number of tertiary interactions thus discovered can be expected to grow with the volume of the sequence database and with the continued development of algorithms for detecting nonconcerted covariations.[64]

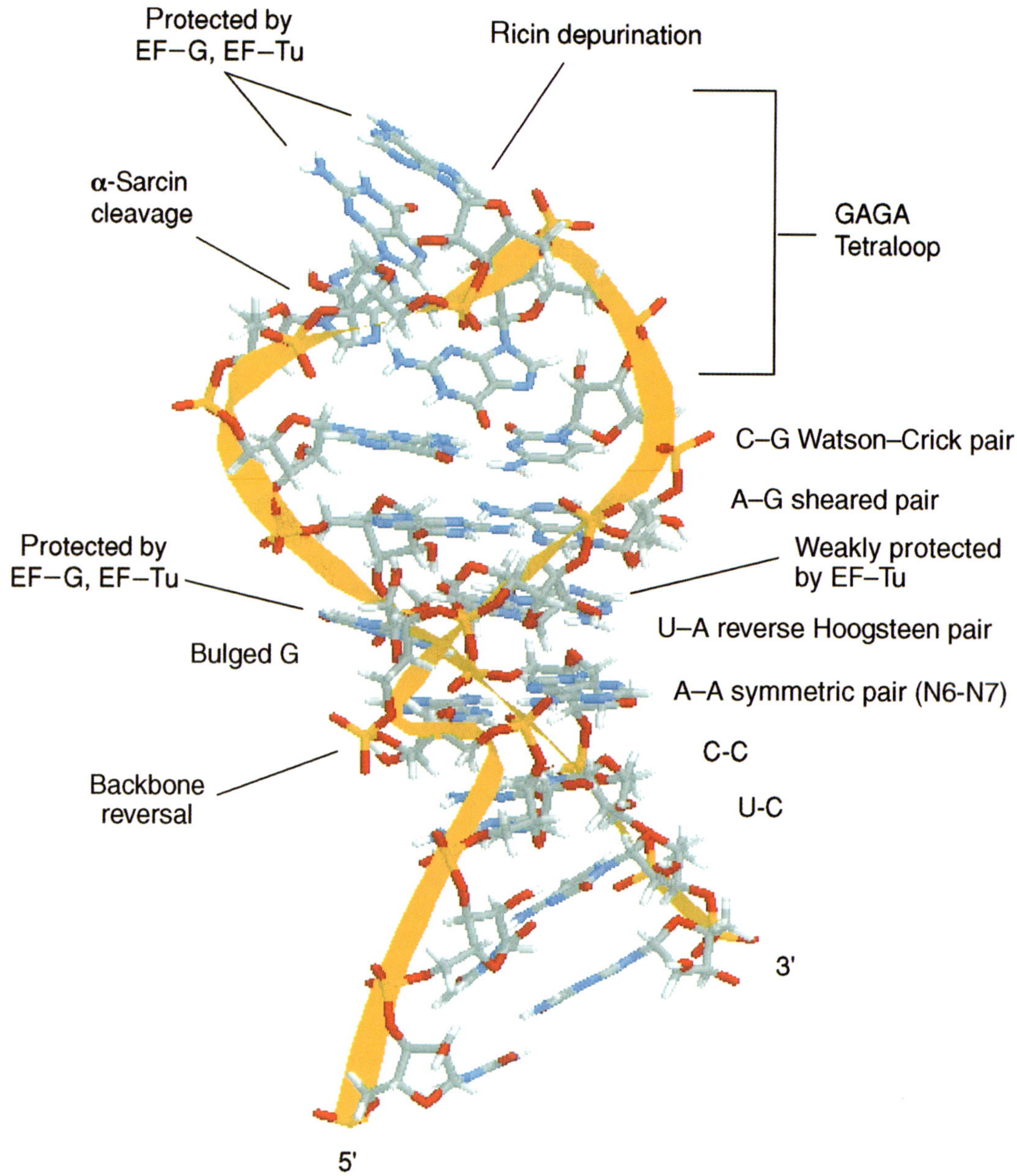

Figure 4 Structure of the sarcin–ricin loop from Rat 28S rRNA,[57] pdb number 1SCL. The loop consists of an E loop motif capped by a GAGA terminal loop. The noncanonical base pairing interactions are indicated, as are sites of protection from chemical probes by elongation factors EF–G and EF–Tu. Also shown are the sites of α-sarcin cleavage and ricin depurination. The path of the phosphodiester backbone is highlighted to emphasize backbone reversal at the bulged G residue and A–A symmetric pair.

Curiously, while many long-range tertiary interactions have been found in 23S rRNA by comparative sequence analysis, few have been found in 16S rRNA (see Figure 1). This is particularly striking considering that the 16S-like rRNA database is an order of magnitude larger than the 23S-like rRNA database.[1] Perhaps this is indicative of some fundamental difference in the principles guiding the global structure of these two molecules.

Cross-linking studies have also indicated the proximity of sites distant from one another in the secondary structure models[4] and, in a number of instances, lend experimental support to phylogenetically-determined interactions. In addition to pseudoknots, lone pairs, and base triples, it is likely that there exists a number of tetraloop–tetraloop receptor interactions and loop–helix interactions of an unknown nature that are important for tertiary structure.

13.4.1 Pseudoknots, Lone Pairs, and Base Triples

A number of pseudoknot structures have been identified by comparative sequence analysis, and several have received experimental support. In 16S rRNA, three pseudoknots have been examined by mutagenesis (see Figure 1). One pseudoknot is formed between the triplets consisting of positions 17–19 and positions 916–918; canonical Watson–Crick pairing in this "central" pseudoknot is necessary for ribosome function.[65] A second pseudoknot in 16S rRNA occurs between the doublets 570/571 and 865/866. As demonstrated by site-directed mutagenesis experiments, the *E. coli* pseudoknot (5′-GU-3′, positions 570 and 571, and 5′-AC-3′, positions 865 and 866) is important for 30S subunit assembly and 70S ribosome formation. Either of the mutations U571A or A865U cause structural instability of the 30S subunits with severe consequences on growth rate. Restoration of Watson–Crick pairing potential by a double mutation, U571A/A865U, alleviates these defects.[66] Interestingly, both the central pseudoknot and the 571–865 pseudoknot are adjacent to the site of a conformational switch structure involving nucleotides 912–914 and 885–890.[67] Mutations perturbing the equilibrium of the switch structure also perturb the reactivity of bases in the 571–865 pseudoknot, suggesting a possible structural and functional relationship between these two regions.

A third pseudoknot found in 16S rRNA involves the 530 loop, an important functional site involved in the decoding of mRNA.[68] A pseudoknot forms between the loop positions 524–526 and positions 505–507 that are part of an asymmetric bulge. Some mutations disrupting these base pairing interactions have severe consequences on cell growth and ribosome function, while others confer resistance to the antibiotic streptomycin.[69]

In 23S rRNA a two base pair pseudoknot occurs in domain II between the doublets at positions 1005, 1006 and 1137, 1138. These nucleotides are located in two internal loops positioned near either end of a seven base pair Watson–Crick helix. Introduction of mismatches at either base pairing position impairs growth, while introduction of mismatches at both positions is lethal.[70] Nucleotides near the sites of these mutations also become more susceptible to nuclease cleavage and chemical modification. Consistent with the phylogenetic data, both growth and structural defects are rescued by compensatory changes that restore the potential for Watson–Crick pairing.

A series of long-range base pairing contacts in the core of 23S rRNA constrain the three-dimensional folding of rRNA.[71] This set of Watson–Crick base pairs, 1262–2017, 1269–2011, and 1270–2010, brings domains II, III, IV, and V closer together in the secondary structure model. Mutagenesis studies combined with enzymatic and chemical probing provide support for the proposed structure and for an extended helical structure including 1262–1270 and 2010–2017, and indicate the importance of this structure for ribosome function.

Base triples represent another class of tertiary interaction that have been observed directly in the crystal structures of tRNAs[49] and the *Tetrahymena* ribozyme.[2] Base triples have been proposed to play an equally important role in rRNA tertiary structure.[64] Some of the candidates identified in rRNAs are short range, whereas others cross secondary structure domains. These interactions are likely to exhibit a high degree of context dependence, and do not display the kind of one-to-one correlation observed for Watson–Crick base pairs.[64] A proposed base triple in 23S rRNA between U746 in domain II and the G2057–C2611 base pair in domain V is supported by a UV cross-link between these two regions[72] and footprinting of antibiotic inhibitors of peptidyl transferase at A2058 and in the 750 loop.[73] This base triple brings together two domains separated by a large distance in the primary and secondary structures of 23S rRNA. Recently, a base triple in the GTPase center in domain II of 23S rRNA, detected by comparative sequence analysis, was established by site-directed mutagenesis.[46]

13.4.2 Secondary Structure Motifs as Potential Sites of Tertiary Interaction

Secondary structure motifs, such as purine–purine tandems, are likely to be sites of tertiary interaction. Most conclusions about the importance of these motifs have been inferred from their conservation at specific sites in rRNA, and by analogy with smaller RNAs where their role in folding and tertiary structure is well established. In purine–purine mismatches, the Watson–Crick faces of both purines are exposed in the major and minor grooves, making them attractive as potential sites for tertiary interactions or as protein recognition sites. The extremely shallow minor groove at purine–purine tandems may facilitate contact between RNA helices.[74] These mismatches are also found often as part of terminal loops, where they may engage in tertiary interactions.[75]

Of the three main classes of tetraloops, the GNRA, UNCG, and UUCG tetraloops, the GNRA tetraloops are now known to engage in tertiary interactions with helices bearing specific tetraloop receptors.[2,17] The high frequency of GNRA loops in a variety of structural RNAs implies that tetraloop–tetraloop receptor interactions may be a common tertiary structure building block. However, GNRA tetraloop receptors have yet to be unambiguously identified in ribosomal RNAs, perhaps due to the lack of highly defined sequence parameters defining potential receptors. Selex experiments have indicated that the secondary structures of GAAA tetraloop receptors may be more variable than originally proposed,[76] making them potentially more difficult to recognize by sequence inspection. Other members of the GNRA class recognize simple motifs such as 5′-CC/ 3′-GG or 5′-CU/3′-AG tandems.[17] Our understanding of the contribution of tetraloops and mismatch tandems to rRNA tertiary structure could benefit from a comprehensive and systematic site-directed mutagenesis approach coupled with structure probing.

13.4.3 Evidence for Tertiary Structure from Functional Studies

Tertiary interactions serve to bring distant secondary structure elements together to form functional sites in the ribosome. Thus, close physical proximity and possibly direct contact can be inferred for structures which participate in a common ribosome function. Evidence for the involvement of specific sites of rRNA in individual steps of translation comes primarily from the experimental approaches of mutagenesis, chemical footprinting, and cross-linking of ribosome ligands (for a series of reviews, see references 4, 37, 68, 77–80).

The importance of functional data in the quest for tertiary structure information is particularly pertinent for regions that are highly conserved and not amenable to dissection by comparative sequence analysis. One such region is the peptidyl transferase center of 23S rRNA (for a review see reference 81). Peptidyl transferase activity is associated primarily with three highly conserved secondary structure elements of domain V: the central loop (a five-helix junction) and the A and P terminal loops. While these three structures are some distance from one another in secondary structure,[1] it is certain that, together with elements of domain IV and possibly domain II, they collectively form part of the active site of the peptidyl transferase. This conclusion is based in part on the large number of mutations in these three structures which produce defects in peptidyl transferase activity[53,68,82] and is supported by chemical footprinting of the CCA terminus of tRNAs in the A and P sites.[55] The challenge remains to determine precisely how these three elements are arranged in three-dimensional space to create the catalytic site of the ribosome.

13.4.4 Evidence for Tertiary Structure from Cross-linking, Chemical Footprinting, and Cleavage Experiments

A great deal of information regarding the overall arrangement of rRNAs in the ribosome has been obtained by a variety of cross-linking techniques (reviewed in references 4 and 83). In addition to direct rRNA–rRNA cross-links, constraints on rRNA folding can also be inferred from cross-links to tRNA, mRNA, or ribosomal proteins. Thus, two nucleotide positions which both cross-link to the same position on a ligand can be inferred to be in close proximity to one another. The limit of resolution of cross-linking methods depends in part upon the size and nature of the cross-linking agent. Site-directed cross-linking can be achieved by incorporation of photoreactive nucleotide analogues such as 4-thiouridine or azidoadenosines which are "zero length" cross-links. Zero length cross-links can also be obtained by utilizing the inherent propensities of some positions to form cross-links upon ultraviolet irradiation.

Chemical footprinting experiments have provided a wealth of information about global conformation. The protection of rRNA by the binding of ribosomal proteins, ligands such as tRNAs or translation factors, and antibiotics, contribute a significant number of constraints. This is particularly true for antibiotics which, by virtue of their small size, can make only a limited number of contacts with nucleotides that are in very close proximity to one another. For example, the antibiotic vernamycin B protects A752 in domain II, as well as A2062 and G2505 in domain V of 23S rRNA.[73] While A2062 and G2505 are both in the central loop of domain V, the proximity of A752 to these nucleotides has been confirmed by cross-linking of the 750 loop to the central loop.

Another powerful approach to defining distance constraints in the ribosome is the use of cleavage reagents tethered to specific sites on various ligands. These include 1, 10-*ortho*-phenanthroline-Cu^{II} [84] and hydroxyl radicals generated by Fe^{II}–EDTA.[68] These reagents catalyze cleavage of the RNA backbone independent of secondary structure. While effects on reactivity of base chemical probes can result from either proximity or long-range conformational effects, cleavage is completely dependent on proximity. rRNA residues neighboring specific nucleotides of tRNA or specific amino acid residues of a ribosomal protein or elongation factor can be identified in this way. These approaches are particularly informative if the crystal structure of the ligand is known, and if several different residues are used as sites of attachment of the cleavage reagent. Generally, results from cleavage experiments agree with those from chemical footprinting experiments and serve to complement the chemical probing approach.

13.4.5 Three-dimensional Reconstructions, X-ray Crystallography, Cryoelectron Microscopy

The accumulated data from cross-linking, mutagenesis, and footprinting experiments have been integrated in the construction of three-dimensional models of the large rRNAs.[3,85–87] One of the key elements of modeling is the placement of ribosomal proteins. Early neutron scattering data were used to position the mass centers of 30S subunit ribosomal proteins,[88] and, in conjunction with protein–RNA cross-linking data, to place constraints on 16S rRNA folding. These models are refined with the incorporation of data from the solution of individual ribosomal protein structures[89] in conjunction with data from experiments utilizing hydroxyl radicals generated from specific sites on ribosomal proteins of known structure,[68] and determination of the precise positions of the cross-linked nucleotide and amino acid.[90]

A 9 Å resolution X-ray crystallographic electron density map of *Haloarcula marismortui* 50S subunits was generated by molecular replacement using cryoelectron microscopic contours of *E. coli* 70S ribosomes.[6] This resolution is sufficient to discern the major and minor grooves of RNA helices. Establishing the path of 23S rRNA through the subunit will next require the identification of landmarks such as ribosomal protein binding sites. Models of 16S rRNA are also fitted to the cryoelectron microscopic contour.[85] It is expected that these approaches will ultimately lead to the derivation of an atomic-resolution structure of the intact ribosome.

13.5 CONFORMATIONAL SWITCHES IN rRNA

The realization of the central role of rRNA in ribosome function has led to the prediction that conformational changes in rRNAs occur during translation. This was inspired in part by the knowledge that conformational changes occur in protein enzymes during catalysis. However, direct evidence for specific changes in rRNA conformation and their association with specific ribosome functions has been difficult to obtain. For example, a reversible interconversion between functionally active and inactive forms of the 30S subunit, dependent upon heat, magnesium, and ammonium ion concentrations, has been reported,[91] and structure probing has provided clues to the nature of the transition.[92] However, its biological significance remains undetermined. Numerous proposals of alternate base pairing conformations during different stages of translation have been proposed,[67] but there is no direct evidence for the existence of any of these switch mechanisms or their association with specific ribosome functions. Currently, only one switch mechanism has been described which is supported by comparative sequence analysis, site-directed mutagenesis, and structure probing, although it is generally believed that other switch mechanisms will eventually be uncovered.

13.5.1 The 912-885/888 Conformational Switch in 16S rRNA

Site-directed mutagenesis and comparative sequence analysis, combined with structure probing of *E. coli* ribosomes, determined that two sets of pairing interactions are physiologically relevant structures

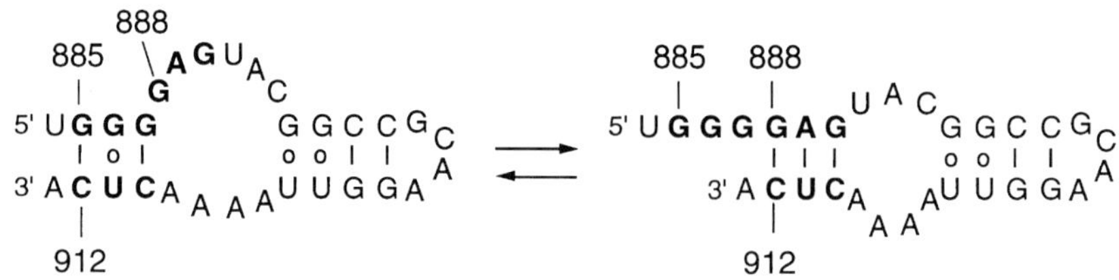

Figure 5 The 912–885/888 conformational switch in 16S rRNA.[67] The left structure represents the proposed base pairing interactions in the 912–885 conformation, the right structure the proposed 912–888 conformation. Mutations predicted to favor the 912–888 conformation are hyperaccurate, are incompatible with hyperaccurate mutations in ribosomal protein S12, and are suppressed by error-prone mutations in ribosomal protein S5. Mutations predicted to favor the alternative rRNA conformation are error prone.

and represent alternate conformations that occur during protein synthesis.[67,93] In the central region of *E. coli* 16S rRNA, three nucleotides, 5′-CUC-3′ (912–910), can pair either with 5′-GAG-3′ (888–890) (the "912-888" conformation), or with 5′-GGG-3′ (885–887) (the "912-885" conformation) (see Figure 5). The dynamic change in rRNA structure involving these two alternate base pairing arrangements is apparently facilitated by ribosomal proteins S5 and S12, which interact with this region of 16S rRNA at the junction of the three major domains. Mutations in ribosomal proteins S5 and S12 have opposite effects on the fidelity of tRNA selection. Similarly, translational fidelity can be manipulated by site-directed mutations in 16S rRNA designed to perturb the equilibrium between the two rRNA conformations, interfering with base pairing in one or the other structure. 16S rRNA mutations favoring the 912-888 conformation increase translational fidelity and are incompatible with S12 hyperaccurate mutations but suppressed by an S5 error-prone mutation. In contrast, mutations predicted to shift the equilibrium toward the 912-885 conformation are error prone and compatible with hyperaccurate S12 mutations. By combining mutations at all three sites, 910–912, 882–885, and 888–890, a balance between the two proposed conformers is re-established and the phenotype reverted to that of the wild-type ribosome.

While the precise function of this conformational switch has yet to be firmly established, ribosomes with mutations favoring the 912-885 conformation have greater affinity for tRNA. This, coupled with the effects on translational fidelity, immediately suggests a role in tRNA selection and proofreading. However, chemical probing reveals significant changes in the rRNA structure at or near sites that have been cross-linked to mRNA, and implies that the switch could also be directly involved in translocation. It is noted that the simultaneous movement of three nucleotides during the conformational switch is consistent with the movement of mRNA by one codon.

The structures participating in the base pairing switch are conserved in all small subunit rRNAs, thus indicating that the switch mechanism is a fundamental process in all ribosomes. The apparent lack of long-range interactions between domains of 16S rRNA supports the idea that this molecule is more adapted to large transformational switches within and between domains than is the case for 23S rRNA.[68]

13.5.2 Potential Conformational Changes in the Ribosomal A Site to Accommodate Different Protein Factors

During translation, the ribosome must interact with many different tRNAs, as well as initiation, elongation, and termination factors. Recent high-resolution structures of some of these translation factors indicate however, that many of these diverse ligands share common structural features.[94–96] This has led to the development of the concept of molecular mimicry, where domains of protein factors or RNA molecules adopt similar conformations so as to utilize a single, common binding site on the ribosome. Consistent with these various proposals is the observation that the EF–Tu, EF–G, the IF1–IF2 (initiation factor) complex, and release factors all compete for binding to the A site of the ribosome. Nevertheless, the ribosome must distinguish among all these factors at the appropriate stage of translation. An intriguing possibility is that the binding site for these factors changes conformation to accommodate structural features unique to the individual proteins, shifting the binding equilibrium in favor of the appropriate factor. This hypothesis must await verification by high-resolution structural studies.

13.5.3 Other Potential Conformational Switches in rRNA

Additional switch structures are to be anticipated in functional centers of both 16S and 23S rRNA in light of the multiple hybrid transition states now known to occur during the various stages of translation.[97] For example, during the translation cycle, EF–Tu and EF–G alternate in binding to the ribosome and interacting with the sarcin–ricin loop in 23S rRNA. It has been proposed that the cytotoxins α-sarcin and ricin recognize differing conformations of the sarcin–ricin loop that are present during differing stages of the elongation cycle, and that the interconversion of these two conformations is promoted by (or at least associated with) the sequential action of EF–Tu and EF–G.[98]

Finally one must consider as a shift structure the intermolecular base pairing between the anti-Shine–Dalgarno (ASD) sequence (5′-CCUCC-3′) at the 3′ end of 16S rRNA and mRNA. During translation initiation, the ASD sequence base pairs with the Shine–Dalgarno sequence (5′-GGAGG-3′) in mRNA upstream of the AUG initiation codon.[99,100] While this specific base pairing ceases once elongation begins, the ASD sequence continues to scan the translated codons and there is evidence that it can base pair with downstream sequences in mRNA.[101] Obviously, we are just beginning to appreciate the complex dynamic nature of rRNA during translation. Given the current rate of progress in the structural analysis of rRNA, a comprehensive understanding of its mechanism seems imminent.

13.6 REFERENCES

1. R.R. Gutell, in "Ribosomal RNA: Structure, Evolution, Processing, and Function in Protein Biosynthesis", eds. R.A. Zimmermann and A.E. Dahlberg, CRC Press, Boca Raton, FL, 1996, 111
2. J.H. Cate, A.R. Gooding, E. Podell, K. Zhou, B.L. Golden, C.E. Kundrot, T.R. Cech and J.A. Doudna, *Science*, 1996, **273**, 1678.
3. H.F. Noller, R. Green, G. Heilek, V. Hoffarth, A. Huttenhofer, S. Joseph, I. Lee, K. Lieberman, A. Mankin, C. Merryman, T. Powers, E.V. Puglisi, R.R. Samaha and B. Weiser, *Biochem. Cell Biol.*, 1995, **73**, 997.
4. R. Brimacombe, P. Mitchell and F. Muller, in "Ribosomal RNA: Structure, Evolution, Processing, and Function in Protein Biosynthesis", eds. R.A. Zimmermann and A.E. Dahlberg, CRC Press, Boca Raton, FL, 1996, 129
5. F. Mueller, T. Doring, T. Erdemir, B. Greuer, N. Junke, M. Osswald, J. Rinke-Appel, K. Stade, S. Thamm and R. Brimacombe, *Biochem. Cell Biol.*, 1995, **73**, 767.
6. N. Ban, B. Freeborn, P. Nissen, P. Penczek, R.A. Grassucci, R. Sweet, J. Frank, P.B. Moore and T.A. Steitz, *Cell*, 1998, **93**, 1105.
7. R.R. Samaha, B. O'Brien, T.W. O'Brien and H.F. Noller, *Proc. Natl. Acad. Sci. USA*, 1994, **91**, 7884.
8. I. Nitta, Y. Kamada, H. Noda, T. Ueda and K. Watanabe, *Science*, 1998, **281**, 666.
9. D. Gautheret, D. Konings and R.R. Gutell, *J. Mol. Biol.*, 1994, **242**, 1.
10. J. Santa Lucia Jr. and D.H. Turner, *Biochemistry*, 1993, **32**, 12–612.
11. H.W. Pley, K.M. Flaherty and D.B. McKay, *Nature*, 1994, **372**, 68.
12. B. Wimberly, G. Varani and I. Tinoco Jr., *Biochemistry*, 1993, **32**, 1078.
13. D. Gautheret, D. Konings and R.R. Gutell, *RNA*, 1995, **1**, 807.
14. B. Wimberly, *Nature Struct. Biol.*, 1994, **1**, 820.
15. C.C. Correll, B. Freeborn, P.B. Moore and T.A. Steitz, *Cell*, 1997, **91**, 705.
16. C.R. Woese, S. Winker and R.R. Gutell, *Proc. Natl. Acad. Sci. USA*, 1990, **87**, 8467.
17. M. Costa and F. Michel, *EMBO J.*, 1995, **14**, 1276.
18. Z.P. Cai, A. Gorin, R. Frederick, X.M. Ye, W.D. Hu, A. Majumdar, A. Kettani and D.J. Patel, *Nature Struct. Biol.*, 1998, **5**, 203.
19. D.L. Abramovitz and A.M. Pyle, *J. Mol. Biol.*, 1997, **266**, 493.
20. C. Cheong, G. Varani and I. Tinoco Jr., *Nature*, 1990, **346**, 680.
21. H.A. Heus and A. Pardi, *Science*, 1991, **253**, 191.
22. F.M. Jucker and A. Pardi, *Biochemistry*, 1995, **34**, 14–416.
23. J.B. Prince, B.H. Taylor, D.L. Thurlow, J. Ofengand and R.A. Zimmermann, *Proc. Natl. Acad. Sci. USA*, 1982, **79**, 5450.
24. J. Rinke-Appel, N. Junke, R. Brimacombe, S. Dokudovskaya, O. Dontsova and A. Bogdanov, *Nucleic Acids Res.*, 1993, **21**, 2853.
25. D. Moazed and H.F. Noller, *Cell*, 1986, **47**, 985.
26. D. Moazed and H.F. Noller, *Nature*, 1987, **327**, 389.
27. E.A. DeStasio and A.E. Dahlberg, *J. Mol. Biol.*, 1990, **212**, 127.
28. P. Purohit and S. Stern, *Nature*, 1994, **370**, 659.
29. H. Miyaguchi, H. Narita, K. Sakamoto and S. Yokoyama, *Nucleic Acids Res.*, 1996, **24**, 3700.
30. D. Fourmy, M.I. Recht, S.C. Blanchard and J.D. Puglisi, *Science*, 1996, **274**, 1367.
31. S.T. Gregory and A.E. Dahlberg, *Nucleic Acids Res.*, 1995, **23**, 4234.
32. D. Fourmy, S. Yoshizawa and J.D. Puglisi, *J. Mol. Biol.*, 1998, **277**, 333.
33. P. Mitchell, M. Osswald and R. Brimacombe, *Biochemistry*, 1992, **31**, 3004.

34. C. Ehresmann, H. Moine, M. Mougel, J. Dondon, M. Grunberg-Manago, J.-P. Ebel and B. Ehresmann, *Nucleic Acids Res.*, 1986, **14**, 4803.
35. T.L. Helser, J.E. Davies and J.E. Dahlberg, *Nature New Biol.*, 1972, **235**, 6.
36. P.H. van Knippenberg, in "Structure, Function and Genetics of Ribosomes", eds. B. Hardesty and G. Kramer, Springer, New York, 1986, 413
37. P.R. Cunningham, C.J. Weitzmann, D. Negre, J.G. Sinning, V. Frick, K. Nurse and J. Ofengand, in "he Ribosome: Structure, Function, and Evolution", eds. A. Dahlberg, W.E. Hill, A. Dahlberg, R.A. Garrett, P.B. Moore, D. Schlessinger and J.R. Warner, ASM Press, Washington, DC, 1990, 243
38. D. Lafontaine, J. Delcour, A.L. Glasser, J. Desgres and J. Vandenhaute, *J. Mol. Biol.*, 1994, **241**, 492.
39. J.P. Rife and P.B. Moore, *Structure*, 1998, **6**, 747.
40. J.P. Rife, C.S. Cheng, P.B. Moore and S.A. Strobel, *Nucleic Acids Res.*, 1998, **26**, 3640.
41. S.E. Butcher, T. Diekmann and J. Feigon, *J. Mol. Biol.*, 1997, **268**, 348.
42. S. Douthwaite, B. Vester, C. Aagaard and G. Rosendahl, in "The Translation Apparatus: Structure, Function, Regulation, Evolution", eds. K.H. Nierhaus, F. Franceschi, A.R. Subramanian, V.A. Erdmann and B. Wittman-Liebold, Plenum, New York, 1993, 339
43. G. Rosendahl and S. Douthwaite, *Nucleic Acids Res.*, 1994, **22**, 357.
44. D. Moazed, J.M. Robertson and H.F. Noller, *Nature*, 1988, **334**, 362.
45. K.S. Wilson and H.F. Noller, *Cell*, 1998, **92**, 131.
46. G.L. Conn, R.R. Gutell and D.E. Draper, *Biochemistry*, 1998, **37**, 11–980.
47. S. Huang, Y.-X. Wang and D.E. Draper, *J. Mol. Biol.*, 1996, **258**, 308.
48. D.K. Jemiolo, F.T. Pagel and E.J. Murgola, *Proc. Natl. Acad. Sci. USA*, 1995, **92**, 12–309.
49. W. Saenger, "Principles of Nucleic Acid Structure", Springer, New York, 1984.
50. F.M. Jucker and A. Pardi, *RNA*, 1995, **1**, 219.
51. R.R. Samaha, R. Green and H.F. Noller, *Nature*, 1995, **377**, 309.
52. S.T. Gregory, K.R. Lieberman and A.E. Dahlberg, *Nucleic Acids Res.*, 1994, **22**, 279.
53. K.R. Lieberman and A.E. Dahlberg, *J. Biol. Chem.*, 1994, **269**, 16–163.
54. E.V. Puglisi, R. Green, H.F. Noller and J.D. Puglisi, *Nature Struct. Biol.*, 1997, **4**, 775.
55. D. Moazed and H.F. Noller, *Cell*, 1989, **57**, 585.
56. D. Moazed and H.F. Noller, *Proc. Natl. Acad. Sci. USA*, 1991, **88**, 3725.
57. A.A. Szewczak, P.B. Moore, Y.-L. Chan and I.G. Wool, *Proc. Natl. Acad. Sci. USA*, 1993, **90**, 9581.
58. A. Munishkin and I.G. Wool, *Proc. Natl. Acad. Sci. USA*, 1997, **94**, 12–280.
59. Y. Endo and I.G. Wool, *J. Biol. Chem.*, 1982, **257**, 9054.
60. Y. Endo, K. Mitsui, M. Motizuki and K. Tsurugi, *J. Biol. Chem.*, 1987, **262**, 5908.
61. W.E. Tapprich and A.E. Dahlberg, *EMBO J.*, 1990, **9**, 2649.
62. P. Melancon, W.E. Tapprich and L. Brakier-Gingras, *J. Bacteriol.*, 1992, **174**, 7896.
63. M. O'Connor and A.E. Dahlberg, *Nucleic Acids Res.*, 1996, **24**, 2701.
64. D. Gautheret, S.H. Damberger and R.R. Gutell, *J. Mol. Biol.*, 1995, **248**, 27.
65. R.A. Poot, S.H.E. vandenWorm, C.W.A. Pleij and J. van Duin, *Nucleic Acids Res.*, 1998, **26**, 549.
66. A. Vila, J. Viril-Farley and W.E. Tapprich, *Proc. Natl. Acad. Sci. USA*, 1994, **91**, 11–148.
67. J.S. Lodmell and A.E. Dahlberg, *Science*, 1997, **277**, 1262.
68. R. Green and H.F. Noller, *Ann. Rev. Biochem.*, 1997, **66**, 679.
69. T. Powers and H.F. Noller, *EMBO J.*, 1991, **10**, 2203.
70. G. Rosendahl, L.H. Hansen and S. Douthwaite, *J. Mol. Biol.*, 1995, **249**, 59.
71. C. Aagaard and S. Douthwaite, *Proc. Natl. Acad. Sci. USA*, 1994, **91**, 2989.
72. P. Mitchell, M. Osswald, D. Schueler and R. Brimacombe, *Nucleic Acids Res.*, 1990, **18**, 4325.
73. D. Moazed and H.F. Noller, *Biochimie*, 1987, **69**, 879.
74. S.A. Strobel, L. Ortoleva-Donnelly, S.P. Ryder, J.H. Cate and E. Moncoeur, *Nature Struct. Biol.*, 1998, **5**, 60.
75. S.A. Strobel and J.A. Doudna, *Trends Biochem. Sci.*, 1997, **22**, 262.
76. M. Costa and F. Michel, *EMBO J.*, 1997, **16**, 3289.
77. L. Brakier-Gingras, R. Pinard and F. Dragon, *Biochem. Cell Biol.*, 1995, **73**, 907.
78. M. O'Connor, C.A. Brunelli, M.A. Firpo, S.T. Gregory, K.R. Lieberman, J.S. Lodmell, H. Moine, D.I. Van Ryk and A.E. Dahlberg, *Biochem. Cell Biol.*, 1995, **73**, 859.
79. R.A. Zimmermann, in "Ribosomal RNA: Structure, Evolution, Processing, and Function in Protein Biosynthesis", eds. R.A. Zimmermann and A.E. Dahlberg, CRC Press, Boca Raton, FL, 1996, 277
80. R.A. Garrett and C. Rodriguez-Fonseca, in "Ribosomal RNA: Structure, Evolution, Processing, and Function in Protein Biosynthesis", eds. R.A. Zimmermann and A.E. Dahlberg, CRC Press, Boca Raton, FL, 1996, 327
81. K.R. Lieberman and A.E. Dahlberg, *Prog. Nucleic Acids Res.*, 1995, **50**, 1.
82. B.T. Porse and R.A. Garrett, *J. Mol. Biol.*, 1995, **249**, 1.
83. D. Mundus and P. Wollenzien, *Nucleic Acids Symp. Ser.*, 1997, **36**, 171.
84. W.E. Hill, D.J. Bucklin, J.M. Bullard, A.L. Galbraith, N.V. Jammi, C.C. Rettberg, B.S. Sawyer and M.A. van Waes, *Biochem. Cell Biol.*, 1995, **73**, 1033.
85. F. Mueller and R. Brimacombe, *J. Mol. Biol.*, 1997, **271**, 524.
86. F. Mueller and R. Brimacombe, *J. Mol. Biol.*, 1997, **271**, 545.
87. F. Mueller, H. Stark, M. van Heel, J. Rinke-Appel and R. Brimacombe, *J. Mol. Biol.*, 1997, **271**, 566.
88. M.S. Capel, D.M. Engelman, B.R. Freeborn, M. Kjeldgaard, J.A. Langer, V. Ramakrishnan, D.G. Schindler, D.K. Schneider, B.P. Schoenborn and I.Y. Sillers *et al. et al.*, *Science*, 1987, **238**, 1403.

89. V. Ramakrishnan and S.W. White, *Trends Biochem. Sci.*, 1998, **23**, 208.
90. B. Thiede, H. Urlaub, H. Neubauer, G. Grelle and B. Wittman-Liebold, *Biochem. J.*, 1998, **334**, 39.
91. A. Zamir, R. Miskin and D. Elson, *J. Mol. Biol.*, 1971, **60**, 347.
92. D. Moazed, B.J. Van Stolk, S. Douthwaite and H.F. Noller, *J. Mol. Biol.*, 1986, **191**, 483.
93. J.S. Lodmell, R.R. Gutell and A.E. Dahlberg, *Proc. Natl. Acad. Sci. USA*, 1995, **92**, 10–555.
94. P. Nissen, M. Kjeldgaard, S. Thirup, G. Polekhina, L. Reshetnikova, B.F. Clark and J. Nyborg, *Science*, 1995, **270**, 1464.
95. K. Ito, K. Ebihara, M. Uno and Y. Nakamura, *Proc. Natl. Acad. Sci. USA*, 1996, **93**, 5443.
96. S. Brock, K. Szkaradkiewicz and M. Sprinzl, *Mol. Microbiol.*, 1998, **29**, 409.
97. D. Moazed and H.F. Noller, *Nature*, 1989, **342**, 142.
98. A. Gluck, Y. Endo and I.G. Wool, *J. Mol. Biol.*, 1992, **226**, 411.
99. J. Shine and L. Dalgarno, *Proc. Natl. Acad. Sci. USA*, 1974, **71**, 1342.
100. J.A. Steitz and K. Jakes, *Proc. Natl. Acad. Sci. USA*, 1975, **72**, 4734.
101. R.B. Weiss, D.M. Dunn, A.E. Dahlberg, J.F. Atkins and R.F. Gesteland, *EMBO J.*, 1988, **7**, 1503.

14

Turnover of mRNA In Eukaryotic Cells

SUNDARESAN THARUN and ROY PARKER
University of Arizona, Tucson, AZ, USA

14.1 INTRODUCTION

Ribonucleic acids (RNA) are polymeric molecules made of ribonucleotide building blocks. There are three major kinds of RNA molecules found in cells. These are the messenger RNAs (mRNA), ribosomal RNAs (rRNA) and the transfer RNAs (tRNA). mRNAs encode the information for building proteins and can be considered informational RNAs. The rRNAs and tRNAs are the critical parts of the protein synthesis machinery required to convert the nucleotide sequence information in each mRNA into the corresponding protein molecule.

Since most cellular functions are performed by proteins, every eukaryotic cell has thousands of different kinds of mRNAs encoding the diversity of cellular proteins. For example, it is estimated that in the simple yeast *Sacchormyces cerevisiae*, approximately 4000 different mRNA molecules are present during growth in rich media.[1] Different mRNA molecules can be found at low, medium, or high concentrations within the cell depending on the biological role of the encoded polypeptide. Moreover, in any given cell the rate of synthesis of each protein is directly related to the concentration of the corresponding mRNA, and its translation rate. Given this fact, the proper regulation of the levels of the individual mRNAs is crucial for the normal functioning of any cell.

mRNA molecules, like all other macromolecules in the cell are in constant turnover, being synthesized and degraded at specified rates at any given time. Thus, the decay rate of a mRNA is a major determinant of its concentration. Decay rates of mRNA molecules — generally expressed as half-lives ($t_{1/2}$) — vary over a wide range in eukaryotic cells (from few minutes to several hours or even few days). In a given cell, different mRNAs have different decay rates and for a given mRNA the decay rate can vary in response to a variety of physiological cues. Not surprisingly, the properties of each mRNA molecule are finely tuned to meet the biological demands on that particular gene. The precise nucleotide sequences of different mRNAs can contain specific information that dictates its intracellular location, translation rate, rate of degradation, and the ability of that mRNA to respond to specific physiological cues.

There are several striking differences between the metabolism of prokaryotic and eukaryotic mRNAs. Many of these can be explained by the fact that in eukaryotic cells the site of mRNA biogenesis (transcription and RNA processing) and its function in protein synthesis are temporally and physically separated by the nuclear membrane, since these two processes occur in nucleus and cytoplasm, respectively. Conversely, in prokaryotes cells the process of mRNA biogenesis and function are coupled both temporally and spatially. The separation of mRNA synthesis from function in eukaryotic has allowed the cell to substantially process the mRNA before translation occurs in manners such as the splicing of intervening sequences.

The coupling of transcription and translation in prokaryotic cells has led to a mechanism for mRNA decay where the primary degradation mechanism is by 3′ to 5′ exonucleases.[2] This can be rationalized since by the time the 3′ end of mRNA is finished the transcript has been largely translated and could be considered for degradation. In contrast, the separation of synthesis and function in eukaryotic cells has demanded that mRNAs be stabilized for a period of time before translation occurs to allow time for RNA processing and transport to the cytoplasmic compartment. The mechanism of mRNA stabilization appears to be related to specialized structures found at the 5′ and 3′ ends of eukaryotic transcripts. The eukaryotic transcripts have a "cap" structure at their 5′ end. This cap structure is a methylated guanosine residue that is attached to the first transcribed nucleotide by a 5′–5′ triphosphate linkage. At the 3′ end, eukaryotic mRNAs have extensive polyadenylate tails that are added posttranscriptionally. Prokaryotic mRNAs lack the cap structure and in general, lack poly(A) tails also although mRNA adenylation may play a role in mRNA degradation in prokaryotes.[3] As discussed further below, both the cap and poly(A) tail play critical roles in translation and decay of eukaryotic mRNAs.

It has been clear for a number of years that the differential degradation of eukaryotic mRNAs plays an important role in the modulation of gene expression. There are a large number of examples where the regulation of gene expression occurs by the modulation of the decay rate of a mRNA or class of mRNAs.[4] Thus, in order to understand the modulation of gene expression it will be important to determine the mechanisms and control of mRNA degradation.

In the last few years several different, and somewhat related, mechanisms by which eukaryotic mRNAs are degraded have been defined. One mRNA decay pathway is initiated by shortening of the 3′ poly(A) tail (often referred to as deadenylation) followed by decapping and 5′ to 3′ exonucleolytic degradation of the transcript. Alternatively, transcripts can undergo 3′ to 5′ decay after poly(A) shortening. Decay of mRNAs can also initiate independently of the shortening of the poly(A) tail. For example, transcripts can be decapped in a deadenylation-independent manner leading to 5′ to 3′ degradation. Similarly, mRNA decay can also be initiated by endonucleolytic cleavages. In this chapter we will discuss these pathways of degradation, the enzymes which catalyze the specific nucleolytic events, and how the activity of the nucleases may be modulated on individual transcripts to give rise to differential mRNA turnover.

14.2 ROLE OF DEADENYLATION IN THE DECAY OF EUKARYOTIC mRNAs

14.2.1 Deadenylation Is a Key Step In mRNA Turnover

Following the demonstration that the 3′ poly(A) tails of eukaryotic mRNAs are shortened in the cytoplasm[5] several observations implicated the removal of the poly(A) tail as an early step in the decay of many mRNAs. One set of observations came from analyzing the poly(A) shortening rates and the decay rates of different mRNAs (for review, see Peltz et al.).[6] For example, in the comparison of two adenovirus mRNAs, the unstable mRNA deadenylated significantly faster than the more stable transcript.[7] This observation suggested that the faster poly(A) shortening might be involved in faster mRNA decay. Correlations between the stability of a transcript and the metabolism of its poly(A) tail have also been observed by examining the poly(A) tail of an individual mRNA under different conditions that alter its decay rate. For example, when the transcripts for human growth hormone, or avian liver apoVLDLII mRNA are stabilized by the presence of glucocorticoids and estrogen, respectively, the average poly(A) tail length is significantly longer.[8,9] Finally, additional evidence that poly(A) tail removal might be required for mRNA degradation came from the examination of the stability of transcripts lacking poly(A) tails, which were either introduced into cells by microinjection or electroporation (reviewed in Bernstein and Ross)[10] or synthesized *in vivo* in the presence of cordycepin (3′ deoxyadenosine).[11] Since in these cases, unadenylated mRNAs were generally less stable than adenylated control mRNAs, the poly(A) tail was inferred to be required for maintaining the cytoplasmic stability of a mRNA.

Additional evidence that poly(A) tail shortening can be required for mRNA decay came from following the degradation of "pulses" of newly synthesized transcripts produced by rapid induction and repression of regulatable promoters. In mammalian cells this approach has been performed by utilizing the c-fos promoter, which after serum stimulation becomes transiently active. By using constructs coding for c-fos and β-globin mRNAs driven by c-fos promoter, the fate of those two mRNAs have been studied in a time-dependent manner.[12,13] These experiments indicated that the unstable c-fos transcript deadenylates much faster than the stable β-globin mRNA and does not begin to be degraded until its poly(A) tail is shortened.

Further evidence that the c-fos mRNA requires poly(A) shortening before degradation came from the analysis of chimeric transcripts. Replacing the coding region of β-globin mRNA with that of c-fos mRNA results in both more rapid deadenylation and rapid decay of the chimeric mRNA as compared to the parental β-globin transcript. Moreover, a deletion in the c-fos coding region of this chimeric mRNA that leads to its stabilization also slows down its deadenylation rate.[12] Similar results have been seen when an AU-rich element, termed an ARE, within the c-fos 3′ UTR is transferred to the 3′ UTR of the β-globin transcript. In this case, the resulting chimeric mRNA is highly unstable and has a rapid rate of deadenylation, suggesting that the c-fos ARE can promote decay, at least in part, by accelerating deadenylation.[12] Similar, AU-rich sequences are found in the 3′ UTRs of several unstable mammalian oncogene and lymphokine mRNAs.[14] Evidence suggests that at least some of the other transcripts containing these elements are dependent on rapid deadenylation for their short half-lives.[15–17]

Similar "transcriptional pulse-chase" experiments have been performed on several yeast mRNAs. In this case, cells are grown in medium containing a "neutral" carbon source, such as raffinose, or a mixture of raffinose and sucrose, in which expression from the *GAL1* promoter is neither induced nor repressed.[18] Addition of galactose to these cultures causes a rapid induction of transcription from the *GAL1* promoter. After a brief induction, typically 10–15 min, transcription is quickly repressed by the addition of glucose. These yeast experiments have been particularly informative since the decay rate of the transcript, the poly(A) shortening rate, and the appearance of decay intermediates have all been examined. The examination of the stable *PGK1* and the unstable *MFA2* transcripts in this manner have led to the following observations.[19] (1) Following transcription repression there is a temporal lag before the decay of these mRNAs initiates. (2) In each case, there is a good correlation between the length of this lag period and the time it takes for the poly(A) tail to be shortened from its initial length (from 50 to 80 residues long) to a short oligo(A) length (10–15 residues). Analysis of chimeric and mutant transcripts provide additional evidence that the lag period was required to allow deadenylation before decay of these transcripts could begin. For example, replacement of the 3′ UTR of the stable *PGK1* mRNA with the 3′ UTR of the unstable *MFA2* mRNA, which contains sequences capable of stimulating rapid deadenylation,[20] results in both an increased deadenylation rate and a corresponding decrease in the length of the lag period.[19] In addition, *cis*-acting mutations in *MFA2* mRNA that stabilize the transcript also change its deadenylation rate to an extent that partially or fully accounts for the increase in stability.[20]

Additional evidence that deadenylation precedes the decay of the *PGK1* and *MFA2* transcripts has come from the examination of mRNA decay products.[19,21] As described below, degradation of the transcript body in yeast is performed by the 5′ to 3′ *XRN1* gene exonuclease. Importantly, intermediates in the degradative process can be trapped by inserting strong RNA secondary structures, which block the exonuclease, into the mRNA.[19,21] Study of such decay intermediates (produced as a result of insertions) during a transcriptional pulse chase experiment showed that they accumulated only after the poly(A) tails of the full-length mRNAs were shortened to an oligo(A) length.[19,22,23] Moreover, these intermediates have short poly(A) tails even when the full-length mRNA consists of a mixture of short- and intermediate-length poly(A)-tailed species.[19] Similar results are seen with decay intermediates trapped by deletion of the *XRN1* gene, where only deadenylated transcripts are seen to be decapped.[22,24] These results argue that decay products are generated only from full-length mRNAs that have been deadenylated.[19]

The data described above suggest that deadenylation is a prerequisite for degradation of the body of many eukaryotic mRNAs. An important issue is at what point during deadenylation the transcript becomes a substrate for subsequent decay events. The data from experiments in yeast argue that shortening to an oligo(A) tail is sufficient for later decay and that the poly(A) tail does not need to be fully removed. For example, degradation of at least three of the four yeast mRNAs examined initiates once the poly(A) tail is shortened to the oligo(A) length (5 to 15 adenosine residues).[19] Moreover, the decay intermediates that are observed have oligo(A) tails.[19] Since *in vitro* studies have shown that the poly(A) binding protein (Pab1p) is unable to bind strongly to oligo(A) tracts shorter than 12 adenylate residues,[25] deadenylation to an oligo(A) tail could lead to the loss of the last Pab1p associated with the transcript and therefore may be functionally equivalent to complete removal of the poly(A) tail. Analysis of chimeric transcripts between the unstable c-fos and stable β-globin mRNAs suggests that deadenylation to a short poly(A) tail can be sufficient to initiate decay of mammalian mRNAs as well.[12] However, since the shortest poly(A) tails observed on these transcripts were approximately 25 to 60 A residues long, decay events following deadenylation may be triggered at a longer poly(A) tail length than in yeast.

Since the rate at which a mRNA is deadenylated contributes to its overall stability it is important to determine how different rates of deadenylation are controlled. In addition, some *cis*-acting sequences that stimulate mRNA turnover do so, at least in part, by accelerating mRNA deadenylation (see below). An important issue concerns how these sequences exert control over mRNA deadenylation, and what nucleases are the targets of this regulation.

14.2.2 Gene Products Involved In Deadenylation

One protein that affects deadenylation is the poly(A) binding protein, Pab1p, which binds to the 3′ poly(A) tail. The poly(A) binding protein has been cloned from several organisms and has a conserved organization consisting of four RNA binding domains and an extended c-terminus.[26–28] Experiments in yeast suggest that Pab1p is required for the proper rates of deadenylation. These experiments are slightly complicated since deletion of the *PAB1* gene is lethal.[26] Given this limitation, the role of Pab1p in deadenylation has been examined either in yeast strains that are conditional *pab1* mutants, or in strains carrying a suppressor mutation that allows growth of *pab1*Δ strains. In such *pab1* mutant strains there is an increase in the length of bulk cellular poly(A).[29] Moreover, the *PGK1* and *MFA2* mRNAs are deadenylated 3- and 6-fold slower, respectively, in *pab1*Δ mutants than in *PAB1* strains.[30] Two general types of models can explain the reduced rates of deadenylation observed in *pab1* mutant yeast strains. In one view, proteins bound to the poly(A) tail in the absence of Pab1p could inhibit poly(A) shortening. Alternatively, Pab1p could directly influence deadenylation by interacting with the nuclease(s) responsible for shortening of the poly(A) tail (see below).

In contrast to the results above, examination of the role of Pab1p in cell-free extracts from human cells reveals that presence of a 3′ poly(A) tail can protect added transcripts from 3′ to 5′ exonucleolytic decay presumably when bound to the poly(A) binding protein.[31,32] At the present time the difference between these results and the yeast experiments remains unclear. Although unlikely, it is possible that the role of Pab1p in deadenylation differs between organisms. Alternatively, the *in vitro* system may be evaluating the function of an alternative 3′ to 5′ exonuclease that does not normally perform poly(A) shortening *in vivo*. This issue will be resolved as the poly(A) nucleases are identified and characterized in these different organisms.

Several studies have been conducted with the goal of identifying the deadenylating enzyme. For example, a 3′ to 5′ poly(A) nuclease has been purified from yeast extracts.[33] Consistent with a role for Pab1p in stimulating deadenylation in yeast, this activity, termed PAN, is largely Pab1p-dependent and exhibits little or no activity in the absence of Pab1p in crude extracts or following further purification.[33,34] Mutations in the two subunits of this enzyme, referred to as PAN2 and PAN3, do lead to an overall increase in the length of poly(A) tails *in vivo* indicating that this activity does play a role somehow in deadenylation *in vivo*. Later results suggested that this enzyme acts at a very early stage of deadenylation, perhaps while still in the nucleus.[35] In any case, since the actual rate of poly(A) shortening appears to be normal in pan2Δ and pan3Δ mutants, this enzyme does not appear to be the major enzyme responsible for cytoplasmic deadenylation. Very recent studies have demonstrated that a complex of proteins containing Ccr4p and Caf1p forms a part of the mRNA deadenylating machinery in yeast. Rates of mRNA deadenylation and decay are severely decreased in ccr4Δ cells. Further, purified Ccv4p complex deadenylates synthetic mRNA substrates *in vitro*.[108]

Candidates for an enzyme involved in cytoplasmic deadenylation have also been described in extracts of mammalian cells. For example, several 3′ to 5′ exonucleases that degrade poly(A) with high preference have been either partially purified,[36,37] or wholly purified to a 74 kDa protein from HeLa cell extracts.[38] Substrates containing non-poly(A) sequences attached to the 3′ end of the poly(A) were poorly degraded by this activity. Similarly, poly(A) molecules with a cordycepin moiety at their 3′ ends were also poor substrates. Future experiments should reveal the potential roles of this activity in deadenylation *in vivo*.

14.2.3 Cis-Acting Sequences Affecting Deadenylation

Several *cis*-acting elements affecting the mRNA deadenylation rate have been identified in yeast and more complex eukaryotes. These sequences are often localized to the 3′ untranslated regions, but can also be found within coding regions. For example, the 3′ UTR of the yeast *MFA2* mRNA is both necessary and sufficient for rapid deadenylation.[19,20] Sequences promoting deadenylation have also been localized to the 3′ untranslated regions of mammalian transcripts. One class of these elements is a set of AU-rich elements, termed AREs, found in the 3′ UTR of the transcripts for several oncogenes and lymphokines (for a review, see Chen and Shyu).[17] These AREs are characterized by the presence of one or multiple copies of the pentanucleotide AUUUA and a high U content. By the analysis of chimeric transcripts, the AREs from many mRNAs, including the *c-fos*, GM-CSF, *c-myc* and *jun*B mRNAs, have been shown to have the ability to accelerate deadenylation and decay rates of otherwise stable β-globin mRNA.[16,39]

In order to understand how these AREs function to promote mRNA decay, several groups have identified proteins that interact with the AREs of various transcripts and therefore might modulate deadenylation rate.[40–45] However, only one of these proteins, termed AUBF, has been shown to stimulate decay in an *in vitro* system.[43] Progress in this area may be aided by the identification of proteins which bind to a small sequence UUAUUUA(U/A)(U/A) determined to be important for a functional ARE to stimulate mRNA decay.[46,47]

Examples of *cis*-acting sequences affecting deadenylation localized to the coding regions are found in the mammalian c-fos and yeast *MATα1* transcripts.[12,48] One of the striking similarities in these cases is that the function of both of these elements requires proper translation of the transcript. For example, in the yeast *MATα1* mRNA, the ability of a 32 nucleotide region to stimulate deadenylation and degradation is stimulated by a region of rare codons just 5′ to this 32 nucleotide element.[49,50] One interpretation of this observation is that the presence of a paused ribosome is required for this element to function.[50] Similarly, in order for the c-fos coding region to promote decay the mRNA needs to be translated.[51] Since this coding region element is known to be recognized as RNA,[52] there may be a similar requirement for translating ribosomes to affect the function of this destabilizing element.

There are several additional cases wherein translation of the mRNA affects poly(A) shortening. For example, deadenylation, but not degradation following deadenylation, of the unstable mammalian c-myc mRNA is slowed when cells are treated with translation inhibitors.[53] In contrast, the insertion of a strong hairpin into the 5′ UTR of the yeast *PGK1* mRNA inhibits translation and leads to an increase in the rate of deadenylation.[23] Similar results where inhibition of translation in *cis* accelerates deadenylation have also been seen with mammalian transcripts.[54] Although how translation and control of deadenylation rate interact is unclear, one hypothesis is based on observations suggesting an important functional interaction between the 5′ and 3′ termini of eukaryotic transcripts. These observations include the fact that the poly(A) tail/Pab1p complex serves as both an enhancer of translation and an inhibitor of mRNA decapping[30] and that sequences in the 3′ UTR of the *MFA2* mRNA can influence mRNA decay following deadenylation.[19,20] In this model a mRNP structure involving the 5′ and 3′ ends of a mRNA would control rates of both translation and deadenylation. This mRNP structure would in turn be influenced by signals received from specific elements within a mRNA and respond by changing rates of deadenylation and/or translation. This model is attractive because it provides an explanation for how sequences located in the coding region of an mRNA are able to influence nucleolytic events at the 3′ ends of an mRNA since the element need only affect a central mRNP structure rather than the actual nucleolytic events themselves. Ultimately, the identification of the specific *trans*-acting factors that mediate the effects of instability elements, and the determination of whether these factors directly or indirectly impact the deadenylation machinery will facilitate an understanding of how poly(A) shortening is controlled.

14.3 DEADENYLATION TRIGGERS DECAPPING AND 5′ TO 3′ DECAY

One important issue is how deadenylation leads to the mRNA degradation. In the absence of any strong experimental data it had been assumed that the loss of the poly(A) tail, and the associated poly(A) binding protein, would lead to degradation of a transcript by exposing the 3′ end to non-specific nucleases.[6] However, evidence has now accumulated in yeast indicating that the major effect of deadenylation is to lead to a decapping reaction that exposes the transcript to 5′ to 3′ exonucleolytic decay.[22,23] One of the key approaches used to determine the events following deadenylation was to insert strong RNA secondary structures, such as a poly(G) tract, which can form a very stable structure,[55] into the mRNA to block processive exonucleases.[19,21] In these experiments, mRNA fragments appear that are trimmed at their 5′ end to the site of the RNA secondary structure, even when the site of insertion is near the 5′ end of the transcript.[22,23] Since these fragments only appear after deadenylation, this observation suggests that poly(A) shortening triggers a cleavage event at the extreme 5′ end. Additional evidence for this conclusion came from the analysis of yeast strains lacking the major cytoplasmic 5′ to 3′ exonuclease, encoded by the *XRN1* gene.[56] In *xrn1*Δ strains, decapped forms of both the *MFA2* and *PGK1* transcripts accumulate following deadenylation.[22,23] Since many 5′ to 3′ exonucleases, including the product of the *XRN1* gene, are blocked by the cap structure located at the 5′ end of mRNAs,[56,57] cleavage of the cap linkage or one of the first few phosphodiester bonds would expose the transcript to 5′ to 3′ exonucleolytic degradation.

Experimental evidence suggests that the 5′ cleavage reaction is likely to be within the cap linkage itself and should be considered a decapping reaction. This possibility was first suggested by the analysis of the decapped transcripts that accumulate in *xrn1*Δ strains. For example, the decapped *MFA2* transcripts that accumulate in *xrn1*Δ strains appear to be full length as judged by primer extension analysis.[22] Consistent with the reaction being cleavage of the cap linkage, an enzymatic activity capable of releasing m7GDP from the 5′ end of capped transcripts had been identified.[58] Evidence that this activity is actually required for mRNA decapping *in vivo* has come from the identification of a gene, termed *DCP1*, encoding the catalytic component of this decapping enzyme.[59,60] Strains deleted for the *DCP1* gene are unable to complete mRNA decapping *in vivo* resulting in increased stability of several mRNAs tested and show no decapping activity in cell-free extracts.[59] These results suggest that the nucleolytic event that exposes the mRNA to 5′ to 3′ decay is cleavage of the cap linkage releasing m7GDP and the mRNA body with a 5′ phosphate. Interestingly, the *PGK1* and *RP51A* transcripts accumulate in *xrn1*Δ strains as decapped species shortened at the 5′ end by a few nucleotides.[23,24] Although these species could arise by cleavage of one of the first few phosphodiester bonds, it is more likely that they arise by decapping followed by a 5′ to 3′ exonucleolytic nibbling that utilizes a nuclease different from the Xrn1p. This is supported by the observations which suggest that there is a second less efficient 5′ to 3′ exonuclease in yeast[23] and that decapping of the *PGK1* mRNA requires the Dcp1p *in vivo* and occurs at the same site as the MFA2 transcript *in vitro*.[59,60]

These observations define a decay pathway common to stable and unstable mRNAs in yeast in which deadenylation triggers decapping, thereby leading to 5′ to 3′ decay of the transcript. The available evidence suggests that this pathway acts on most, if not all, yeast mRNAs. For example, it has been shown that the *MAT*α*1* mRNA is also degraded by this same mechanism.[48] Moreover in strains lacking the decapping enzyme (*dcp1*Δ) several yeast mRNAs including the unstable *MFA2*, *STE2*, *GAl10*, and *HIS3* mRNAs as well as the stable *ACT1* and *PGK1* transcripts are stabilized.[59] Similarly, in cells lacking Xrn1p, full-length deadenylated forms of the *Rp51A*, *CYC1*, *ACT1*, *MF*α*1* and *MAT*α*1* transcripts accumulate.[24,48] The deadenylated forms of several of these mRNAs (*ACT1*, *RP51A* and *CYC1*) are degradable by purified Xrn1p suggesting that they lack a 5′ cap.[24] Lastly, Xrn1p is very abundant and primarily cytoplasmic.[61] These observations are consistent with the hypothesis that deadenylation triggers decapping of many yeast transcripts.

Since poly(A) tails and cap structures are common features of eukaryotic transcripts an appealing model is that mRNA decay by deadenylation-dependent decapping and 5′ to 3′ digestion could be a conserved mechanism of mRNA turnover. Several observations suggest that decapping and 5′ to 3′ decay may occur in other eukaryotes. For example, mRNAs lacking the cap structure are rapidly degraded in many eukaryotic cells.[62] In addition, enzymes that could catalyze the removal of the cap structure and subsequent 5′ to 3′ degradation of the transcript have been described in mammalian cells.[63] More recently, decay intermediates consistent with mRNA decapping have been observed in several other eukaryotic cell types. For example, full-length mRNAs that are devoid of both the cap and most of the poly(A) tail, have been detected from murine liver cells.[64] Similarly, mRNA decay intermediates that are shortened at their 5′ ends have been identified in both plant and animal cells.[65,66] Furthermore, homologs of the yeast Xrn1p 5′ to 3′ exonuclease that degrade mRNA following decapping have been identified in several organisms.[61,67] Taken together, these observations suggest that deadenylation-dependent decapping followed by 5′ to 3′ exonucleolytic decay is a conserved mRNA decay mechanism in eukaryotic cells.

How would deadenylation stimulate decapping of the mRNA? One possibility is that decapping is an indirect result of deadenylation. For example, since the poly(A) tail of mRNAs can associate with the cytoskeleton,[68,69] poly(A) tail shortening could alter a mRNA's subcellular localization and thereby expose the mRNA to a decapping. An alternative model is that the poly(A) tail and the poly(A) binding protein might inhibit the decapping reaction directly by forming, or stabilizing, an mRNP structure involving the 5′ and 3′ termini of the mRNA. Such an mRNP structure has been suggested previously based on the stimulation of translational initiation by poly(A) tails,[29,70–72] and on the presence of circular polysome structures in electron micrographs.[73] Such an interaction is also supported by observations that suggest that base-pairing between the 5′ and 3′ UTR sequences can occur *in vivo*.[74] More recent results have provided insight into how such an interaction could form. This is based on the observation that the poly(A) binding protein, presumably bound to the 3′ poly(A) tail can associate with the eIF-4G component of the cap binding complex found at the 5′ end of the transcript.[75] One simple model is that the same mRNP structure that stimulates translational initiation could inhibit decapping, perhaps by efficiently recruiting translation initiation factors to the 5′ UTR and cap structure.

14.4 DEADENYLATION INDEPENDENT DECAPPING OF mRNAs

In some cases mRNA decay is brought about by decapping and 5′ to 3′ decay of the transcript independent of poly(A) shortening. To date the only definitive example of this decay mechanism has come from the analysis of yeast transcripts that contain an early nonsense codon (Muhlrad and Parker).[76] The degradation of mRNAs with nonsense codons is part of a conserved process, termed mRNA surveillance, which promotes the rapid turnover of aberrant mRNAs. These aberrant transcripts contain early nonsense codons,[77–80] unspliced introns[81] or have extended 3′ UTRs.[80] However, it should be expected that some other features of transcripts that promote rapid decay will do so by stimulating deadenylation-independent decapping.

mRNA surveillance may exist, in part, to increase the fidelity of gene expression by degrading aberrant mRNAs that if translated would produce truncated proteins. This process would be relevant since truncated proteins often have dominant negative phenotypes. A striking example of this phenomenon is seen in *C. elegans* where loss of function mutations in the *SMG* genes, which are required for mRNA surveillance, convert recessive nonsense mutations in the myosin gene unc-54 into dominant negatives.[80] Since mRNA surveillance degrades unspliced and aberrantly processed transcripts, this decay mechanism might be most important in organisms with a large number of introns where processing errors might lead to aberrant transcripts giving rise to truncated proteins.

The evidence that premature nonsense codons trigger deadenylation-independent decapping consists of the following observations.[76] First, in transcriptional pulse-chase experiments in yeast, a *PGK1* transcript with an early nonsense codon starts decaying without any detectable temporal lag, in contrast to the wild-type transcript which only decays after a lag period during which deadenylation occurs.[19] Second, decay products trapped by the insertion of a poly(G) tract in the 3′ UTR of the nonsense codon containing transcripts have long poly(A) tails. Third, full-length decapped *PGK1* transcripts containing nonsense codons are stabilized in *xrn1*Δ strains. Finally, in strains lacking the decapping enzyme encoded by the *DCP1* gene, transcripts with early nonsense codons are greatly stabilized.[59] These results suggest that the mechanism of degradation of these transcripts converges with the normal deadenylation-dependent decapping pathway at the nucleolytic step of decapping.

Two observations raise the possibility that nonsense codons might trigger deadenylation-independent decapping in mammalian cells. First, introduction of a nonsense mutation into the β-globin mRNA accelerates its decay rate more than 10-fold, the deadenylation rate almost remains unaffected. As a result, in transcriptional pulse-chase experiments, even at a time when 90% of the mutant mRNA is degraded the poly(A) tails are as long as >50 nucleotides, therefore suggesting that decay is independent of poly(A) shortening.[12] In addition, examination of the turnover of a β-globin transcript with an early nonsense codon in murine erythroid tissue revealed the presence of capped decay products that were shortened at the 5′ end.[82] Although the mechanism by which these decay products receive a 5′ cap is unclear, the presence of such products imply that decay is proceeding in a 5′ to 3′ direction for these transcripts.

14.5 CONTROL OF mRNA DECAPPING

As described in previous sections, for the body of the mRNA to be degraded by 5′ to 3′ exonucleolytic process, the mRNA needs to be decapped. Decapping appears to be a central point at which many of the inputs that control mRNA decay rates converge. For example, at least in yeast, the deadenylation rate of the mRNA affects overall transcript stability mainly by virtue of the ability of poly(A) tail to inhibit decapping.[22,23] In addition, specific sequences that modulate mRNA decay rate can do so by affecting the rates of decapping.[20] Finally, as discussed above, the rapid decay of transcripts with early nonsense codons works by triggering extremely rapid mRNA decapping independently of poly(A) tail shortening.[76]

Recent results have demonstrated that in yeast cells a single decapping activity, dependent on, or encoded by, the *DCP1* gene, is responsible for the decapping of stable and unstable mRNAs, including those that are decapped independently of deadenylation. This observation suggests that features of mRNAs that alter the rates of decapping will do so by affecting the activity of the Dcp1p on individual transcripts. As discussed below, current evidence suggests that Dcp1p activity on a transcript will be modulated by proteins that recognize specific mRNA features, or sequences, and thereby affect the rate of Dcp1p cleavage.

The observation that decapping does not normally occur until the poly(A) tail is shortened to an oligo(A) length strongly suggests that the poly(A) tail inhibits this process. The inhibition of the decapping reaction by the poly(A) tail in yeast is mediated by the *PAB1* gene product. This conclusion is supported by the observation that in *pab1* mutant strains, mRNA molecules are decapped independently of deadenylation.[30] This suggests a model in which the shortening of the poly(A) tail to oligo(A) length triggers decapping by promoting the dissociation of the last Pab1p molecule.

In addition to the poly(A) tail, there are specific sequences within mRNAs that can affect decapping rates of mRNA. For example, the unstable *MFA2* mRNA ($t_{1/2} = 4'$) is rapidly decapped following deadenylation while the stable *PGK1* transcript ($t_{1/2} = 35'$) is decapped slowly after deadenylation.[19,22,23] These differences in decapping rates are at least partially determined by specific sequences within these mRNAs,[20,23] although these sequences are currently poorly defined. Similarly, as discussed above, the presence of an early nonsense codon can trigger rapid decapping.[76,83]

In each case where decapping rate responds to some mRNA feature, work has identified *trans*-acting factors that are required for the modulation of decapping rate. For example, as discussed above, the inhibition of decapping by the poly(A) tail requires the associated poly(A) binding protein[30] to communicate the presence of a 3′ poly(A) tail. Similarly, the deadenylation-independent decapping caused by premature translational termination requires the *UPF1*, *UPF2*, and *UPF3* gene products to signal that termination is premature, thereby triggering rapid decapping.[78,84,85] In a similar manner, recent work has identified lesions in several genes including *LSM1* thru' *LSM1*, *PAT1* and *MRT3*, that appear to modulate the rates of decapping on substrates after deadenylation and therefore may function analogously to the *UPF* gene products.[86,109] Moreover, some dcp1 mutants have also been identified that are defective in deadenylation-dependent mRNA decapping *in vivo* but yield decapping enzymes exhibiting wild-type-specific activity *in vitro*[87] which again supports the idea that decapping may be regulated *in vivo*.

There are two general types of hypotheses to explain how the decapping rate will be modulated on individual transcripts, either by the poly(A) tail or by specific sequences. In one possibility, the decapping rate would be modulated by affecting the interaction, or recruitment, of the decapping enzyme with each mRNA. The central feature of this model is that the decapping enzyme itself would be the sight of regulatory inputs. An alternative is that the modulation of decapping activity on individual mRNAs occurs by alterations in mRNP structure that affect the decapping rate. For example, all of the features that modulate the decapping rate could be envisioned to affect the interaction of proteins, such as translation initiation factors, with the 5′ cap structure thereby sterically competing with the decapping activity. Such a model is appealing for two reasons. First, it would provide a common explanation for how the poly(A) tail could affect both translation initiation rates and decapping. In addition, such a model would provide a basis for the observation that the translation and turnover of eukaryotic mRNAs are often mechanistically coupled.

14.6 DEADENYLATION CAN ALSO LEAD TO 3′ TO 5′ DEGRADATION

Eukaryotic transcripts can also be degraded in a 3′ to 5′ direction following deadenylation. Evidence for this mechanism of degradation has primarily come from examining mRNA turnover in yeast when the 5′ to 3′ decay pathway is inhibited. For example, fragments of the yeast *PGK1* mRNA shortened at the 3′ end are observed when the 5′ to 3′ decay pathway is inhibited by deletion of either the *XRN1* or *DCP1* genes.[23,59,76] Similarly, mRNA fragments with poly(G) tracts to block the 5′ to 3′ exonuclease are ultimately degraded 3′ to 5′, although at a relatively slow rate.[88] Interestingly, in wild-type yeast, the 3′ trimmed fragments of the *PGK1* mRNA can also be observed at low levels. These observations argue that the 3′ to 5′ degradation is generally a slower process than 5′ to 3′ degradation. Decay intermediates that are consistent with 3′ to 5′ decay following deadenylation have also been observed for the oat phytochrome A mRNA *in vivo*.[65] This observation, and the conservation of gene products required for this decay pathway (see below), suggest that 3′ to 5′ degradation of eukaryotic mRNAs is a conserved process.

In yeast, several observations now indicate that 3′ to 5′ nucleolytic degradation of the transcript body is a general mechanism of mRNA turnover capable of acting on many mRNAs. This evidence includes the analysis of mRNA decay when decapping and 5′ to 3′ degradation are blocked, which indicated that both the stable *PGK1* mRNA,[23] and the unstable *MFA2* mRNA were degraded 3′ to 5′.[88] Moreover, in strains defective for both 5′ to 3′ and 3′ to 5′ decay mechanisms several mRNAs are extremely stable, indicating that these mRNAs are all substrates for 3′ to 5′ degradation.[88] These observations suggest that there are two general mechanisms for degrading the mRNA body following poly(A) shortening in yeast, decapping leading to 5′ to 3′ degradation, or 3′ to 5′ degradation. Moreover, since transcripts were extremely long-lived in the absence of these two mechanisms of degradation,[88] there are unlikely to be other *major* nucleolytic activities that can act to degrade mRNAs at a reasonable rate.

The generality of the 3′ to 5′ decay mechanism is strengthened by the observation that any combination of mutations that inactivate the 5′ to 3′ decay pathway with mutations that inactivate the 3′ to 5′ pathway leads to inviability.[88] This observation argues that efficient mRNA degradation, by either one of these pathways, is essential for viability. Thus, mRNA turnover is a redundant and essential process in yeast. Moreover, several of the proteins required for these mRNA decay mechanisms (see below), including *XRN1*, *SKI2*, *SKI6/RRP41*, and *RRP4* have homologs in other eukaryotic cells, including mammals.[67,89–91] The existence of these homologs argues that these pathways of mRNA turnover occur in all eukaryotic cells and are likely to be the two general mechanisms of mRNA decay.

An interesting issue is the relationship between 5′ to 3′ degradation and 3′ to 5′ degradation of mRNA and their respective roles in eukaryotic cells. Currently, the available evidence suggests that the major mechanism of mRNA decay in yeast is by decapping and 5′ to 3′ degradation.[19,22–24,59] However, the 3′ to 5′ mechanism of degradation is likely to have unique functions. For example, it is likely that particular mRNAs are preferentially degraded by the 3′ to 5′ mechanism even in wild-type cells. Similarly, since the pathways of mRNA degradation have only been examined under an extremely limited set of growth conditions, there may be specific conditions where the 3′ to 5′ mechanism is primary. In addition, the 3′ to 5′ mechanism of degradation may play an antiviral role by reducing expression from viral poly(A)$^-$ transcripts.[92] Finally, it should be considered that in other eukaryotes the relative importance of these two mechanisms may be different. For example, in oat seedlings the phytochrome A mRNA is degraded by both 5′ to 3′ and 3′ to 5′ decay mechanisms at similar rates.[65]

Several proteins have been identified that are required for the 3′ to 5′ degradation of mRNAs.[88] Additionally the components of a multiprotein complex termed the exosome, which is known to be required for rRNA processing[91] are also involved. Three of the proteins from this complex (Rrp4p, Ski6/Rrp41p, and Rrp44p) have been shown to have 3′ to 5′ exoribonuclease activity *in vitro*, and the remaining two (Rrp42 and Rrp43) have sequence similarity to bacterial 3′ to 5′ exonucleases.[91] While the complex is known to be required for proper 5.8S rRNA processing in the nucleus in yeast, a homologous complex in HeLa cells is also found in the cytoplasm,[91] suggesting a potential role in 3′ to 5′ mRNA degradation. The essential observation suggesting that the exosome is involved in 3′ to 5′ degradation of mRNA is that mutations in components of the exosome lead to defects in 3′ to 5′ mRNA degradation. This observation has led to the working hypothesis that the exosome is the degradative activity that acts on mRNAs in a 3′ to 5′ direction.[88]

In addition to the exosome three other factors are known to be required for 3′ to 5′ degradation. These are the products of the *SKI2*, *SKI3*, and *SKI8* genes.[88] Although the exact role of these factors in mRNA decay is not yet clear, mutations in these genes have other phenotypes that are consistent with a defect in 3′ to 5′ degradation of poly(A)$^-$ mRNAs. For example, these mutants lead to an overexpression of the mRNAs from the double-stranded RNA killer virus, which could be due to a stabilization of the poly(A)$^-$ viral mRNAs. In addition, poly(A)$^-$ mRNAs introduced into yeast by electroporation showed a longer functional stability in *ski2*Δ and *ski8*Δ strains as compared to wild-type.[92] Interestingly, the *ski2*Δ and *ski8*Δ strains also showed increased initial rates of protein production from electroporated poly(A)$^-$ transcripts.[92] This observation was interpreted to indicate that these proteins function to repress translation of poly(A)$^-$ mRNAs due to an alteration in the biogenesis of the 60S ribosomal subunit and that the longer functional mRNA stability was a consequence of differences in translation rates.[92] However, several observations now suggest that the *SKI2*, *SKI3* and *SKI8* proteins affect 3′ to 5′ mRNA degradation more directly. First, polysome profiles in *ski2*Δ, *ski3*Δ, and *ski8*Δ mutants are identical to wild-type[92] and examination of 5.8S processing indicated that at least this aspect of rRNA processing was normal in these mutants.[88] Second, since the *ski2*, *ski3*, *ski6/rrp41*, *ski8*, and *rrp4* mutants affected the 3′ to 5′ degradation of a mRNA fragment, which was *not* being translated, it is unlikely that an increase in translation rate in the mutant strains could be indirectly protecting the RNA from 3′ to 5′ degradation. Given this, there are two possible explanations for the results with electroporated mRNAs. First, if there is a competition between 3′ to 5′ degradation and translation initiation for electroporated mRNAs, when the RNAs are first introduced into cells, more transcripts would be getting translated, but at the same initiation rate, in the mutant strains. Alternatively, the *SKI2*, *SKI3*, and *SKI8* proteins might function in remodeling mRNP structure, perhaps by promoting the disassociation of proteins from the 3′ UTR, which might decrease the translation rate and also make the 3′ end more accessible to the exosome.

A simpler model is that the *SKI2*, *SKI3* and *SKI8* proteins function to adapt, or recruit, the exosome to mRNA substrates. This is a particularly appealing model for the Ski2p, which is a member of the DEVH box family of proteins and thus a putative RNA helicase, because some 3′ to 5′ exonuclease complexes have been shown to have associated RNA helicases of this type.[93,94] In this view, other proteins would serve as "adapters" for other exosome substrates, such as the 5.8S pre-rRNA. This hypothesis would explain why the *ski2*Δ, *ski3*Δ, and *ski8*Δ mutations do not affect 5.8S processing.[88] Strikingly, mutations in another DEVH protein closely related to Ski2p, Dob1p, show a defect in processing of the 5.8S pre-rRNA.[95] This observation suggests that the Dob1p might serve as the exosome "adapter" for 5.8S pre-rRNA. This view makes the testable predictions that Ski2p, and perhaps Ski3p and Ski8p, will show interactions with the exosome and will directly affect its ability to degrade mRNA substrates.

14.7 ROLE OF ENDONUCLEOLYTIC CLEAVAGES IN THE DECAY OF EUKARYOTIC mRNAs

There are now several examples of eukaryotic mRNAs whose degradation includes a sequence-specific endonucleolytic cleavage. Evidence for this mechanism comes from the analysis of the several mRNAs including the mammalian 9E3, IGFII, transferrin receptor (TfR) apoII and albumin mRNAs and *Xenopus* Xlhbox2B mRNA. In these cases, mRNA fragments are detected *in vivo* that correspond to the 5′ and 3′ portion of the transcript and are consistent with endonucleolytic cleavages.[96–101] In addition, endonucleolytic cleavages have been defined *in vitro* for the albumin mRNA[102] and in the coding region of the c-myc mRNA.[103] One interesting feature of several of these endonuclease cleavage sites is that there are multiple cleavages within a limited region of the transcript. For example, in the mouse albumin mRNA, *Xenopus* homeobox mRNA and chicken liver apoII mRNA several endonucleolytic cleavages have been observed in a small region and the various cleavage sites were found to have some sequence homology.[99–101,104] However, the cleavage site consensus sequences observed in each of these cases is different. This suggests that there may be a wide variety of endonucleases with different cleavage specificities. However, it is also possible that the specificity of a common endonuclease could be modified by association with different specificity determining factors.

Since sequence-specific endonuclease target sites are likely to be limited to individual mRNAs or classes of mRNAs, their presence allows for specific control of the decay rate of these transcripts. In some cases, the rate of endonucleolytic cleavage is modulated by the activity of protective factors that bind at, or near, the cleavage site and compete with the endonuclease.[98,103,105] For example, the binding of the iron response element binding protein in the TfR 3′ UTR in response to low intracellular iron concentrations inhibits the endonucleolytic cleavage of this mRNA.[98] Therefore, some endonucleases may be constitutively active and the accessibility of the cleavage site is regulated.

In other cases, the endonuclease activity may be directly regulated. For example, the mammalian endonuclease RNase L is normally inactive and is only activated by oligomers of 2′, 5′ phosphodiester bonded adenylate residues, which are produced in response to the presence of double stranded RNA.[106] Although RNase L is important in mediating interferon responses[107] it has yet to be established whether this enzyme normally degrades any cellular mRNAs.

Given the role of deadenylation in mRNA turnover, it has been of interest to determine whether or not these endonuclease cleavage reactions are dependent on prior poly(A) shortening. At least in some cases it is clear that deadenylation to an oligo(A) tail length is not required for endonucleolytic cleavage since the 3′ fragment of the 9E3 and TfR mRNAs is polyadenylated.[96,98] In addition, the cleavage of the Xlhbox2B mRNA is not affected by the adenylation state of the mRNA.[105] However, it is possible that endonucleolytic cleavage of some mRNAs could be dependent on the length of the poly(A) tail. For example, in the case of the mouse liver albumin mRNA, the downstream degradation intermediates resulting from the *in vivo* endonucleolytic cleavages occurring in the middle region of mRNA are enriched in the poly(A)$^-$ fraction but present in very low or undetectable levels in the poly(A)$^+$ fraction.[101] Similarly, in the case of chicken apoII mRNA, the downstream fragments were also found to be poly(A)$^-$.[100] These results raise the possibility that these cleavages could be deadenylation dependent.

14.8 SUMMARY

It is now clear that there is a diversity of different pathways for mRNA turnover in eukaryotic cells. One simple integrated view is that all polyadenylated mRNAs would be degraded by the deadenylation-dependent pathway at some rate. For different transcripts, or potentially in different cell types, the relative importance of 5′ to 3′ and 3′ to 5′ decay may differ. In addition to this deadenylation-dependent "default" pathway, another layer of complexity would come from degradation mechanisms specific to individual mRNAs, or classes of mRNAs. Such mRNA-specific mechanisms would include sequence-specific endonuclease cleavage and deadenylation-independent decapping. Thus, the overall decay rate of an individual transcript will be a function of its susceptibility to these turnover pathways. In addition, *cis*-acting sequences that specify mRNA decay rate, and regulatory inputs that control mRNA turnover, are likely to affect all the steps of these decay pathways. Future work identifying and characterizing the gene products that catalyze and modulate the rates of these nucleolytic steps should provide a basis for the understanding of differential mRNA decay.

14.9 REFERENCES

1. V.E. Velculescu, L. Zhang, W. Zhou, J. Vogelstein, M.A. Basrai, D.E. Bassett Jr., P. Hieter, B. Vogelstein and K.W. Kinzler, *Cell*, 1997, **88**, 243.
2. P. Alifano, C.B. Bruni and M.S. Carlomagno, *Genetica*, 1994, **94**, 157.
3. S.N. Cohen, *Cell*, 1995, **80**, 829.
4. G. Caponigro and R. Parker, *Microbiol. Rev.*, 1996, **60**, 233.
5. D. Sheiness and J.E. Darnell, *Nat. New Biol.*, 1973, **241**, 265.
6. S.W. Peltz, G. Brewer, P. Bernstein, P. Hart and J. Ross, *Crit. Rev. Euk. Gene Exp.*, 1991, **1**, 99.
7. M.C. Wilson, S.G. Sawicki, P.A. White and J.E. Darnell, *J. Mol. Biol.*, 1978, **126**, 23.
8. I. Paek and R. Axel, *Mol. Cell. Biol.*, 1987, **7**, 1496.

9. A.W. Cochrane and R.G. Deeley, *J. Mol. Biol.*, 1988, **203**, 555.
10. P. Bernstein and J. Ross, *Trends Biochem. Sci.*, 1989, **14**, 373.
11. M. Zeevi, J.R. Nevins and J.E. Darnell, *Mol. Cell. Biol.*, 1982, **2**, 517.
12. A.B. Shyu, J.G. Belasco and M.E. Greenberg, *Genes Dev.*, 1991, **5**, 221–234.
13. T. Wilson and R. Treisman, *EMBO J.*, 1988, **7**, 4193.
14. D. Caput, B. Beutler, K. Hartog, R. Thayer, S. Brown-Shimer and A. Cerami, *Proc. Natl. Acad. Sci. USA*, 1986, **83**, 1670.
15. G. Brewer and J. Ross, *Mol. Cell. Biol.*, 1988, **8**, 1697.
16. C.-Y. Chen and A.-B. Shyu, *Mol. Cell. Biol.*, 1994, **14**, 8471.
17. C.-Y. Chen and A.-B. Shyu, *Trends Biochem. Sci.*, 1995, **20**, 465.
18. M. Johnston, *Microbiol. Rev.*, 1987, **51**, 458.
19. C.J. Decker and R. Parker, *Genes Dev.*, 1993, **7**, 1632.
20. D. Muhlrad and R. Parker, *Genes Dev.*, 1992, **6**, 2100.
21. P. Vreken and H.A. Raue', *Mol. Cell. Biol.*, 1992, **12**, 2986.
22. D. Muhlrad, C.J. Decker and R. Parker, *Genes Dev.*, 1994, **8**, 855.
23. D. Muhlrad, C.J. Decker and R. Parker, *Mol. Cell. Biol.*, 1995, **15**, 2145.
24. C.L. Hsu and A. Stevens, *Mol. Cell. Biol.*, 1993, **13**, 4826.
25. A.B. Sachs, R.W. Davis and R.D. Kornberg, *Mol. Cell. Biol.*, 1987, **7**, 3268.
26. A.B. Sachs, M.W. Bond and R.D. Kornberg, *Cell*, 1986, **45**, 827.
27. S.A. Adam, T. Nakagawa, M.S. Swanson, T.K. Woodruff and G. Dreyfuss, *Mol. Cell. Biol.*, 1986, **6**, 2932.
28. T. Grange, C. Martin de Sa, J. Oddos and R. Pictet, *Nucleic Acids Res.*, 1987, **15**, 4771.
29. A.B. Sachs and R.W. Davis, *Cell*, 1989, **58**, 857.
30. G. Caponigro and R. Parker, *Genes Dev.*, 1995, **9**, 2421.
31. P. Bernstein, S.W. Peltz and J. Ross, *Mol. Cell. Biol.*, 1989, **9**, 659.
32. L.P. Ford, P.S. Bagga and J. Wilusz, *Mol. Cell. Biol.*, 1997, **17**, 398.
33. A. Sachs and J. Deardorff, *Cell*, 1992, **70**, 961.
34. J. Lowell, D. Rudner and A. Sachs, *Genes Dev.*, 1992, **6**, 2088.
35. C.E. Brown and A.B. Sachs, *Mol. Cell. Biol.*, , 1998, **18**, 6548.
36. J. Astrom, A. Astrom and A. Virtanen, *J. Biol. Chem.*, 1992, **267**, 18154.
37. J. Astrom, A. Astrom and A. Virtanen, *EMBO J.*, 1991, **10**, 3067.
38. C.G. Korner and E. Wahle, *J. Biol. Chem.*, 1997, **272**, 10448.
39. C.-Y. Chen, N. Xu and A.-B. Shyu, *Mol. Cell. Biol.*, 1995, **15**, 5777.
40. J.S. Malter, *Science*, 1989, **246**, 664.
41. P.R. Bohjanen, B. Petryniak, C.H. June, C.B. Thompson and T. Lindsten, *Mol. Cell. Biol.*, 1991, **11**, 3288.
42. D.A. Katz, N.G. Theodorakis, D.W. Cleveland, T. Lindsten and C.B. Thompson, *Nucleic Acids Res.*, 1994, **22**, 238.
43. G. Brewer, *Mol. Cell. Biol.*, 1991, **11**, 2460.
44. C.-Y. Chen, Y. You and A.-B. Shyu, *Mol. Cell. Biol.*, 1992, **12**, 5748.
45. E. Vakalopoulou, J. Schaack and T. Shenk, *Mol. Cell. Biol.*, 1991, **11**, 3355.
46. C.A. Lagnado, C.Y. Brown and G.J. Goodall, *Mol. Cell. Biol.*, 1994, **14**, 7984.
47. A.M. Zubiaga, J.G. Belasco and M.E. Greenberg, *Mol. Cell. Biol.*, 1995, **15**, 2219.
48. G. Caponigro and R. Parker, *Nucleic Acids Res.*, 1996, **24**, 4304.
49. R. Parker and A. Jacobson, *Proc. Natl. Acad. Sci. USA*, 1990, **87**, 2780.
50. G. Caponigro, D. Muhlrad and R. Parker, *Mol. Cell. Biol.*, 1993, **13**, 5141.
51. S.C. Schiavi, C.L. Wellington, A.-B. Shyu, A.-Y.A. Chen, M.E. Greenberg and J.G. Belasco, *J. Biol. Chem.*, 1994, **269**, 3441.
52. C.L. Wellington, M.E. Greenberg and J.G. Belasco, *Mol. Cell. Biol.*, 1993, **13**, 5034.
53. I.A. Laird-Offringa, C.L. de Wit, P. Elfferich and A.J. van der Eb, *Mol. Cell. Biol.*, 1990, **10**, 6132.
54. M. Muckenthaler, N. Gunkel, R. Stripecke and M.W. Hentze, *RNA*, 1997, **3**, 983.
55. J.R. Williamson, M.K. Raghuraman and T.R. Cech, *Cell*, 1989, **59**, 871.
56. F.W. Larimer and A. Stevens, *Gene*, 1990, **95**, 85.
57. A. Stevens, *Biochem. Biophys. Res. Comm.*, 1978, **81**, 656.
58. A. Stevens, *Mol. Cell. Biol.*, 1988, **8**, 2005.
59. C.A. Beelman, A. Stevens, G. Caponigro, T.E. La Grandeur, L. Hatfield, D. Fortner and R. Parker, *Nature*, 1996, **382**, 642.
60. T.E. LaGrandeur and R. Parker, *EMBO J.*, 1998, **17**, 1487.
61. W.D. Heyer, A.W. Johnson, U. Reinhart and R.D. Kolodner, *Mol. Cell. Biol.*, 1995, **15**, 2728.
62. D.R. Drummond, J. Armstrong and A. Colman, *Nucleic Acids Res.*, 1985, **13**, 7375.
63. M. Coutts and G. Brawerman, *Biochim. Biophys. Acta*, 1993, **1173**, 57.
64. P. Couttet, M. Fromont-Racine, D. Steel, R. Pictet and T. Grange, *Proc. Natl. Acad. Sci. USA*, 1997, **94**, 5628.
65. D. Higgs and J. Colbert, *Plant Cell*, 1994, **6**, 1007.
66. S.-K. Lim and L.E. Maquat, *EMBO J.*, 1992, **11**, 3271.
67. V.I. Bashkirov, H. Scherthan, J.A. Solinger, J.-M. Buerstedde and W.-D. Heyer, *J. Cell Biol.*, 1997, **136**, 761.
68. R. Lenk, L. Ransom, Y. Kaufmann and S. Penman, *Cell*, 1977, **10**, 67.
69. K.L. Taneja, L.M. Lifshitz, F.S. Fay and R.H. Singer, *J. Cell Biol.*, 1992, **119**, 1245.
70. M.T. Doel and N.H. Carey, *Cell*, 1976, **8**, 51.
71. G. Galili, E. Kawata, L.D. Smith and F.A. Larkins, *J. Biol. Chem.*, 1988, **263**, 5764.
72. D. Munroe and A. Jacobson, *Mol. Cell. Biol.*, 1990, **10**, 3441.
73. A.K. Christensen, L.E. Kahn and C.M. Bourne, *Am. J. Anat.*, 1987, **178**, 1.

74. J.J. Van den Heuvel, R.J. Planta and H.A. Raue, *Yeast*, 1990, **6**, 473.
75. S.Z. Tarun Jr. and A.B. Sachs, *EMBO J.*, 1996, **15**, 7168.
76. D. Muhlrad and R. Parker, *Nature*, 1994, **370**, 578.
77. R. Losson and F. Lacroute, *Proc. Natl. Acad. Sci. USA*, 1979, **76**, 5134.
78. P. Leeds, S.W. Peltz, A. Jacobson and M.R. Culbertson, *Genes Dev.*, 1991, **5**, 2303.
79. S.W. Peltz, A.H. Brown and A. Jacobson, *Genes Dev.*, 1993, **7**, 1737.
80. R. Pulak and P. Anderson, *Genes Dev.*, 1885, **1993**, 7.
81. F. He, S.W. Peltz, J.L. Donahue, M. Rosbash and A. Jacobson, *Proc. Natl. Acad. Sci. USA*, 1993, **90**, 7034.
82. S.-K. Lim, C.D. Sigmund, K.W. Gross and L.E. Maquat, *Mol. Cell. Biol.*, 1992, **12**, 1149.
83. K.W. Hagan, M.J. Ruiz-Echevarria, Y. Quan and S.W. Peltz, *Mol. Cell. Biol.*, 1995, **15**, 809.
84. B.-S. Lee and M.R. Culbertson, *Proc. Natl. Acad. Sci. USA*, 1995, **92**, 10354.
85. P. Leeds, J.M. Wood, B.-S. Lee and M.R. Culbertson, *Mol. Cell. Biol.*, 1992, **12**, 2165.
86. L.K. Hatfield, C.A. Beelman, A. Stevens and R. Parker, *Mol. Cell. Biol.*, 1996, **16**, 5830.
87. S. Tharun and R. Parker, *Genetics*, 1999, **151**, 1273.
88. J.S. Jacobs Anderson and R. Parker, *EMBO J.*, 1998, **17**, 1497.
89. S.G. Lee, I. Lee, S.H. Park, C. Kang and K. Song, *Genomics*, 1995, **10**, 660.
90. A.W. Dangel, L. Shen, A.R. Mendoza, L.C. Wu and C.Y. Yu, *Nucleic Acids Res.*, 1995, **23**, 2120.
91. P. Mitchell, E. Petfalski, A. Shevchenko, M. Mann and D. Tollervey, *Cell*, 1997, **91**, 457.
92. D.C. Masison, A. Blanc, J.C. Ribas, K. Carroll, N. Sonenberg and R.B. Wickner, *Mol. Cell. Biol.*, 1995, **15**, 2763.
93. B. Py, C.f. Higgins, H.M. Krisch and A.J. Carpousis, *Nature*, 1996, **381**, 169.
94. S.P. Margossian, H. Li, H.P. Zassenhaus and R.A. Butow, *Cell*, 1996, **84**, 199.
95. J. de la Cruz, D. Kressler, D. Tollervey and P. Linder, *EMBO J.*, 1998, **17**, 1128.
96. M.Y. Stoekle and H. Hanafusa, *Mol. Cell. Biol.*, 1989, **9**, 4738.
97. F.C. Nielsen and J. Christiansen, *J. Biol. Chem.*, 1992, **267**, 19404.
98. R. Binder, J.A. Horowitz, J.P. Basilion, D.M. Koeller, R.D. Klausner and J.B. Harford, *EMBO J.*, 1969, **1994**, 13.
99. B. Brown and R. Harland, *Genes Dev.*, 1925, **1990**, 4.
100. A. Cochrane and R.G. Deeley, *J. Biol. Chem.*, 1989, **264**, 6495.
101. S. Tharun and R. Sirdeshmukh, *Nucleic Acids Res.*, 1995, **23**, 641.
102. R.L. Pastori, J.E. Moskaitis and D.R. Schoenberg, *Biochemistry*, 1991, **30**, 10490.
103. P.L. Bernstein, D.J. Herrick, R.D. Prokipcak and J. Ross, *Genes Dev.*, 1992, **6**, 642.
104. R. Binder, S.-P.L. Hwang, R. Ratnasabapathy and D.L. Williams, *J. Biol. Chem.*, 1989, **264**, 16910.
105. B.D. Brown, I.D. Zipkin and R.M. Harland, *Genes Dev.*, 1993, **7**, 1620.
106. R.H. Silverman, in "Ribonucleases: Structure and Function", eds. G. D'Alessio and J.F. Riordan, Academic Press, New York, 1996.
107. B.A. Hassel, A. Zhou, C. Sotomayor, A. Maran and R.H. Silverman, *EMBO J.*, 1993, **12**, 3297.
108. M. Tucker, M.A. Valencia-Sanchez, R.R. Staples, J. Chen, C.L. Denis and R. Parker, *Cell*, 2001, **104**, 1.
109. S. Tharun, W. Hi, A.E. Mayes, P. Lennertz, J.D. Beggs and R. Parker, *Nature*, 2000, , 404, 515.

15

Applications of Ribonucleotide Analogues in RNA Biochemistry

SANDEEP VERMA, NARENDRA K. VAISH
and FRITZ ECKSTEIN
Max-Planck-Institut für Experimentelle Medizin, Göttingen, Germany

15.1 INTRODUCTION

Ribonucleotide analogues are beneficial for understanding catalytic RNA mechanisms and to study RNA–RNA, RNA–DNA, and RNA–protein interactions. Such studies often focus on the role of the phosphate group and thus employ a phosphate analogue, most often the phosphorothioate. However, as the difference between the function of DNA and RNA might be located at the 2′ position, analogues which replace the 2′-OH group are often useful in understanding the role of this group in recognition processes or its involvement in reactions. Additionally, base analogues can be used for photoactivated cross-linking or to study the involvement of base exocyclic functional groups. This review will discuss some of the more commonly used analogues in the RNA field, concentrating on oligoribonucleotides and RNA, not considering nucleosides by themselves. A more comprehensive review of oligonucleotide analogues covering both oligodeoxynucleotides and oligoribonucleotides has appeared.[1] Another extensive review on nucleoside analogues and modified oligonucleotides is presented elsewhere.[2].

15.2 INCORPORATION OF ANALOGUES INTO OLIGORIBONUCLEOTIDES

15.2.1 Chemical Synthesis

Automated, solid-phase DNA synthesizers can also be conveniently used for the synthesis of short oligoribonucleotides. The solid-phase synthesizers make use of fully protected ribonucleosides, anchored to a controlled pore glass support, at the 3′ end, via a chemically cleavable linker. A series of reactions couple ribonucleoside phosphoramidites sequentially to the support-bound nucleoside and after the last step, base hydrolysis is used to cleave the linker from the anchored nucleoside to yield a 2′-*O*-silyl protected oligomer. Desilylation is generally achieved by using tetrabutylammonium fluoride or similar reagents.

Development of effective 2′-OH protecting groups, use of efficient activators and oligomer purification methods have greatly improved RNA synthesis.[3–5] (see also Chapter 6). At the time of writing, synthesis of oligoribonucleotides using the solid-phase approach, is limited to a length of ~50 nucleotides. For the preparation of longer oligoribonucleotides, enzymatic ligation of chemically synthesized oligonucleotide fragments is preferred.

15.2.2 Enzymatic Ligation of Oligoribonucleotides

Enzymatic ligation is the method of choice for the synthesis of oligoribonucleotides that are too long to be obtained by chemical synthesis. Originally, RNA ligase was used for the ligation of the 5′ and 3′ termini of RNA fragments through the formation of a phosphodiester bond. This enzyme requires single-stranded ends for ligation and is thus restricted to the ligation of fragments in loop regions such as in the anticodon loop of tRNA.[6] An alternative method which employs T4 DNA ligase has been developed.[7] This enzyme requires the presence of the ends to be ligated on a duplex, which is achieved by annealing two oligoribonucleotide fragments to a complementary oligodeoxynucleotide bridge. This is a versatile method which permits efficient ligation and allows for the incorporation of nucleotide analogues. The methodology has been described in detail by Moore and Query.[8]

15.2.3 Enzymatic Incorporation by Polymerization

Template-directed synthesis of oligoribonucleotides and long transcripts is achieved by the use of T7 or SP6 RNA polymerases. This process is also well suited for the incorporation of modified nucleotides, provided their triphosphates are good substrates for these enzymes. This requirement, however, limits the number of modified ribonucleotides which can be used in transcription reactions and thus only a few modifications can be enzymatically introduced into oligoribonucleotides.

The phosphorothioate internucleotidic linkage is the most common enzymatically introduced modification as all of the four nucleoside α-thiotriphosphates are good substrates for RNA polymerases ((**1**)–(**4**)).[9] Substitution of one of the nonbridging oxygen atoms of a phosphate group by sulfur introduces chirality at the phosphorus center and consequently two diastereomers exist for the nucleoside α-thiotriphosphates. Only the Sp-diastereomers are good substrates for the polymerases. The enzymatic incorporation proceeds with inversion of configuration at phosphorus resulting in the formation of

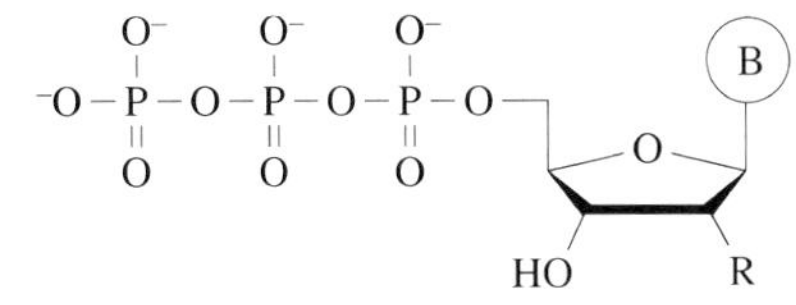

R = F, NH_2

(**1**) 2'-Modified nucleoside 5'-triphosphates

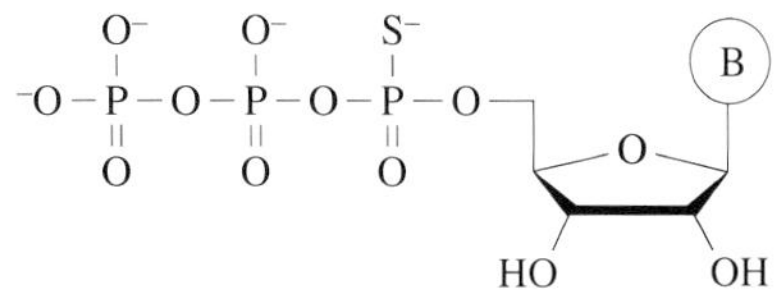

(**2**) Nucleoside 5'-α-thiotriphosphates

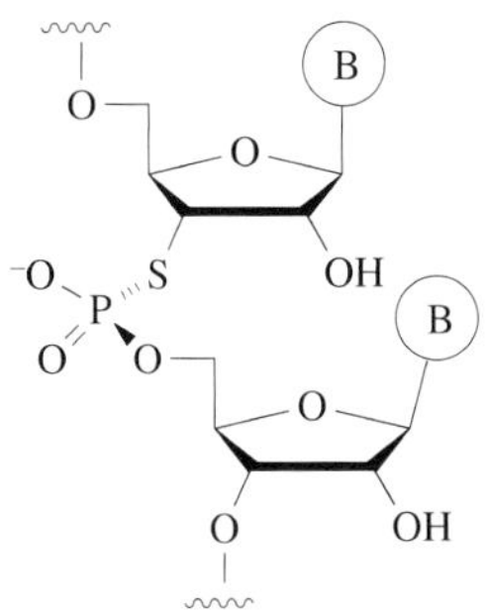

(**3**) 3'-Phosphorothiolate

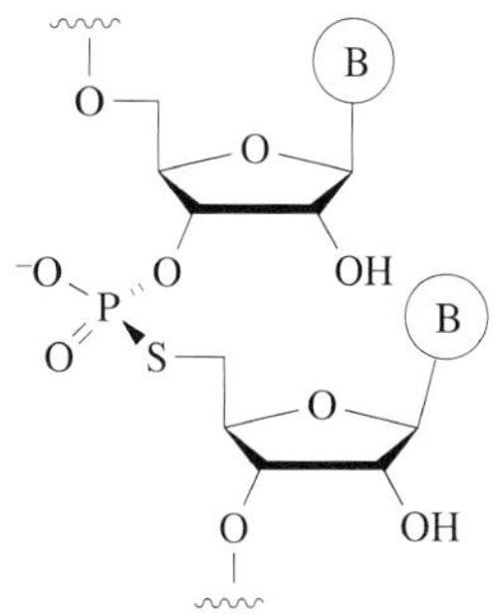

(**4**) 5'-Phosphorothiolate

the Rp internucleotidic linkage. Thus, the enzymatic synthesis is limited to the incorporation of the Rp-phosphorothioate linkage.[9,10]

Of the ribose analogues, only 2′-amino- and 2′-fluoro-2′-deoxynucleoside triphosphates are substrates for the wild-type T7 RNA polymerase ((**1**)–(**4**)). In particular, 2′-amino nucleoside triphosphates are better substrates as observed from the synthesis of a luciferase transcript of 2500 nt in length.[11] In comparison, 2′-fluoro-2′-deoxy analogues are less readily incorporated. However, a mutant T7 RNA polymerase which incorporates 2′-deoxynucleotides better than the wild-type enzyme, also accepts 2′-fluoro analogues quite well.[12]

T7 RNA polymerase normally initiates transcription with a guanosine triphosphate, but it can also accept other guanosine derivatives such as guanosine 5′-monophosphorothioate[13] or guanosine 5′-γ-thiotriphosphate[14] for initiation. The applications of 5′-terminal phosphorothioate-containing constructs will be discussed in Section 15.3.3.

15.3 PHOSPHOROTHIOATE INTERNUCLEOTIDIC LINKAGES

The exchange of a nonbridging oxygen with a sulfur in the phosphorothioate is a minimal change since the negative charge of the phosphate group is retained and the van der Waals radius of sulfur (1.85 Å) is only 30% larger than that of oxygen (1.40 Å).[15] Sulfur is also more hydrophobic than oxygen, but the contribution of this factor to the difference in the properties of phosphates and phosphorothioates has not yet been determined. As mentioned before, the sulfur/oxygen exchange makes the phosphorus center chiral, resulting in two diastereomers of the phosphorothioate internucleotide linkage. The chirality of the phosphorothioate linkage permits the determination of the stereospecificity and of the stereochemical course of reactions occurring at phosphorus.[9] In addition, sulfur is a "soft" atom and it coordinates preferentially with "soft" metal ions, whereas oxygen is a "hard" atom and coordinates to "hard" metal ions.[16] Therefore, incorporation of phosphorothioates also allows for the identification of metal ion-binding phosphate group oxygen atoms in RNA or oligoribonucleotides.[17] The literature on nucleoside phosphorothioates, up to 1984, was reviewed by Eckstein.[9]

The oligonucleotide phosphorothioates can also be synthesized by chemical synthesis, using a sulfurizing reagent in the oxidation step. Chemical synthesis again produces a mixture of diastereomers, which in some cases can be separated, while the enzymatic incorporation results exclusively in the formation of the Rp-diastereomeric internucleotidic linkage.[9,10,18,19]

15.3.1 Phosphorothioates for Stereochemical Analysis

The chirality of the phosphorothioate group as an internucleotidic linkage in oligonucleotides permits analysis of the stereochemical course of transesterification reactions as long as the configurations of the starting material and the product are known. In most cases, the starting material for such studies is a transcript obtained by polymerization of a nucleoside α-thiotriphosphate with an RNA polymerase and therefore the configuration of the phosphorothioate internucleotidic linkage is known. Scheme 1 shows the stereospecificity of nucleoside phosphorothioate diastereomers with enzymes. (Literature may be consulted for individual reactions: polymerization;[9,20] reaction with RNase A;[21] reaction with RNase T_1;[22] reaction with SVPDE (snake venom phosphodiesterase);[20,23] reaction with nuclease Pl;[24] reaction with CNPase.[25]).

The configuration of the phosphorothioate linkage in the product of a transesterification reaction can be determined in several ways depending on the type of product. If the product is a nucleoside 2′,3′-cyclic phosphorothioate, comparison can be made with an authentic sample or by using cyclic nu-

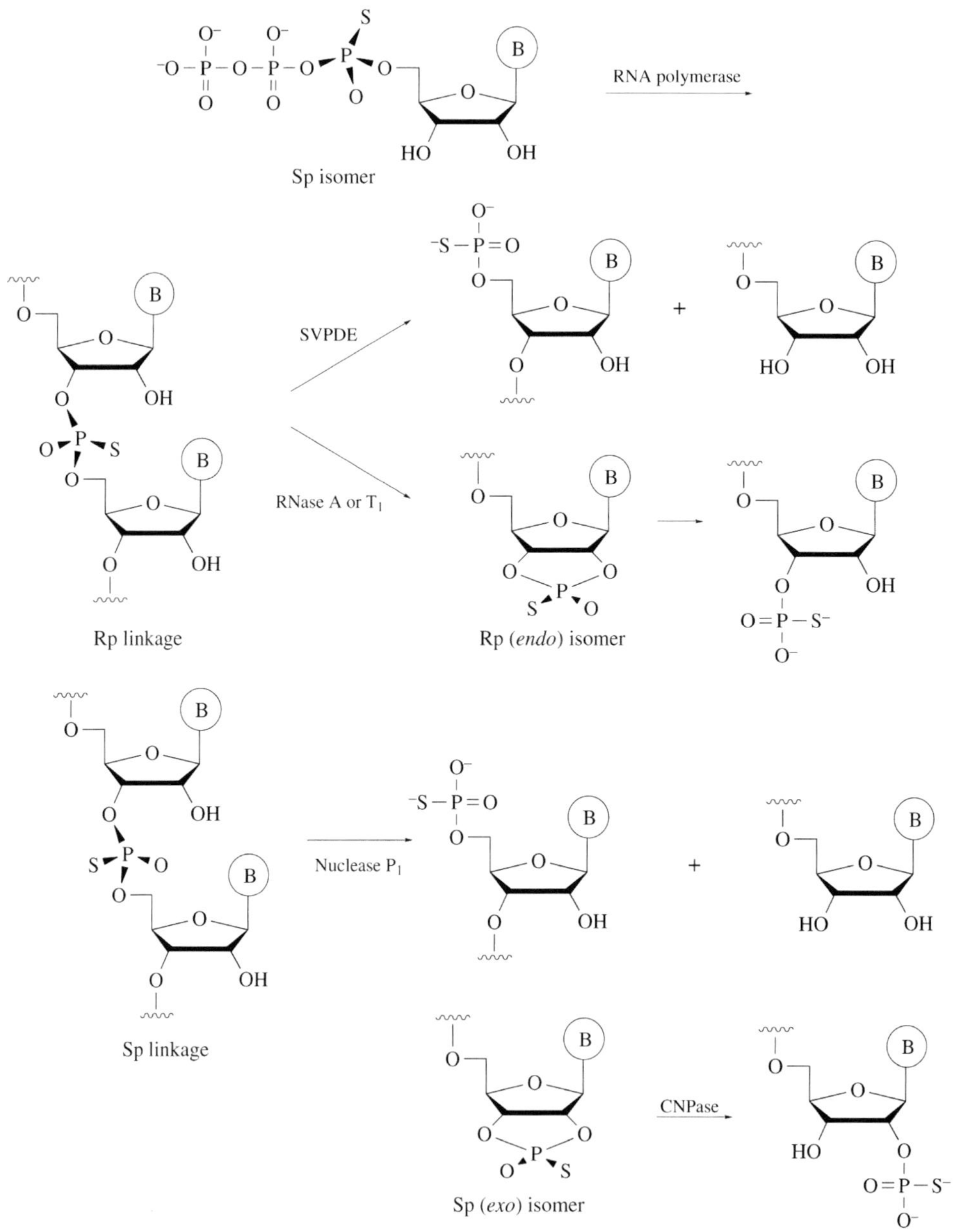

Scheme 1

cleotide phosphodiesterase which stereospecifically cleaves the Sp-isomer of cyclic phosphorothioates.[25] In cases where the product also contains a phosphorothioate internucleotidic linkage, the stereochemical preference of certain nucleases can be exploited (Scheme 1). Using such an approach, the stereochemical course of ribozyme-catalyzed reactions such as RNA cleavage by the hammerhead and the hairpin ribozymes have been determined.[26,27] These reactions proceed with inversion of configuration at phosphorus providing convincing evidence for an in-line mechanism in the absence of an intermediate. However, the ground state conformation of the hammerhead ribozyme, as established by X-ray crystallography,[28,29] is not compatible with such a mechanism and strongly suggests that a conformational change is needed to attain the transition state geometry.[30,31] The stereochemical course of the reactions catalyzed by the group I intron,[32,33] pre-mRNA splicing,[34,35] and group II intron self-splicing,[36,37] have been elucidated by stereopreferential degradation of the products. All these reactions also proceed with an inversion of configuration at phosphorus. Using phosphorothioates, a comparison of the stereochemical requirements of the individual steps has allowed similarities to be established between different splicing systems as determined for pre-mRNA splicing and group II splicing reaction.[38]

However, caution should be exercised in the mechanistic interpretation of rate differences observed between phosphate and phosphorothioate substrates in enzyme-catalyzed reactions.[39] From a purely chemical viewpoint, reactivity of the phosphorothioate group should only be marginally lower than that of the phosphate diester. Steric clashes in the transition state because of the larger van der Waals radius of sulfur, rather than the electronegativity difference, have been invoked to explain the reduction in reaction rate for the phosphorothioate.[40,41]

15.3.2 Phosphorothioate Interference

Phosphorothioates facilitate identification of the phosphate groups in an oligoribonucleotide involved in RNA–RNA or RNA–protein interactions. These studies exploit either the difference in size or a change in binding specificity when oxygen is substituted by sulfur. Binding of a metal ion to a specific phosphate group might be altered if it is replaced by a phosphorothioate, which makes the exact position of the phosphate group involved in metal ion binding easier to identify. Phosphorothioates are generally incorporated randomly by transcription for interference studies with oligoribonucleotides, so that on average only one phosphorothioate is present per molecule. The enzymatic incorporation results in the formation of an Rp-phosphorothioate linkage and thus an observed interference can be attributed to the pro-Rp oxygen of the replaced phosphate group. The success of this method depends on separation of the active species from the inactive species. Once separated, the precise position of the interfering phosphorothioate can be determined by iodine cleavage (*vide infra*).[42,43] Finally, a comparison of the phosphorothioate cleavage pattern of the active and inactive species allows identification of interfering positions in the reaction or interaction under investigation.

The phosphorothioate modification can also probe for phosphate–metal ion coordination in a reaction. This is based on the concept of "hard" and "soft" metal ions. Hard metal ions interact preferentially with the phosphorothioate oxygen atom, whereas soft ones interact with the phosphorothioate sulfur atom.[16] If a reaction is metal ion dependent and shows stereoselectivity for one diastereomer of the phosphorothioate in the presence of the hard metal ion Mg^{2+}, then use of a soft metal ion, such as Mn^{2+}, often reverses the diastereomer selectivity by making the other diastereomer the prefered substrate. Such a switch in stereospecificity indicates metal ion coordination to one of the oxygens of the phosphate group in the reaction. It is important to understand that this effect is only manifested if the phosphate–metal ion interaction is important in the rate-limiting step of the reaction. Also, the absence of such reversion does not necessarily indicate a lack of metal ion binding as pointed out for the 3′-5′-exonuclease reaction of the Klenow polymerase.[44]

15.3.2.1 Interference analysis in catalytic RNA

A study using phosphorothioate interference has identified 44 phosphate groups important for Mg^{2+} ion binding in the group I intron reaction.[45,46] For most of these positions, catalysis of the cleavage reaction could be restored with Mn^{2+} which indicates that the pro-Rp oxygen atoms are coordinated to the metal ion in the splicing reaction. The P4–P6 domain of this ribozyme has been crystallized and

Mg^{2+} binding sites have been identified[47] (see also Chapter 3). In order to determine the importance of domains in RNA tertiary structure, they were individually replaced by the Rp-phosphorothioate and the folding was followed in the presence of Mg^{2+} or Mn^{2+}. This approach allowed for the identification of four phosphates critical for folding.

A development in the area of interference analysis is "nucleotide analogue interference mapping", which has been applied to identify the functional groups essential for group I intron catalysis[48] (see also Chapter 11). In this approach, a functional group modification in the nucleobase was coupled to the phosphorothioate linkage for easy identification. Modified group I intron constructs were prepared by substituting inosine α-thiotriphosphate in place of GTP, in the transcription reactions. Iodine-mediated phosphorothioate cleavage[42,43] was used to identify the positions of essential guanosine residues. In a parallel experiment, interference by the incorporation of GTPαS alone was used to separate the phosphorothioate effect from the inosine effect. This assay successfully mapped almost every guanosine residue essential for the catalytic activity of the group I intron ribozyme.

Four phosphate groups in the RNase P RNA were identified by phosphorothioate interference as important for an intramolecular, pre-tRNA cleavage reaction.[49] Beside these four positions, another study identified an additional twelve phosphate positions which were found to be important for the binding of the pre-tRNA.[50] Phosphorothioate substitution at the cleavage site of the pre-tRNA resulted in a 1000-fold reduction of the cleavage rate for both diastereomers, but it was found that the rate inhibition of the Rp-diastereomer could be largely restored by Cd^{2+}, a soft metal ion.[51] In yet another study, phosphorothioate substitutions of the pro-Rp oxygens in a pre-tRNA transcript revealed that direct coordination of Mg^{2+} ion to the pro-Rp oxygen of the scissile bond is required for RNase P catalysis.[52]

Rp-phosphorothioate interference also identified three phosphate groups, besides the one at the cleavage site, as important for catalysis by the hammerhead ribozyme.[53] The importance of one of these sites, where a metal ion is coordinated to the pro-Rp oxygen of the A9 phosphate, has been confirmed by X-ray structural analyses.[28,29] As suggested by the X-ray structure, a metal ion rescue could be observed with the Rp isomer at A9, which confirms the observation of metal ion coordination to this site. A Mg^{2+} ion has been found coordinated to the pro-Rp oxygen of the phosphate group at the cleavage site in a recent X-ray structure analysis.[31] These results are consistent with the cleavage of the Rp-phosphorothioate diastereomer in the presence of Mn^{2+} and that of the Sp-diastereomer in the presence of Mg^{2+} ion (Scheme 2).[27,54] However, the results of the phosphorothioate experiments have been challenged.[55] Zhou *et al.* observed a rate enhancement with Mn^{2+}, irrespective of the diastereomer used. Synthetic hammerhead ribozymes containing a mixture of phosphorothioate diastereomers have also been prepared.[56] The presence of the phosphorothioate Sp isomer at positions A6 and U16.1 severely reduced the rate of cleavage, indicating other important metal ion–phosphate interactions for this ribozyme.

Similar interference studies have identified important phosphate groups for the extended hammerhead ribozyme present in the satellite 2 transcripts[57] and for the hepatitis delta virus ribozyme.[58]

15.3.2.2 Interference analysis of phosphate–protein interaction

The phosphorothioate interference analysis and footprinting with iodine can also probe for the phosphate groups involved in RNA–protein interactions. This method was first applied to study the affinity of an oligoribonucleotide for the R17 coat protein.[59] It was extended to probe the interaction of *Escherichia coli* $tRNA^{Ser}$ with its cognate aminoacyl synthetase.[43] The results suggested that contact with the tRNA variable loop might be located on the extended arm of the synthetase. All of these contacts, with one exception, were later confirmed by the X-ray crystal structure analysis of the related *Thermus thermophilus* complex.[60] Results obtained using the same approach with the yeast $tRNA^{Asp}$ system were consistent with the interaction of the identity elements of the tRNA with the synthetase protein.[61] Phosphorothioate footprinting to investigate the interaction between $tRNA^{Phe}$ transcripts from *T. thermophilus* and its synthetase revealed a complex and novel pattern of $tRNA^{Phe}$–synthetase interaction.[62] This strategy has also been used to study mitochondrial group I intron RNA splicing, which involves interaction with tyrosyl-tRNA synthetase.[63] These observations have led to the construction of a model which suggests that the recognition of the synthetase occurs by a tRNA-like structure in the group I intron.[64] The interaction of tRNA with the ribosome A and P sites was also investigated

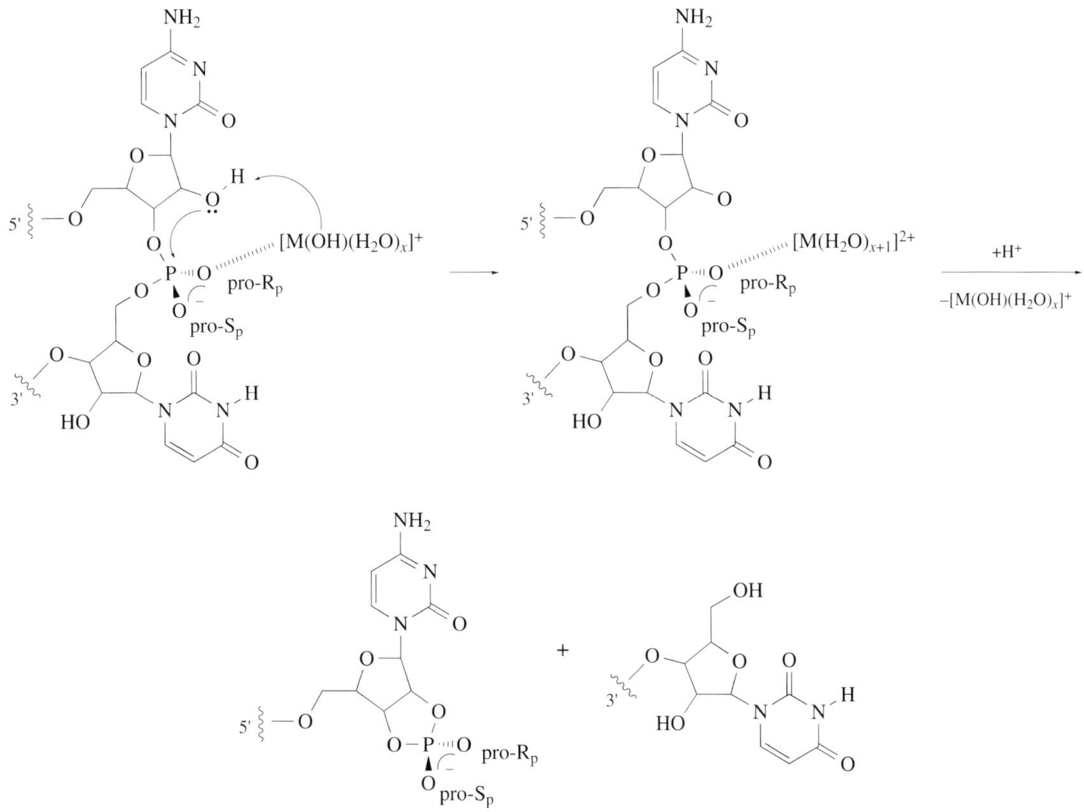

Scheme 2

by phosphorothioate interference to map the essential "identity elements" involved in ribosome-tRNA recognition.[65]

15.3.2.3 Interference analysis for importance of 2′-hydroxy groups

A similar method was used to identify essential 2′-OH groups in the binding of tRNA to RNase P RNA.[66] dNTPαS was introduced into an RNase P RNA transcript by T7 RNA polymerase using a mixture of Mg^{2+} and Mn^{2+} in the reaction mixture to facilitate incorporation of the dNTPαS. Separation of the tRNA/RNA complex and comparison of the phosphorothioate iodine footprinting with the unbound RNA, revealed ~20 positions for which 2′-deoxy substitution in the RNase P RNA impaired binding to the tRNA.

In the authors' laboratory, a similar study was undertaken to identify 2′-OH positions in the *E. coli* $tRNA^{Asp}$ responsible for charging by the cognate synthetase.[67] A T7 RNA polymerase mutant[68] was used for the random incorporation of dNTPαS into the tRNA transcript. In a separate experiment, NTPαS was incorporated to obtain tRNA transcripts to assess the effect of phosphorothioates alone on charging. Charged tRNAs were separated from the uncharged ones and iodine footprinting was performed on the two species. For several nucleotide positions, interference with charging could be attributed to the absence of the 2′-OH group. However, for some positions, the phosphorothioate effect could not be separated from the 2′-deoxy effect. Individual tRNAs, where the identified positions were singly replaced by the corresponding deoxynucleotide, were prepared and the kinetics of charging was determined. In general, the reduction in charging rate was between 2–10-fold.[67] This study differs from the one performed on yeast $tRNA^{Asp}$ using dNTP incorporation with a double mutant of the T7 RNA polymerase.[69] In the transcription reactions, one NTP was totally replaced by the corresponding dNTP. The tRNAs with all the uridines replaced by deoxyuridine and those with all guanosines replaced by deoxyguanosine, could not be charged. These results were interpreted on the basis of the X-ray structure of the native complex rather than evaluating tRNAs with individual deoxynucleotide replacements.

15.3.3 Terminal phosphorothioates for attachment

Oligonucleotides with a terminal phosphorothioate can be prepared by chemical synthesis,[70,71] by enzymatic 5′-thiophosphorylation using ATPγS,[14,72] by the attachment of pdCpS with RNA ligase to the 3′-end of oligoribonucleotides,[73] or by initiating transcription with guanosine derivatives such as GMPS or GTPγS.[13,14] Sodium periodate oxidation of a terminal ribonucleotide, linked to the oligonucleotide by a phosphorothioate internucleotidic linkage, followed by β-elimination is yet another alternative for postsynthetic labeling.[74] In oligonucleotides, terminal phosphorothioate groups have been used for a variety of purposes such as the separation of oligonucleotides or transcripts on mercury gels or columns,[14] reaction with haloacetyl derivatives of fluorescent dyes,[73] and with photoaffinity labels.[13]

15.4 S-BRIDGING PHOSPHOROTHIOLATES

Replacement of either the 3′- or the 5′-bridging oxygen of the phosphodiester internucleotidic linkage by sulfur, is another important modification used in oligoribonucleotides to probe mechanisms of ribozyme catalysis (Structures (**1**)–(**4**)). In the group I intron ribozyme-catalyzed transesterification reaction, cleavage of a substrate containing a 3′-*S*-phosphorothiolate internucleotidic substitution was about 1000-fold slower in the presence of Mg^{2+} when compared to an unmodified phosphodiester substrate.[75] However, in the presence of soft metal ions such as Mn^{2+} and Zn^{2+}, the cleavage activity was restored. This observation indicates a possible role of metal ions in stabilizing the negative charge on 3′ oxygen in the transition state. The presence of a second metal ion involved in group I ribozyme catalysis was demonstrated using 3′-(thioinosinyl)-(3′-5′)-uridine (IspU) dinucleotide as a substrate in the splicing reaction.[76] In a reaction mimicking the exon ligation step, IspU substrate required a second, thiophilic metal ion in optimum activity. Based on these observations, involvement of a second metal ion in activating the 3′-hydroxyl group of guanosine in the first step of splicing was proposed. The metal ion specificity switch from Mg^{2+} to Mn^{2+} using substrate analogues containing a 3′-*S*-phosphorothiolate was also determined for the first and second step of splicing of pre-mRNA. In contrast to the second step, the first step of splicing needs metal ion assistance.[77]

The mechanism of the hammerhead ribozyme-catalyzed reaction has also been probed using 5′-*S*-phosphorothiolates. This modification was introduced at the cleavage site in a DNA substrate.[78] However, an appreciable thio effect by changing the metal ion from Mg^{2+} to Mn^{2+} was not observed. There was concluded to be a lack of metal ion coordination with the 5′-oxygen of the leaving group in the transition state. Another group, which used an all-RNA substrate for similar experiments, came to a different conclusion.[79] They reasoned that attack at the 2′-OH is rate limiting for the modified substrate, whereas the departure of the leaving group is rate limiting for the unmodified one. Rate differences observed for the hydrolysis of 3′- and 5′-*S*-phosphorothiolate susbtrates with different metal ions should be interpreted with caution as even the metal ion-dependent hydrolysis of simple dinucleotides is not well understood.[80]

These examples elegantly document the power of sulfur substitution in the bridging oxygens of a phosphate internucleotidic linkage to study metal ion involvement in RNA-catalyzed reactions and complement the nonbridging oxygen substitution discussed earlier. However, it would be desirable to have more quantitative data on softness and hardness of the coordination partners in these systems which has been described only for ATPβS and related compounds.[16] Also, the absence of rescue effect with a thiophilic metal ion should not be interpreted as a lack of interaction in the native system, since metal ion rescue can be prevented or obscured by many factors.

15.5 MODIFICATION OF THE NUCLEOBASES

The heterocyclic ring of purine and pyrimidine bases provides the hydrogen bonding functional groups in nucleic acids. Thus base analogues, when introduced into oligonucleotides, can provide information on the importance of specific functional groups present in natural bases. In interpreting results obtained with base analogues, it should be realized that even a subtle change in the analogue can have dramatic effects due to the change in size, electronic distribution, nucleoside sugar conformation, tautomeric structure, or functional group pK_a values. The representative structures (**5**)–(**19**) of modified nucleosides

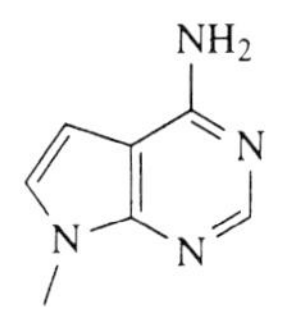

(5) 7-Deazaadenine

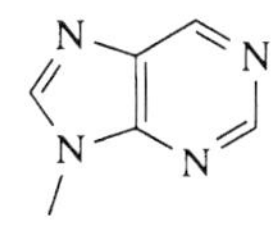

(6) Purine

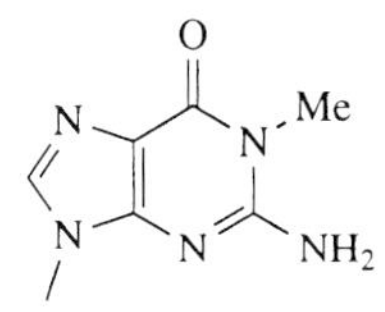

(7) *N*′-Methylguanosine

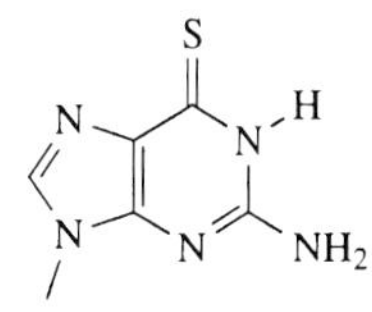

(8) 6-Thioguanosine

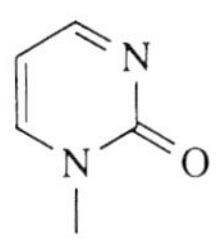

(9) 2-Pyrimidinone

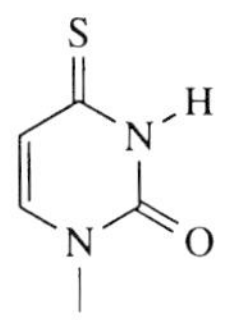

(10) 4-Thiouridine

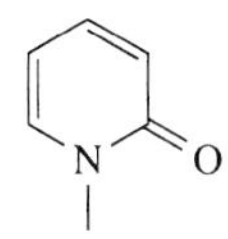

(11) 2-Pyridinone

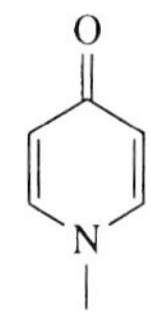

(12) 4-Pyridinone

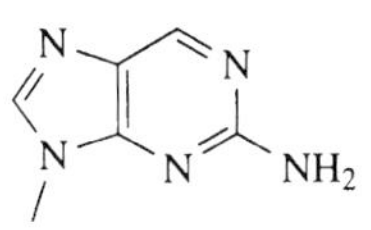

(13) 2-Aminopurine

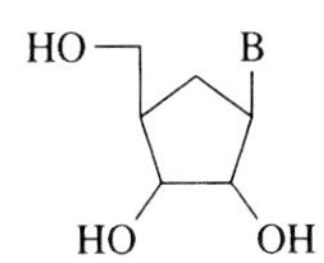

(14) 4'-Carbocyclic analogue

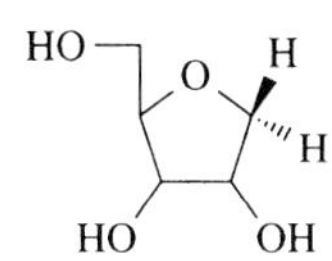

(15) Abasic ribonucleoside

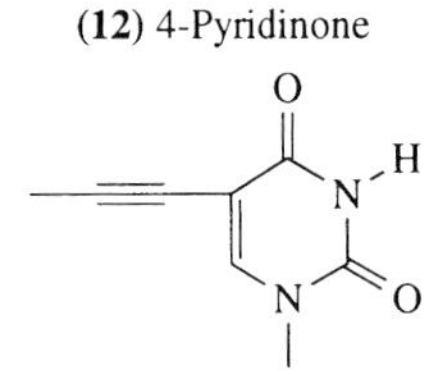

(16) 5-Propynyluridine

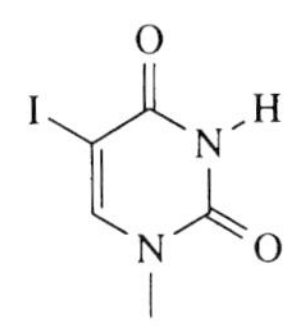

(17) 5-Iodouridine

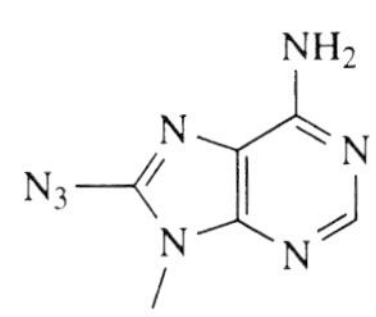

(18) 8-Azidoadenosine

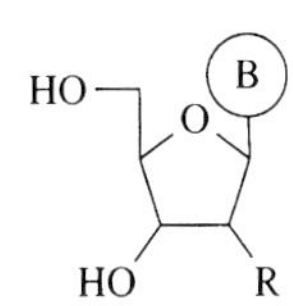

(19)
R = OMe, 2'-*O*-Methylnucleoside
R = SH, 2'-Thionucleoside
R = $OCH_2CH{=}CH_2$, 2'-*O*-Allylnucleoside

discussed in subsequent sections are shown. Several synthetic approaches for the introduction of C-5 modifications in uridine and at C-8 of adenosine and guanosine have been reviewed.[81]

7-Deazaadenosine and purine ribonucleoside, lacking the N-7 nitrogen and the exocyclic amino group, respectively, have been incorporated into the branchpoint in a nuclear pre-mRNA and a group II intron to compare the effect in the two systems.[82] The changes in both systems were alike and further supported the functional similarity between pre-mRNA and group II intron splicing.

Another purine analogue, *N*′-methylguanosine (**7**), when substituted in the core region of the hammerhead ribozyme, completely inhibited its catalytic activity.[83] This analogue has also been incorporated into $tRNA^{Asp}$, where it abolished mischarging.[84]

2-Aminopurine (**13**) is often used for fluorescence measurements and will be discussed in Section 15.7. Other analogues such as purine (**6**), isoguanine, 2,6-diaminopurine, 6-thioguanine, and hypoxanthine have also helped to identify purine nucleosides essential for RNA catalysis.[85–88]

2-Pyrimidinone (**9**) and 4-thiouridine (**10**) ribonucleosides as well as 2- and 4-pyridinone ribonucleosides (**11**) and (**12**) have been used to assess the participation of conserved pyrimidine bases in making hydrogen bonding contacts in the catalytic core of the hammerhead ribozyme and to provide evidence for a Mg^{2+} ion-binding site in the catalytically active ribozyme.[89,90]

The hammerhead and the hairpin ribozymes remain the most studied of RNA molecules with base analogues. A detailed discussion of the results of these studies with the hammerhead ribozyme and their correlation with its X-ray structure has been published.[30] A summary of the studies with the hairpin ribozyme has also been discussed.[91]

A new area of application of base-modified ribonucleotide analogues involves *in vitro* selection of aptamers and ribozymes. Such analogues can provide new functional groups which might aid in the desired interaction or enhance catalytic properties of selected RNA molecules, by enlarging the functional space in the selection process. A precondition, of course, is that the corresponding analogue triphosphates should act as substrates for the polymerases employed in the selection procedure. The first report of such an application was the incorporation of 5-(propynyl)-2′-deoxy- uridine into a thrombin aptamer.[92] Another group has reported the incorporation of 5-pyridylmethyl- and 5-imidazolylmethyluridine during the selection processes designed to generate RNA sequences which can catalyze amide bond and carbon–carbon bond formation, respectively.[93,94]

Deletion of the purine or pyrimidine bases from the ribose sugar moiety gives rise to abasic nucleosides. These analogues preserve the sugar hydroxy groups and therefore any biochemical effect observed from their use results directly from the loss of the heterocyclic ring. Substitution of abasic ribonucleosides (**15**) in the catalytic core of the hammerhead ribozyme significantly impairs its catalytic activity.[95] However, it was possible to restore the activity of four abasic ribozymes by a simple addition of the missing base, indicating that the ribozyme structure can be altered to create a binding site for small ligands. Other studies using the ribo-abasic nucleosides include the recognition of branch-point adenosine in the group II intron,[88] the identification of essential base residues in loop B of the hairpin ribozyme,[96] and the importance of the stem-loop II and the conserved U residues in the catalytic core of the hammerhead ribozyme.[97,98]

15.6 RIBOSE MODIFICATION AT THE 2′-POSITION

Several sugar modifications, primarily at the 2′-position, have been employed to study the structure–function relationship in RNA. Such analogues can be used to investigate the conformational importance of the sugar pucker in RNA interactions. The furanose ring in oligoribonucleotides maintains a $C_{3'}$-*endo* sugar ring conformation.[99] In 2′-substituted nucleosides, the electronegativity of the substituent exerts a profound effect on the sugar conformation and as a result, the population of $C_{3'}$-*endo* conformer increases with an increase in the electronegativity. However, the conformational effect is often difficult to separate from the effect of the 2′-substituent on the hydrogen bonding at this position. For example, a 2′-fluoro-2′-deoxyribose sugar has a very high percentage of the $C_{3'}$-*endo* conformer, more than the deoxyribose, but is unable to support hydrogen bonding. Conversely, the 2′-amino derivative has an extremely low percentage of the $C_{3'}$-*endo* conformer compared to deoxyribose, but retains both hydrogen donating and accepting abilities similar to the 2′-OH group. Additionally, the 2′-amino group is an excellent nucleophile for substitution reactions and this property has been exploited in cross-linking experiments discussed in a subsequent section, and also in the attachment of cholesterol to the hammerhead ribozyme.[100] The synthesis of several 2′-modified nucleosides and their triphosphates has been reviewed.[81]

Sugar-modified analogues have been extensively used to study ribozymes and RNA aptamers to probe the importance of specific 2′-hydroxyl groups in catalysis and to provide resistance against nucleases. The 2′-hydroxyl modifications introduced to probe hammerhead ribozyme catalysis include fluoro and amino,[101,102] methoxy and allyloxy,[103] and 2′-*C*-allyl substituents.[104] 2′-*O*-methyl sugar modifications have been introduced to identify the positions of 2′-hydroxyl groups necessary for the hairpin ribozyme activity.[105] Interactions involving 2′-hydroxyl groups and the importance of having a ribose moiety at the cleavage site in the *Tetrahymena* ribozyme reaction, has been studied using 2′-deoxy- and 2′-fluoro-substituted nucleosides.[106–109] 2′-Deoxynucleosides have also been used to investigate the role of active site 2′-hydroxyl groups in group II intron molecular recognition and catalysis.[110]

A 2′-thio-2′-deoxynucleoside has been incorporated into a dinucleotide and into an oligodeoxynucleotide, but mechanistic studies using this modification have not been reported.[111,112] However, it is expected that they will also serve as interesting probes for establishing metal ion assistance for RNA in catalysis.

By employing a variety of selection protocols, high affinity nucleic acid aptamers that recognize a wide range of molecular targets such as antibiotics, amino acids, nucleoside triphosphates, cofactors, organic dyes, porphyrins, and proteins have been isolated.[113] In addition to specificity and affinity, improved chemical and enzymatic stability is also essential for the development and use of RNA-based aptamers in therapeutics and diagnostics. The substitution of 2′-fluoro- and 2′-amino-2′-deoxy pyrimidine nucleosides in RNA imparts resistance to nucleases[102] and moreover these modifications can be easily introduced during aptamer selection since the corresponding nucleotide triphosphates are good substrates for T7 RNA polymerase.[11,12] Using these analogues, aptamers have been isolated against basic fibroblast growth factor,[114] human thyroid stimulating hormone,[115] a monoclonal antibody which recognizes the main immunogenic region of human acetylcholine receptor,[116] and the keratinocyte growth factor.[117]

2′-*O*-methyl-modified, biotinylated oligoribonucleotides have been used as antisense probes to map U2 snRNP-pre-mRNA interactions and to study structure and function of U4/U6 snRNPs.[118,119] Modified nucleosides such as 2′-*O*-methyl have also been placed at either the 3′- or 5′-splice site of a nuclear pre-mRNA to understand chemical steps involved in the spliceosomal assembly.[7] *In vitro* studies have revealed that the presence of 2′-hydroxyl group is more important at the 3′-splice site for the second step in the splicing reaction which results in the ligation of two exons with a concomitant release of an intron.

Another sugar modification involves replacement of the *O*-4′ furanose ring oxygen by a methylene group to give carbocyclic analogues. Ribocarbocyclic analogues of cytidine and adenosine have been chemically incorporated into the hammerhead ribozyme and were found to confer RNase resistance.[120]

15.7 FLUORESCENT LABELS FOR OLIGORIBONUCLEOTIDES

2-Aminopurine (2AP), a fluorescent nucleoside analogue, has been extensively used to detect changes in oligonucleotide conformation and its applications have been reviewed.[121] It can substitute for adenosine in base pairing with uridine. The absorption and excitation maximum for the nucleoside is at 330 nm and has an emission at 380 nm. Importantly, the quantum yield of 2AP fluorescence, when substituted in oligonucleotides, depends on the degree of base stacking.[122] In one study, Mg^{2+} ion induced conformational perturbations in the tertiary structure of the hammerhead ribozyme were followed by monitoring the change in fluorescence of 2AP and the affinity constants for Mg^{2+} ion binding were determined.[123]

The guanosine-specific fluorescence quenching of 3′-fluorescein-labeled oligoribonucleotides was used to determine rate constants for substrate binding, cleavage, and dissociation of the hairpin ribozyme.[124] Two examples from the hammerhead ribozyme structural studies further emphasize the power and versatility of fluorescence resonance energy transfer (FRET) studies. In both the studies, helical termini of the ribozyme were labeled with a donor and an acceptor dye. In one study, FRET efficiency was used to determine the distances between these ends to propose a three-dimensional model for the ribozyme[125] and in another study, FRET as a function of Mg^{2+} concentration was used to follow the folding of the ribozyme.[126] There is considerable literature on fluorescent dyes attached at the termini of oligonucleotides for FRET measurements and their application in structural studies has been reviewed.[121]

15.8 NUCLEIC ACID CROSS-LINKING

The conformational flexibility of RNA–RNA duplexes or RNA–protein complexes can be restricted by site-specific cross-links using chemically modified oligonucleotides. Such cross-links can also determine the proximity of nucleotides or amino acids, since they would not be formed at distances greater than the length of the spacer. The reagents for cross-linking are usually attached to either the base or the sugar moiety and in most cases the sulfhydryl group is used for introducing cross-links.

15.8.1 Chemical Cross-links

A versatile strategy for the cross-linking of RNA oligomers results from the extension of the convertible nucleoside approach to oligoribonucleotides.[127] In this method, uridine and inosine derivatives that possess good leaving groups at the 4 or 6 position, respectively, are site-specifically incorporated in oligoribonucleotides (Figure 1). Subsequent reaction of the convertible nucleoside, at the oligonucleotide level, with a symmetrical ω,ω-dithiobis(alkylamine), results in the formation of cytidine and adenosine derivatives bearing an alkyl disulfide tether. Reduction of the disulfide linkage results in the formation of a free sulfhydryl group, which can be reoxidized to form a cross-link. Oligoribonucleotides can be cross-linked in this manner as exemplified by the formation of an intramolecular disulfide in an RNA ministem loop.[128] Thioethyl groups can also be attached to C-5 of pyrimidines and oxidized to the disulfides as described for the cross-linking of $tRNA^{Phe}$.[129]

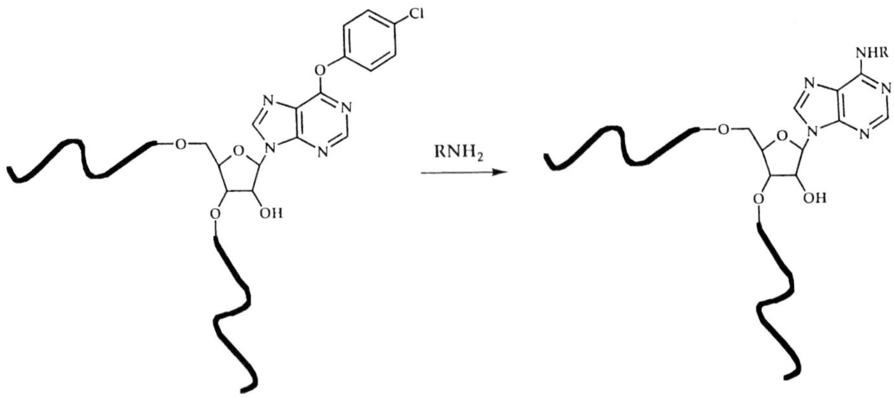

Figure 1 Convertible ribonucleoside approach for chemical modification of RNA. RNH_2 represents the attacking amine nucleophile which may contain a disulfide group for further cross-linking studies. Similarly, the C-4 position of uridine can also be activated. The ribbon structure represents a fully deprotected oligoribonucleotide, containing a convertible inosine derivative.

However, there are systems where cross-linking via base modification is undesirable and, therefore, attachment of the cross-linking reagent via the sugar ring is preferred. This approach has been used to discriminate between the two models of the hammerhead ribozyme, one based on X-ray crystallographic analysis and the other based on FRET results.[130] In this method, the nucleotides to be linked were replaced by the corresponding 2′-amino nucleotides which were then reacted with 2-pyridyl 3-isothiocyanatobenzyl disulfide. After reduction, an intrastrand disulfide was introduced by oxidation and the catalytic activity of cross-linked ribozymes was determined. More reactive aliphatic isocyanates have been used for the interstrand cross-linking of the hairpin ribozyme as part of an effort to build a first three-dimensional model (Figure 2).[131] This method has been further extended to introduce interstrand disulfide cross-links in a group I ribozyme.[132] In this study, cross-links were formed by disulfide exchange reaction rather than oxidation as this former reaction occurs at a much faster rate (Figure 2). The cross-linked ribozymes were catalytically competent even though some cross-links bridged distances considerably farther apart than expected from the current model. These results reveal interesting dynamic properties of large catalytic RNAs. Alternatively, 2′-hydroxyl groups have also been derivatized with thioethyl groups to introduce intra- and interhelical disulfide cross-links in $tRNA^{Phe}$.[129]

15.8.2 Photocross-linking

15.8.2.1 Nucleobase cross-links

Natural nucleosides can be photoactivated to form new bonds within approachable distances. A good example is the formation of a cross-link between a cytidine and a uridine residue, which was observed upon irradiation of yeast $tRNA^{Phe}$ and was found to be consistent with the X-ray structure.[133] Another example is described for the hairpin ribozyme, where a uridine and a guanosine in a loop were photochemically cross-linked in high yields.[134]

Usually, the cross-linking yields obtained with the natural nucleotides are low. Moreover, the wavelength required for activation is around 260 nm, which is not specific for a particular nucleotide. Thus, analogues which are more reactive and absorb at higher wavelength are usually preferred for photoactivated cross-linking. 4-Thiothymidine and 4-thiouridine (**10**), 5-bromouridine and 5-iodouridine (**17**) and 6-thioguanosine (**8**) meet this requirement and they also form stable Watson–Crick base pairs. The most frequently used analogues are 4-thiouridine (**10**) or 4-thiothymidine, which absorb at 331 nm and can be irradiated between 300 nm and 400 nm for activation.[135,136] Cross-linking can occur under anaerobic conditions by the addition at the 5, 6 double bond or at C-4 with loss of sulfur, by invoking a radical mechanism. Under aerobic conditions, sulfur becomes oxidized to sulfonic acid, which can subsequently be displaced by a nucleophile.

The first intrastrand RNA site-specific cross-link with 4-thiouridine was achieved upon irradiation of yeast $tRNA^{Val}$, which contains 4-thiouridine as a natural constituent.[137] 4-Thiouridine was also

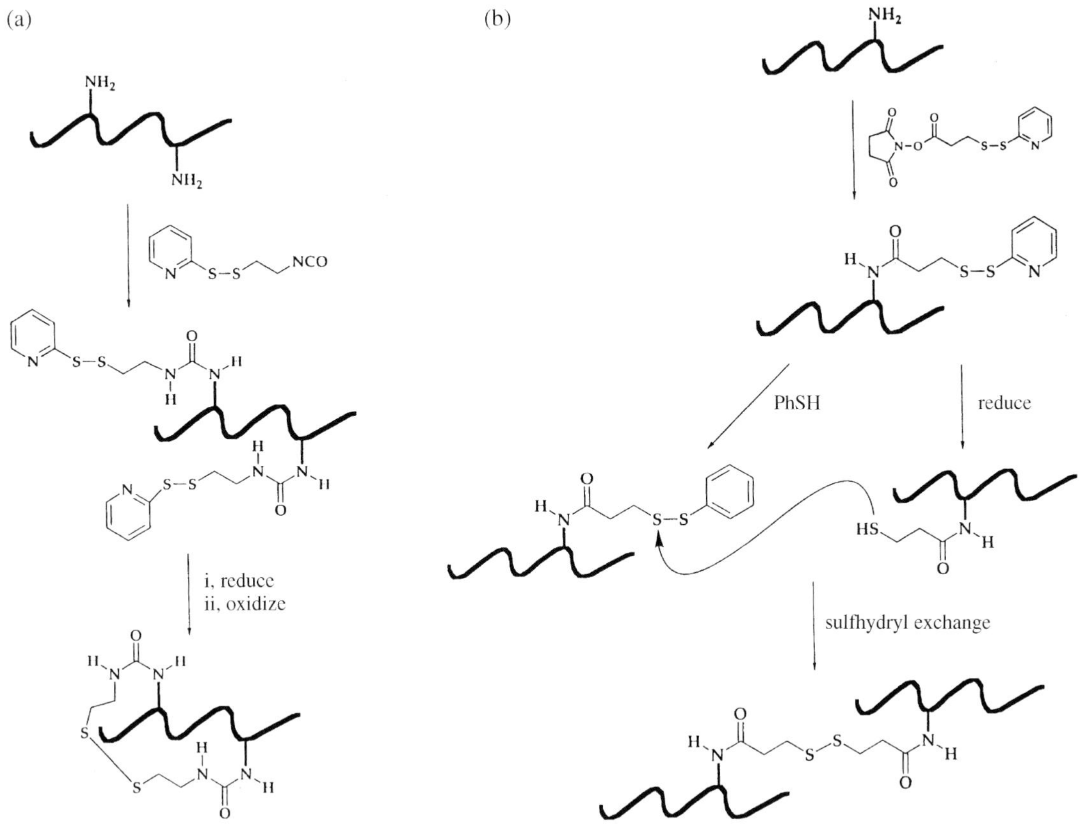

Figure 2 (a) Chemical cross-linking via the reaction of a sugar 2′-amino group and 2-pyridyl-3-isocyanatobenzyl disulfide. (b) Chemical crosslinking via the reaction of a 2′-amino group and *N*-hydroxysuccinimidyl-activated disulfide esters. The ribbon structure represents 2′-amino-modified RNA.

incorporated into oligoribonucleotides by enzymatic methods and depending on its proximal position, photocross-links were observed to ribosomal proteins and 16S RNA.[138]

4-Thiouridine has been incorporated at specific positions in an adenovirus pre-mRNA, using the Moore and Sharp ligation approach[7,8] to identify contacts to snRNAs during the splicing reaction.[139] Cross-links were observed with a loop sequence in U5 snRNA and with an invariant sequence of U6. Using a similar strategy, strand-specific cross-linking between 4-thiouridine-modified human $tRNA^{Lys}$ with HIV-1 reverse transcriptase subunits p66 and p51 have also been investigated.[140] 4-Thiouridine was used to cross-link small nuclear ribonucleoprotein particles U11 and U6 to the 5′-splice site of an AT-AC intron found in metazoan genes.[141]

6-Thioinosine and 4-thiouridine have been used as photoaffinity probes to study hammerhead and hairpin ribozyme conformations.[136] Multiple cross-links were found in both of the cases and the results were difficult to explain on the basis of models. It was concluded that these ribozymes can adopt multiple conformations making it difficult to identify catalytically competent conformations by this approach alone.

5-Bromouridine and 5-iodouridine (**17**), which can be activated at 308 nm and 325 nm, respectively, can also serve as convenient photoaffinity labels as exemplified by the cross-linking of an RNA transcript to the bacteriophage R17 coat protein.[142,143] The protein photodamage was considerably reduced with the iodo derivative.

The 5-position of uridine can be conveniently functionalized, without affecting the hydrogen bonding interaction, to attach various nonradioactive probes. In one of the first reports, biotin was covalently attached at the 5-position of uridine triphosphate for incorporation in RNA by transcription for a subsequent streptavidin-affinity isolation.[144] Many biotin phosphoramidites are now commercially available for solid-phase chemical coupling to oligonucleotides. A similar approach can also be used for the incorporation of noradioactive probes such as digoxigenin or fluorescein, instead of biotin.

In one study, multiple 5-methyleneaminouridine triphosphate residues were introduced in oligoribonucleotides by T7 RNA polymerase transcription of synthetic DNA templates.[145] The free amino group in the modified uridine was postsynthetically reacted with an azirinyl group-containing reagent. The photocross-linking results with the azirinyl–uridine derivative and 6-thioguanosine were used to study the binding of radiolabeled mRNA analogues to *E. coli* ribosomes. Precise cross-link positions were determined by a combination of ribonuclease H and T1 digestion and by primer extension analysis.

8-Azidoadenosine (**18**) and 8-azidoguanosine are frequently used for the photoaffinity labeling of enzymes.[146,147] Due to incompatibility of the azido group with phosphoramidite-based chemical synthesis, these analogues are not suitable for automated synthesis. However, enzymatic methods have been used to incorporate 8-azidoadenosine into the acceptor stem of tRNAs to probe its interaction with ribosomes.[148] 5-Azidouridine and 5-azidodeoxyuridine triphosphates are also suitable for photoactivated cross-linking and are substrates for RNA polymerases.[149] However, affinity labeling studies have not been reported for these analogues.

5-Sulfhydryluridine triphosphate has been incorporated enzymatically into RNA followed by subsequent derivatization with azidophenacyl bromide, for cross-linking to several RNA polymerases.[150]

15.8.2.2 Terminal cross-linking

Posttranscriptional reaction of a 5′-terminal guanosine phosphorothioate is an elegant method of attaching a photoactivatable azidophenacyl group to the transcripts. So far, it has mainly been used to investigate RNase P RNA interactions. Interestingly, the strategy involved transcription initiation at numerous positions and thus placed the terminal guanosine phosphorothioate along the RNase P RNA molecule. Upon irradiation of azidophenacyl-derivatized transcripts, intra- and intermolecular cross-links were obtained in the absence or presence of pre-tRNA.[151,152] The library of cross-links has facilitated significant refinement of the three-dimensional model of this ribozyme. A similar cross-linking study was also used to identify "neighbors" in the group I intron ribozyme.[153]

15.8.2.3 Metal derivatives

There have been few investigations regarding metal derivatives of oligoribonucleotides. In one study, 2′-*O*-methyl oligoribonucleotides were postsynthetically modified with transplatin to give a 1,3-intrastrand cross-linked oligomer.[154,155] When hybridized with complementary RNA strands, these intrastrand platinated oligomers triggered the formation of specific interstrand cross-links. The utility of transplatin-modified oligoribonucleotides was further demonstrated by introducing cross-links in Ha-ras mRNA and by concomitant inhibiton of cell proliferation.

15.9 CONCLUSIONS

The introduction of modified nucleotides into oligoribonucleotides or transcripts has greatly aided the understanding of mechanistic and structural aspects of many biochemical reactions involving RNA. Ribonucleotide analogues which can be enzymatically incorporated have found wider application because of their ease of introduction. Additionally, the analogues which confer resistance to modified RNA from nucleases are of particular interest for their possible use in the development of RNA-based therapeutics.

15.10 ACKNOWLEDGMENTS

Work in this laboratory was supported by the Deutsche Forschungsgemeinschaft and the Fonds der Chemischen Industrie. A fellowship from the Alexander von Humboldt-Stiftung (N.K.V.) is gratefully acknowledged.

15.11 REFERENCES

1. S. Verma and F. Eckstein, *Ann. Rev. Biochem*, 1998, **67**, 99.
2. S. Verma, N.K. Vaish and F. Eckstein, *Comp. Nat. Prod. Chem.*, 1999, **6**, 217.
3. M.J. Gait, C. Pritchard and G. Slim, in "Oligonucleotides and Analogs-A Practical Approach", ed. F. Eckstein, IRL Press, Oxford, 1991, 25
4. F. Wincott, A. Direnzo, C. Shaffer, S. Grimm, D. Tracz, C. Workman, D. Sweedler, C. Gonzalez, S. Scaringe and N. Usman, *Nucleic Acids Res.*, 1995, **23**, 2677.
5. R.H. Davis, *Curr. Opin. Biotechnol.*, 1995, **6**, 213.
6. O.C. Uhlenbeck and R.I. Gumport, in "The Enzymes", ed. P.D. Boyer, Academic Press, New York, 1982, 31
7. M.J. Moore and P.A. Sharp, *Science*, 1992, **256**, 992.
8. M.J. Moore and C.C. Query, in "RNA-Protein Interactions: A Practical Approach", ed. C. Smith, Oxford University Press, 1998, 75
9. F. Eckstein, *Annu. Rev. Biochem.*, 1985, **54**, 367.
10. A.D. Griffiths, B.V.L. Potter and I.C. Eperon, *Nucleic Acids Res.*, 1987, **15**, 4145.
11. H. Aurup, D.M. Williams and F. Eckstein, *Biochemistry*, 1992, **31**, 9636.
12. Y. Huang, F. Eckstein, R. Padilla and R. Sousa, *Biochemistry*, 1997, **36**, 8231.
13. A.B. Burgin and N.R. Pace, *EMBO J.*, 1990, **9**, 4111.
14. G.L. Igloi, *Biochemistry*, 1988, **27**, 3842.
15. L. Pauling, "The Nature of the Chemical Bond, 3rd edn.", Cornell University Press, Ithaca, NY, 1960.
16. R.G. Pearson, *Science*, 1966, **151**, 172.
17. V.L. Pecoraro, J.D. Hermes and W.W. Cleland, *Biochemistry*, 1984, **23**, 5262.
18. G. Zon and W.J. Stec, in "Oligonucleotides and Analogs-A Practical Approach", ed. F. Eckstein, IRL Press, Oxford, 1991, 87
19. W.J. Stec, A. Grajkowski, A. Kobylanska, B. Karwowski, M. Koziolkiewicz, K. Misiura, A. Okruszek, A. Wilk, P. Guga and M. Boczkowska, *J. Am. Chem. Soc*, 1995, **117**, 12–019.
20. P.M.J. Burgers and F. Eckstein, *Proc. Natl. Acad. Sci. USA*, 1978, **75**, 4798.
21. P.M.J. Burgers and F. Eckstein, *Biochemistry*, 1979, **18**, 592.
22. F. Eckstein, H.H. Schulz, H. Ruterjans, W. Haar and W. Maurer, *Biochemistry*, 1972, **11**, 3507.
23. F.R. Bryant and S.J. Benkovic, *Biochemistry*, 1979, **18**, 2825.
24. B.V.L. Potter, B.A. Connolly and F. Eckstein, *Biochemistry*, 1983, **22**, 1369.
25. P.A. Heaton and F. Eckstein, *Nucleic Acids Res.*, 1996, **24**, 850.
26. H. van Tol, J.M. Buzayan, P.A. Feldstein, F. Eckstein and G. Bruening, *Nucleic Acids Res.*, 1990, **18**, 1971.
27. G. Slim and M.J. Gait, *Nucleic Acids Res.*, 1991, **19**, 1183.
28. H.W. Pley, K.M. Flaherty and D.B. McKay, *Nature*, 1994, **372**, 68.
29. W.G. Scott, J.T. Finch and A. Klug, *Cell*, 1995, **81**, 991.
30. D.B. McKay, *RNA*, 1996, **2**, 395.
31. W.G. Scott, J.B. Murray, J.R.P. Arnold, B.L. Stoddard and A. Klug, *Science*, 1996, **274**, 2065.
32. J.A. McSwiggen and T.R. Cech, *Science*, 1989, **244**, 679.
33. J. Rajagopal, J.A. Doudna and J.W. Szostak, *Science*, 1989, **244**, 692.
34. M.J. Moore and P.A. Sharp, *Nature*, 1993, **365**, 364.
35. K.L. Maschhoff and R.A. Padgett, *Nucleic Acids Res.*, 1993, **21**, 5456.
36. G. Chanfreau and A. Jacquier, *Science*, 1994, **266**, 1383.
37. R.A. Padgett, M. Podar, S.C. Boulanger and P.S. Perlman, *Science*, 1994, **266**, 1685.
38. P.A. Sharp, *Cell*, 1994, **77**, 805.
39. D. Herschlag, J.A. Piccirilli and T.R. Cech, *Biochemistry*, 1991, **30**, 4844.
40. A.H. Polesky, M.E. Dahlberg, S.J. Benkovic, N.D.F. Grindley and C.M. Joyce, *J. Biol. Chem.*, 1992, **267**, 8417.
41. J.P. Noel, H.E. Hamm and P.B. Sigler, *Nature*, 1993, **366**, 654.
42. G. Gish and F. Eckstein, *Science*, 1988, **240**, 1520.
43. D. Schatz, R. Leberman and F. Eckstein, *Proc. Natl. Acad. Sci. USA*, 1991, **88**, 6132.
44. C.A. Brautigam and T.A. Steitz, *J. Mol. Biol.*, 1998, **277**, 363.
45. R.B. Waring, *Nucleic Acids Res.*, 1989, **17**, 10–281.
46. E.L. Christian and M. Yarus, *Biochemistry*, 1993, **32**, 4475.
47. J.H. Cate, R.L. Hanna and J.A. Doudna, *Nature Struct. Biol*, 1997, **4**, 553.
48. S.A. Strobel and K. Shetty, *Proc. Natl. Acad. Sci. USA*, 1997, **94**, 2903.
49. M.E. Harris and N.R. Pace, *RNA*, 1995, **1**, 210.
50. W.D. Hardt, J.M. Warnecke, V.A. Erdmann and R.K. Hartmann, *EMBO J.*, 1995, **14**, 2935.
51. J.M. Warnecke, J.P. Furste, W.D. Hardt, V.A. Erdmann and R.K. Hartmann, *Proc. Natl. Acad. Soc. USA*, 1996, **93**, 8924.
52. Y. Chen, X.Q. Li and P. Gegenheimer, *Biochemistry*, 1997, **36**, 2425.
53. D.E. Ruffner and O.C. Uhlenbeck, *Nucleic Acids Res.*, 1990, **18**, 6025.
54. S.C. Dahm and O.C. Uhlenbeck, *Biochemistry*, 1991, **30**, 9464.
55. D.-M. Zhou, P.K.R. Kumar, L.-H. Zhang and K. Taira, *J. Am. Chem. Soc*, 1996, **118**, 8969.
56. R. Knoll, R. Bald and J.P. Furste, *RNA*, 1997, **3**, 132.
57. O. Mitrasinovic and L.M. Epstein, *Nucleic Acids Res.*, 1997, **25**, 2189.
58. Y.-H. Jeoung, P.K.R. Kumar, Y.-A. Suh, K. Taira and S. Nishikawa, *Nucleic Acids Res.*, 1994, **22**, 3722.
59. J.F. Milligan and O.C. Uhlenbeck, *Biochemistry*, 1989, **28**, 2849.

60. V. Biou, A. Yaremchuk, M. Tukalo and S. Cusack, *Science*, 1994, **263**, 1404.
61. J. Rudinger, J.D. Puglisi, J. Putz, D. Schatz, F. Eckstein, C. Florentz and R. Giege, *Proc. Natl. Acad. Soc.USA*, 1992, **89**, 5882.
62. R. Kreutzer, D. Kern, R. Giege and J. Rudinger, *Nucleic Acids Res.*, 1995, **23**, 4598.
63. M.G. Caprara, G. Mohr and A.M. Lambowitz, *J. Mol. Biol.*, 1996, **257**, 512.
64. M.G. Caprara, V. Lehnert, A.M. Lambowitz and E. Westhof, *Cell*, 1996, **87**, 1135.
65. M. Dabrowski, C.M.T. Spahn and K.H. Nierhaus, *EMBO J.*, 1995, **14**, 4872.
66. W.D. Hardt, V.A. Erdmann and R.K. Hartmann, *RNA*, 1996, **2**, 1189.
67. T. Persson, C.S. Vortler, O. Fedorova and F. Eckstein, *RNA*, 1998, **4**, 1444.
68. R. Sousa and R. Padilla, *EMBO J*, 1995, **14**, 4609.
69. R. Aphasizhev, A. Theobald-Dietrich, D. Kostyuk, S.N. Kochetkov, L. Kisselev, R. Giege and F. Fasiolo, *RNA*, 1997, **3**, 893.
70. J.-C. Francois, T. Saison-Behmoaras, C. Barbier, M. Chassignol, N.T. Thuong and C. Helene, *Proc. Natl. Acad. Sci. USA*, 1989, **86**, 9702.
71. S.M. Gryaznov and R.L. Letsinger, *Nucleic Acids Res.*, 1993, **21**, 1403.
72. B.C.F. Chu and L.E. Orgel, *DNA Cell Biol.*, 1990, **9**, 71.
73. R. Cosstick, L.W. McLaughlin and F. Eckstein, *Nucleic Acids Res.*, 1984, **12**, 1791.
74. S. Alefelder, B.K. Patel and F. Eckstein, *Nucleic Acids Res*, 1998, **26**, 4983.
75. J.A. Piccirilli, J.S. Vyle, M.H. Caruthers and T.R. Cech, *Nature*, 1993, **361**, 85.
76. L.B. Weinstein, B.C.N.M. Jones, R. Cosstick and T.R. Cech, *Nature*, 1997, **388**, 805.
77. E.J. Sontheimer, S.G. Sun and J.A. Piccirilli, *Nature*, 1997, **388**, 801.
78. R.G. Kuimelis and L.W. McLaughlin, *Biochemistry*, 1996, **35**, 5308.
79. D.-M. Zhou, N. Usman, F.E. Wincott, J. Matulic-Adamic, M. Orita, L.-H. Zhang, M. Komiyama, P.K.R. Kumar and K. Taira, *J. Am. Chem. Soc.*, 1996, **118**, 5862.
80. J.B. Thomson, B.K. Patel, V. Jimenez, K. Eckart and F. Eckstein, *J. Org. Chem.*, 1996, **61**, 6273.
81. B.E. Eaton and W.A. Pieken, *Annu. Rev. Biochem.*, 1995, **64**, 837.
82. R.K. Gaur, L.W. McLaughlin and M.R. Green, *RNA*, 1997, **3**, 861.
83. S. Limauro, F. Benseler and L.W. McLaughlin, *Bioorg. Med. Chem. Lett.*, 1994, **4**, 2189.
84. J. Putz, C. Florentz, F. Benseler and R. Giege, *Nature Struct. Biol.*, 1994, **1**, 580.
85. S. Bevers, G.B. Xiang and L.W. McLaughlin, *Biochemistry*, 1996, **35**, 6483.
86. T. Tuschl, M.M.P. Ng, W. Pieken, F. Benseler and F. Eckstein, *Biochemistry*, 1993, **32**, 11–658.
87. J.A. Grasby, K. Mersmann, M. Singh and M.J. Gait, *Biochemistry*, 1995, **34**, 4068.
88. Q.L. Liu, J.B. Green, A. Khodadi, P. Haeberli, L. Beigelman and A.M. Pyle, *J. Mol. Biol.*, 1997, **267**, 163.
89. J.B. Murray, C.J. Adams, J.R.P. Arnold and P.G. Stockley, *Biochem. J.*, 1995, **311**, 487.
90. A.B. Burgin Jr., C. Gonzalez, J. Matulic-Adamic, A.M. Karpeisky, N. Usman, J.A. McSwiggen and L. Beigelman, *Biochemistry*, 1996, **35**, 14–090.
91. D.J. Earnshaw and M.J. Gait, *Antisense Nucleic Acid Drug Dev*, 1997, **7**, 403.
92. J.A. Latham, R. Johnson and J.J. Toole, *Nucleic Acids Res*, 1994, **22**, 2817.
93. T.W. Wiegand, R.C. Janssen and B.E. Eaton, *Chem. Biol.*, 1997, **4**, 675.
94. T.M. Tarasow, S.L. Tarasow and B.E. Eaton, *Nature*, 1997, **389**, 54.
95. A. Peracchi, L. Beigelman, N. Usman and D. Herschlag, *Proc. Natl. Acad. Sci. USA*, 1996, **93**, 11–522.
96. S. Schmidt, L. Beigelman, A. Karpeisky, N. Usman, U.S. Sorensen and M.J. Gait, *Nucleic Acids Res.*, 1996, **24**, 573.
97. L. Beigelman, A. Karpeisky and N. Usman, *Bioorg. Med. Chem. Lett.*, 1994, **4**, 1715.
98. L. Beigelman, A. Karpeisky, J. Matulic-Adamic, C. Gonzalez and N. Usman, *Nucleosides Nucleotides*, 1995, **14**, 907.
99. W. Saenger "Principles of Nucleic Acid Structure", Springer-Verlag, New York, 1984.
100. B. Bramlage, S. Alefelder, P. Marschall and F. Eckstein, *Nucleic Acids Res.*, 1999, **27**, 3159.
101. D.B. Olsen, F. Benseler, H. Aurup, W.A. Pieken and F. Eckstein, *Biochemistry*, 1991, **30**, 9735.
102. W.A. Pieken, D.B. Olsen, F. Benseler, H. Aurup and F. Eckstein, *Science*, 1991, **253**, 314.
103. G. Paolella, B.S. Sproat and A.I. Lamond, *EMBO J*, 1992, **11**, 1913.
104. T.C. Jarvis, F.E. Wincott, L.J. Alby, J.A. McSwiggen, L. Beigelman, J. Gustofson, A. DiRenzo, K. Levy, M. Arthur, J. Matulic-Adamic, A. Karpeisky, C. Gonzalez, T.M. Woolf, N. Usman and D.T. Stinchcomb, *J. Biol. Chem.*, 1996, **271**, 29–107.
105. B.M. Chowrira, A. Berzal-Herranz, C.F. Keller and J.M. Burke, *J. Biol. Chem.*, 1993, **268**, 19–458.
106. A.M. Pyle and T.R. Cech, *Nature*, 1991, **350**, 628.
107. A.M. Pyle, F.L. Murphy and T.R. Cech, *Nature*, 1992, **358**, 123.
108. D. Herschlag, F. Eckstein and T.R. Cech, *Biochemistry*, 1993, **32**, 8299.
109. D. Herschlag, F. Eckstein and T.R. Cech, *Biochemistry*, 1993, **32**, 8312.
110. D.L. Abramovitz, R.A. Friedman and A.M. Pyle, *Science*, 1996, **271**, 1410.
111. C.L. Dantzman and L.L. Kiessling, *J. Am. Chem. Soc*, 1996, **118**, 11–715.
112. M.L. Hamm and J.A. Piccirilli, *J. Org. Chem*, 1997, **62**, 3415.
113. S.E. Osborne and A.D. Ellington, *Chem. Rev.*, 1997, **97**, 349.
114. D. Jellinek, L.S. Green, C. Bell, C.K. Lynott, N. Gill, C. Vargeese, G. Kirschenheuter, D.P.C. McGee, P. Abesinghe, W.A. Pieken, R. Shapiro, D.B. Rifkin, D. Moscatelli 103nd N. Janjic, *Biochemistry*, 1995, **34**, 11–363.
115. Y. Lin, D. Nieuwlandt, A. Magallanez, B. Feistner and S.D. Jayasena, *Nucleic Acids Res.*, 1996, **24**, 3407.
116. S.-W. Lee and B.A. Sullenger, *Nature Biotech.*, 1997, **15**, 41.
117. N.C. Pagratis, C. Bell, Y.-F. Chang, S. Jennings, T. Fitzwater, D. Jellinek and C. Dang, *Nature Biotech.*, 1997, **15**, 68.

118. S.M.L. Barabino, B.S. Sproat, U. Ryder, B.J. Blencowe and A.I. Lamond, *EMBO J.*, 1989, **8**, 4171.
119. B.J. Blencowe, B.S. Sproat, U. Ryder, S. Barabino and A.I. Lamond, *Cell*, 1989, **59**, 531.
120. F. Burlina, A. Favre, J.-L. Fourrey and M. Thomas, *J. Chem. Soc., Chem. Commun.*, 1996, , 1623.
121. D.P. Millar, *Curr. Opin. Struct. Biol.*, 1996, **6**, 322.
122. D.C. Ward, E. Reich and L. Stryer, *J. Biol. Chem.*, 1969, **244**, 1228.
123. M. Menger, T. Tuschl, F. Eckstein and D. Porschke, *Biochemistry*, 1996, **35**, 14–710.
124. N.G. Walter and J.M. Burke, *RNA*, 1997, **3**, 392.
125. T. Tuschl, C. Gohlke, T.M. Jovin, E. Westhof and F. Eckstein, *Science*, 1994, **266**, 785.
126. G.S. Bassi, A.I.H. Murchie, F. Walter, R.M. Clegg and D.M.J. Lilley, *EMBO J.*, 1997, **16**, 7481.
127. C.R. Allerson, S.L. Chen and G.L. Verdine, *J. Am. Chem. Soc.*, 1997, **119**, 7423.
128. C.R. Allerson and G.L. Verdine, *Chem. Biol.*, 1995, **2**, 667.
129. J.T. Goodwin, S.E. Osborne, E.J. Scholle and G.D. Glick, *J. Am. Chem. Soc.*, 1996, **118**, 5207.
130. S.T. Sigurdsson, T. Tuschl and F. Eckstein, *RNA*, 1995, **1**, 575.
131. D.J. Earnshaw, B. Masquida, S. Muller, S.T. Sigurdsson, F. Eckstein, E. Westhof and M.J. Gait, *J. Mol. Biol.*, 1997, **274**, 197.
132. S.B. Cohen and T.R. Cech, *J. Am. Chem. Soc.*, 1997, **119**, 6259.
133. L.S. Behlen, J.R. Sampson and O.C. Uhlenbeck, *Nucleic Acids Res.*, 1992, **20**, 4055.
134. J.M. Burke, S.E. Butcher and B. Sargueil, in "Nucleic Acids & Molecular Biology", ed. F. Eckstein and D.M.J. Lilley,, Springer-Verlar, Berlin Heidelberg, 1996, 129
135. T.T. Nikiforov and B.A. Connolly, *Nucleic Acids Res.*, 1992, **20**, 1209.
136. A. Favre and J.-L. Fourrey, *Acc. Chem. Res.*, 1995, **28**, 375.
137. A. Favre, A.M. Michelson and M. Yaniv, *J. Mol. Biol.*, 1971, **58**, 367.
138. O. Dontsova, A. Kopylov and R. Brimacombe, *EMBO J.*, 1991, **10**, 2613.
139. E.J. Sontheimer and J.A. Steitz, *Science*, 1993, **262**, 1989.
140. Y. Mishima and J.A. Steitz, *EMBO J.*, 1995, **14**, 2679.
141. Y.-T. Yu and J.A. Steitz, *Proc. Natl. Acad. Sci. USA*, 1997, **94**, 6030.
142. J.M. Gott, M.C. Willis, T.H. Koch and O.C. Uhlenbeck, *Biochemistry*, 1991, **30**, 6290.
143. M.C. Willis, B.J. Hicke, O.C. Uhlenbeck, T.R. Cech and T.H. Koch, *Science*, 1993, **262**, 1255.
144. P.R. Langer, A.A. Waldrop and D.C. Ward, *Proc. Natl. Acad. Sci. USA*, 1981, **78**, 6633.
145. P.V. Sergiev, I.N. Lavrik, V.A. Wlasoff, S.S. Dokudovskaya, O.A. Dontsova, A.A. Bogadonov and R. Brimacombe, *RNA*, 1997, **3**, 464.
146. B. Jayaram and B.E. Haley, *J. Biol. Chem.*, 1994, **269**, 3233.
147. A.J. Chavan, Y. Nemoto, S. Narumiya, S. Kozaki and B.E. Haley, *J. Biol. Chem.*, 1992, **267**, 14–866.
148. J. Wower, S.S. Hixson and R.A. Zimmermann, *Proc. Natl. Acad. Sci. USA*, 1989, **86**, 5232.
149. R.K. Evans and B.E. Haley, *Biochemistry*, 1987, **26**, 269.
150. B.K. He, D.L. Riggs and M.M. Hanna, *Nucleic Acids Res.*, 1995, **23**, 1231.
151. M.E. Harris, J.M. Nolan, A. Malhotra, J.W. Brown, S.C. Harvey and N.R. Pace, *EMBO J.*, 1994, **13**, 3953.
152. M.E. Harris, A.V. Kazantsev, J.-L. Chen and N.R. Pace, *RNA*, 1997, **3**, 561.
153. J.-F. Wang, W.D. Downs and T.R. Cech, *Science*, 1993, **260**, 504.
154. M. Boudvillain, M. Guerin, R. Dalbies, T. Saison-Behmoaras and M. Leng, *Biochemistry*, 1997, **36**, 2925.
155. C. Colombier, M. Boudvillain and M. Leng, *Antisense Nucleic Acid Drug Dev.*, 1997, **7**, 397.

16

RNA in Biotechnology: Towards a Role for Ribozymes in Gene Therapy

MASAKI WARASHINA[1], TOMOKO KUWABARA[1], HIROAKI KAWASAKI[1], JUN OHKAWA[1] and KAZUNARI TAIRA[1]
The University of Tokyo, Tokyo, Japan

[1]Also: Gene Discovery Research Center, National Institute of Advanced Industrial Science and Technology (AIST), Tsukuba Science City, Japan

16.1 INTRODUCTION

Catalytic RNAs include: hammerhead, hairpin, and hepatitis delta virus (HDV) ribozymes; group I and II introns; the RNA subunit of RNase P; and ribosomal RNA (Figure 1).[1–12] Among these catalytic RNAs, the first two ribozymes to be discovered, by Altman and Cech, respectively, were the RNA subunit of RNase P and a group I intron.[1,2] Within five years of these discoveries, small ribozymes, such as hammerhead, hairpin and HDV ribozymes, were discovered in studies of the replication, via a rolling-circle mechanism, of certain viroids, satellite RNAs and an RNA virus.[3–5,8,10,12] The hammerhead ribozyme is the smallest of all these catalytic RNAs.[13,14]

The reaction mechanisms of large and small ribozymes are different. Large ribozymes, such as group I introns and the catalytic RNA subunit of RNase P, use external nucleophiles. By contrast, small ribozymes, such as hammerheads, hairpins and HDV ribozymes, use an internal nucleophile, namely, the 2′-oxygen at the cleavage site, with resultant formation of a cyclic phosphate. Since large ribozymes do not require the 2′-hydroxyl group as a nucleophile at the cleavage site, the ribozymes of *Tetrahymena* and of RNase P can cleave DNA substrates in addition to RNA substrates.[15–17] Ribozymes are now recognized to be metalloenzymes.[10,12,18–28] In the reaction mediated by the ribozyme from *Tetrahymena*, a metal ion catalyst coordinates directly to and stabilizes the developing negative charge of the leaving 3′-oxygen, acting as a Lewis acid.[19] Moreover, very recent evidence suggests that a metal ion activates the nucleophilic 3′-hydroxyl group of guanosine in the same reaction, lending support to the proposed double-metal-ion mechanism of catalysis.[20,28] In reactions mediated by hammerhead ribozymes, base catalysis mediated by Mg^{2+}-hydroxide was first proposed on the basis of profiles of pH *versus* the rate of the reaction.[18] However, it has also been noted that a general double-metal-ion mechanism, in which metal ions act as Lewis acids and coordinate directly to the 2′-hydroxyl moiety and the leaving 5′-oxygen to activate a nucleophile and to stabilize a developing negative charge, respectively, might well explain reactions catalyzed by hammerhead ribozymes.[24,25,29–34] By contrast, the absence of metal-ion-mediated catalysis has been reported in the reactions catalyzed by hairpin ribozymes.[35–37] Thus, hairpin ribozymes

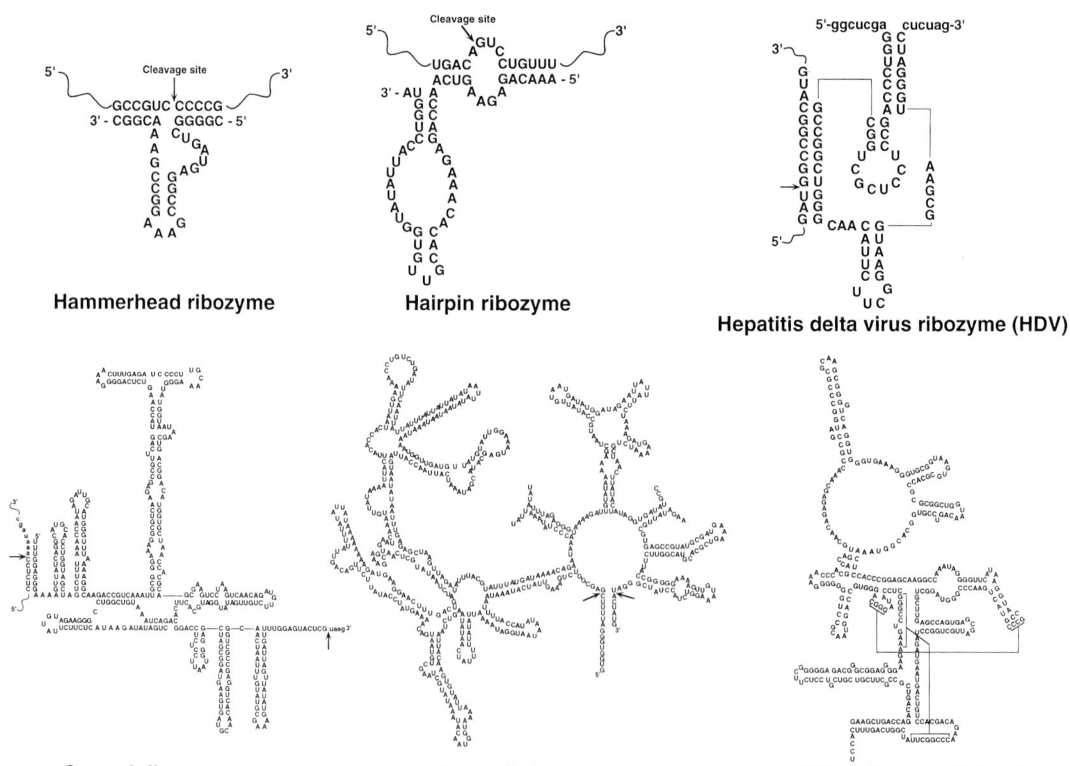

Figure 1. Secondary structures of six types of ribozyme, namely, a hammerhead ribozyme, a hairpin ribozyme, and the hepatitis delta virus ribozyme; group I and group II introns; and the RNA subunit of RNase P. The cleavage sites in substrates are indicated by an arrow. The RNA subunit of RNase P cleaves precursors to tRNAs so no cleavage site is shown.

can be classified as a distinct class of ribozymes that do not require metal ions as catalysts. It should also be noted that, under extreme conditions (in the presence of 1 to 4 M monovalent cations such as Li^+, Na^+, and NH_4^+), hammerhead ribozymes do not require divalent metal ions for catalysis.[38]

The extensive efforts of many research groups have revealed some details of the mechanisms of ribozyme-mediated cleavage of RNA and studies of ribozymes have become very exciting. It is considered likely that ribozymes are fossil molecules that originated in the primitive "RNA world" and it is likely that elucidation of their mechanisms of action will enhance our understanding of the life processes of primitive organisms. The rapidly developing field of RNA catalysis is of particular current interest not only because of the intrinsic catalytic properties of ribozymes but also because of their potential utility as therapeutic agents and specific regulators of gene expression.[12,39–46] To date, numerous studies directed towards the application of ribozymes *in vivo* have been performed and many successful experiments, aimed at the use of ribozymes for the suppression of gene expression in different organisms, have been reported.[47–68] In this chapter, we shall discuss a variety of ribozymes and various methods that have been tested for the efficient inhibition of gene expression by ribozymes. We recently succeeded in constructing an efficient ribozyme-expression system under the control of our improved version of the promoter of a human gene for $tRNA^{Val}$.[61–63,66–68] Using this ribozyme-expression system, we succeeded in analyzing the intracellular functions of endogenous transcriptional co-activators[55,61] and in creating a novel and highly specific method for the suppression of gene expression.[62,63,66–68] We shall focus on these examples in this chapter. RNA is not the only nucleic acid with catalytic activity. The recent discovery of DNA enzymes, created by a selection procedure *in vitro*, has created still more excitement in the field of catalytic nucleic acids that might have important applications *in vivo*, as well as in the field of mechanisms of catalysis.[69–74]

16.2 HAMMERHEAD RIBOZYMES AS TOOLS FOR THE SUPPRESSION OF GENE EXPRESSION

16.2.1 General Features of Hammerhead Ribozymes

Hammerhead ribozymes are among the smallest catalytic RNAs. They are called "hammerheads" because of their two-dimensional structure.[8] The sequence motif, with three duplex stems and a conserved "core" of two non-helical segments that are responsible for the self-cleavage reaction (*cis* action), was first recognized in the satellite RNAs of certain viruses.[8] However, hammerhead ribozymes have been generated in the laboratory that are able to act "in *trans*",[13,14] and *trans*-acting hammerhead ribozymes, consisting of an antisense section (stems I and stem III) and a catalytic core with a flanking stem-loop II section, have been tested as potential therapeutic agents and used extensively in mechanistic studies (Figure 2(A)).[14] Such RNAs can cleave oligoribonucleotides at specific sites (namely, after the sequence NUX, where N can be A, G, C or U and X can be A, C or U, with the most efficient cleavage occurring after a GUC triplet).[75–77] In the case of *trans*-acting hammerhead ribozymes, most of the conserved nucleotides that are essential for the cleavage reaction are included in the catalytic core.[14] Therefore, RNA molecules consisting of only about 30 nucleotides can be generated for use as artificial endonucleases that can cleave specific RNA molecules. Because of their small size and considerable potential as inhibitors of gene expression, hammerhead ribozymes are the ribozymes that are used most frequently in studies directed towards therapeutic applications *in vivo*.

16.2.2 Sequence Requirements of Hammerhead Ribozymes

Phylogenetic analysis, extensive mutagenesis experiments, and selection procedures *in vitro* have been used in attempts to identify the sequence requirements of an active hammerhead ribozyme.[8,12,70,75,78–81] In the catalytic core of the hammerhead ribozyme in Figure 2(A), replacement of certain nucleotides, with the exception of residue of U7, dramatically reduces cleavage activity. Selection *in vitro* for active hammerhead ribozymes revealed that active sequences correspond broadly to the consensus core sequence, and no other sequences support efficient cleavage,[80] an indication that the consensus sequence derived from viruses and virusoids is probably the optimal sequence. Nonetheless, chemical modification at U7 produced a non-natural ribozyme that had higher activity than the wild-type hammerhead

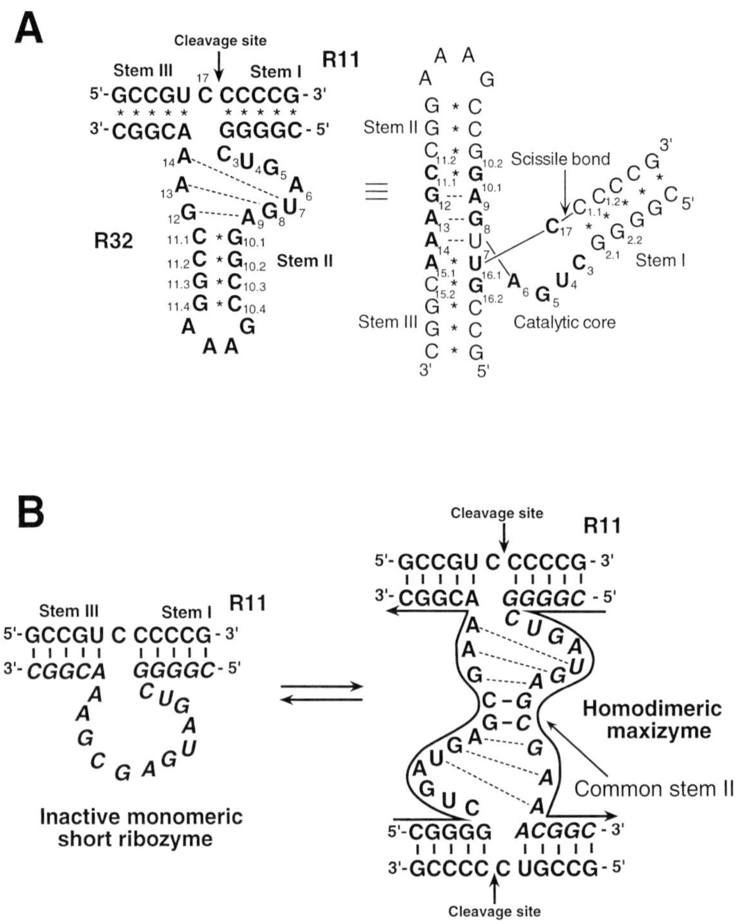

Figure 2. Secondary structures of (A) a wild-type hammerhead ribozyme (R32) and (B) a maxizyme that is capable of forming a homodimer. A schematic representation of the overall folding of the hammerhead ribozyme is also shown in (A) on the right.

ribozyme.[82] We also isolated a ribozyme with an extra G inserted within the catalytic core and the mutant ribozyme had higher activity than that of the parental ribozyme, an indication that there exist RNA motifs with higher catalytic activity than the natural motifs.[83,84] In stem I, any residue can be tolerated without substantial loss of cleavage activity. With the exception of the cleavage triplet, substitutions of nucleotides in stem III are also tolerated.

With respect to the important trinucleotide at the cleavage site, the NUX rule (where N can be A, U, G or C, and X can be A, U or C) is generally accepted. This rule states that any oligonucleotide with a NUX triplet can be cleaved by a hammerhead ribozyme.[75,76] The first residue, N, can fundamentally be A, U, G or C but the actual nucleotide has a considerable effect on the efficiency of cleavage by a hammerhead ribozyme. Among NUX triplets, GUC is cleaved the most efficiently under k_{cat}/K_M conditions namely, when the encounter of ribozymes and their substrates is part of the rate-limiting step, with CUC and UUC being cleaved somewhat less efficiently.[76] Therefore, when a target site in a *trans*-acting system (an intermolecular reaction) is to be chosen, GUC or CUC are likely to be preferred. However, in *cis*-acting systems (intramolecular reactions), in which K_M values are irrelevant, other triplets, such as AUC, GUA, and AUA, can be chosen since these triplets are associated with high values of k_{cat}. In fact, the minus strand of the virusoid of Lucerne transient streak virus, (−)vLTSV, and the plus strand of the satellite RNA of barley yellow dwarf virus, (+)sBYDV, cleave the GUA triplet and the AUA triplet, respectively, in the hammerhead-catalyzed cleavage reactions that occur during their replication.[76] Eckstein and his coworkers demonstrated recently that the GAC triplet can, in fact, also be cleaved by a hammerhead ribozyme. In addition, Eckstein's group generated a hammerhead-like ribozyme that cleaved RNAs, after an AUG triplet, that were not cleaved by conventional hammerhead ribozymes. They used an *in vitro* selection procedure to identify this ribozyme and they found that it had a catalytic core sequence unlike those of "conventional" hammerhead ribozymes.[80,81]

In stem II, the $G_{10.1}$–$C_{11.1}$ pair adjacent to the catalytic core is essential for efficient catalysis.[75,79,85,86] The nucleotides in the loop at the end of stem II can be varied, although, in many cases, a stable GAAA tetra nucleotide loop is used. Stem II is the only helix in the hammerhead ribozyme that is not involved in binding of the substrate and it can be shortened without complete loss of activity. The minimal length of helix II was found to be two base pairs and hammerhead ribozymes with a short stem II are designated minizymes. The activities of most such minizymes are very low.[78,86–91] However, we found that some short ribozymes, with short oligonucleotide linkers instead of the stem-loop II region, can form homodimers or heterodimers that are very active (Figure 2(B)).[62,63,66,92,93] In order to distinguish monomeric forms of conventional minizymes that have extremely low activity from our novel dimers with high-level activity, we chose the name "maxizyme" for the latter, very active, short ribozymes that are capable of forming dimers. Since the maxizymes tested to date have been more effective, in mammalian cells, in reducing gene expression than "conventional" hammerhead ribozymes, the homodimeric or heterodimeric maxizymes appear to have considerable potential utility as gene-inactivating agents (see below for details).[62,63,66] More recently, Conaty *et al.* succeeded in selection *in vitro* of active minimized ribozymes with a 6-nt linker instead of the stem-loop II region.[94] Their minimized ribozymes had significant cleavage activity at low Mg^{2+} concentrations, suggesting a potential utility *in vivo* of the minimized ribozymes where the concentration of Mg^{2+} is lower than that commonly used *in vitro* kinetic analysis.

16.2.3 Expression Systems for Ribozymes

The specific association of nucleic acid-based enzymes, such as ribozymes, with their targets *via* base pairing and the subsequent cleavage of the RNA substrate suggest that these catalytic molecules might be useful for gene therapy. There are basically two ways to introduce ribozymes into cells. One way involves a drug-delivery system (DDS) in which a chemically synthesized ribozyme is encapsulated in liposomes or some other similar vehicle and delivered to target cells.[95] The other way involves generating ribozymes within cells by transcription from the corresponding DNA template (gene therapy).[39–68] Current gene-therapy technology is limited primarily by the necessity for manipulations of target tissues outside the body (*ex vivo*). In other words, target cells must be removed from the body, manipulated, and returned. Therefore, the limitations that determine the genetic diseases that can potentially be treated are linked directly to the limitations of current cell biology.[96,97]

Many successful attempts at using ribozymes for the suppression of gene expression have been reported.[39–68] However, the efficacy of ribozymes *in vitro* is not necessarily correlated with their functional activity in vivo.[67] When a ribozyme is generated within a cell by transcription from an expression vector, various factors, such as the transcription rate, the stability and localization of the transcript, and the cleavage activity, are likely to influence the functional activity of the ribozyme in vivo.[44,50,67,98] Since such factors depend, in large part, on the activities of cellular proteins, it is important that the ribozymes' characteristics should be appropriate for optimal functioning in the intracellular environment. Thus, the successful intracellular action of a ribozyme depends significantly on the choice of expression system. It is also important that active ribozymes be selected from the pool of random sequences within the intracellular environment.[56,83,99] At present, RNA polymerase system II (pol II system) and RNA polymerase system III (pol III system) are the main systems used for the expression of ribozymes in the eukaryotic systems. Of course, these expression systems are not limited to the expression of hammerhead ribozymes and can be exploited for the expression of other ribozymes, such as hairpin ribozymes, and antisense RNAs. In the following sections, we will discuss examples of the successful intracellular application of ribozymes, including examples from our laboratory, and we will identify some of the problems that remain to be solved.

16.2.3.1 RNA polymerase II system

The RNA polymerase II system is the system that is normally used for the expression of proteins in cells. Numerous constructs have been designed that use this system for the efficient production of proteins and this type of expression system was adopted in early experiments for the expression of ribozymes. In attempts to express functionally active ribozymes, strong and constitutive promoters, such as the cytomegalovirus (CMV) immediate-early-gene enhancer/promoter, the simian virus 40 (SV40) early enhancer/promoter, and the promoter of the human gene for β-actin have been used successfully. Cameron and Jennings demonstrated the functional activity of a ribozyme in a mammalian cell system.[100] In their study, the ribozyme sequence targeted to the mRNA for chloramphenicol acetyl transferase (CAT) was introduced into the 3′ noncoding region of the mRNA for luciferase, and transcription was controlled by the SV40 early promoter. In this system, the expression of CAT was suppressed to 30% of the control level. By contrast, Sarver *et al.* used the promoter of the human gene for β-actin for expression of a ribozyme targeted to HIV-1 RNA.[47] They achieved a considerable reduction in the rate of viral replication in cultured cells.

Although strong and constitutive promoters of the type mentioned above are effective in cultured cells, they are not always suitable for use in transgenic animal systems. In addition, they may not always be useful under conditions when analysis depends on, for example, the stage of development of germ cells or the phase of the cell cycle. Under such conditions, a temporally controlled promoter is required. Thus, for example, Zhao and Pick used the promoter of a heat-shock gene to control expression of a ribozyme targeted to the transcript of the *fushi tarazu* gene.[101] They generated transgenic flies that carried a heat-shock-inducible ribozyme-expression cassette and were able to examine the function of the *fushi tarazu* gene at each stage of development by appropriately timed induction of the ribozyme. Efrat *et al.* used the promoter of a gene for insulin, namely, a tissue-specific promoter that operates specifically in pancreatic cells, to regulate the expression of a ribozyme targeted to the transcript of a gene for glucokinase.[102] Mutations in this gene are found in patients with maturity-onset diabetes of the young (MODY). To generate an animal model of MODY, they established a line of transgenic mice and characterized the relationship between the expression of the gene for glucokinase and this kind of diabetes. However, such successful examples are rare. Since the efficiency of transcription from a conditionally controlled promoter is lower than that from a constitutive-type promoter, such as the CMV or SV40 promoter, it is often difficult to obtain an effective intracellular concentration of the ribozyme of interest.

The pol II system consists basically of an enhancer/promoter sequence and a terminator sequence, and the DNA between these sequences is transcribed by RNA polymerase II (Figure 3(A)). The transcribed RNA (usually referred to as mRNA) is processed in several ways, such as RNA capping, addition of poly-A, and removal of introns. These modifications are essential for transport from the nucleus to the cytoplasm, for stability, and for the effective translation of the transcribed RNA. However, such modifications are likely to facilitate the binding of ribosomes to the mRNA for translation and it is possible that such binding of ribosomes might interfere with the association of the transcribed ribozyme with its target RNA. Moreover, the pol II promoter is not suitable for production of short RNAs.[103] In the pol II system, several hundred nucleotides at least are necessary between the promoter and the terminator for effective transcription and termination at the correct site. Thus, either extra sequences must be added at both ends of the ribozyme sequence or, alternatively, the ribozyme sequence must be inserted into the noncoding region of a stable mRNA, such as CAT mRNA. However, the long 5′ and 3′ extra sequences have undesirable effects on cleavage activity.[98,104] Therefore, a 100- or 1000-fold molar excess of ribozyme over the target RNA is required for successful suppression of the expression of the target gene.

We developed a novel system for the expression of ribozymes (Figure 4).[48,53,60,105–107] Our system consists of a promoter, one *trans*-acting and two *cis*-acting ribozymes and terminator sequences, with the *trans*-acting ribozyme sequence inserted between the two *cis*-acting ribozyme sequences (Figure 4(A)). During or after transcription, the two *cis*-acting ribozymes catalyze the liberation of the short *trans*-acting ribozyme. This system allows elimination of *cis*-appended sequences that might otherwise affect the functional activity of the ribozyme. Another advantage of this system is that the yield of ribozyme can be increased by connecting several copies of the same ribozyme-expression unit in tandem (multi-homo type; Figure 4(B-i)). This strategy might be useful when a conditionally regulated promoter is used, in particular since the transcriptional capacity of such a promoter is usually low. With such a strategy, the level of ribozyme transcripts can be raised by increasing the number of ribozyme units ("n" in Figure 4).

An alternative version of the above-described strategy involves connecting different ribozyme-expression units, such that several different ribozymes, directed against different portions of a specific RNA molecule, can be expressed under the control of a single promoter from a single vector (multi-hetero type; Figure 4(B-ii)). This scenario might allow more efficient cleavage and decreases the likelihood of selection of ribozyme-resistant mutant sequences (for example, in HIV-1 RNA). We have designated

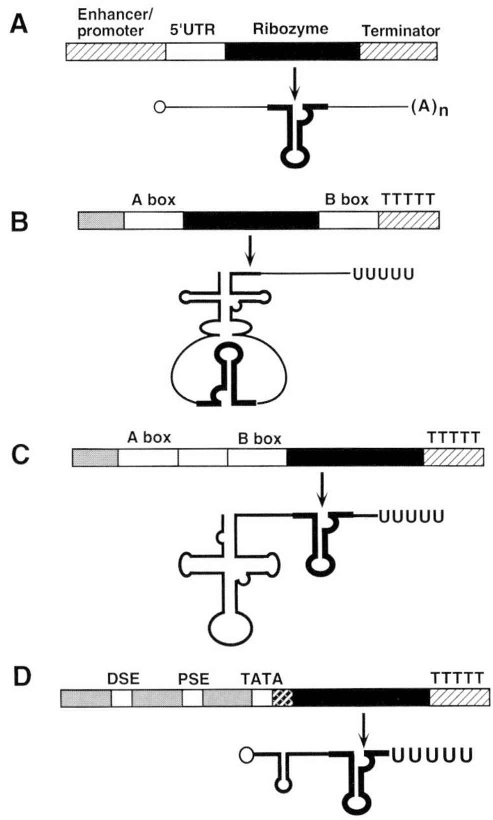

Figure 3. Expression systems for ribozymes *in vivo*. (A) The pol II expression cassette of a ribozyme. Transcription of the RNA that includes a ribozyme sequence starts within the 5′ untranslated region (UTR) and terminates at the terminator sequence. The open circle represents the RNA cap structure and $(A)_n$ represents a polyA tail, both of which are generated after transcription. (B–D) Typical pol III-type cassettes for expression of ribozymes. (B) and (C) represent pol III promoter cassettes driven from the gene for a tRNA. The ribozyme sequence is inserted within the anticodon loop (B) or replaces the amino-acceptor stem (C). In both cases, the A and B boxes act as the promoter sequence. Five T residues (U residues in RNA) terminate transcription. (D) Schematic representation of the promoter of the human gene for U6 snRNA. In this promoter, the enhancer/promoter sequence is located in the mature coding region. The ribozyme sequence is located immediately after a small-stem structure and a short stretch of adjacent nucleotides, which are essential for the capping reaction and intracellular stability. Transcription starts after the TATA box and terminates at a string of 5 T residues. The open circle represents the RNA cap structure. DSE represents the distal-sequence enhancer and PSE represents the proximal-sequence element.

these two kinds of strategy "shotgun-type expression". The second strategy, in particular, might be suitable for gene therapy against AIDS since HIV is infamous for its high rates of mutation. This mutability hinders the application of ribozymes to AIDS therapy. Each ribozyme has extremely high specificity for its target sequence. Thus, the mutation of a single nucleotide around the cleavage site is sufficient to inhibit a ribozyme's activity.[108–110] It might be possible to overcome the problems posed by the mutability of HIV by using several ribozymes simultaneously, each targeted to a different site in HIV RNA.[48] Using our trimming system, we have succeeded in preventing replication of HIV-1 in cell cultures with much greater efficiency than has been observed with more usual ribozyme-expression systems.[53,107] In our study, several *trans*-acting tRNA-embedded ribozymes, targeted to different sites in HIV-1 RNA, were expressed under the control the SRα promoter, which is a modified form of the early promoter of SV40.[111] We examined the effects on the replication of HIV-1 of the ribozymes in cultured cells in a co-transfection transient-expression assay with an HIV-1 infectious DNA clone. In order to stabilize the *trans*-acting ribozyme in cells, we inserted it into the anticodon of a tRNA loop (Figure 4). All constructs caused significant inhibition of the replication of HIV-1, and one of them was 90% effective in inhibiting replication when the molar ratio of the DNA template that encoded the target HIV-1 RNA to that for the ribozyme was 1 : 8.

Our ribozyme-expression system, consisting of a promoter, a *trans*-acting ribozyme sandwiched between two *cis*-acting ribozymes, and a terminator, was successfully used by Clawson and Norris and their colleagues to demonstrate the effectiveness of the trimmed ribozyme not only in cultured cells

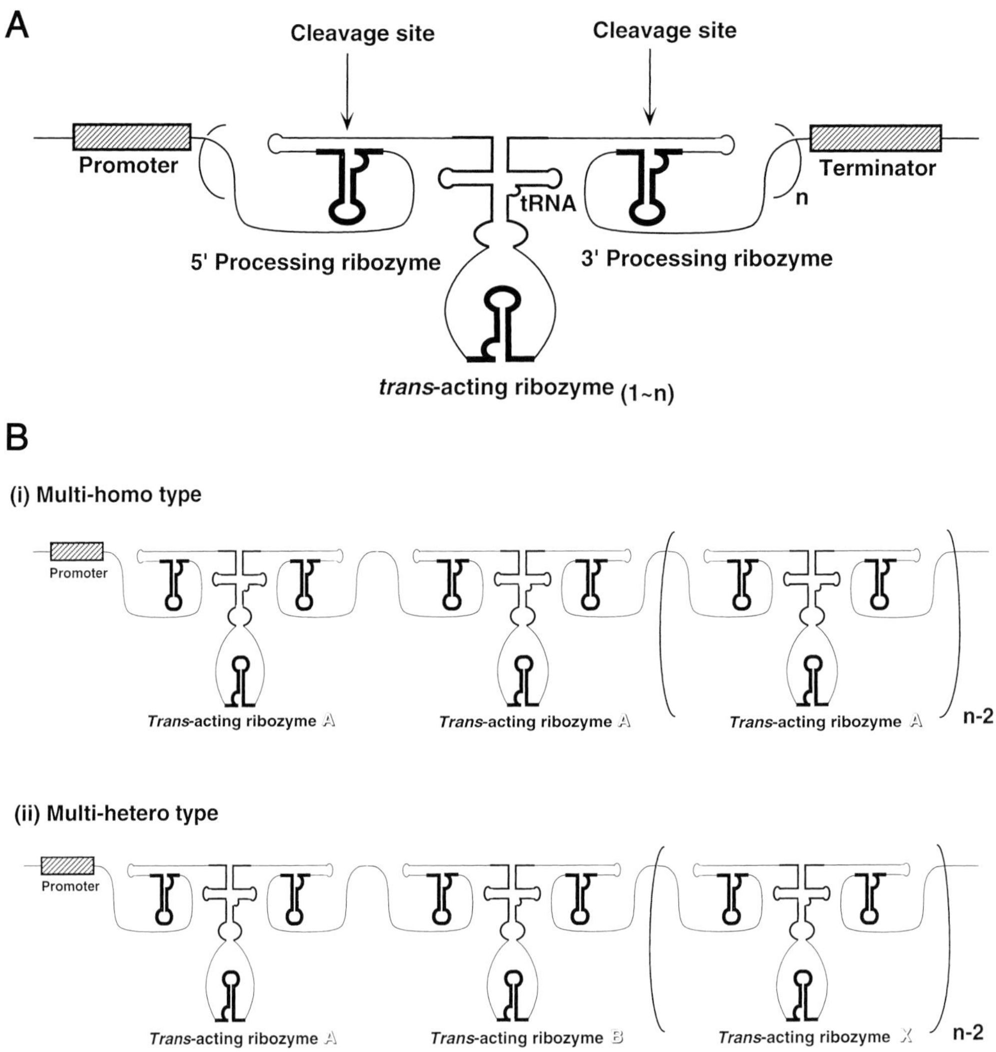

Figure 4. A novel ribozyme-expression system with *cis*-acting (trimming) ribozymes. (A) A novel ribozyme-expression system with two *cis*-acting ribozymes that liberate a tRNA-embedded *trans*-acting ribozyme during or after transcription. Here, the tRNA promoter is inactivated and is used only to increase the stability of the liberated *trans*-acting ribozyme *in vivo*. (B) Shotgun-type expression strategies. (i) The "multi-homo" type of ribozyme-expression system. Identical ribozyme cassettes composed of two *cis*-acting ribozyme sequences and one *trans*-acting ribozyme sequence are connected in tandem. (ii) The "multi-hetero" type of ribozyme-expression system. Different ribozyme cassettes that have different *trans*-acting ribozyme sequences are connected in tandem. Upon transcription, several independent *trans*-acting ribozymes are liberated by the actions of the *cis*-acting ribozymes.

but also in animal studies.[112–114] Of course, the *trans*-acting ribozyme can be trimmed not only by a hammerhead ribozyme but also by other types of ribozyme.[115–118]

16.2.3.2 RNA polymerase III system

The polymerase III system is used for transcription of small RNA molecules, such as tRNAs, U6 snRNA, and adenoviral-VA 1 RNA. Two important features of this system are that (1) minimal numbers of extra nucleotides are required at the 3′- and 5′-ends for production of properly initiated and terminated transcripts, and (2) the level of expression is at least one order of magnitude higher than that obtained with the pol II system.[119] Indeed, in our studies, pol III-driven ribozymes[62,63,66,67] but not pol II-driven ribozymes[107] could be detected by Northern blotting analysis, providing evidence of the higher level of the former transcripts. Thus, the pol III system appears to be particularly useful for the expression of ribozymes. In addition, in the case of a tRNA promoter and the VA 1 RNA promoter, since each expression unit is compact (less than 200 bp including the ribozyme sequence), the promoters are suitable for the expression of multiple ribozymes from a single vector, such as a retroviral vector. However, in many cases, the expected effects of ribozymes have not been achieved, in spite of the apparently favorable features of the pol III system.

Among the pol III-type promoters, the promoters of genes for tRNAs (tRNA promoters) have been well studied as controllers of the expression of ribozymes. Cotten and Birnstiel used a $tRNA^{Met}$ promoter from *Xenopus* and they inserted the coding sequence of a ribozyme, targeted to U7 snRNA, into the anticodon loop of the tRNA (Figure 3(B)).[119] They coinjected radioisotopically labeled U7 snRNA and the ribozyme-expression plasmid into the nucleus of frog oocytes. Twenty hours after the injection, the target RNA had completely vanished. The transcribed ribozyme had remained localized in the nucleus for the most part, whereas the target RNA had been rapidly exported to the cytoplasm. In this case, large amounts of ribozymes, as compared to the amount of target RNA (about 500- to 1000-fold molar excess of ribozyme over target RNA), were needed for cleavage of the target RNA. Wong-Staal's group used another strategy that exploited a tRNA promoter.[49,120] They used the human $tRNA^{Val}$ promoter and replaced the amino-acceptor stem sequence with a hairpin ribozyme sequence targeted against an HIV-1 (Figure 3(C)). Replication of various strains of HIV-1 was efficiently inhibited by a single ribozyme in cell culture. Bertrand and Rossi's groups used a human $tRNA_i^{Met}$ promoter for expression of a hammerhead ribozyme directed against HIV-1 (the target site was identical to that in the study by Wong-Staal's group). However, the effective inhibition of viral replication was not observed.[57,98,121] The ribozymes had remained mainly in the cell nuclei. Immature mRNA in the nucleus is decorated with splicesomes and other RNA-binding proteins, and mature mRNA is exported rapidly to the cytoplasm by the protein factors that serve as an export system. Therefore, a ribozyme might not be able to find its target site within an mRNA or even the target mRNA itself in the nucleus. Furthermore, the structure of mRNA that has already been exported to the cytoplasm can be loosened by the helicase activity of eIF4A (eukaryotic initiation factor) prior to binding of ribosomes. Thus, we might expect more efficient suppression of a specific gene if a ribozyme could be designed to attack its target mRNA in the cytoplasm (Figure 5).

We used the pol III system for expression of a ribozyme, that was targeted to the same sequence in HIV-1 as mentioned in the previous paragraph. In our system, a human $tRNA^{Val}$ promoter was used to control expression of a hammerhead ribozyme.[67] The ribozyme sequence was connected to the $tRNA^{Val}$ promoter sequence with a short intervening linker sequence. The length and the sequence of the linker were varied so that we obtained a set of $tRNA^{Val}$-ribozymes (Figure 6). Computer-generated analysis of folding predicted that the insertion of different short linkers would change the overall structure of the

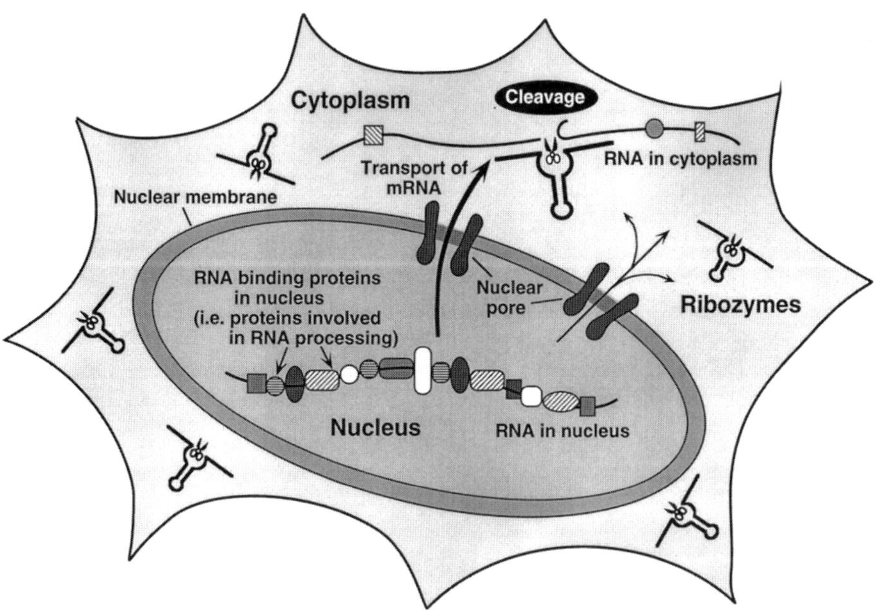

Figure 5. Export of $tRNA^{Val}$-driven ribozymes to the cytoplasm that ensures co-localization with target sequences. mRNAs are expected to be more accessible to ribozymes in the cytoplasm than in the nucleus because many proteins, including enzymes involved in processing pre-mRNA, might render the target site inaccessible to ribozymes in the nucleus.

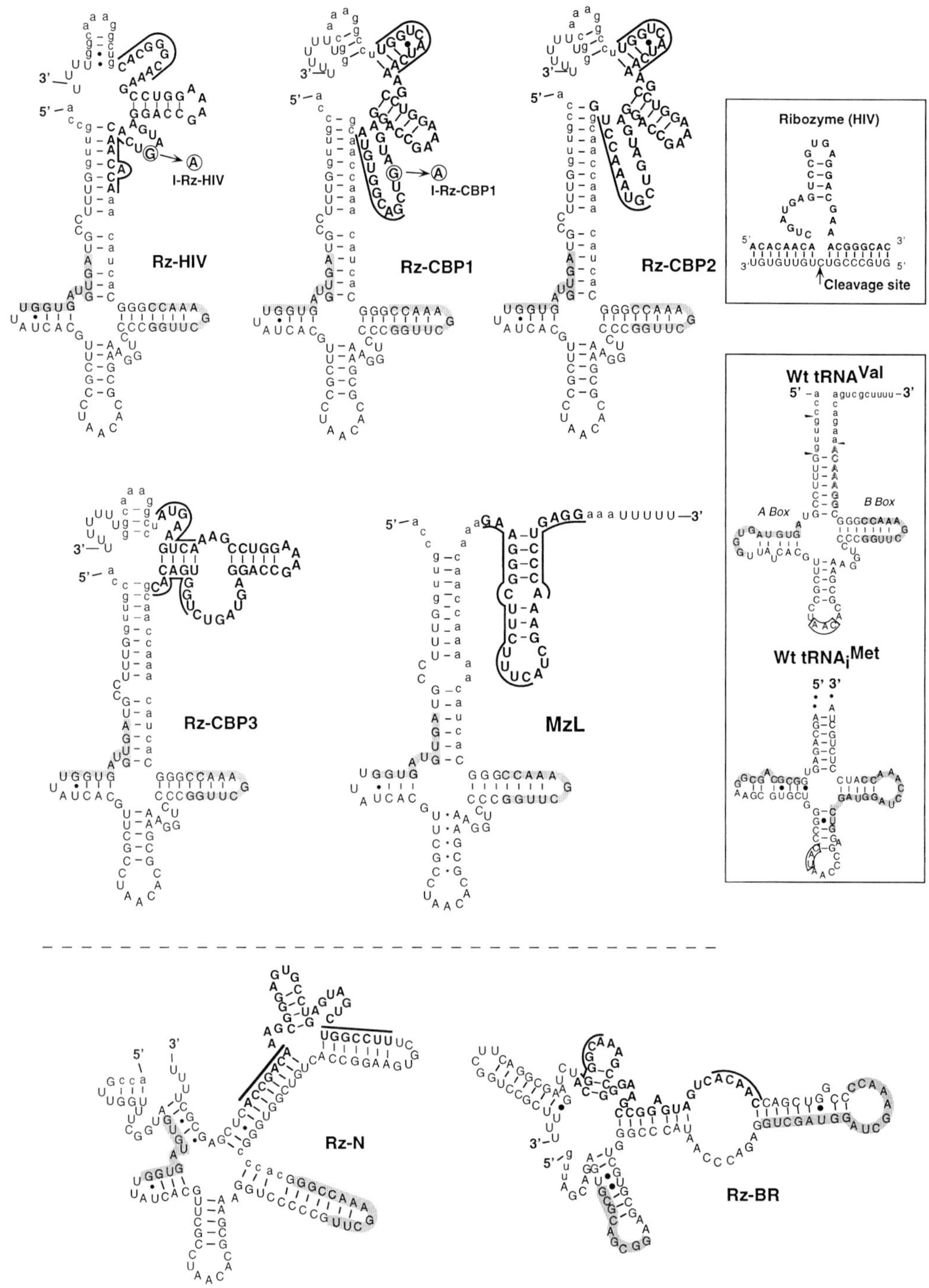

Figure 6. The secondary structures of tRNA^{Val}-ribozymes, as predicted by computer-folding analysis. The sequence of the hammerhead ribozyme (bold capital letters; also shown in the upper right panel) was ligated downstream of that of a seven-base-deleted tRNA^{Val} (capital letters) with various linker sequences (lowercase letters). The sequences that correspond to the internal promoter of tRNA^{Val}, namely the A and B boxes, are indicated by shaded boxes. The recognition arms of ribozymes are indicated by underlining. The Rz-HIV and Rz-BR (Bertrand's tRNA$_i^{Met}$-ribozyme; see Ref. [57]) target the LTR(R-U5) region of HIV-1. The inactive tRNA^{Val}-ribozyme HIV (I-Rz-HIV) was made from Rz-HIV by changing G_5 in the core region to A. The predicted secondary structures of human placental tRNA^{Val} and tRNA$_i^{Met}$ are shown in the lower right panel. The tRNA is processed at three sites (arrowheads) to yield the mature tRNA^{Val} (capital letters). Rz-CBP1 through Rz-CBP3 are ribozymes targeted to CBP mRNA. MzL represents maxizyme left.

transcripts. We found that the putative structural differences influenced the cleavage activity *in vitro*, as well as the stability of the transcripts in cultured cells. The steady-state concentration of the ribozyme differed over a 26-fold range.[67] It should be useful to identify the structural factors that determine stability *in vivo* in order to enhance the functional activity of such ribozymes.

When the expression cassette that encoded the most stable ribozyme (Rz-HIV in Figure 6) was introduced into H9 cells by an HIV-1 vector and then the cells were challenged by infection with HIV-1, replication of the virus was almost completely inhibited for more than 10 days.[67] The most stable ribozyme (Rz-HIV), as well as other related ribozymes (structures not shown), that had similar computer-predicted secondary structures (a cloverleaf configuration), were exported to the cytoplasm. Nevertheless, cytoplasmic localization did not, by itself, guarantee the effectiveness of the ribozyme. When the intracellular stability of the ribozyme was low or when its cleavage activity was low, the ribozyme failed to inhibit viral replication.[67] It is appropriate, however, to emphasize here that, in our hands, all intracellularly active ribozymes transcribed by our expression system were effectively exported to the cytoplasm (see below).[61–63,66,67] Thus, cytoplasmic localization is one of the key determinants of the functional activity of ribozymes in vivo (Figure 5).

The importance of the cytoplasmic localization of ribozymes is also supported by the results of several studies in which the promoter of the gene for U6 snRNA (an RNA that is known to remain in the nucleus) was used for expression of ribozymes (Figure 3(D)).[67,122,123] In all cases examined, even though large amounts of the U6 snRNA-ribozymes were present in cells, no inhibition of expression of target genes occurred. The failure of the ribozymes to cleave their targets in cells might have been due to the propensity for nuclear localization of the U6 snRNA portion of the ribozyme transcript and, indeed, it was confirmed that the U6 snRNA-ribozyme remained localized mainly in the nucleus.[57,123]

A protein, designated Exportin-t (Xpo-t), which transports tRNA from the nucleus to the cytoplasm was identified recently.[124–126] Xpo-t binds RanGTP in the absence of tRNA but it does not bind tRNA in the absence of RanGTP. Therefore, a model for the transport of tRNAs was proposed wherein Xpo-t first associates with RanGTP in the nucleus and then the complex binds a mature tRNA molecule. The entire complex is then translocated through a nuclear pore to the cytoplasm, where the Ran-bound GTP is hydrolyzed, releasing the tRNA into the cytoplasm and allowing Xpo-t to be recycled to the nucleus. The minimal sequence or structure within a tRNA that can be recognized by Xpo-t is being clarified.[125,127–129] The 3′-terminal CCA region of the tRNA appears to be important for recognition by Xpo-t. With regard to the export of the $tRNA^{Val}$-ribozymes transcribed from our expression system, we cannot exclude the possibility that these $tRNA^{Val}$-ribozymes might be recognized by an export receptor similar to Xpo-t[124–129] since all $tRNA^{Val}$-ribozymes that had a cloverleaf tRNA-like structure (Figure 6) were found in the cytoplasm. In the case of the tRNA-ribozymes, such as Rz-N and Rz-BR in Figure 6, that remained localized in the nucleus, the results of computer-folding analysis did not support a cloverleaf tRNA-like structure.[67] Studies with *Xenopus* oocytes indicate that only tRNAs with mature 5′ and 3′ ends are recognized by Xpo-t and can be exported to the cytoplasm.[125,127,128] Clearly, our $tRNA^{Val}$-ribozymes have a ribozyme sequence at their 3′ ends, but nevertheless, it appears that they can be exported to the cytoplasm in mammalian cells as long as their secondary structures retain a cloverleaf tRNA-like configuration.[61–63,66,67] At present, we are trying to identify the export receptor that recognizes $tRNA^{Val}$-ribozymes in mammalian cells. The results from our laboratory indicate that a tRNA-like structure is important for the production of intracellularly active $tRNA^{Val}$-ribozymes.[61–63,66–68]

16.2.4 Functional Analysis of the Transcriptional Co-Activator CBP Using $tRNA^{Val}$-Ribozymes

The mechanism responsible for the transport of our $tRNA^{Val}$-ribozymes to the cytoplasm remains to be clarified but it is clearly important that properly designed $tRNA^{Val}$-ribozymes be transported to the cytoplasm and that they have high-level activity *in vivo*. In this section, we will discuss our evidence that $tRNA^{Val}$-ribozymes are very useful tools in molecular biology, including a specific example of the use of such ribozymes in the functional analysis of a co-activator CBP (CREB *b*inding *p*rotein, where CREB stands for cAMP-response-element-binding protein). The co-activators CBP and p300, a related protein, are functionally conserved proteins that act in concert with other factors to regulate transcription.[130,131] Three target sites that are not conserved in the mRNA for the related protein p300 were selected within CBP mRNA and the individual ribozymes targeted to these sites were designated Rz-CBP1 through Rz-CBP3 (Figure 6) in upstream to the downstream direction. In our pol III-driven ribozyme-expression cassette, in order to block 3′-end processing of the transcript without any negative effects on transcription, we first removed the last seven nucleotides from the mature $tRNA^{Val}$ (Wt $tRNA^{Val}$ in Figure 6; removed nucleotides are indicated by outlined capital letters in the lower box on the right). The removed nucleotides were replaced by a linker (lowercase letters in Figure 6).

In our analysis, we focused on three types of ribozyme that we cloned adjacent to the sequence of a $tRNA^{Val}$ promoter, with a short linker between them (linker and ribozyme sequences are indicated by lowercase and bold capital letters, respectively, in Figure 6). The overall secondary structures were adjusted to maintain the cloverleaf tRNA-like configuration by the selection of an appropriate linker sequence. The selected target sequence might also be expected to influence the overall structure of the transcripts and, thus, to affect the accessibility of the recognition arms of the ribozyme (recognition arms are underlined). Naturally, it is important that the 5′ and 3′ substrate-recognition arms of the ribozyme be available to the substrate so that the ribozyme and substrate RNA together can form the stem structures that ensure subsequent cleavage of the substrate. The inactive ribozyme (designated I-Rz) differed from the active ribozymes by a single G_5 to A mutation in the catalytic core (the numbering system follows the rule for hammerhead ribozymes).[132] This single-nucleotide change would be expected to diminish cleavage activity while antisense effects, if any, should remain unaffected.[51,75]

16.2.4.1 Suppression of expression of the gene for CBP by $tRNA^{Val}$-ribozyme

The effects of $tRNA^{Val}$-ribozymes on the specific target mRNA, namely, the transcript of the CBP gene, were examined by measuring the level of CBP itself in F9 cells that had been transfected with the plasmids that encoded the various $tRNA^{Val}$-ribozymes. To determine the effectiveness of the ribozyme-mediated inhibition of expression of the CBP gene in a transient expression assay, we compared relative levels of CBP by an amplified sandwich ELISA. As shown in Figure 7, all the active ribozymes decreased the level of CBP *in vivo*. The extent of inhibition by the various ribozymes, expressed from the pol III promoter, ranged from 27% to 86% under the conditions of our studies. Stronger inhibition can, of course, be achieved by the introduction of a larger quantity of the ribozyme-coding plasmid.

Our $tRNA^{Val}$-ribozymes, targeted to CBP mRNA, had high activity, which allows us to investigate the function of the transcriptional co-activator CBP in F9 cells. From among our active $tRNA^{Val}$-ribozymes (Figure 7), we chose the most active one, Rz-CBP1, to generate stable lines of F9 cells that expressed a $tRNA^{Val}$-ribozyme. The concentrations of CBP mRNA in these cell lines were measured by a competitive reverse transcription-polymerase chain reaction (RT-PCR; Figure 8(A)). The level of CBP mRNA fell dramatically from 1×10^{-2} attomole/μl to less than 1×10^{-4} attomole/μl in F9 cells that expressed Rz-CBP1 (in lanes 1–10 in Figure 8(A), note that no CBP cDNA was detectable in cells

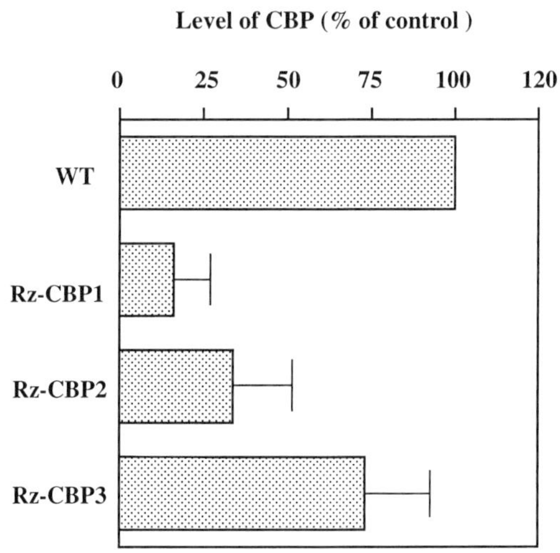

Figure 7. Effects of ribozyme-expression plasmids specific for cleavage of CBP mRNA on the levels of CBP in a transient-expression assay. Relative levels of CBP were compared by an amplified sandwich ELISA. Normalized levels of CBP in wild-type (WT) F9 cells were taken arbitrarily as 100%. All values are the averages of results from at least three experiments and the standard deviation for each value relative to the value for WT cells is indicated.

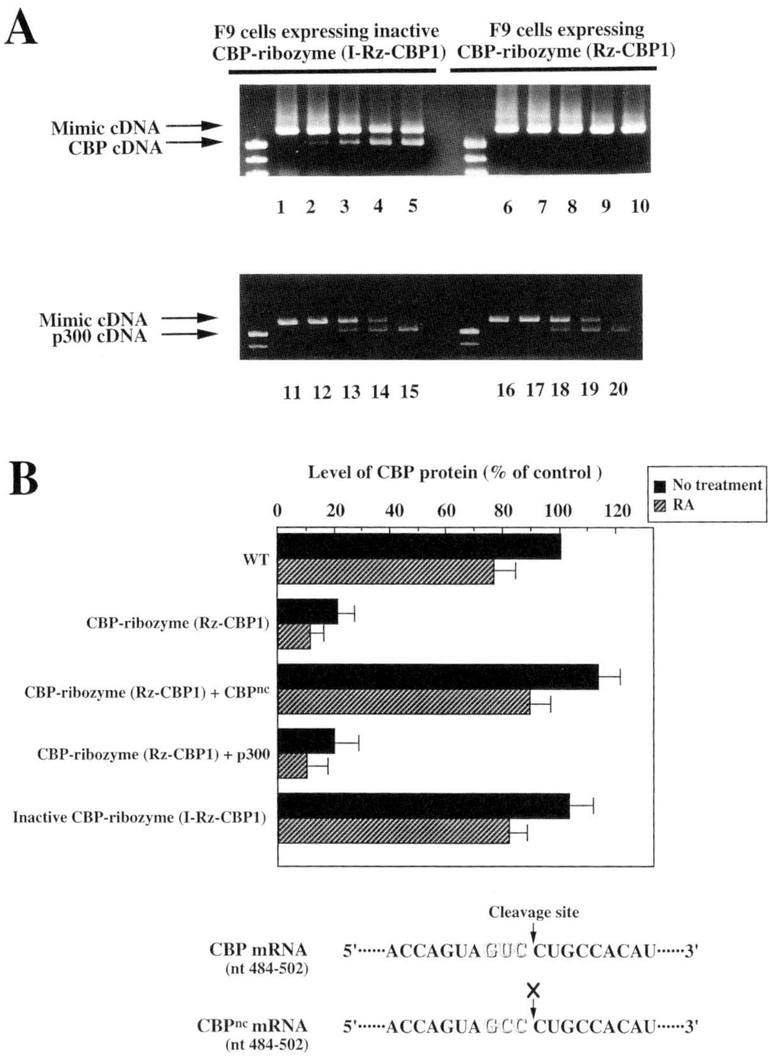

Figure 8. Effects of ribozyme vectors specific for cleavage of CBP mRNA on the levels of p300 and CBP mRNAs and on the levels of the proteins themselves. (A) Analysis by competitive RT-PCR of mRNAs for p300 and CBP. The "mimic" DNA fragment specific for p300 or CBP was prepared as described in Ref. [61]. Products of PCR were subjected to electrophoresis on a 2% agarose gel. The concentrations of p300 or CBP mimic DNA were as follows: lanes 1, 6, 11 and 16, 1 attomole/μl; lanes 2, 7, 12 and 17, 1×10^{-1} attomole/μl; lanes 3, 8, 13 and 18, 1×10^{-2} attomole/μl; lanes 4, 9, 14 and 19, 1×10^{-3} attomole/μl; and lanes 5, 10, 15 and 20, 1×10^{-4} attomole/μl. (B) Relative levels of CBP were compared by an amplified sandwich ELISA. The normalized level of CBP in wild-type (WT) F9 cells was taken arbitrarily as 100%. CBP^{nc} indicates an expression plasmid that encoded a mutant mRNA for CBP that was not cleaved (nc) by the CBP-ribozyme. All values are the averages of results from at least three experiments and the standard deviation for each value relative to the control is indicated. X indicates the absence of cleavage.

that produced the active ribozyme). In the same cells, levels of p300 mRNA were unchanged (lanes 11–20 in Figure 8(A)), indicating that the Rz-CBP1 ribozyme had specifically cleaved the CBP mRNA without damaging the related p300 mRNA. An inactive ribozyme, I-Rz-CBP1, targeted to CBP mRNA had no effect on the levels of either p300 mRNA or CBP mRNA (lanes 1–5 and 11–15 in Figure 8(A)). These data indicate that the $tRNA^{Val}$-ribozyme acted, by catalyzing cleavage and not *via* an antisense effect, on CBP mRNA specifically despite the strong homology of p300 mRNA to CBP mRNA within conserved regions. Compared to levels of CBP mRNA in lines of F9 cells that expressed an inactive $tRNA^{Val}$-ribozyme, levels of CBP mRNA were about 100-fold lower in cells that expressed an active $tRNA^{Val}$-ribozyme.

We next examined the levels of CBP itself. In F9 cells that expressed an active CBP-ribozyme, levels of CBP were at least 5-fold lower than levels in F9 cells that harbored an inactive CBP-ribozyme and than levels in WT F9 cells (Figure 8(B)). In these experiments, in order to rescue the expression of CBP, we also constructed a version of CBP mRNA that was not cleavable (nc) by the active CBP-ribozyme (CBP^{nc}; in which the triplet at the cleavage site was changed from the cleavable GUC to the uncleavable GCC, a mutation that does not change the translated amino acids). As expected, stable introduction of CBP^{nc} mRNA completely restored the production of CBP in F9 cells that expressed an active CBP-ribozyme.

16.2.4.2 A role for the transcriptional co-activator CBP during the RA-induced differentiation of F9 cells

Upon exposure to retinoic acid (RA), F9 cells can be induced to differentiate into endoderm cells.[133–135] Then, we next asked whether F9 cells that lacked CBP could still respond to treatment with RA. When challenged with RA, F9 cells expressing an active CBP-ribozyme showed morphological signs of differentiation. This result suggests that CBP does not have a critical role in this differentiation process. When a population of F9 cells is treated with RA, many cells differentiate but some undergo apoptosis.[136,137] To explore the possible dependence of apoptosis on CBP, we evaluated levels of nicked DNA, using the TUNEL assay (TdT-mediated dUTP nick end labeling assay),[138] since nicked DNA is one of the hallmarks of apoptosis. F9 cells expressing the CBP-ribozyme were much less sensitive to the RA-induced nicking of DNA than were WT cells or those that synthesized the inactive ribozyme.[61] Thus, it appeared that RA-induced apoptosis required the expression of intact CBP.

The experiments described above provide an example of the utility of our ribozyme-expression system. In our functional analysis of the transcriptional co-activator CBP, we found that the CBP-ribozyme was very active in F9 cells. CBP exhibits strong homology to p300, in particular within the functional domains. However, despite the similarity between CBP and p300, the CBP-ribozyme cleaved CBP mRNA specifically, without affecting p300 mRNA, allowing us to elucidate the role of the transcriptional co-activator CBP in F9 cells. The advantages of ribozymes over conventional antisense RNA/DNA technology include the following: (1) inclusion in studies of a ribozyme that is rendered inactive by a point mutation allows discrimination of the effects of cleavage of the target RNA from other secondary effects, such as antisense effects; and (2) as demonstrated in this section, it is usually possible to create an uncleavable version of the target mRNA, by introducing a point mutation into the target mRNA, that can be used to rescue cells from the effects of the ribozyme. The conclusion that we reached using tRNA^{Val}-ribozymes[61] was identical to that reached in an analysis of transgenic mice.[139] Therefore, properly designed tRNA^{Val}-ribozymes with high-level activity appear to be very useful tools in molecular biology, with potential utility in medicine also, as described below.

16.2.5 A Novel Dimeric Ribozyme (Maxizyme) Engineered from the Hammerhead Ribozyme with Considerable Potential as a Gene-Inactivating Agent

After our success in using tRNA^{Val}-ribozymes for functional analysis of the co-activator CBP, we wondered whether we might be able to use the same expression system for our novel dimeric short ribozymes (maxizymes) which had such high level of activity *in vitro*. As discussed below, we found that the bulky tRNA^{Val} portion of the transcript did not inhibit the dimerization process and, moreover, that the tRNA^{Val}-driven maxizyme was more active *in vivo* than the tRNA^{Val}-driven conventional hammerhead ribozyme.

16.2.5.1 Shortened variants of the hammerhead ribozyme (conventional minizymes)

Discussions of the potential utility of the hammerhead ribozyme have been numerous since the discovery of this small ribozyme and many attempts have been made to create ribozymes with improved features. In one case, for example, extra sequences were deleted from the stem/loop II region of the hammerhead ribozyme. Such deletions are acceptable since stem II is the only helix in the hammerhead ribozyme that is not involved in binding of the substrate and the stem/loop II region appeared initially not to be directly involved in catalysis (Figure 2).[87] For the development of chemically synthesized ribozymes as potential therapeutic agents, it would certainly be advantageous to remove any surplus nucleotides that are not essential for catalytic activity. Removal of surplus nucleotides would obviously reduce the cost of synthesis, increase the overall yield of the desired polymer, and simplify purification. These considerations led to the production of short ribozymes (minizymes), namely, conventional hammerhead ribozymes with a deleted stem/loop II region.[78,86,87,140] However, the activities of the minizymes turned out to be two to three orders of magnitude lower than those of the parental hammerhead ribozymes, a result that led to the suggestion that minizymes might not be suitable as gene-inactivating agents.[78]

16.2.5.2 Discovery of maxizymes that act as a dimeric form of short ribozymes

We tried to create variants of hammerhead ribozymes with deletions in the stem/loop II region and, fortunately, we found that some shortened forms of hammerhead ribozymes had high cleavage activity that was similar to that of the wild-type parental hammerhead ribozyme (R32; Figure 2(A)).[92] Moreover, the active species appeared to form dimeric structures with a common stem II (Figure 2(B)). In the active short ribozymes, the linker sequences that replaced the stem/loop II region were palindromic so that two short ribozymes were capable of forming a dimeric structure with a common stem II. In order to distinguish monomeric forms of conventional minizymes that have extremely low activity from our novel dimers with high-level activity, we chose the name "maxizymes" for the very active short ribozymes that were capable of forming dimers.[63,66] The activity of the homodimeric maxizymes (a dimer with two identical binding sequences) depended on Mg^{2+} ions and, in addition, interactions with the substrates also stabilized the dimeric structures. In addition, we tried to obtain more direct evidence for the existence of homodimeric maxizymes by NMR spectroscopy.[62] The observations suggested that the major population of maxizymes was in a dimeric form even in the absence of the substrate.

We next synthesized heterodimeric maxizymes, in which one maxizyme left (MzL) and a one maxizyme right (MzR) were the monomers that together formed one heterodimeric maxizyme, as shown in Figure 9. Such a heterodimeric maxizyme has two independent substrate-recognition arms and can cleave its substrates only when the individual maxizymes form a heterodimer (Figure 9(A)). Increases in the length of the common stem II region were associated with increases in the activity of the heterodimeric maxizymes *in vitro* (Figure 9(B)).[141,142] Moreover, we found that the cleavage rate of one of the substrates by the heterodimeric maxizyme increased linearly with increases in the concentration of the other substrate, a clear demonstration that the formation of the dimer was essential for cleavage of the substrate and, moreover, that the dimeric structure could be stabilized in the presence of the substrates.

16.2.5.3 Construction of tRNA^{Val}-embedded maxizymes

For the application of maxizymes to gene therapy, they must be expressed constitutively *in vivo* under the control of a strong promoter. As described in previous sections, we succeeded in establishing an effective ribozyme-expression system that was based on the pol III promoter. While pol II promoters might allow tissue-specific or regulatable expression,[102,114] we would expect that pol III transcripts would be expressed at significantly higher levels.[119] High-level expression under control of the pol III promoter would be advantageous for maxizymes and enhance the likelihood of their dimerization. Therefore, we chose the expression system with the promoter of a human gene for tRNA^{Val}, which we had used successfully in the suppression of target genes by ribozymes, as discussed above.[55,61,67] We embedded maxizymes similarly in the 3′-modified side of the tRNA^{Val} portion of the human gene and one such sequence is shown in Figure 6 (MzL).[62,63,66]

In the case of the tRNA^{Val}-embedded maxizymes, we worried initially about the possibility that the tRNA^{Val} portion might cause severe steric hindrance that would inhibit dimerization, with resultant production of monomers with extremely low activity. To our surprise, the tRNA^{Val}-embedded maxizymes still had significant cleavage activity *in vitro*.[62] We performed modeling studies, using the coordinates of the crystal structure of a hammerhead ribozyme[143] and those[144] of yeast tRNA^{Phe}, to confirm that the formation of dimeric structures by tRNA^{Val}-embedded homodimeric maxizymes was theoretically feasible (Figure 10). The resultant model structure appeared feasible and steric hindrance by the two tRNA moieties seemed not to be a problem.[62] More direct evidence for the formation of dimers was provided by gel-shift analysis in the absence of the substrate.[62]

16.2.5.4 Intracellular activities of tRNA^{Val}-dimeric maxizymes; simultaneous cleavage of HIV-1 Tat mRNA at two independent sites by heterodimeric maxizymes

The high activity of the tRNA^{Val}-embedded maxizymes that had been transcribed *in vitro* prompted us to examine whether similar transcripts might also be active within cells. We designed heterodimeric maxizymes that should be able to cleave HIV-1 Tat mRNA at two independent sites simultaneously. We confirmed *in vitro* that such heterodimeric maxizymes, with their two independent catalytic cores, were able to cleave HIV-1 Tat mRNA at two independent sites simultaneously (Figure 9(B)).[141,142]

Then, we examined intracellular activities of the HIV-1-maxizymes by using a reporter construct, in which we used LTR-Luc HeLa cells which endogenously encoded a chimeric gene that consisted of the long terminal repeat (LTR) of HIV-1 and a gene for luciferase (Figure 11(A)).[145] The LTR of HIV-1 contains regulatory elements that include a TAR region. The HIV-1 regulatory protein, Tat, binds to TAR and the binding of Tat stimulates transcription substantially. Therefore, luciferase activity originating from the chimeric LTR-Luc gene increases in response to increases in the concentration of Tat.[145] Measurements of luciferase activity allowed us to monitor the effects of tRNA^{Val}-embedded ribozymes on the Tat-mediated transcription of the chimeric LTR-Luc gene. After transient expression of both Tat and tRNA^{Val}-embedded ribozymes by co-transfection of cells with a Tat expression vector (pCD-SRα Tat) and one of our tRNA^{Val}-embedded ribozyme-expression vectors (Figure 11(A)), we estimated the intracellular activity of each tRNA^{Val}-ribozyme by measuring the luciferase activity.

The luciferase activity recorded when we used only the Tat-expression vector was taken as 100%.

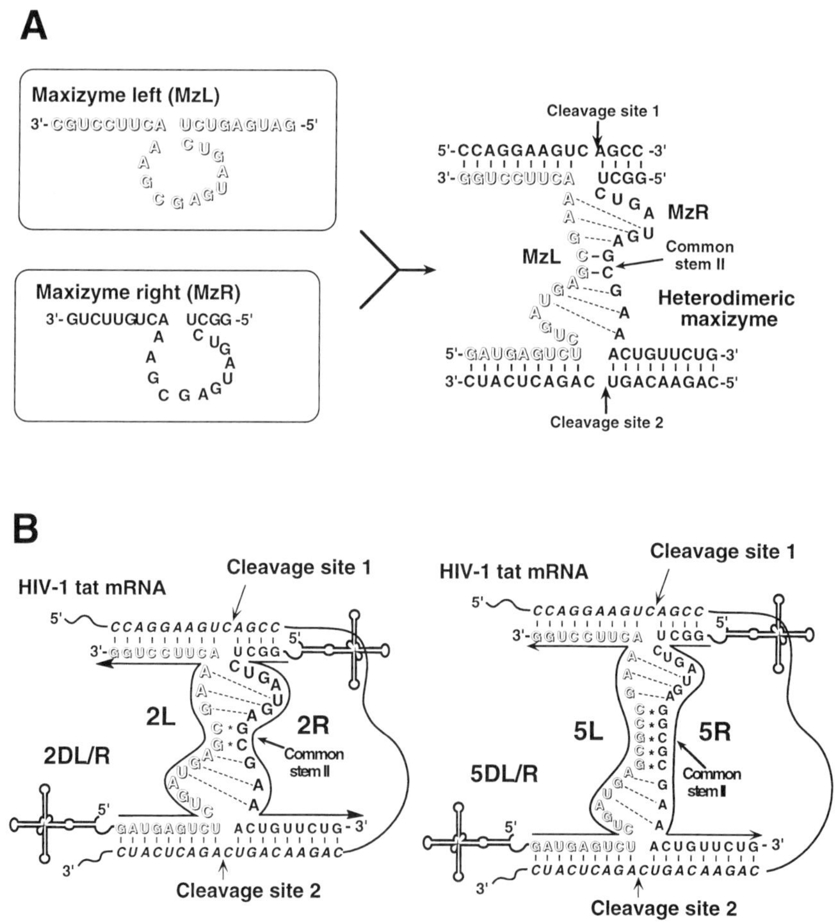

Figure 9. Formation of the heterodimeric maxizyme. (A) Secondary structure of the heterodimeric maxizyme. The MzL and the MzR form a dimeric structure with a common stem II. The MzL–MzR heterodimeric maxizyme can generate two different binding sites. The substrate RNA can be cleaved only after the formation of the heterodimer. (B) tRNA^{Val}-driven heterodimeric maxizymes that can cleave a HIV-1 Tat mRNA at two sites simultaneously. Two types of heterodimeric maxizyme with a common stem II of 2-bp (2DL/R) and 5-bp (5DL/R) are shown.

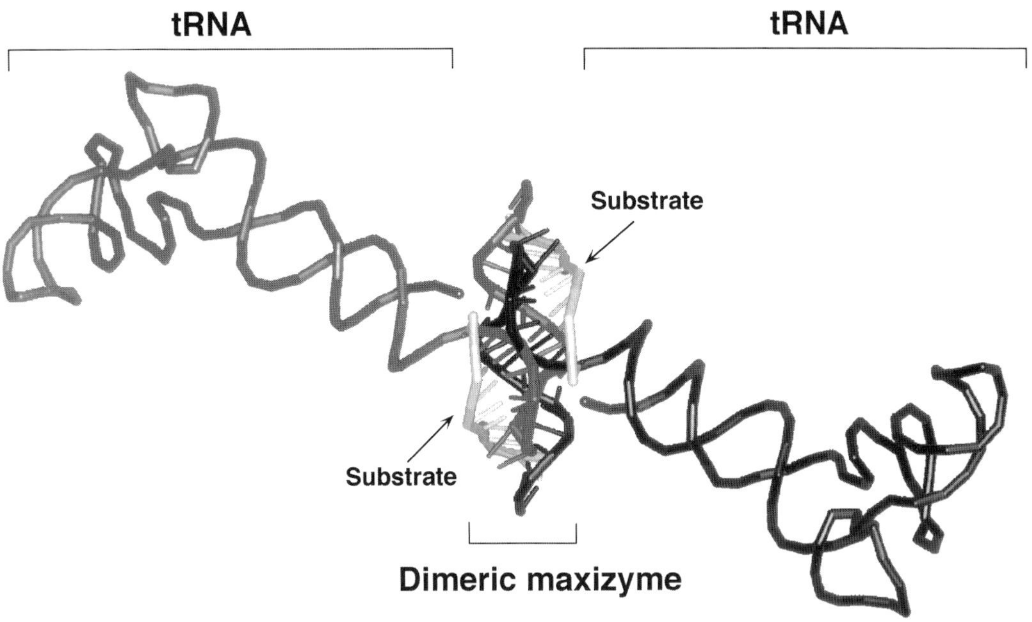

Figure 10. A model of the tRNA-embedded dimeric maxizyme (tRNA^{Val}-Mz). In this model, the internal loop or linker is ignored and is assumed to be an A-form duplex.

Ribozyme- and Tat-expression vectors were used at a molar ratio of 2 : 1 for co-transfection of LTR-Luc HeLa cells. As shown in Figure 11(B), all of the tRNA^{Val}-dimeric maxizymes (lanes 10, 11, 13, and 14) were extremely effective (>90% inhibition), despite the fact that the SRα promoter, which controlled the transcription of the target Tat mRNA, is 10- to 30-fold more active than the SV40 early promoter regardless of the species and origin of cells.[111] Conventional tRNA^{Val}-hammerhead ribozymes (Rz1 and Rz2; lanes 8, 9, 12, and 15) were also effective, albeit to a lesser extent (>60% inhibition). It is important to note that, since the inactive tRNA^{Val}-driven maxizymes (D2-I-L/R; lane 5) that had been created by a single $G_5 \Rightarrow A_5$ mutation within the catalytic core did not have any inhibitory effects, it is clear that the intracellular activities of the tRNA^{Val}-embedded ribozymes originated from their cleavage activities in cultured cells and not from any antisense effects.[66]

The heterodimers of tRNA^{Val}-driven maxizymes cleaved two sites within Tat mRNA more efficiently (lanes 10, 11, 13, and 14) than two independent tRNA^{Val}-driven conventional hammerhead ribozymes, each targeted to one of the two cleavage sites (lanes 12 and 15), despite the fact that, in the case of the maxizymes, a complicated dimerization process is involved in addition to the process of association with the target. The activities in mammalian cells of tRNA^{Val}-maxizymes (in particular, that of the 2-bp heterodimeric maxizyme, D2L/R)[66] relative to those of standard ribozymes (Figure 11) were greater than we had anticipated from kinetic parameters determined *in vitro*.[141,142] In general, the rate-limiting step of a reaction *in vivo* that is mediated by a catalytic RNA has been considered to be the substrate-binding step.[57,146–149] Under such k_{cat}/K_M controlled reaction conditions, once the dimeric maxizyme has bound to the more accessible target site with one of its binding arm (site I in Figure 12; the interaction at this site has a lower K_M value than that of the other interaction at site II), the subsequent second binding step at the second site becomes an intramolecular interaction. This intramolecular annealing process is more favorable in terms of entropy than the intermolecular annealing process in which independent standard ribozymes are involved. The dual-site maxizymes might also have an off-rate advantage. In other words, if binding at one site is hindered by translation or other facilitators of strand displacement, continued binding to the other site might keep the RNAs associated, facilitating re-binding at the first site. In other cases, even if binding to the second target site had a high K_M value because of some undesirable tertiary structure of the target mRNA (a hidden target site), the binding to the first site might change the overall structure of the mRNA and the second site might become more accessible. Such phenomena, in addition to the minimal decrease in activity associated with embedding in a tRNA, might contribute to the strong activities of heterodimeric maxizymes in mammalian cells.

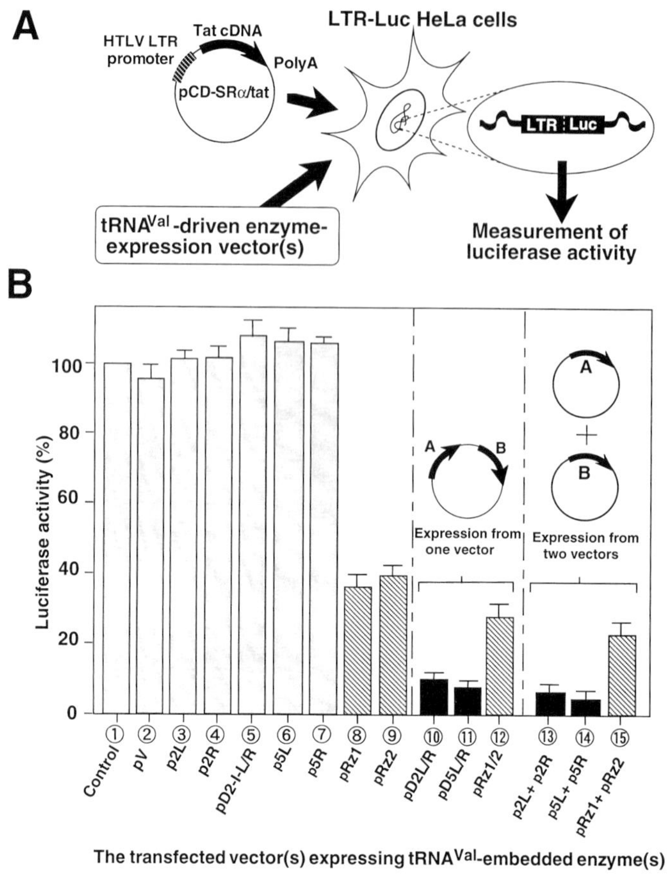

Figure 11. Intracellular activities of tRNA^{Val}-ribozymes in LTR-Luc HeLa cells. (A) Assay system for measurements of activities of tRNA^{Val}-ribozymes in LTR-Luc HeLa cells and (B) the effects of tRNA^{Val}-ribozymes on the Tat-mediated transcription of the chimeric LTR-Luc gene. tRNA^{Val}-ribozyme- and Tat-expression vectors were used at a molar ratio of 2:1 for co-transfection of LTR-Luc HeLa cells. The results shown are the averages of results from five sets of experiments. Luciferase activity was normalized by reference to the efficiency of transfection, which was determined by monitoring activity of a co-transfected gene for β-galactosidase. pV is the second control vector containing only the tRNA^{Val}-expression cassette. In the cases of pD2L/R, pD5L/R and pRz1/2, each monomer unit (L and R or Rz1 and Rz2) was cloned into the same vector in tandem.

16.2.6 A Novel Allosterically Trans-Activated Ribozyme (Maxizyme) That Acts as an Exceptionally Specific Inhibitor of Gene Expression

The studies described above demonstrated that our homodimeric and heterodimeric maxizymes were more active than the parental hammerhead ribozymes *in vivo*. The novel tRNA^{Val}-driven heterodimeric maxizymes, which can cleave one substrate at two independent sites simultaneously, can be designed very easily and, thus, they should be very useful tools *in vivo*. Nevertheless, it remains true that two independent tRNA^{Val}-driven parental ribozymes can also cleave such a substrate at the two different sites, albeit with lower efficiency. In this section, we shall describe tRNA^{Val}-driven heterodimeric maxizymes that specifically cleaved a chimeric mRNA with high selectivity while conventional ribozymes failed to do so.

16.2.6.1 Chronic myelogenous leukemia (CML) and the potential for ribozyme therapy

In some cytogenetic abnormalities, such as certain leukemias, chimeric fusion mRNAs connecting strange exons result from reciprocal chromosomal translocations and cause abnormalities. For the design of ribozymes that can disrupt such chimeric RNAs, it is necessary to target the junction sequence. Otherwise, normal mRNAs that share part of the chimeric RNA sequence will also be cleaved by the ribozyme, with resultant damage to the host cells (Figure 13).

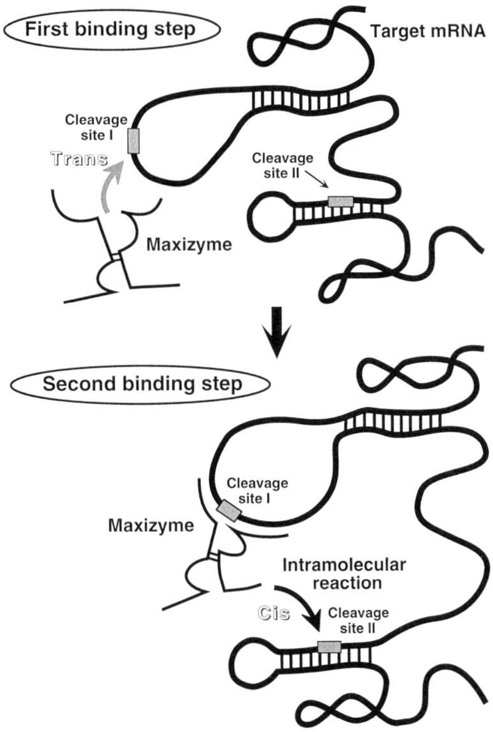

Figure 12. Schematic representation of the cleavage of an mRNA at two sites by a dimeric maxizyme. Site I is more accessible than site II.

Chronic myelogenous leukemia (CML) is a clonal myeloproliferative disorder of hematopoietic stem cells that is associated with the Philadelphia chromosome.[150] The reciprocal chromosomal translocation t(9; 22) (q34; q11) (K28 translocations and L6 translocations, Figure 13) results in the formation of the *BCR–ABL* fusion genes that encode two types of mRNA: b3a2 (consisting of *BCR* exon 3 and *ABL* exon 2) and b2a2 (consisting of *BCR* exon 2 and *ABL* exon 2).[151,152] Both of these mRNAs are translated into a protein of 210 kDa (p210$^{BCR\text{–}ABL}$) which is unique to the malignant phenotype.[153] In the case of the b2a2 sequence, there are no triplet sequences that are potentially cleavable by hammerhead ribozymes within two or three nucleotides of the junction in question (Figure 13(B)).[63,154] GUC triplets are generally the most susceptible to cleavage by a hammerhead ribozyme, and one such triplet is located 45 nucleotides from the junction. If this GUC triplet were cleaved by a ribozyme, normal *ABL* mRNA that shares part of the sequence of the abnormal *BCR–ABL* mRNA would also be cleaved by the ribozyme, with resultant damage to host cells. In designing ribozymes that might cleave b2a2 mRNA, we must be sure to avoid cleavage of normal *ABL* mRNA.

Previous attempts at the cleavage of *BCR–ABL* (b2a2) mRNA have involved a combination of a long antisense arm and a ribozyme sequence for generation of specificity (Figure 13(B), bottom right).[155,156] However, we demonstrated that the antisense-type of ribozyme cleaved normal *ABL* mRNA non-specifically both *in vitro* and *in vivo*,[63,154,157] most probably because hammerhead ribozymes can cleave their substrates even when the binding arms are as little as three nucleotides in length.[12,110]

16.2.6.2 *Design of an allosterically controllable maxizyme and its specificity of the cleavage of the chimeric BCR–ABL substrate in vitro*

In extending our studies of dimeric maxizymes, we tried to create an allosterically controllable ribozyme, based on the heterodimeric maxizyme, that would have cleavage activity only when the target sequence of interest was present. Such a ribozyme should be controlled such that it is active only in the presence of an abnormal RNA sequence, for example, the junction sequence of a chimeric mRNA. We first thought about using one of the two substrate-binding regions of the heterodimeric maxizyme as sensor arms (Figure 14(A), left). Then, since one domain of the maxizyme was to be used solely for

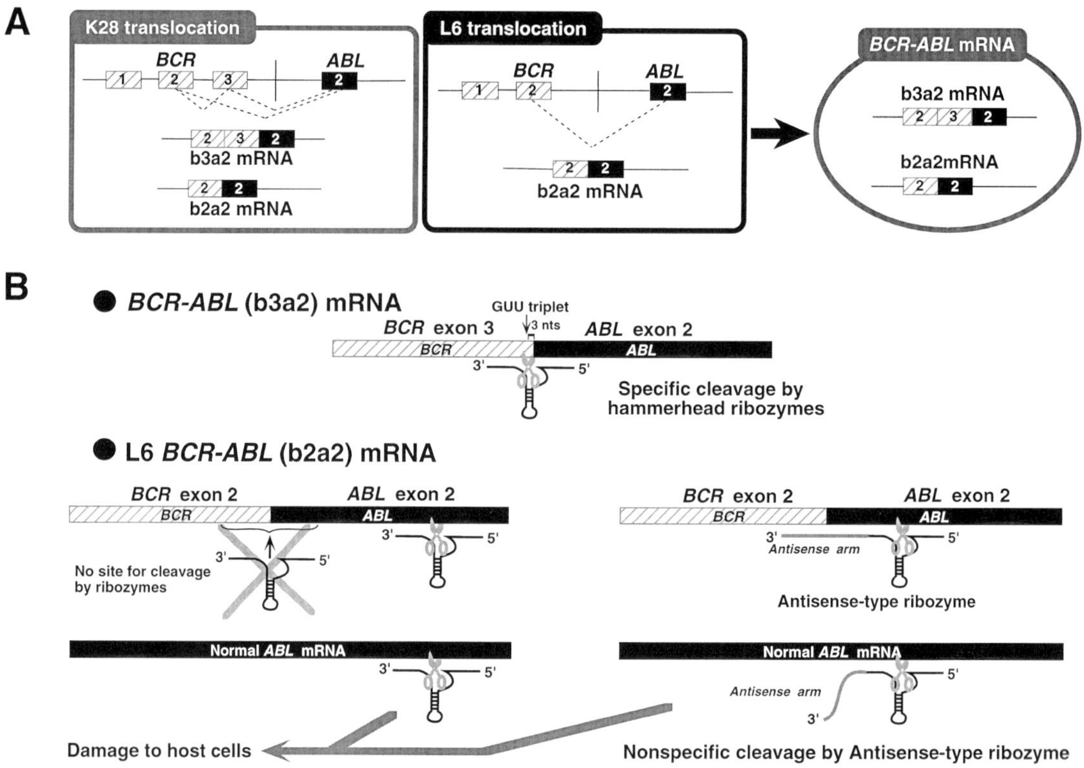

Figure 13. Types of translocation and the possible sites of cleavage within normal *ABL* mRNA and abnormal *BCR–ABL* fusion mRNAs by conventional ribozymes. (A) *BCR–ABL* translocations and fusion mRNAs. The two types of chromosomal translocation (K28-type and L6-type) that are associated with chronic myelogenous leukemia and the corresponding fusion mRNAs are depicted. Dotted lines connecting *BCR* and *ABL* exons indicate alternative splicing pathways. (B) Cleavage of normal and/or abnormal mRNAs by conventional hammerhead ribozymes. In b2a2 mRNA, there are no triplet sequences near the *BCR–ABL* junction that are potentially cleavable by hammerhead ribozymes. A GUC triplet, which is generally the triplet that is most susceptible to cleavage by hammerhead ribozymes, is located 45 nucleotides from the junction. If such a GUC triplet were selected as the site of cleavage by ribozymes, normal *ABL* mRNA that shares part of the abnormal *BCR–ABL* RNA sequence would also be cleaved by the ribozyme, with resultant damage to host cells (bottom left). Previous attempts involving a combination of a long antisense arm and a ribozyme sequence could not show sufficient specificity (bottom right). nts, nucleotides.

recognition of the target sequence of interest, we deleted its catalytic core completely to generate an even smaller monomeric unit (Figure 14(A), right). Our goal was to control the activity of maxizymes allosterically by introducing sensor arms so that, only in the presence of the correct target sequence of interest, would the maxizyme be able to create a cavity for the capture of catalytically indispensable Mg^{2+} ions. Ribozymes are metalloenzymes,[18–34] and we wanted to ensure that the active structure of the maxizyme with a Mg^{2+}-binding pocket would be formed only in the presence of the sequence of interest.

In order to achieve high substrate-specificity, we wanted our maxizyme to adopt an active conformation exclusively in the presence of the abnormal *BCR–ABL* junction (Figure 14(B), top panel), with the maxizyme remaining in its inactive conformations in the presence of normal *ABL* mRNA and in the absence of the abnormal *BCR–ABL* junction (Figure 14(B), bottom panel). The specifically designed sequences, which are shown in Figure 14(B) (note that the lengths and sequences of sensor arms and those of common stem II are the variables), should ensure such conformations in the presence and in the absence, respectively, of the abnormal b2a2 mRNA.[63] This phenomenon would resemble the changes in conformation of allosteric proteinaceous enzymes that occur in response to their effector molecules. The name "*maxizyme*" was initially chosen as an abbreviation of *m*inimized, *a*ctive, *x*-shaped (heterodimeric), and *i*ntelligent (allosterically controllable) ribo*zyme*.[63] Later, the term of maxizyme was used simply to distinguish monomeric forms of conventional minizymes that have extremely low activity from the novel dimeric short ribozyme with high-level activity.[66]

In order to prove *in vitro* that changes in the conformation of our heterodimeric maxizyme depended on the presence or absence of the abnormal b2a2 mRNA, specificity was examined by incubating the

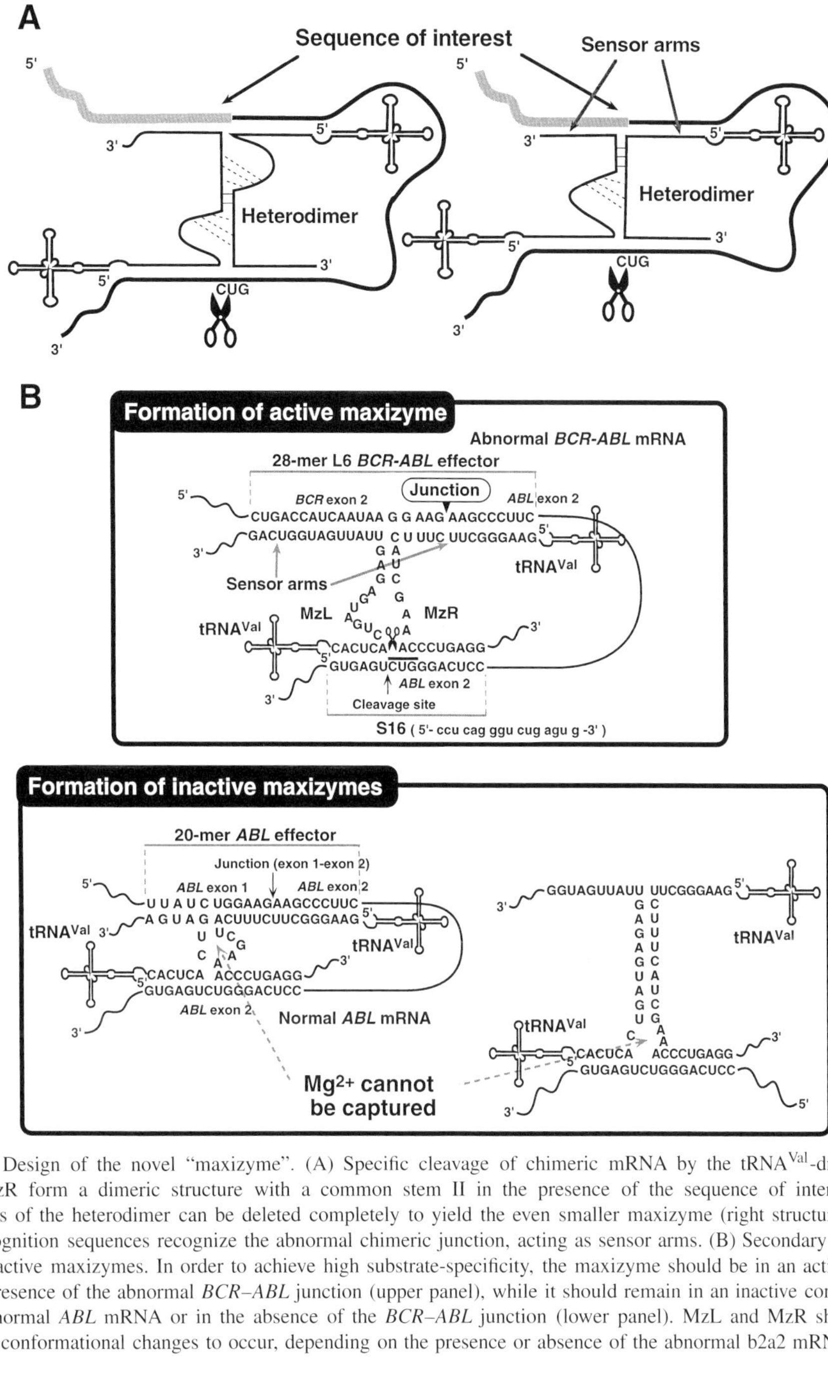

Figure 14. Design of the novel "maxizyme". (A) Specific cleavage of chimeric mRNA by the $tRNA^{Val}$-driven maxizyme. MzL and MzR form a dimeric structure with a common stem II in the presence of the sequence of interest. One of the catalytic cores of the heterodimer can be deleted completely to yield the even smaller maxizyme (right structure), in which the substrate-recognition sequences recognize the abnormal chimeric junction, acting as sensor arms. (B) Secondary structures of the active and inactive maxizymes. In order to achieve high substrate-specificity, the maxizyme should be in an active conformation only in the presence of the abnormal *BCR–ABL* junction (upper panel), while it should remain in an inactive conformation in the presence of normal *ABL* mRNA or in the absence of the *BCR–ABL* junction (lower panel). MzL and MzR should allow such conformational changes to occur, depending on the presence or absence of the abnormal b2a2 mRNA.

maxizyme, which had been transcribed *in vitro*, with the $5'$-^{32}P-labeled short 16-nt substrate (S16) that corresponded to the target (cleavage) site in the presence and in the absence of either a 20-nt normal *ABL* effector molecule or a 28-nt *BCR–ABL* effector molecule (Figure 15).[63] Indeed, no products of cleavage of substrate S16 were detected in the absence of the *BCR–ABL* junction or in the presence of the normal *ABL* sequence (effector molecule), demonstrating the expected high substrate-specificity of the maxizyme (Figure 15). These results proved that the maxizyme was subject to complete allosteric control *in vitro*, in accord with the predicted conformational changes (depicted at the bottom in Figure 15) that should occur in response to the effector molecule (the *BCR–ABL* junction) that was added *in trans*. Furthermore, the results confirmed that the $tRNA^{Val}$ portion of the maxizyme did not interfere with the allosteric control.

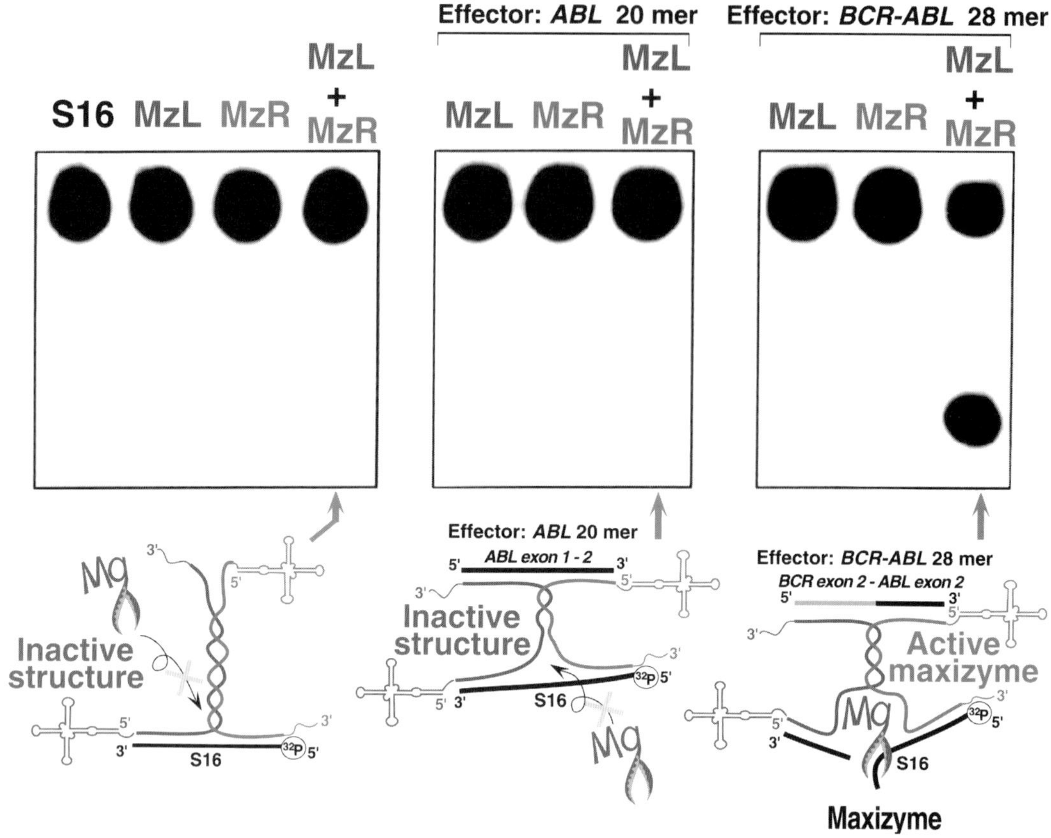

Figure 15. Allosteric control of the activity of the maxizyme *in vitro*. The specificity of maxizyme-mediated cleavage was examined by incubating tRNA^{Val}-driven component(s) with the 5′-^{32}P-labeled short 16-mer substrate (S16) in the presence and in the absence of an allosteric effector molecule, namely, either a short 20-mer normal *ABL* sequence (*ABL* 20-mer) or a short 28-mer *BCR–ABL* sequence (*BCR–ABL* 28-mer). Sequences of these effector molecules are shown in Fig. 14B. MzL and/or MzR were incubated at 0.1 μM with 2 nM 5′-^{32}P-labeled substrate (S16). When applicable, the concentration of the effector, 20-mer *ABL* or 28-mer *BCR–ABL*, was 1 μM. In order to achieve high substrate-specificity, the tRNA^{Val}-driven maxizyme should be in an active conformation only in the presence of the abnormal *BCR–ABL* junction (right), while it should remain in an inactive conformation in the presence of normal *ABL* mRNA or in the absence of the *BCR–ABL* junction (left and middle). The catalytic action of Mg^{2+} ions is indicated by "Mg scissors".

16.2.6.3 The intracellular activity and specificity of the maxizyme against an endogenous BCR–ABL cellular target in mammalian cells

We next examined the activity of the maxizyme against an endogenous *BCR–ABL* (b2a2 mRNA) target in mammalian cells. For this purpose, we established a line of murine cells, BaF3/p210$^{BCR\text{-}ABL}$, that expressed human b2a2 mRNA constitutively, by integrating into the murine genome a plasmid construct that expressed p210$^{BCR\text{-}ABL}$ (p210$^{BCR\text{-}ABL}$ was generated from human b2a2 mRNA). Although the parental BaF3 cell line is an interleukin-3-dependent (IL-3-dependent) hematopoietic cell line,[158,159] our transformed BaF3/p210$^{BCR\text{-}ABL}$ cells were IL-3-independent because of the tyrosine kinase activity of p210$^{BCR\text{-}ABL}$ and, thus, the latter transformed cells were able to grow in the absence of IL-3. However, if the expression of p210$^{BCR\text{-}ABL}$ were to be inhibited, BaF3/p210$^{BCR\text{-}ABL}$ cells should become IL-3-dependent and, in the absence of IL-3, they should undergo apoptosis. In order to examine the specificity of the maxizyme, we also used H9 cells, which originated from human T cells and expressed normal *ABL* mRNA, as control cells.

These cell lines led us to examine whether the maxizyme could play a role in the regulation of apoptosis. We transfected BaF3 cells that stably expressed *BCR–ABL* (b2a2) mRNA with plasmids that encoded the wild-type ribozyme (pVwtRz), the maxizyme (pV-MzL/R), or the parental vector (pV), and we selected cells by exposing them to puromycin 24 h after transfection. After incubation for 60 h in the presence of puromycin, dead cells were removed and puromycin-resistant cells were cultured for various

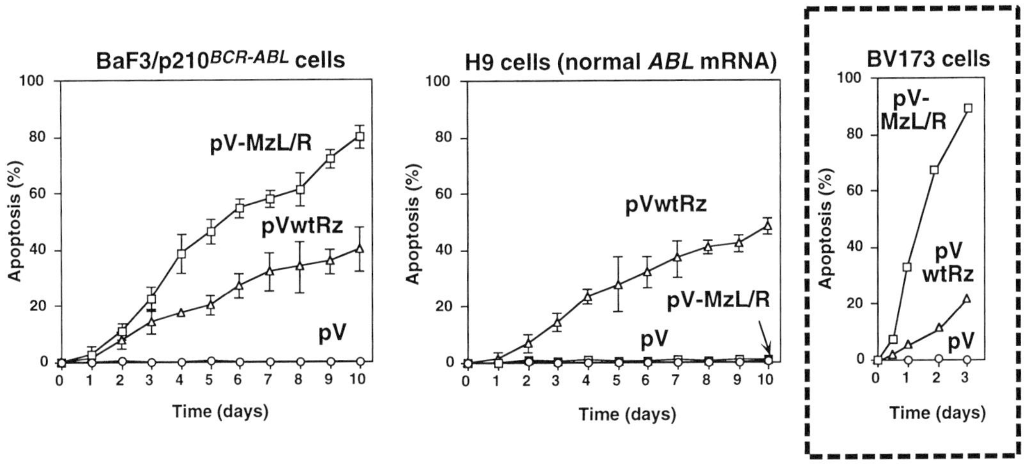

Figure 16. Efficiency of cleavage by the maxizyme of the endogenous *BCR–ABL* mRNA target. The viability of $tRNA^{Val}$-ribozyme-transduced BaF3/p210$^{BCR-ABL}$ cells and H9 cells were measured. The viability of BV173 cells, which were derived from a patient with a Philadelphia chromosome and which transiently expressed $tRNA^{Val}$-ribozymes, is also shown. Each monomer unit of the maxizyme was expressed independently by using tRNA-expression system. The maxizyme-expression vector of pV-MzL/R has both MzL- and MzR-expression cassettes in tandem. The wtRz is a conventional hammerhead ribozyme that was designed to direct the same cleavage site with the maxizyme. pV is a control vector containing only the $tRNA^{Val}$-expression cassette.

times in medium without IL-3. The viability of cells was assessed in terms of their ability to exclude trypan blue dye. As shown in the left panel of Figure 16, BaF3/p210$^{BCR-ABL}$ cells that expressed the maxizyme died rapidly, whereas the control-transfected BaF3/p210$^{BCR-ABL}$ (pV) cells were still alive 10 days after withdrawal of IL-3. Moreover, the maxizyme did not kill any H9 cells that expressed normal *ABL* mRNA (Figure 16, middle), a result that demonstrates the high specificity of the maxizyme for its target, the chimeric *BCR–ABL* gene. By contrast, in the presence of the conventional hammerhead ribozyme wtRz, apoptosis was induced in both BaF3/p210$^{BCR-ABL}$ and H9 cells (Figure 16, left and middle). This result was consistent with the observation that wtRz can target the transcripts of both the *BCR–ABL* gene and the normal *ABL* gene *in vitro*. Furthermore, the maxizyme also killed many more BV173 cells,[160] derived from a leukemic patient with a Philadelphia chromosome, than did the wild-type ribozyme or the parental vector (Figure 16, right).

Moreover, we tried to detect the anticipated products of cleavage directly by Northern blotting analysis. Total RNA from $tRNA^{Val}$-ribozyme/maxizyme-transduced BaF3/p210$^{BCR-ABL}$ cells was extracted after the withdrawal of IL-3. The length of the major cleavage product detected was exactly as anticipated (3.5 kb). No reduction in the level of expressed *BCR–ABL* mRNA was observed in the case of the control $tRNA^{Val}$ RNA (pV). Detection of the cleavage fragment proved that the maxizyme and the conventional ribozyme were catalytically active and cleaved specifically the target mRNA in cultured cells. In one set of control experiments, BaF3/p210$^{BCR-ABL}$ cells that expressed the target RNA substrate but no maxizyme RNA were mixed, just before isolation of the RNA, with transformed BaF3 cells that had expressed the maxizyme but not the substrate RNA, and then total RNA was isolated from the mixture of cells. In this case, no cleavage products were detected by Northern blotting, a clear demonstration that cleavage of the substrate RNA had occurred within cells and not during the isolation of the RNA *in vitro*. Thus, we confirmed that the apoptosis of cells originated from the cleavage of *BCR–ABL* mRNA by the maxizyme or of *BCR–ABL* and *ABL* mRNAs by the ribozyme, with resultant depletion of p210$^{BCR-ABL}$ and/or p145 c-*ABL* proteins in the respective hematopoietic cells. (The p145 c-*ABL* protein is a nuclear protein with low intrinsic tyrosine kinase activity, whereas the p210$^{BCR-ABL}$ protein is a cytoplasmic, membrane-associated protein with a constitutively high level of tyrosine kinase activity that prolongs the survival of hematopoietic cells by inhibiting apoptosis.)

16.3 OTHER RIBOZYMES AS TOOLS FOR GENE THERAPY

16.3.1 Hairpin Ribozymes as Gene-Inactivating Agents

A hairpin ribozyme was first discovered in the (−)-strand of the satellite RNA associated with the tobacco ringspot virus (−)sTRSV.[3] This satellite RNA replicates via a rolling-circle mechanism. During replication, concatenated strands of satellite RNAs are produced. The concatenated strands are self-cleaved to yield monomeric strands and then self-ligation occurs to generate monomeric circles. The hairpin ribozyme is responsible for the self-processing reaction of the (−)-strand.[161–163] Other naturally occurring hairpin ribozymes have also been discovered in the (−)-strands of the satellite RNAs associated with chicory yellow mottle virus (sCYMV1) and arabis mosaic virus (sArMV).[164] During the replication of the (+)-strand of these satellite RNAs, concatenated (+)-strands are self-cleaved by hammerhead ribozymes. However, the ligation activity of hairpin ribozymes is very much higher than that of hammerhead ribozymes, which have almost undetectable ligation activity. Indeed, in some studies, hairpin ribozymes have been used as ligases to connect single-stranded RNAs *in vitro*.[44]

The catalytically essential domain within the 359-nt, negative-strand satellite RNA of TRSV, was identified as two minimal sequences by deletion analysis.[161] Subsequently, hairpin ribozymes were engineered to be able to act *in trans* in the same way as hammerhead ribozymes. A 50-nt catalytic segment can cleave a 10- to 14-nt substrate at a phosphodiester bond between residue A_{-1} and residue G_{+1} and can catalyze ligation of cleaved products with a terminal 2′,3′-cyclic phosphate on the 5′-fragment and a 5′-OH group on the 3′-fragment.[161,165] Calculations of minimal energy for folding, results of mutagenesis experiments and limited phylogenetic comparisons have revealed a possible secondary structure for hairpin ribozymes, and results of selection *in vitro* have lent support to this model.[161–164,166–171] The secondary structure of the hairpin ribozyme appears to consist of four helices and two internal loops (Figure 17). In a *trans*-acting hairpin ribozyme, helices I and II are formed by interactions between the substrate and the ribozyme, while helices III and IV are formed by sequences within the ribozyme itself. Internal loop A, which contains the cleavage and ligation site, is located between helices I and II. The other internal loop B is located between helices III and IV. Helix IV is capped by a hairpin loop that is not important for catalysis.[172] Most of the residues in loops A and B are essential for efficient cleavage. By contrast, all residues in helices I and IV, with the exception of the base pairs that flank loop A, can be varied, provided that the base pairs within helices can be formed. Moreover, the lengths of helices can be increased without loss of activity.[167–170]

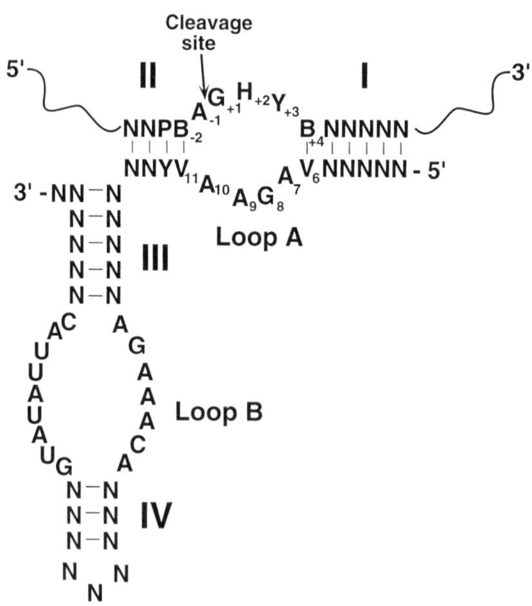

Figure 17. Secondary structure of a hairpin ribozyme. B = U, C, or G; H = A, C, or U; N = any nucleotide, except in helices where there is a requirement for base-pairing; P = A or G; V = A, C, or G; and Y = U or C.

One of the most important features of the hairpin ribozyme is the strong requirement for G_{+1} immediately adjacent to the cleavage site on its 3′ side (Figure 17). Replacement of this residue by A, U or C reduces the catalytic efficiency by a factor of at least 10^5 but it does not affect the binding of the substrate to the ribozyme.[167,169,170,173] These observations indicate that G_{+1} is necessary for a step that occurs after the substrate-binding step. In addition, compared to hammerhead ribozymes, hairpin ribozymes have much stronger preference for particular residues around the cleavage site for the efficient cleavage of RNA (Figure 17). However, since the cleavage activity of hairpin ribozymes is almost the same as that of hammerhead ribozymes, hairpins have also been examined for their ability to suppress gene expression *in vivo*. For the application of hairpin ribozymes *in vivo*, it is obviously important to use an appropriate expression system, as discussed above in the case of hammerhead ribozymes. Wong-Staal and her co-workers reported examples of the expression of hairpin ribozymes in cells, in which they used a human $tRNA^{Val}$ promoter and replaced the amino-acceptor stem sequence of the tRNA with a hairpin ribozyme directed against HIV-1.[49,120,174] They succeeded in inhibiting the replication of various strains of HIV-1 in cultured cells. Medical applications of hairpin ribozymes are being tested. Hairpin ribozymes clearly have considerable potential as gene-inactivating agents *in vivo* because of their strong cleavage activity, which is similar to that of hammerhead ribozymes, even though the potential sites of cleavage by hairpins are more limited.

16.3.2 RNase P-Mediated Regulation of Gene Expression

As mentioned above, small ribozymes, in particular hammerhead and hairpin ribozymes, are very attractive as inactivators of gene expression and their uses have been extensively explored by many scientists. However, the potential utility of large ribozymes, such as the *Tetrahymena* ribozyme and RNase P, has also been investigated.

Successful trials exploiting the activity of endogenous RNase P were reported by Altman's group.[59,65,175–177] The RNA subunit of ribonuclease P (the RNase P ribozyme) was one of the first two ribozymes to be discovered.[2,6] RNase P catalyzes the processing of the precursors to all kinds of tRNA (pre-tRNAs) in eukaryotic and prokaryotic cells. The enzyme cleaves pre-tRNAs, removing 5′ precursor-specific sequences and generating the mature 5′ ends of the tRNAs. A unique feature of the RNase P ribozyme is that, in nature, it catalyzes intermolecular reactions. By contrast, all other known ribozymes catalyze intramolecular reactions in nature. RNase P from both eukaryotic and prokaryotic sources consists of an RNA subunit and a protein subunit. For example, RNase P from *Escherichia coli* consists of a large RNA component of about 400 nucleotides and a relatively small protein component of about 120 amino acids. In nature, the RNA subunit, the ribozyme, acts together with the protein component. However, the catalytic center is in the RNA subunit. The protein subunit of RNase P was initially thought to be responsible for the catalytic activity. However, the RNA subunit of bacterial RNase P was shown to be able to cleave pre-tRNA in the absence of the protein component at high ionic strength *in vitro*, and, thus, the RNA component was identified as a catalytically active RNA or ribozyme.[2,178] The substrate-binding domain is located in the RNA subunit.[179] A putative model of the secondary structure of the ribozyme of RNase P was deduced from comparative phylogenetic analysis and the results of crosslinking studies, and it is shown in Figure 1.

The mechanism for the RNase P-mediated inhibition of gene expression is based on findings made by Altman's group.[180] They found that RNAs that form complexes that resemble the aminoacyl stem of a tRNA and that terminate similarly in the sequence CCA can be efficiently cleaved by RNase P both *in vitro* and *in vivo* (Figure 18). These RNA complexes consist of a substrate molecule and an effector molecule, which was designated the external guide sequence (EGS).[180] Each specific EGS RNA contains a sequence complementary to the substrate and a 3′-proximal CCA sequence. The EGS and the substrate form an aminoacyl stem-like structure which is recognized by RNase P as a substrate and also identifies the site of cleavage. Therefore, in principle, a specific EGS can target any specific RNA, such as an mRNA, for specific cleavage by RNase P. Indeed, Altman and coworkers succeeded in the specific inhibition of expression of several kinds of mRNA *in vitro*, in *Escherichia coli* and in mammalian cells.[59,65,175–177]

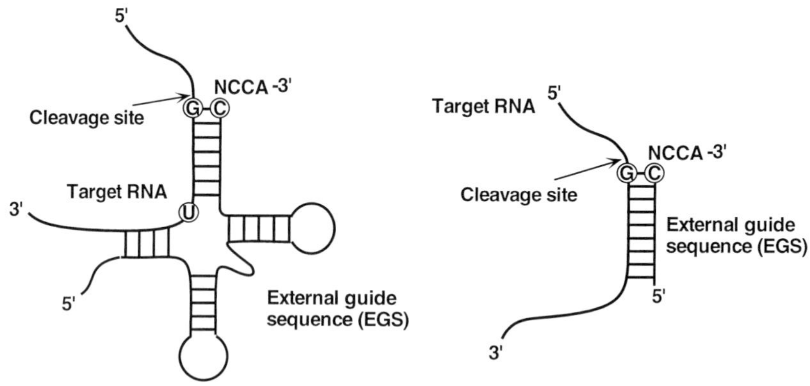

Figure 18. Schematic representation of the complexes between a target RNA and an EGS RNA. RNase P can cleave the target RNA when the latter forms a complex with an appropriate EGS RNA. The EGS RNA with a tRNA-like structure (left) and the shortened variant of EGS RNA (right). The arrows indicate the sites of cleavage by RNase P. The indicated nucleotides are essential for the effective cleavage by RNase P.

16.3.3 Group I Ribozymes as Tools for Gene Therapy

RNA splicing is a fundamental feature of the processing of RNA in many organisms. The pre-rRNA of *Tetrahymena thermophila* was found to undergo "self-splicing" *in vitro* without the need for a protein catalyst and it was one of the first RNA molecules to be discovered to have enzymatic activity.[1] RNA splicing by group I introns is extremely widespread and occurs during the generation of mature mRNAs, rRNAs and tRNAs. Such introns have been found in the mitochondrial, chloroplast and nuclear genomes of numerous eukaryotes and they have also been found in prokaryotic and eubacterial genomes. However, their distribution is irregular. For example, some species of *Tetrahymena* have group I introns while closely related species do not. The *Tetrahymena* and other "group I" self-splicing ribozymes promote two phosphoester-transfer reactions (Figure 19(A)), which result in the removal of an intervening sequence and the splicing of adjacent RNA domains.[7,9,44,70,181] Group I ribozymes were first defined on the basis of their common structural features and Figure 1 shows their general secondary structure, with lowercase letters indicating exons and uppercase letters indicating the intron, and with the cleavage and splicing sites indicated by thick arrows. The first and the second steps in RNA splicing by a group I ribozyme are shown in Figure 19. In the initial transesterification reaction, the 3′-OH group of guanosine or one of its 5′-phosphorylated derivatives is used as the nucleophile for S_N2 attack on the phosphate of the target internucleotide linkage (5′ side of the exon/intron junction).[182–187] In the second transesterification, the newly formed 3′-OH group of the 5′ exon is used as the nucleophile in the subsequent attack at the second splice-site junction (3′ side of the exon/intron junction). Each reaction results in inversion at the phosphorus center.

Exploiting the splicing ability of group I intron, Sullenger and Cech and their coworkers succeeded in applying such ribozymes *in vivo* in an unusual way.[54,64,188,189] Unlike the methodologies described above, their methodology does not make use of the ribozyme as an inhibitor of gene expression via disruption of a target mRNA. In their system, defective mRNA is repaired by a *trans*-splicing ribozyme (Figure 19(B)). While group I ribozymes naturally catalyze self-splicing reactions, they also appear to be able to mediate *trans*-splicing reactions, resembling other ribozymes that can act *in trans*.[190,191] Sullenger and Cech created a *trans*-splicing ribozyme with the appropriate coding sequence in the exon region that was connected downstream of the catalytic intron (Figure 19(B)).[188] Such a *trans*-splicing ribozyme should potentially be able to replace a defective portion of a mutant mRNA with a functional sequence.

A slightly shortened version of a *trans*-splicing ribozyme, designated L21,[192] was able to repair truncated transcripts of the *lacZ* gene in *Escherichia coli*, as well as in the cytoplasm of mammalian cells.[54,188] More recently, a *trans*-splicing ribozyme was used to repair mutant transcripts associated with a common genetic disorder, namely, to amend mutant β-globin transcripts in erythroid lineage cells derived from the peripheral blood of patients with sickle cell disease.[64] These successes suggest that *trans*-splicing ribozymes for repair of mutant RNAs might also turn out to be uniquely useful tools for gene therapy, in addition to other mRNA-cleaving ribozymes of the types described above.

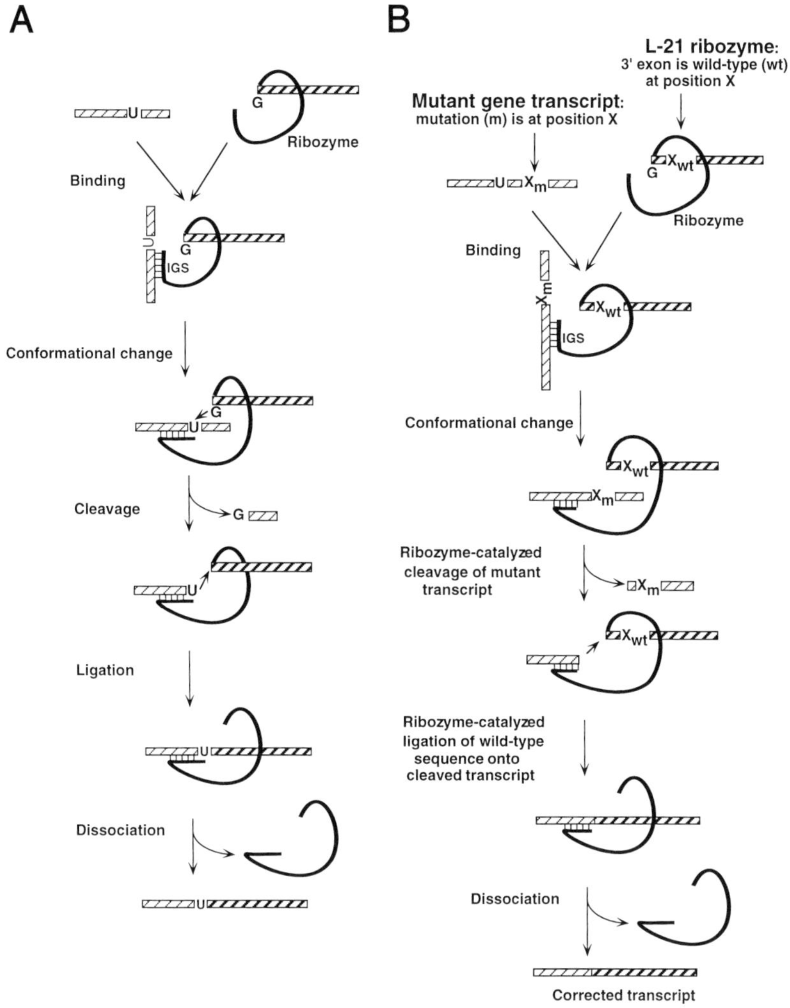

Figure 19. Reaction scheme for the splicing reaction mediated by a *trans*-acting group I intron. (A) The splicing reaction mediated by a group I intron. (B) Schematic representation of group I intron-mediated repair of RNA.

16.3.4 Recently Discovered Artificial DNA Enzymes as Gene-Inactivating Agents

There have been several successful attempts to generate DNA enzymes by selection *in vitro* and DNA molecules with catalytic activity have been isolated. Such DNA enzymes include a Pb^{2+}-dependent DNA enzyme with RNA-cleavage activity isolated by Joyce's group,[193] a DNA enzyme with DNA ligase activity isolated by Szostak's group,[194] and a DNA enzyme with self-cleaving activity isolated by Breaker's group.[195] DNA enzymes are single-stranded and, naturally, they cannot be produced *in vivo* by the use of the conventional expression vectors that are used for RNA enzymes. Nevertheless, compared with RNA molecules, DNA molecules have several advantageous features, provided that they can be introduced into cells by external delivery systems. For example, DNA is amenable to molecular design by rational and combinatorial methods; DNA can be synthesized with ease; DNA is more stable than RNA *in vivo*; and various chemical-modification procedures are available for further stabilization of DNA *in vivo*. These properties of DNA suggest that it might have considerable potential value in therapeutic and industrial situations.

In a recent study, Santoro and Joyce successfully selected DNA enzymes *in vitro* that were able to cleave RNA molecules with any sequence.[69] Joyce's DNA enzymes, shown in Figure 20, are somewhat similar to conventional hammerhead ribozymes. They consist of a catalytic domain of fifteen deoxyribonucleotides and metal ions, such as Mg^{2+} ions, are necessary for their catalytic activity,[69,71,74] as is the case for hammerhead ribozymes, which are recognized as metalloenzymes.[18–34] The catalytic domain is flanked by two substrate-recognition domains of seven or eight deoxyribonucleotides each,

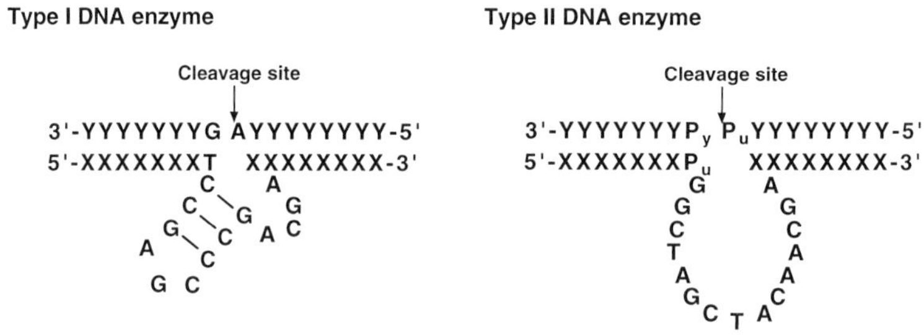

Figure 20. Secondary structure of the DNA enzymes that were isolated by Joyce's group.[69] In the substrate, base Y can form a base pair with base X. P_u denotes nucleotide A or G, and P_y denotes nucleotide U or C. Cleavage occurs at the site indicated by an arrow.

and the RNA substrate is bound by Watson–Crick base pairing (Figure 20). These DNA enzymes can be divided into two types. Type I DNA enzymes can cleave an RNA sequence at a phosphodiester bond that is located between an A and a G residue. The catalytic domain consists of a 4-nt loop adjacent to the cleavage site and a stem-loop region that resembles the stem-loop II region of the hammerhead ribozyme. However, the stem-loop region in the DNA enzyme is essential for catalysis.[69] Type II DNA enzymes can cleave an RNA sequence at a phosphodiester bond between a purine and pyrimidine residue. In this case, the catalytic domain consists of 15 nt (Figure 20). It should be possible to engineer such DNA enzymes, which have considerable flexibility in terms of the choice of cleavage site, so that they can cleave almost any target RNA substrate specifically. With respect to substrate specificity, Joyce's DNA enzymes might have still more advantages as compared to ribozymes. In general, DNA–RNA duplexes are less stable than the corresponding RNA–RNA duplexes. Therefore, in the case of a DNA enzyme, mismatches along binding arms that might be generated during binding with a non-target mRNA would interfere with subsequent cleavage to a greater extent than would be the case with a ribozyme.

Since we are interested in the specific cleavage of *BCR–ABL* (b2a2) mRNA, as described above, and since there are several potential sites of cleavage by Joyce's DNA enzymes within three nucleotides of the *BCR–ABL* junction in this mRNA, we tested these DNA enzymes for their ability to cleave *BCR–ABL* mRNA specifically. As expected, these DNA enzymes did cleave the *BCR–ABL* chimeric mRNA specifically *in vitro*.[157] We also demonstrated the usefulness of DNA enzymes for the specific suppression of expression of L6 *BCR–ABL* mRNA in mammalian cells, such as BV173 cells from a leukemic patient.[73] By contrast to conventional antisense DNAs, which suppressed expression not only of abnormal *BCR–ABL* mRNA but also of normal *ABL* mRNA, most of the DNA enzymes that we tested were highly specific and cleaved only abnormal *BCR–ABL* mRNA without damaging the normal message.

According to our studies,[73] although the efficacy of DNA enzymes with natural linkages decreased 12 h after their introduction into cells, partially modified DNA enzymes, with either phosphorothioate or 2′-*O*-methyl groups at both their 5′ and 3′ ends, remained active for much longer times in mammalian cells. Moreover, a DNA enzyme with 2′-*O*-methyl modifications, but not a DNA enzyme with phosphorothioate linkages, was also highly specific.[73] Therefore, DNA enzymes with the former but not the latter modifications appear to have potential utility as gene-inactivating agents, in particular for the treatment of CML.

Since DNA enzymes are easier to synthesize, easier to handle, and more stable *in vivo* than ribozymes, appropriately modified DNA enzymes should turn out to be powerful catalytic nucleic-acid drugs if appropriate exogenous delivery systems can be developed.[73,196]

16.4 FINAL REMARKS ON THE EXPRESSION SYSTEM

It is clear that ribozymes are potentially useful tools for suppression of the expression of specific genes. Although many trials have been successful,[39–68] it remains difficult to design an effective ribozyme-expression system that can be used *in vivo*. The activity of ribozymes *in vivo* appears to depend to a very considerable extent on the expression system. When ribozymes are properly expressed, higher activity can be expected. Therefore, one major challenge related to the use of ribozymes, as well as of antisense RNAs, as therapeutic or genetic agents is the development of suitable expression vectors.

Two kinds of expression system have been used to date, as discussed in this chapter, namely, the pol II system and the pol III system. We have focused on the pol III system and the promoter of a human gene for $tRNA^{Val}$ for the transcription of hammerhead ribozymes.[55,61–63,66–68] Not only is this promoter suitable for transcription of small RNAs but its use also facilitates predictions of secondary structure by computer folding. More importantly, if properly designed, this promoter system allows export of transcribed $tRNA^{Val}$-ribozymes from the nucleus to the cytoplasm so that the $tRNA^{Val}$-ribozymes can find their mRNA targets in an accessible state.

The $tRNA^{Val}$-vector may prove useful for the expression of functional RNAs, other than ribozymes, whose target molecules are localized in the cytoplasm. Although the co-localization of a ribozyme and its substrate in the cytoplasm does not by itself guarantee effectiveness, it clearly increases the probability of success. In our hands, properly designed $tRNA^{Val}$-ribozymes that can be exported to the cytoplasm have consistently high activities and, thus, they appear to be very useful as tools in molecular biology, with potential utility in medicine as well.[55,61–63,66–68] It should be noted, however, that for the RNase P-based gene-inactivation approach (Figure 18), the expression cassettes employed to date were designed to ensure high levels of EGS expression in the nucleus since EGSs are likely to work best in the nucleus, the natural location of RNase P.

Our novel maxizymes were more effective than similarly transcribed standard ribozymes in cells. To the best of our knowledge, our novel maxizyme is superior to other nucleic acid-based drugs reported to date because of its extremely high substrate-specificity and high cleavage activity. Novel maxizymes, whose activity can be controlled allosterically by sensor arms that recognize abnormal mRNAs specifically should be powerful tools for the disruption of abnormal chimeric targets and might provide the basis for future gene therapy for the treatment of CML and other diseases.

We trust that several small and large ribozymes and DNA enzymes described in this chapter will demonstrate their usefulness in the future animal studies and in medical settings.

ACKNOWLEDGMENTS

The authors thank their collaborators, in particular, Drs. S. Koseki, K. K. Yokoyama, T. Tanabe, K. Tani and S. Asano, for many helpful discussions.

16.5 REFERENCES

1. T.R. Cech, A.J. Zaug and P.J. Grabowski, *Cell*, 1981, **27**, 487.
2. C. Guerrier-Takada, K. Gardiner, T. Marsh, N. Pace and S. Altman, *Cell*, 1983, **35**, 849.
3. J.M. Buzayan, W.L. Gerlach and G. Bruening, *Nature*, 1986, **323**, 349.
4. L. Sharmeen, M.Y.-P. Kuo, G. Dinter-Gottlieb and J. Taylor, *J. Virol.*, 1988, **62**, 2674.
5. H.-N. Wu, Y.-J. Lin, F.-P. Lin, S. Makino, M.-F. Chang and M.M.C. Lai, *Proc. Natl. Acad. Sci. USA*, 1989, **86**, 1831.
6. S. Altman, *Adv. Enzymol.*, 1989, **62**, 1.
7. T.R. Cech, *Angew. Chem. Int. Ed. Engl.*, 1989, **29**, 759.
8. R.H. Symons, *Annu. Rev. Biochem.*, 1992, **61**, 641.
9. R.F. Gesteland and J.F. Atkins, eds., "The RNA World", Cold Spring Harbor Laboratory Press, New York, 1993.
10. J. Bratty, P. Chartrand, G. Ferbeyre and R. Cedergren, *Biochim. Biophys. Acta*, 1993, **1216**, 345.
11. N.R. Pace and J.W. Brown, *J. Bacteriol.*, 1919, **1995**, 177.
12. K.R. Birikh, P.A. Heaton and F. Eckstein, *Eur. J. Biochem.*, 1997, **245**, 1.
13. O.C. Uhlenbeck, *Nature*, 1987, **328**, 596.
14. J. Haseloff and W.L. Gerlach, *Nature*, 1988, **334**, 585.
15. D. Herschlag and T.R. Cech, *Nature*, 1990, **344**, 405.
16. D.L. Robertson and G.F. Joyce, *Nature*, 1990, **344**, 467.
17. D. Smith and N.R. Pace, *Biochemistry*, 1993, **32**, 5273.
18. S.C. Dahm, W.B. Derrick and O.C. Uhlenbeck, *Biochemistry*, 1993, **32**, 13040.
19. J.A. Piccirilli, J.S. Vyle, M.H. Caruthers and T.R. Cech, *Nature*, 1993, **361**, 85.
20. T.A. Steitz and J.A. Steitz, *Proc. Natl. Acad. Sci. USA*, 1993, **90**, 6498.
21. M. Yarus, *FASEB J.*, 1993, **7**, 31.
22. A.M. Pyle, *Science*, 1993, **261**, 709.
23. T. Uchimaru, M. Uebayasi, K. Tanabe and K. Taira, *FASEB J.*, 1993, **7**, 137.
24. M. Uebayasi, T. Uchimaru, T. Koguma, S. Sawata, T. Shimayama and K. Taira, *J. Org. Chem.*, 1994, **59**, 7414.
25. S. Sawata, M. Komiyama and K. Taira, *J. Am. Chem. Soc.*, 1995, **117**, 2357.

26. D.-M. Zhou, N. Usman, F.E. Wincott, J. Matulic-Adamic, M. Orita, L.-H. Zhang, M. Komiyama, P.K.R. Kumar and K. Taira, *J. Am. Chem. Soc.*, 1996, **118**, 5862.
27. P.K.R. Kumar, D.M. Zhou, K. Yoshinari and K. Taira, in "Catalytic RNA", eds. F. Eckstein and D.M.J. Lilley, Nucleic Acids and Molecular Biology 10, Springer-Verlag, Berlin, 1996, p. 217.
28. L.B. Weinstein, B.C.N.M. Jones, R. Cosstick and T.R. Cech, *Nature*, 1997, **388**, 805.
29. B.W. Pontius, W.B. Lott and P.H. von Hippel, *Proc. Natl. Acad. Sci. USA*, 1997, **94**, 2290.
30. D.-M. Zhou, L.-H. Zhang and K. Taira, *Proc. Natl. Acad. Sci. USA*, 1997, **94**, 14343.
31. D.-M. Zhou and K. Taira, *Chem. Rev.*, 1998, **98**, 991.
32. W.B. Lott, B.W. Pontius and P.H. von Hippel, *Proc. Natl. Acad. Sci. USA*, 1998, **95**, 542.
33. S. Kuusela and H. Lönnberg, *Curr. Top. Solut. Chem.*, 1997, **2**, 29.
34. M. Warashina, D.-M. Zhou, T. Kuwabara and K. Taira, in "Comprehensive Natural Product Chemistry", eds. D. Söll and S. Nishimura, Elsevier, Oxford, p. 235.
35. A. Hampel and J.A. Cowan, *Chem. Biol.*, 1997, **4**, 513.
36. S. Nesbitt, L.A. Hegg and M.J. Fedor, *Chem. Biol.*, 1997, **4**, 619.
37. K.J. Young, F. Gill and J.A. Grasby, *Nucleic Acids Res.*, 1997, **25**, 3760.
38. J.B. Murray, A.A. Seyhan, N.G. Walter, J.M. Burk and W.G. Scott, *Chem. Biol.*, 1998, **5**, 587.
39. J.A.H. Murray, ed., "Antisense RNA and DNA", Wiley-Liss, New York, 1992.
40. R.P. Erickson and J. Izant, eds., "Gene Regulation: Biology of Antisense RNA and DNA", Raven Press, New York, 1992.
41. T.R. Cech, *Curr. Opin. Struct. Biol.*, 1992, **2**, 605.
42. S. Altman, *Proc. Natl. Acad. Sci. USA*, 1993, **90**, 10898.
43. J.J. Rossi, *Trends Biotechnol.*, 1995, **13**, 301.
44. F. Eckstein and D.M.J. Lilley, eds., "Catalytic RNA", Nucleic Acids and Molecular Biology 10, Springer-Verlag, Berlin, 1996.
45. P.C. Turner, ed., "Ribozyme Protocols", Methods in Molecular Biology 74, Humana Press Inc., New Jersey, 1997.
46. K.J. Scanlon, ed., "Therapeutic Applications of Ribozymes", Methods in Molecular Medicine 11, Humana Press Inc., New Jersey, 1998.
47. N. Sarver, E.M. Cantin, P.S. Chang, J.A. Zaia, P.A. Ladne, D.A. Stephens and J.J. Rossi, *Science*, 1990, **247**, 1222.
48. J. Ohkawa, N. Yuyama, Y. Tanabe, S. Nishikawa and K. Taira, *Proc. Natl. Acad. Sci. USA*, 1993, **90**, 11302.
49. M. Yu, J. Ojwang, O. Yamada, A. Hampel, J. Rapapport, D. Looney and F. Wong-Staal, *Proc. Natl. Acad. Sci. USA*, 1993, **90**, 6340.
50. B.A. Sullenger and T.R. Cech, *Science*, 1993, **262**, 1566.
51. Y. Inokuchi, N. Yuyama, A. Hirashima, S. Nishikawa, J. Ohkawa and K. Taira, *J. Biol. Chem.*, 1994, **269**, 11361.
52. P. Marschall, J.B. Thomson and F. Eckstein, *Cell. Mol. Neurobiol.*, 1994, **14**, 523.
53. J. Ohkawa, T. Koguma, T. Kohda and K. Taira, *J. Biochem. (Tokyo)*, 1995, **118**, 251.
54. J.T. Jones, S.W. Lee and B.A. Sullenger, *Nat. Med.*, 1996, **2**, 643.
55. H. Kawasaki, J. Ohkawa, N. Tanishige, K. Yoshinari, T. Murata, K. Yokoyama and K. Taira, *Nucleic Acids Res.*, 1996, **24**, 3010.
56. S. Fujita, T. Koguma, J. Ohkawa, K. Mori, T. Kohda, H. Kise, S. Nishikawa, M. Iwakura and K. Taira, *Proc. Natl. Acad. Sci. USA*, 1997, **94**, 391.
57. E. Bertrand, D. Castanotto, C. Zhou, C. Carbonnelle, G.P. Lee, S. Chatterjee, T. Grange, R. Pictet, D. Kohn, D. Engelke and J.J. Rossi, *RNA*, 1997, **3**, 75.
58. H. Chen, G. Ferbeyre and R. Cedergren, *Nat. Biotechnol.*, 1997, **15**, 432.
59. C. Guerrier-Takada, R. Salavati and S. Altman, *Proc. Natl. Acad. Sci. USA*, 1997, **94**, 8468.
60. J. Ohkawa, Y. Takebe and K. Taira, in "Therapeutic Applications of Ribozymes", ed. K.J. Scanlon, Methods in Molecular Medicine 11, Humana Press Inc., New Jersey, 1998, p. 83.
61. H. Kawasaki, J. Ohkawa, R. Eckner, T.P. Yao, K. Taira, R. Chiu, D.M. Livingston and K.K. Yokoyama, *Nature*, 1998, **393**, 284.
62. T. Kuwabara, M. Warashina, M. Orita, S. Koseki, J. Ohkawa and K. Taira, *Nat. Biotechnol.*, 1998, **16**, 961.
63. T. Kuwabara, M. Warashina, T. Tanabe, K. Tani, S. Asano and K. Taira, *Mol. Cell*, 1998, **2**, 617.
64. N. Lan, R.P. Howrey, S.-W. Lee, C.A. Smith and B.A. Sullenger, *Science*, 1998, **280**, 1593.
65. D. Plehn-Dujowich and S. Altman, *Proc. Natl. Acad. Sci. USA*, 1998, **95**, 7327.
66. T. Kuwabara, M. Warashina, A. Nakayama, J. Ohkawa and K. Taira, *Proc. Natl. Acad. Sci. USA*, 1886, **1999**, 96.
67. S. Koseki, T. Tanabe, K. Tani, S. Asano, T. Shioda, Y. Nagai, T. Shimada, J. Ohkawa and K. Taira, *J. Viol.*, 1999, **73**, 1868.
68. H. Kawasaki, S. Koseki, J. Ohkawa, K.K. Yokoyama and K. Taira, in "Ribozymes: Biology and Biotechnology", eds. G. Krupp and R.K. Gaur, Eaton Publisher, MA, in press.
69. S.W. Santoro and G.F. Joyce, *Proc. Natl. Acad. Sci. USA*, 1997, **94**, 4262.
70. R.R. Breaker, *Chem. Rev.*, 1997, **97**, 371.
71. S.W. Santoro and G.F. Joyce, *Biochemistry*, 1998, **37**, 13330.
72. N. Ota, M. Warashina, K. Hirano, K. Hatanaka and K. Taira, *Nucleic Acids Res.*, 1998, **26**, 3385.
73. M. Warashina, T. Kuwabara, Y. Nakamatsu and K. Taira, *Chem. Biol.*, 1999, **6**, 237.
74. Q.-C. He, D.-M. Zhou, J.-M. Zhou, Y. Nakamatsu and K. Taira, submitted for publication.
75. D.E. Ruffner, G.D. Stormo and O.C. Uhlenbeck, *Biochemistry*, 1990, **29**, 10695.
76. T. Shimayama, S. Nishikawa and K. Taira, *Biochemistry*, 1995, **34**, 3649.
77. A.R. Kore, N.K. Vaish, U. Kutzke and F. Eckstein, *Nucleic Acids Res.*, 1998, **26**, 4116.
78. D.M. Long and O.C. Uhlenbeck, *Proc. Natl. Acad. Sci. USA*, 1994, **91**, 6977.
79. K.L. Nakamaye and F. Eckstein, *Biochemistry*, 1994, **33**, 1271.

80. N.K. Vaish, P.A. Heaton and F. Eckstein, *Biochemistry*, 1997, **36**, 6495.
81. N.K. Vaish, P.A. Heaton, O. Fedorova and F. Eckstein, *Proc. Natl. Acad. Sci. USA*, 1998, **95**, 2158.
82. A.B.J. Burgin, C. Gonzalez, J. Matulic-Adamic, A.M. Karpeisky, N. Usman, J.A. McSwiggen and L. Beigelman, *Biochemistry*, 1996, **35**, 14090.
83. M. Hamada, H. Kawasaki, S. Fujita and K. Taira, in "Intracellular Ribozyme Technology: Protocols and Applications", ed. J.J. Rossi, Horizon Scientific Press, Norfolk, in press
84. Y. Nakamatsu, T. Kuwabara, M. Warashina, A. Shibata, H. Kawasaki and K. Taira, *Nucleic Acids Symp. Ser.*, 1998, **39**, 143.
85. C.C. Sheldon and R.H. Symons, *Nucleic Acids Res.*, 1989, **17**, 5679.
86. T. Tuschl and F. Eckstein, *Proc. Natl. Acad. Sci. USA*, 1993, **90**, 6991.
87. M.J. McCall, P. Hendry and P.A. Jennings, *Proc. Natl. Acad. Sci. USA*, 1992, **89**, 5710.
88. D.J. Fu and L.W. McLaughlin, *Proc. Natl. Acad. Sci. USA*, 1992, **89**, 3985.
89. J.B. Thomson, T. Tuschl and F. Eckstein, *Nucleic Acids Res.*, 1993, **21**, 5600.
90. P. Hendry, M.J. McCall, F.C. Santiago and P.A. Jennings, *Nucleic Acid Res.*, 1995, **23**, 3922.
91. H. Sugiyama, K. Hatano, I. Saito, S.V. Amontov and K. Taira, *FEBS Lett.*, 1996, **392**, 215.
92. S.V. Amontov and K. Taira, *J. Am. Chem. Soc.*, 1996, **118**, 1624.
93. T. Kuwabara, S.V. Amontov, M. Warashina, J. Ohkawa and K. Taira, *Nucleic Acids Res.*, 1996, **24**, 2302.
94. J. Conaty, P. Hendry and T. Lockett, *Nucleic Acids Res.*, 1999, **27**, 2400.
95. V. Budker, V. Gurevich, J.E. Hagstrom, F. Bortzov and J.A. Wolff, *Nat. Biotechnol.*, 1998, **14**, 760.
96. R.A. Morgan and W.F. Anderson, *Annu. Rev. Biochem.*, 1993, **62**, 191.
97. T. Boulikas, ed., "Gene Therapy and Molecular Biology: From Basic Mechanisms to Clinical Applications", Gene Therapy Press, CA, 1998, Vols. 1 and 2.
98. E. Bertrand, R. Pictet and T. Grange, *Nucleic Acids Res.*, 1994, **22**, 293.
99. M. Hamada, S. Fujita, H. Kise, Y. Jigami and K. Taira, *J. Biochem. (Tokyo)*, 1998, **123**, 684.
100. S.H. Cameron and P.A. Jennings, *Proc. Natl. Acad. Sci. USA*, 1993, **86**, 9139.
101. J.J. Zhao and L. Pick, *Nature*, 1993, **365**, 448.
102. S. Efrat, M. Leiser, Y.J. Wu, D. Fusco-DeMane, O. Emran, M. Surana, T.L. Letton, M.A. Magnuson, G. Weir and N. Fleischer, *Proc. Natl. Acad. Sci. USA*, 1994, **91**, 2051.
103. H. Sanfacon and T. Hohn, *Nature*, 1990, **346**, 81.
104. B.M. Chowrira, P.A. Pavco and J.A. McSwiggen, *J. Biol. Chem.*, 1994, **269**, 25856.
105. K. Taira, M. Oda, H. Shinshi, H. Maeda and K. Furukawa, *Protein Eng.*, 1990, **3**, 733.
106. K. Taira, K. Nakagawa, S. Nishikawa and K. Furukawa, *Nucleic Acids Res.*, 1991, **19**, 5125.
107. J. Ohkawa, N. Yuyama, S. Koseki, Y. Takebe, M. Homann, G. Sczakiel and K. Taira, *Gene Ther. Mol. Biol.*, 1998, **2**, 69.
108. D. Herschlag, *Proc. Natl. Acad. Sci. USA*, 1991, **88**, 6921.
109. M. Werner and O.C. Uhlenbeck, *Nucleic Acids Res.*, 1995, **23**, 2092.
110. K. Hertel, D. Herschlag and O.C. Uhlenbeck, *EMBO J.*, 1996, **15**, 3751.
111. Y. Takebe, M. Seiki, J. Fujisawa, P. Hoy, K. Yokota, K. Arai, M. Yoshida and N. Arai, *Mol. Cell. Biol.*, 1988, **8**, 466.
112. D.J. Voeks, G.A. Clawson and J.S. Norris, *Gene Ther. Mol. Biol.*, 1998, **1**, 407.
113. G.A. Clawson, W. Pan, S.E. Loy, B. Hoel, D.J. Voeks and J.S. Norris, *Gene Ther. Mol. Biol.*, 1998, **2**, 494.
114. C.M. Benedict, W. Pan, S.E. Loy and G.A. Clawson, *Carcinogenesis*, 1998, **19**, 1223.
115. A. Dzianott and J. Bujarski, *Proc. Natl. Acad. Sci. USA*, 1989, **86**, 4823.
116. M. Altschuler, R. Tritz and A. Hampel, *Gene*, 1992, **122**, 85.
117. S.R. Price, N. Ito, C. Oubridge, J.M. Avis and K. Nagai, *J. Mol. Biol.*, 1995, **249**, 398.
118. Y. Komatsu, I. Kanzaki, M. Shirai and E. Ohtsuka, *Biochemistry*, 1997, **36**, 9935.
119. M. Cotten and M. Birnstiel, *EMBO J.*, 1989, **8**, 3861.
120. O. Yamada, M. Yu, J.-K. Yee, G. Kraus, D. Looney and F. Wong-Staal, *Gene Ther.*, 1994, **1**, 38.
121. P.D. Good, A.J. Krikos, S.X. Li, E. Bertrand, N.S. Lee, L. Giver, A. Ellinton, J.A. Zaia, J.J. Rossi and D.R. Engelke, *Gene Ther.*, 1997, **4**, 45.
122. J. Beck and M. Nassal, *Nucleic Acids Res.*, 1995, **23**, 4954.
123. J. Ohkawa and K. Taira, unpublished data.
124. G.J. Arts, M. Fornerod and I.W. Mattaj, *Curr. Biol.*, 1998, **8**, 305.
125. G.J. Arts, S. Kuersten, P. Romby, B. Ehresman and I.W. Mattaj, *EMBO J.*, 1998, **17**, 7430.
126. U. Kutay, G. Lipowsky, E. Izaurralde, F.R. Bischoff, P. Schwarzmaier, E. Hartman and D. Görlich, *Mol. Cell*, 1998, **1**, 359.
127. E. Lund and J.E. Dahlberg, *Science*, 1998, **282**, 2082.
128. G. Lipowsky, F.R. Bischoff, E. Izaurralde, U. Kutay, S. Schäfer, H.J. Gross, H. Beier and D. Görlich, *RNA*, 1999, **5**, 539.
129. G. Simos and E. Hurt, *Curr. Biol.*, 1999, **9**, 238.
130. N. Shikama, J. Lyon and N.B. LaThangue, *Trends Cell Biol.*, 1997, **7**, 230.
131. H. Kawasaki, J. Song, R. Eckner, H. Ugai, R. Chiu, K. Taira, Y. Shi, N. Jones and K.K. Yokoyama, *Genes Dev.*, 1998, **12**, 233.
132. K.J. Hertel, A. Pardi, O.C. Uhlenbeck, M. Koizumi, E. Ohtsuka, S. Uesugi, R. Cedergren, F. Eckstein, W.L. Gerlach, R. Hodgson and R.H. Symons, *Nucleic Acids Res.*, 1992, **20**, 3252.
133. E.G. Bernstine and B. Ephrussi, in "Teratomas and Differentiation", eds. M.I. Sherman and D. Solter, Academic Press, New York, 1976, p. 271.
134. S. Strickland and V. Mahdavi, *Cell*, 1978, **15**, 393.
135. S. Strickland, K.K. Smith and K.R. Marotti, *Cell*, 1980, **21**, 347.
136. R. Atencia, M. Garcia-Sanz, M. Unda and J. Arechaga, *Exp. Cell Res.*, 1994, **214**, 663.

137. J. Clifford, H. Chiba, D. Sobieszczuk, D. Metzger and P. Chambon, *EMBO J.*, 1996, **15**, 4142.
138. Y. Gavrieli, Y. Sherman and S.A. Ben-Sasson, *J. Cell Biol.*, 1992, **119**, 493.
139. T.-P. Yao, S.P. Oh, M. Fuchs, N.-D. Zhou, L.-E. Ch'ng, D. Newsome, R.T. Bronson, E. Li, D.M. Livingston and R. Eckner, *Cell*, 1998, **93**, 361.
140. D.J. Fu, F. Benseer and L.W. McLaughlin, *J. Am. Chem. Soc.*, 1994, **116**, 4591.
141. T. Kuwabara, S.V. Amontov, M. Warashina, J. Ohkawa and K. Taira, *Nucleic Acids Res.*, 1996, **24**, 2302.
142. A. Nakayama, T. Kuwabara, M. Warashina and K. Taira, *FEBS Lett.*, 1999, **448**, 67.
143. H.W. Pley, K.M. Flaherty and D.B. McKay, *Nature*, 1994, **372**, 68.
144. J.L. Sussman, S.R. Holbrook, R.W. Warrant, G.M. Church and S.H. Kim, *J. Mol. Biol.*, 1978, **123**, 607.
145. S. Koseki, J. Ohkawa, R. Yamamoto, Y. Takebe and K. Taira, *J. Control.*, 1998, **53**, 159.
146. B.W. Pontius and P. Berg, *Proc. Natl. Acad. Sci. USA*, 1991, **88**, 8237.
147. J. Ellis and J. Rogers, *Nucleic Acids Res.*, 1993, **21**, 5171.
148. R. Kronenwett, R. Haas and G. Sczakiel, *J. Mol. Biol.*, 1996, **259**, 632.
149. M. Sioud and L. Jesperson, *J. Mol. Biol.*, 1997, **257**, 775.
150. P.C. Nowell and D.A. Hungerford, *Science*, 1960, **132**, 1497.
151. J. Groffen, J.R. Stephenson, N. Heisterkamp, A. de Klein, C.R. Bartram and G. Grosveld, *Cell*, 1984, **36**, 93.
152. E. Shtivelman, B. Lifschitz, R.P. Gale, B.A. Roe and J. Canaani, *Cell*, 1986, **47**, 277.
153. J.B. Konopka, S.M. Watanabe and O.N. Witte, *Cell*, 1984, **37**, 1035.
154. T. Kuwabara, M. Warashina, A. Nakayama, M. Hamada, S.V. Amontov, Y. Takasuka, Y. Komeiji and K. Taira, *Gene Ther. Mol. Biol.*, 1998, **1**, 435.
155. C.J. Pachuk, K. Yoon, K. Moelling and L.R. Coney, *Nucleic Acids Res.*, 1994, **22**, 301.
156. H. James, K. Mills and I. Gibson, *Leukemia*, 1996, **10**, 1054.
157. T. Kuwabara, M. Warashina, T. Tanabe, K. Tani, S. Asano and K. Taira, *Nucleic Acids Res.*, 1997, **25**, 3074.
158. G.Q. Daley and D. Baltimore, *Proc. Natl. Acad. Sci. USA*, 1988, **85**, 9312.
159. Y. Choo, I. Sánchez-García and A. Klug, *Nature*, 1994, **372**, 642.
160. L. Pegorato, L. Matera, J. Ritz, A. Levis, A. Palumbo and G. Biagini, *J. Natl. Cancer Inst.*, 1983, **70**, 447.
161. A. Hampel and R. Tritz, *Biochemistry*, 1989, **28**, 4929.
162. J. Haseloff and W.L. Gerlach, *Gene*, 1989, **82**, 43.
163. P.A. Feldstein, J.M. Buzayan and G. Bruening, *Gene*, 1989, **82**, 53.
164. L. Rubino, M.E. Tousignnant, G. Steger and J.M. Kaper, *J. Gen. Virol.*, 1897, **1990**, 71.
165. J.M. Buzayan, A. Hampel and G. Bruening, *Nucleic Acids Res.*, 1986, **14**, 9729.
166. A. Hampel, R. Tritz, M. Hicks and P. Cruz, *Nucleic Acids Res.*, 1990, **18**, 299.
167. A. Berzal-Herranz, S. Joseph and J.M. Burke, *Genes Dev.*, 1992, **6**, 129.
168. S. Joseph, A. Berzal-Herranz, B.M. Chowrira, S.E. Butcher and J.M. Burke, *Genes Dev.*, 1993, **7**, 130.
169. A. Berzal-Herranz, S. Joseph, B.M. Chowrira, S.E. Butcher and J.M. Burke, *EMBO J.*, 1993, **12**, 2567.
170. P. Anderson, J. Monforte, R. Tritz, S. Nesbitt, J. Hearst and A. Hampel, *Nucleic Acids Res.*, 1994, **22**, 1096.
171. M.B. De Young, A.M. Siwkowski, Y. Lian and A. Hampel, *Biochemistry*, 1995, **34**, 15785.
172. B.M. Chowrira and J.M. Burke, *Nucleic Acids Res.*, 1992, **20**, 2835.
173. B.M. Chowrira, A. Berzal-Herranz and J.M. Burke, *Nature*, 1991, **354**, 320.
174. M. Yu, M.C. Leavitt, M. Maruyama, O. Yamada, D. Young, A.D. Ho and F. Wong-Staal, *Proc. Natl. Acad. Sci. USA*, 1995, **92**, 699.
175. Y. Li, C. Guerrier-Takada and S. Altman, *Proc. Natl. Acad. Sci. USA*, 1992, **89**, 3185.
176. Y. Yuan and S. Altman, *Science*, 1994, **263**, 1269.
177. C. Guerrier-Takada, Y. Li and S. Altman, *Proc. Natl. Acad. Sci. USA*, 1995, **92**, 11115.
178. C. Guerrier-Takada and S. Altman, *Science*, 1984, **223**, 285.
179. C. Reich, G.J. Olsen, B. Pace and N.R. Pace, *Science*, 1988, **239**, 178.
180. A.C. Forster and S. Altman, *Science*, 1990, **249**, 783.
181. T.R. Cech, *Annu. Rev. Biochem.*, 1990, **59**, 543.
182. G. Garriga and A.M. Lambowitz, *Cell*, 1984, **39**, 631.
183. G. van der Horst and H.F. Tabak, *Cell*, 1985, **40**, 759.
184. J.M. Gott, D.A. Shub and M. Belfort, *Cell*, 1986, **47**, 81.
185. J.A. McSwiggen and T.R. Cech, *Science*, 1989, **244**, 679.
186. J. Rajagopal, J.A. Doudna and J.W. Szostak, *Science*, 1989, **244**, 692.
187. M.Q. Xu, S.D. Kathe, H. Goodrich-Blair, S.A. Nierzwicki-Bauer and D.A. Shub, *Science*, 1990, **250**, 1566.
188. B.A. Sullenger and T.R. Cech, *Nature*, 1994, **371**, 619.
189. J.T. Jones and B.A. Sullenger, *Nat. Biotechnol.*, 1997, **15**, 902.
190. T. Inoue, F.X. Sullivan and T.R. Cech, *Cell*, 1985, **43**, 431.
191. M.D. Been and T.R. Cech, *Cell*, 1986, **47**, 207.
192. A.J. Zaug, C.A. Grosshans and T.R. Cech, *Biochemistry*, 1988, **27**, 8924.
193. R.R. Breaker and G.F. Joyce, *Chem. Biol.*, 1994, **1**, 223.
194. B. Cuenoud and J.W. Szostak, *Nature*, 1995, **375**, 611.
195. N. Carmi, L.A. Shulz and R.R. Breaker, *Chem. Biol.*, 1996, **3**, 1039.
196. L.Q. Sun, M.J. Cairns, W.L. Gerlach, C. Witherington, L. Wang and A. King, *J. Biol. Chem.*, 1999, **274**, 17236.

Appendix

Modified Nucleosides from RNA

J.A. McCLOSKEY
University of Utah, Salt Lake City, UT, USA

Structures shown below and on the following pages represent all reported posttranscriptionally modified ribonucleosides of known structure from RNA. Included are nucleosides from established sequence positions as well as those characterized from RNA hydrolysates without knowledge of sequence location. Also shown in the accompanying table are commonly used symbols[1] and corresponding one-character representations. Many of the latter are ASCII symbols which have found use in tRNA sequence database compilations[2] in which presentation of aligned sequences requires one character per nucleotide. The original one-character list[2] has been extended to all RNA nucleosides using the Extended ASCII code and similar representations.

These data are continuously updated and maintained, and are accessible through the worldwide web at http://medlib.med.utah.edu/RNAmods. Also included at this site are Chemical Abstracts registry numbers which permit computer-based literature searches for each nucleoside, and designation of the RNA type (tRNA, SSU rRNA, etc.) in which the nucleoside has been found and the corresponding phylogenetic distribution. Initial literature citations are also provided for structural characterization of the nucleoside and its chemical synthesis.

The reader is referred to reference 3 as a comprehensive treatise covering a range of topics related to structure, function and biosynthesis of modified nucleosides from RNA.

Structure	*Common name*	*Symbol*	*One-character abbreviation*
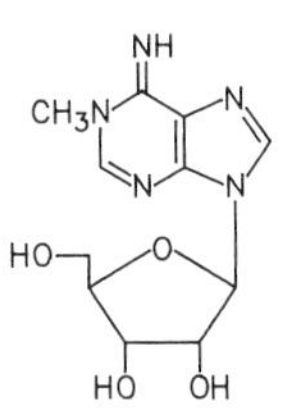 m1A	1-Methyladenosine	m^1A	"
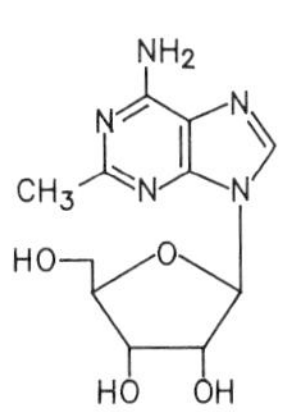 m2A	2-Methyladenosine	m^2A	/

Structure	*Common name*	*Symbol*	*One-character abbreviation*
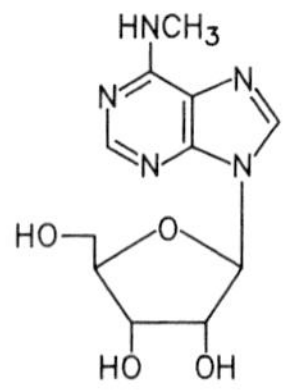 m^{6}A	N^6-Methyladenosine	m^{6}A	=
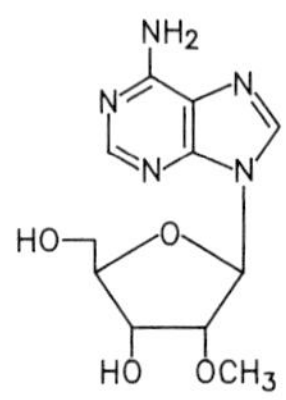 Am	2′-*O*-Methyladenosine	Am	:
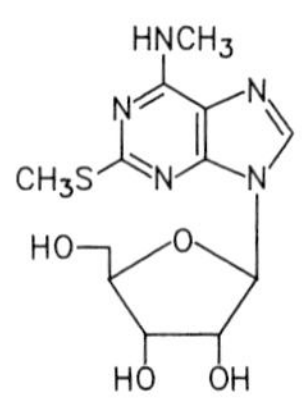 ms^{2}m^{6}A	2-Methylthio-N^6-methyladenosine	ms^{2}m^{6}A	δ
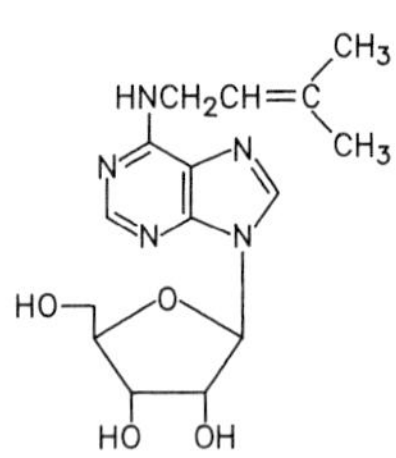 i^{6}A	N^6-Isopentenyladenosine	i^{6}A	+
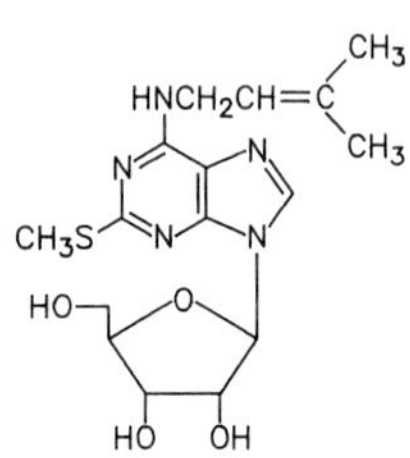 ms^{2}i^{6}A	2-Methylthio-N^6-isopentenyladenosine	ms^{2}i^{6}A	⋆
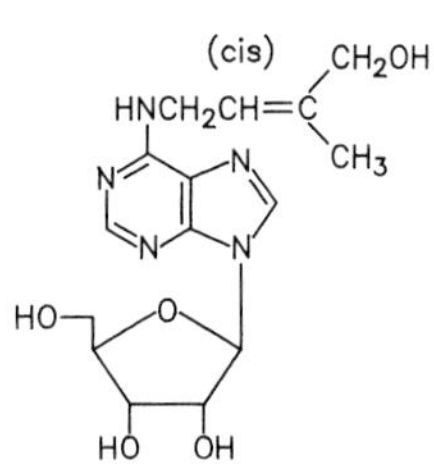 io^{6}A	N^6-(*cis*-Hydroxyisopentenyl)adenosine	io^{6}A	`

Structure	*Common name*	*Symbol*	*One-character abbreviation*
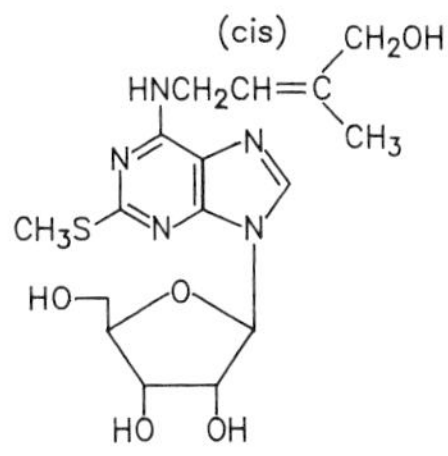 ms^2io^6A	2-Methylthio-N^6-(*cis*-hydroxyisopentenyl)adenosine	ms^2io^6A	σ
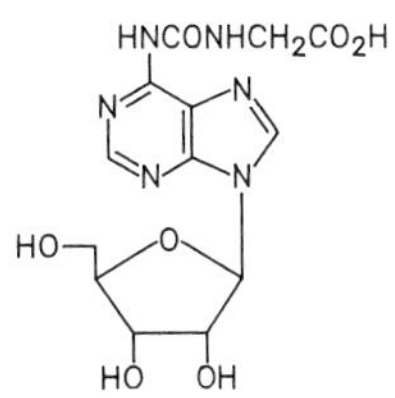 g^6A	N^6-Glycinylcarbamoyladenosine	g^6A	à
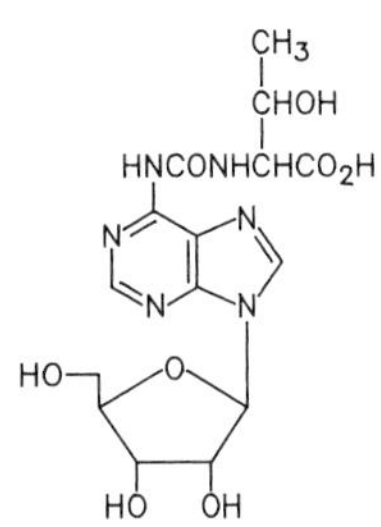 t^6A	N^6-Threonylcarbamoyladenosine	t^6A	6
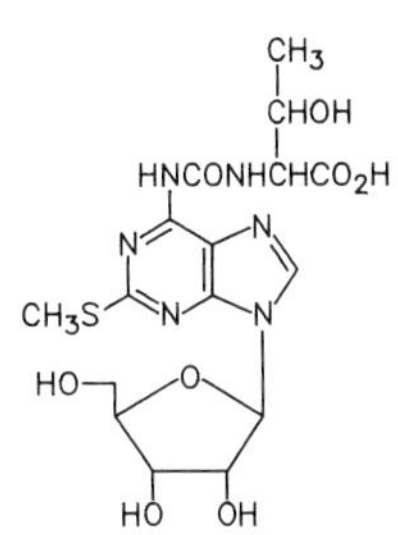 ms^2t^6A	2-Methylthio-N^6-threonylcarbamoyladenosine	ms^2t^6A	[
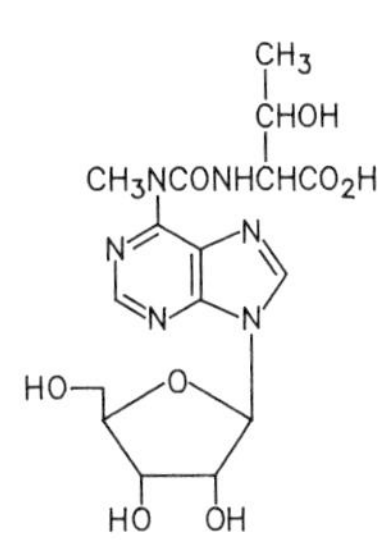 m^6t^6A	N^6-Methyl-N^6-threonylcarbamoyladenosine	m^6t^6A	E

Structure	*Common name*	*Symbol*	*One-character abbreviation*
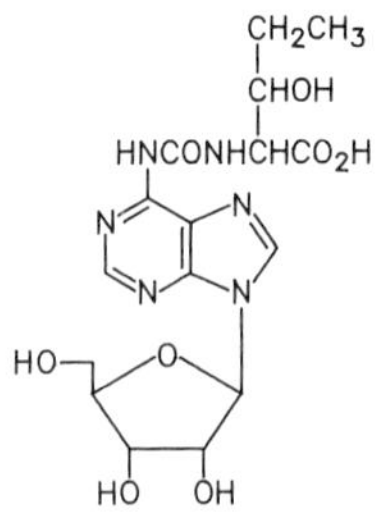 hn⁶A	N^6-Hydroxynorvalylcarbamoyladenosine	hn^6A	Ä
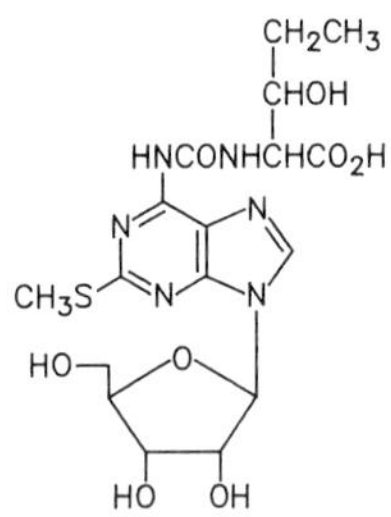 ms²hn⁶A	2-Methylthio-N^6-hydroxynorvalylcarbamoyladenosine	ms^2hn^6A	â
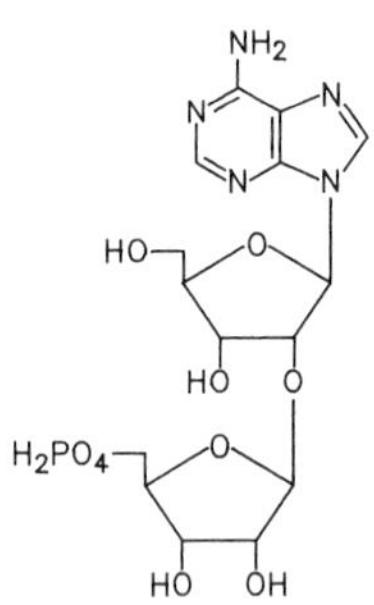 Ar(p)	2′-*O*-Ribosyladenosine (phosphate)	Ar(p)	^
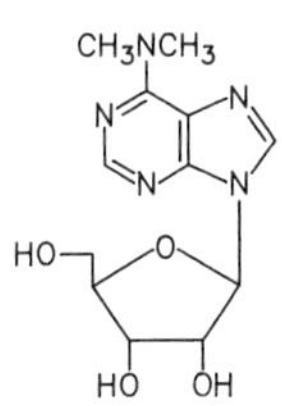 m^6_2A	N^6, N^6-Dimethyladenosine	m^6_2A	α
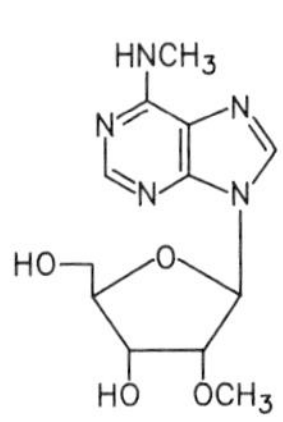 m⁶Am	N^6,2′-*O*-Dimethyladenosine	m^6Am	Å

Structure	*Common name*	*Symbol*	*One-character abbreviation*
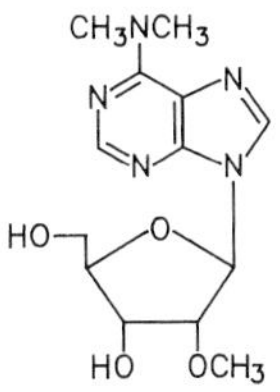 m^6_2Am	N^6, N^6, O-2′-Trimethyladenosine	m^6_2Am	ä
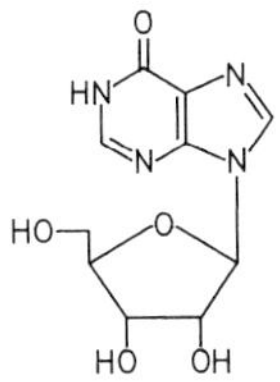 I	Inosine	I	I
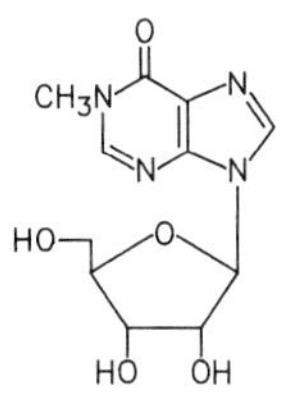 m^1I	1-Methylinosine	m^1I	O
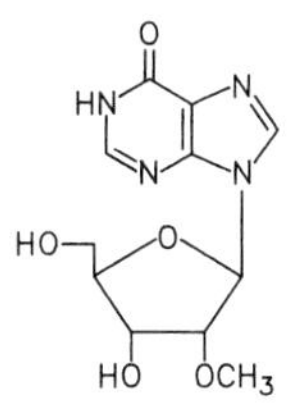 Im	2′-O-Methylinosine	Im	ô
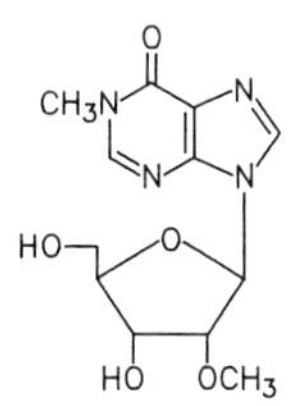 m^1Im	1,2′-O-Dimethylinosine	m^1Im	≤
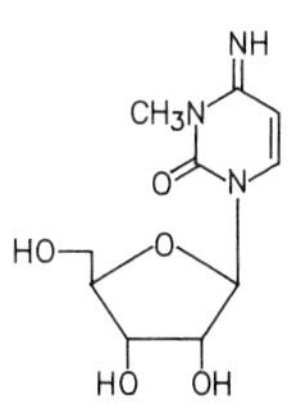 m^3C	3-Methylcytidine	m^3C	'

Structure	*Common name*	*Symbol*	*One-character abbreviation*
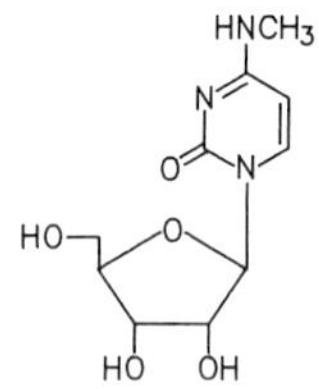 m^{4}C	N^4-Methylcytidine	m^{4}C	Γ
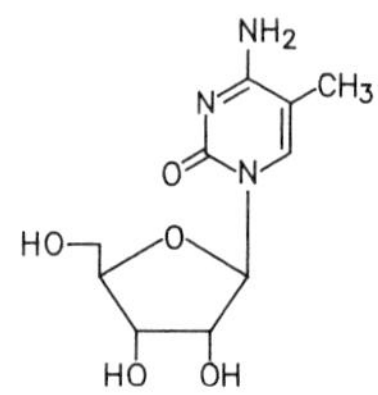 m^{5}C	5-Methylcytidine	m^{5}C	?
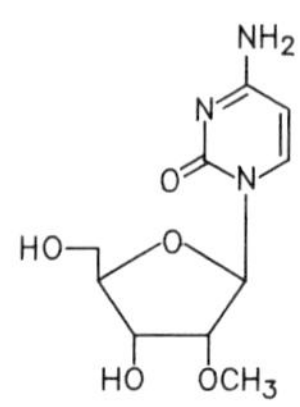 Cm	2′-*O*-Methylcytidine	Cm	B
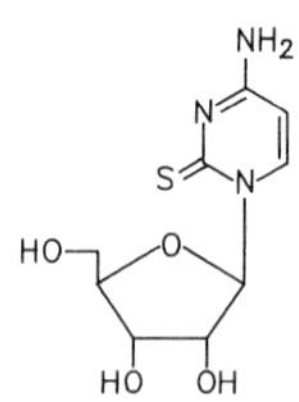 s^{2}C	2-Thiocytidine	s^{2}C	%
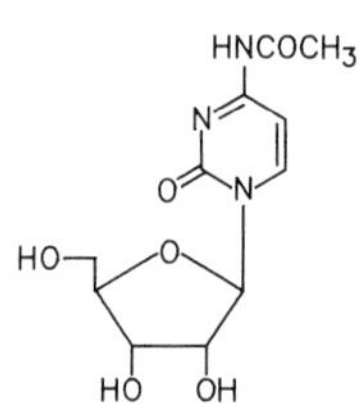 ac^{4}C	N^4-Acetylcytidine	ac^{4}C	M
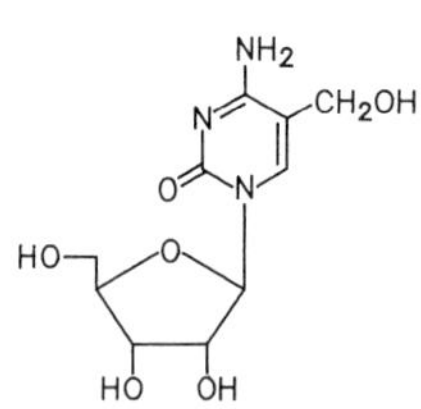 hm^{5}C	5-Hydroxymethylcytidine	hm^{5}C	«

Structure	*Common name*	*Symbol*	*One-character abbreviation*
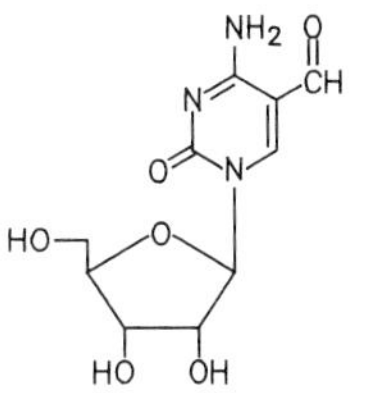 f^5C	5-Formylcytidine	f^5C	>
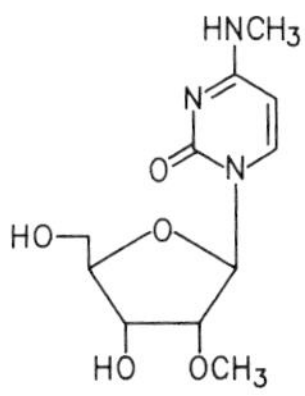 m^4Cm	N^4,2′-*O*-Dimethylcytidine	m^4Cm	Δ
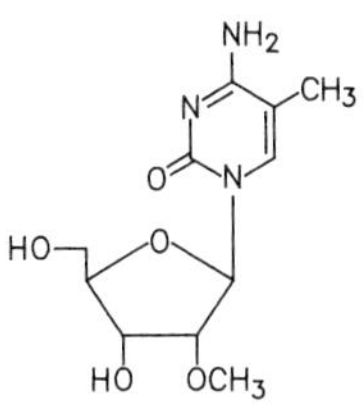 m^5Cm	5,2′-*O*-Dimethylcytidine	m^5Cm	£
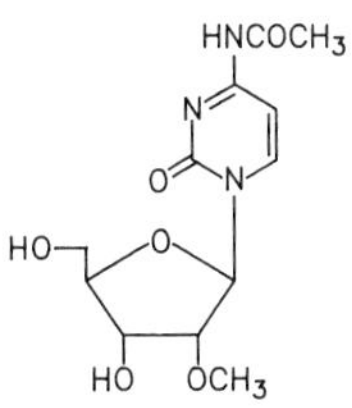 ac^4Cm	N^4-Acetyl-2′-*O*-methylcytidine	ac^4Cm	Σ
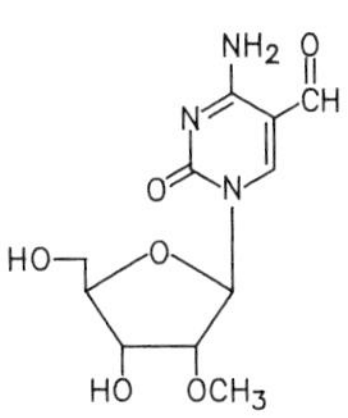 f^5Cm	5-Formyl-2′-*O*-methylcytidine	f^5Cm	°
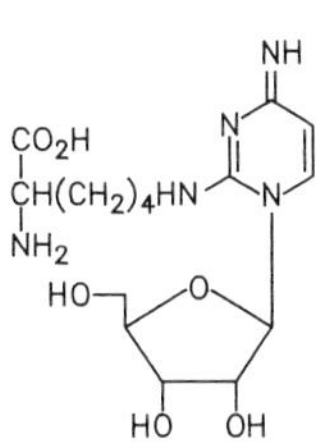 k^2C	Lysidine	k^2C	}

Structure	*Common name*	*Symbol*	*One-character abbreviation*
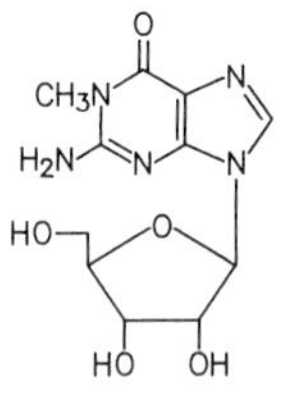 m^1G	1-Methylguanosine	m^1G	K
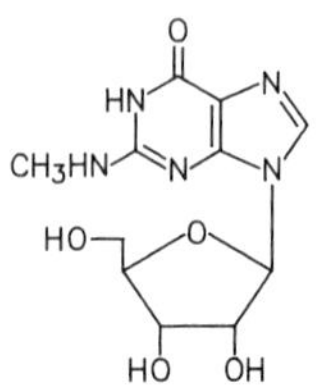 m^2G	N^2-Methylguanosine	m^2G	L
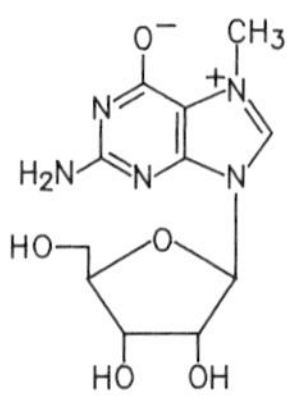 m^7G	7-Methylguanosine	m^7G	7
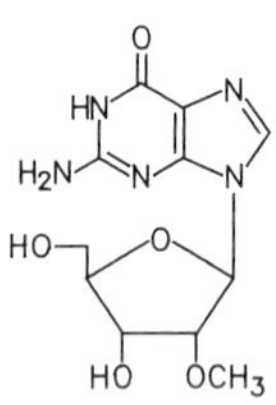 Gm	2′-*O*-Methylguanosine	Gm	#
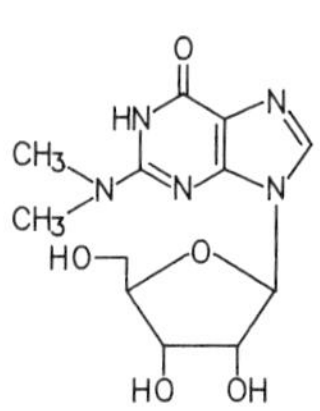 m^2_2G	N^2, N^2-Dimethylguanosine	m^2_2G	R
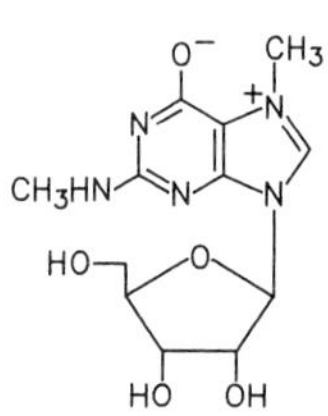 $m^{2,7}G$	N^2,7-Dimethylguanosine	$m^{2,7}G$	é

<table>
<tr><th>Structure</th><th>Common name</th><th>Symbol</th><th>One-character abbreviation</th></tr>
<tr><td>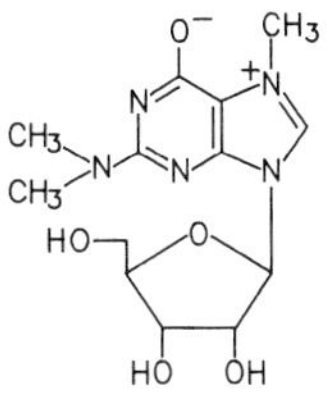
$m^{2,2,7}G$</td><td>N^2, N^2,7-Trimethylguanosine</td><td>$m^{2,2,7}G$</td><td>ë</td></tr>
<tr><td>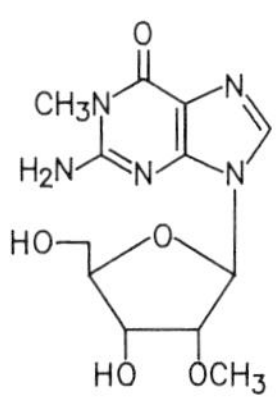
m^1Gm</td><td>1,2′-*O*-Dimethylguanosine</td><td>m^1Gm</td><td>≥</td></tr>
<tr><td>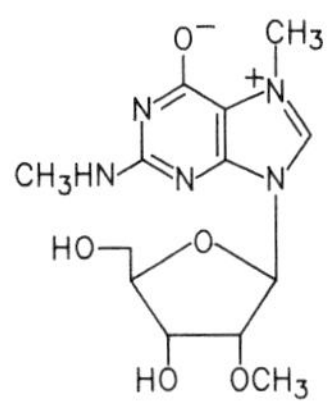
$m^{2,7}Gm$</td><td>N^2,7,2′-*O*-Trimethylguanosine</td><td>$m^{2,7}Gm$</td><td>í</td></tr>
<tr><td>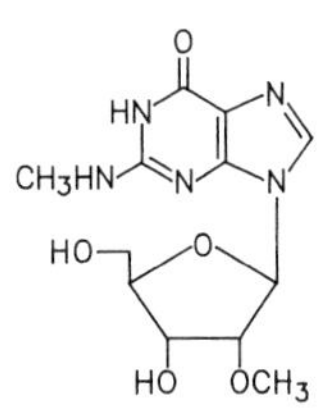
m^2Gm</td><td>N^2,2′-*O*-Dimethylguanosine</td><td>m^2Gm</td><td>ï</td></tr>
<tr><td>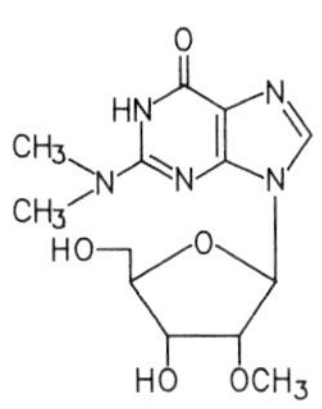
m^2_2Gm</td><td>N^2, N^2,2′-*O*-Trimethylguanosine</td><td>m^2_2Gm</td><td>|</td></tr>
<tr><td>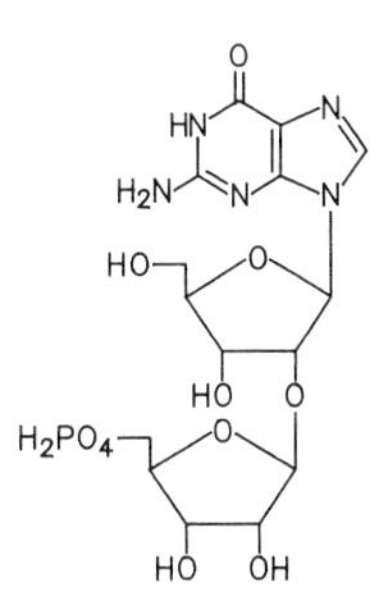
Gr(p)</td><td>2′-*O*-Ribosylguanosine (phosphate)</td><td>Gr(p)</td><td>±</td></tr>
</table>

Structure	*Common name*	*Symbol*	*One-character abbreviation*
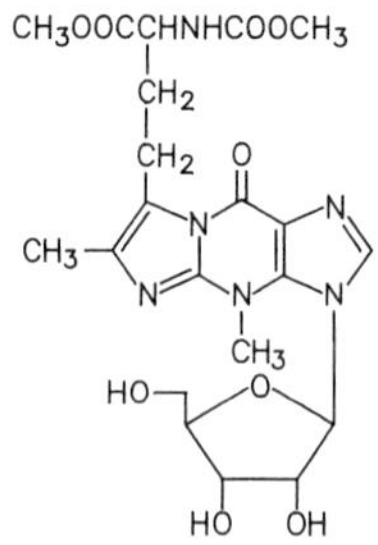 yW	Wybutosine	yW	Y
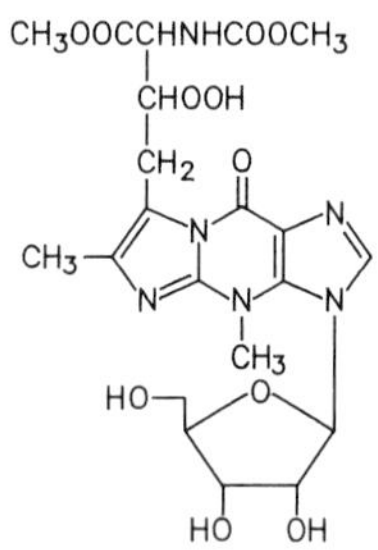 o_2yW	Peroxywybutosine	o_2yW	W
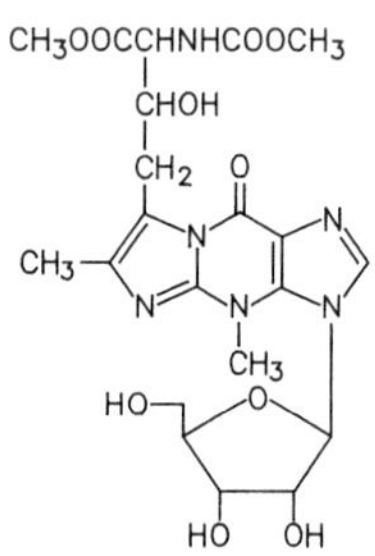 OHyW	Hydroxywybutosine	OHyW	β
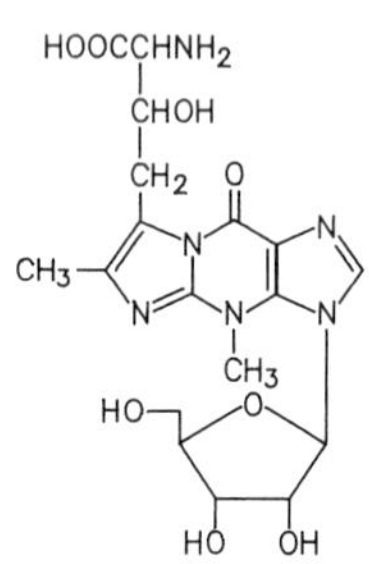 OHyW*	Undermodified hydroxywybutosine	OHyW*	ε
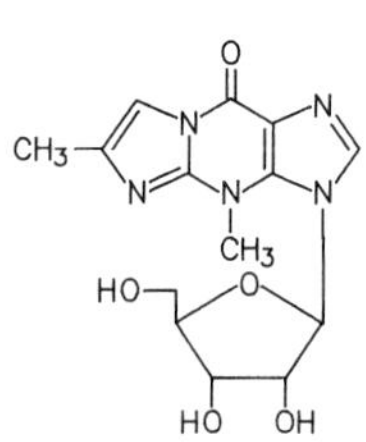 imG	Wyosine	imG	θ

Structure	*Common name*	*Symbol*	*One-character abbreviation*
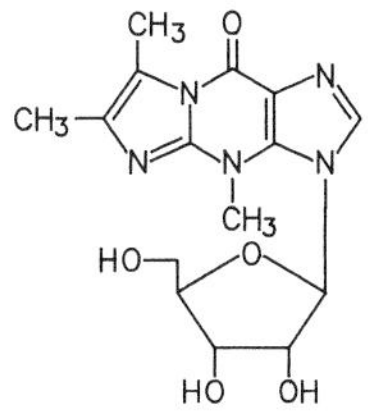 mimG	Methylwyosine	mimG	μ
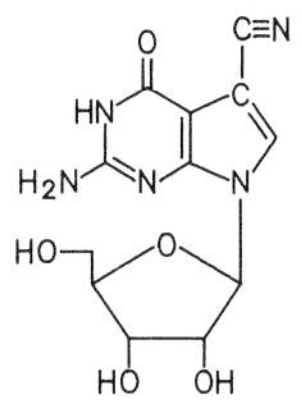 $preQ_0$	7-Cyano-7-deazaguanosine	$preQ_0$	ê
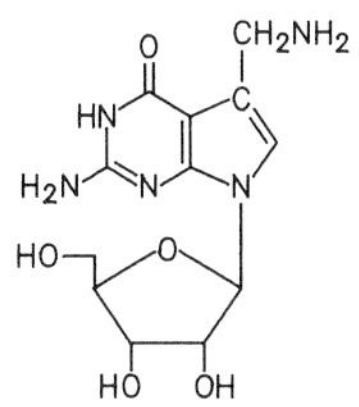 $preQ_1$	7-Aminomethyl-7-deazaguanosine	$preQ_1$	î
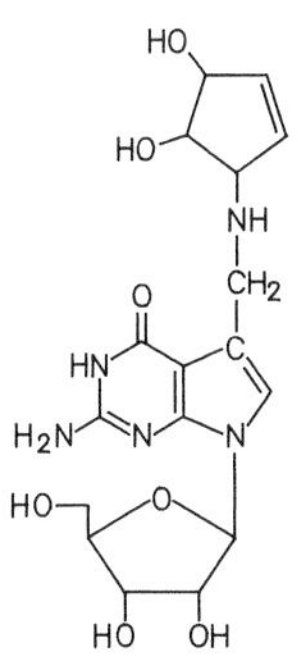 Q	Queuosine	Q	Q
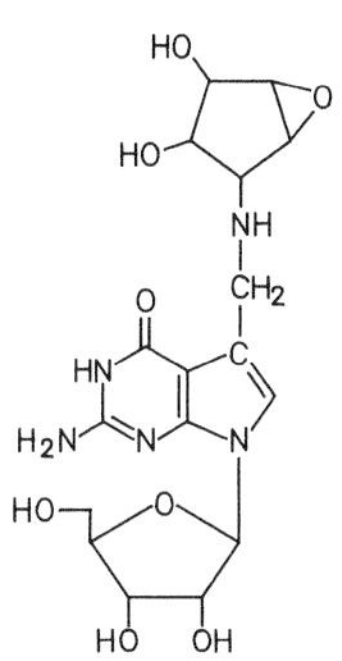 oQ	Epoxyqueuosine	oQ	≈

Structure	*Common name*	*Symbol*	*One-character abbreviation*
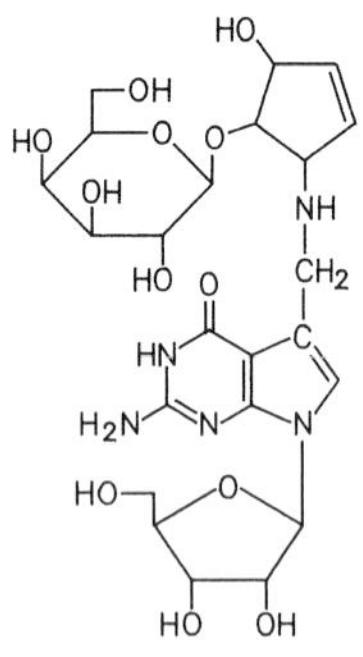 galQ	Galactosyl-queuosine	galQ	9
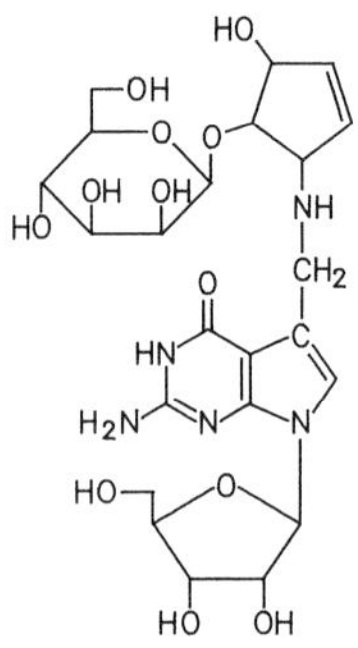 manQ	Mannosyl-queuosine	manQ	8
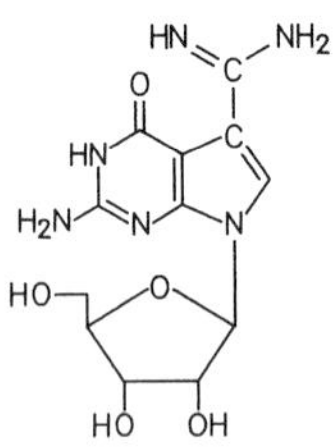 G^+	Archaeosine	G^+	(
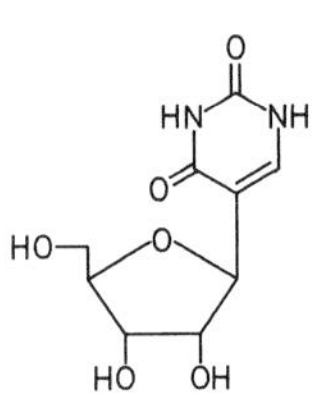 Ψ	Pseudouridine	Ψ	P
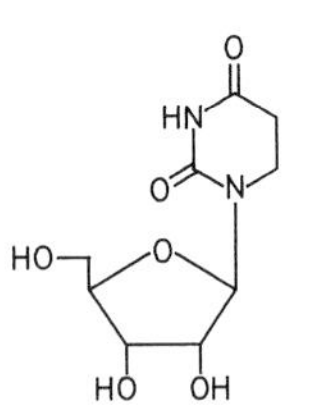 D	Dihydrouridine	D	D

Structure	*Common name*	*Symbol*	*One-character abbreviation*
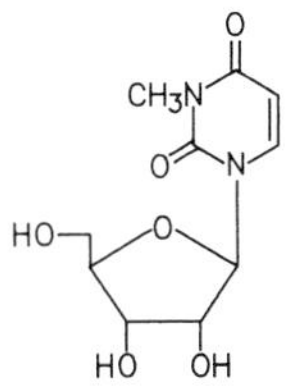 m^3U	3-Methyluridine	m^3U	ü
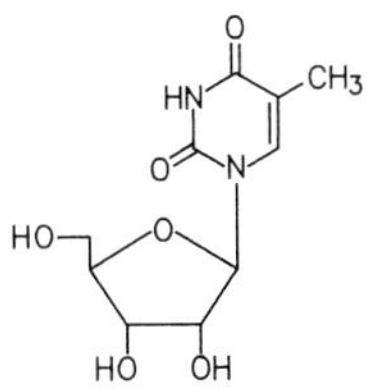 m^5U	5-Methyluridine	m^5U	T
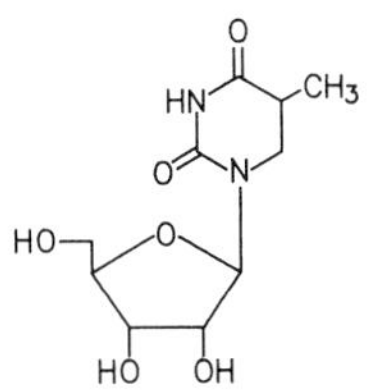 m^5D	5-Methyldihydrouridine	m^5D	û
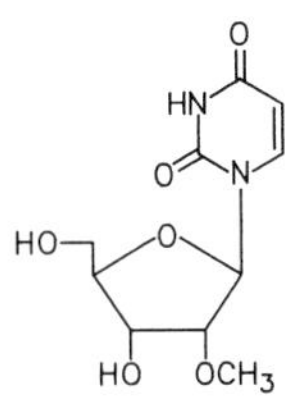 Um	2′-*O*-Methyluridine	Um	J
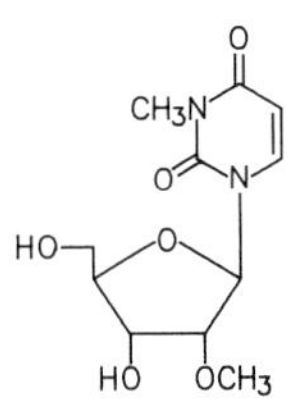 m^3Um	3,2′-*O*-Dimethyluridine	m^3Um	Ω
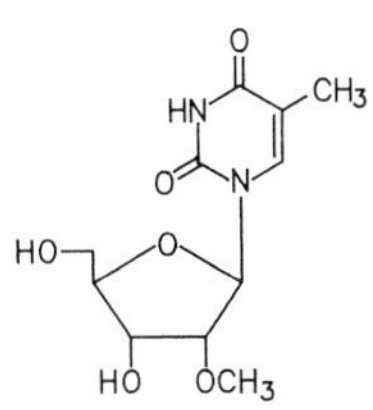 m^5Um	5,2′-*O*-Dimethyluridine	m^5Um	\
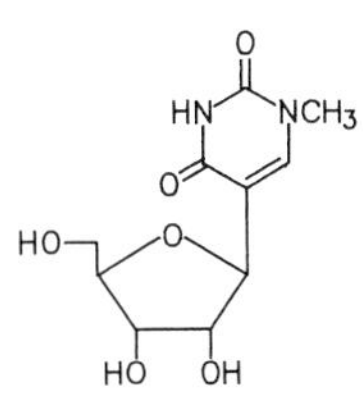 $m^1\Psi$	1-Methylpseudouridine	$m^1\Psi$	]

Structure	*Common name*	*Symbol*	*One-character abbreviation*
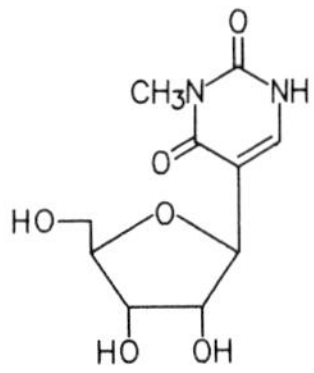 $m^3\Psi$	3-Methylpseudouridine	$m^3\Psi$	÷
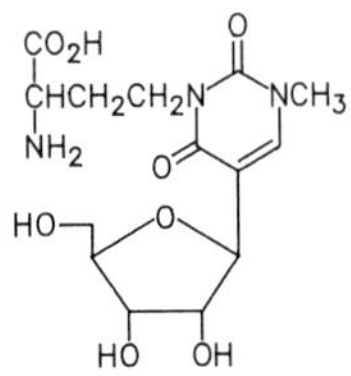 $m^1acp^3\Psi$	1-Methyl-3-(3-amino-3-carboxypropyl)pseudouridine	$m^1acp^3\Psi$	π
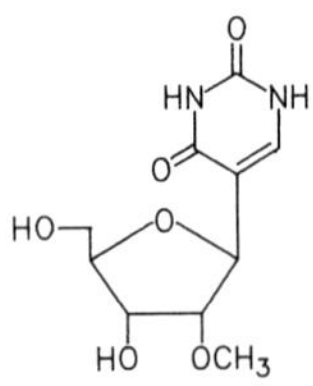 Ψm	2′-*O*-Methylpseudouridine	Ψm	Z
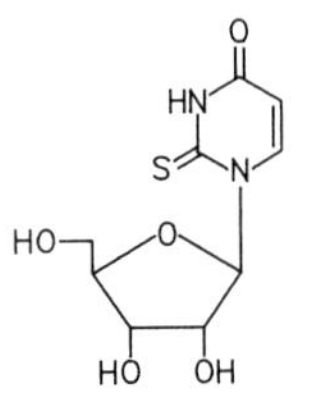 s^2U	2-Thiouridine	s^2U	2
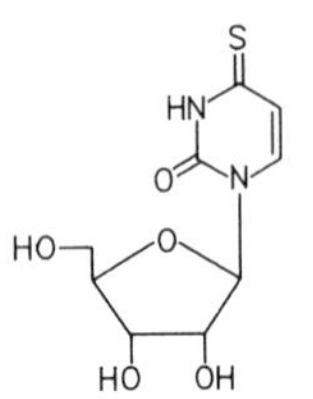 s^4U	4-Thiouridine	s^4U	4
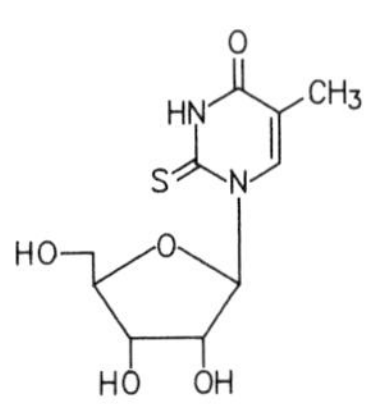 m^5s^2U	5-Methyl-2-thiouridine	m^5s^2U	F
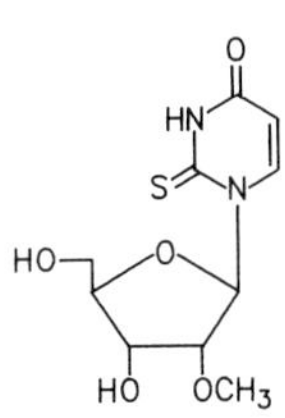 s^2Um	2-Thio-2′-*O*-methyluridine	s^2Um	≡

Structure	*Common name*	*Symbol*	*One-character abbreviation*
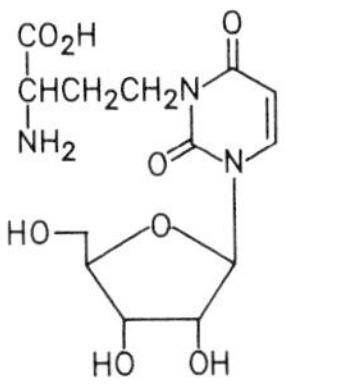 acp^{3}U	3-(3-Amino-3-carboxypropyl)uridine	acp^{3}U	X
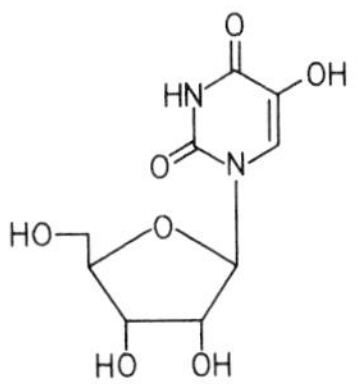 ho^{5}U	5-Hydroxyuridine	ho^{5}U	ú
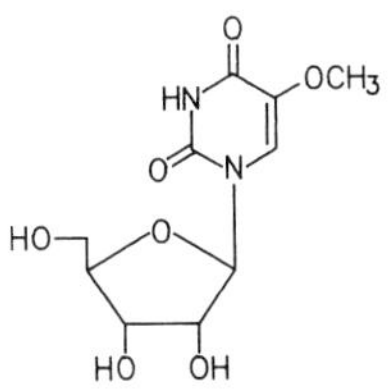 mo^{5}U	5-Methoxyuridine	mo^{5}U	5
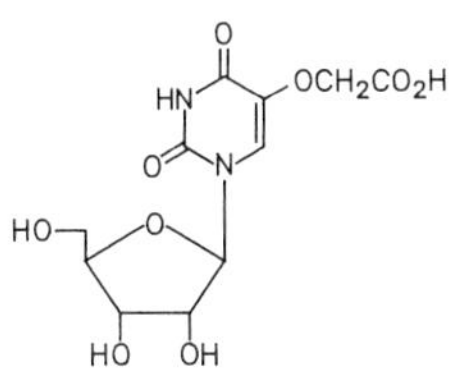 cmo^{5}U	Uridine 5-oxyacetic acid	cmo^{5}U	V
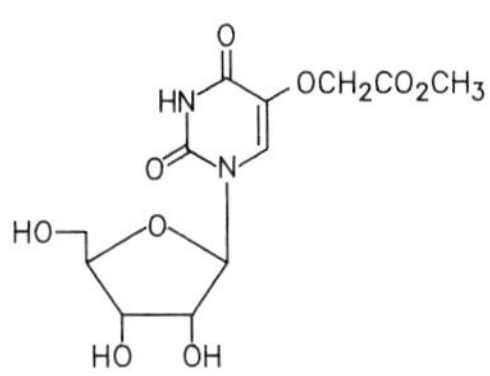 mcmo^{5}U	Uridine 5-oxyacetic acid methyl ester	mcmo^{5}U	ì
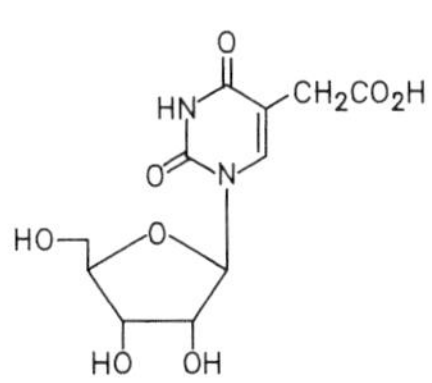 cm^{5}U	5-Carboxymethyluridine	cm^{5}U	ö
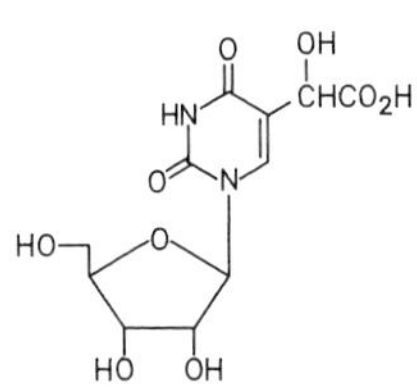 chm^{5}U	5-(Carboxyhydroxymethyl)uridine	chm^{5}U	»

Structure	*Common name*	*Symbol*	*One-character abbreviation*
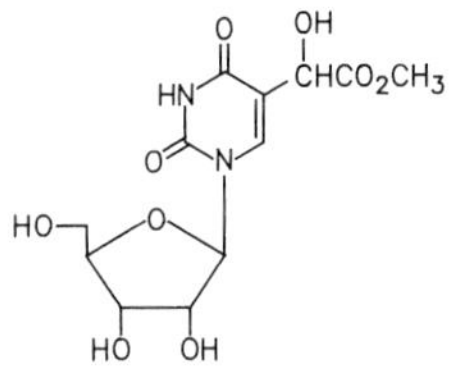 mchm^{5}U	5-(Carboxyhydroxymethyl)uridine methyl ester	mchm^{5}U	′
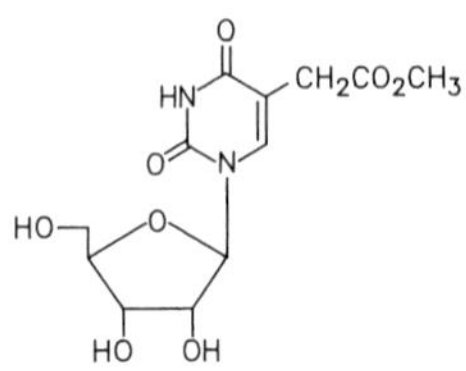 mcm^{5}U	5-Methoxycarbonylmethyluridine	mcm^{5}U	1
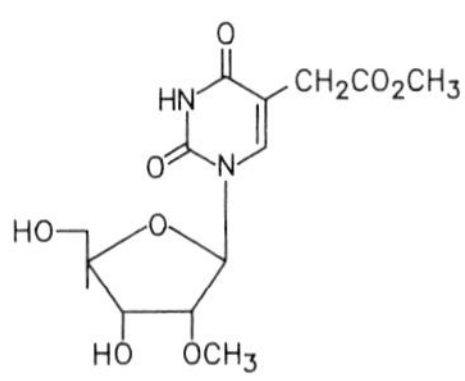 mcm^{5}Um	5-Methoxycarbonylmethyl-2′-*O*-methyluridine	mcm^{5}Um	ò
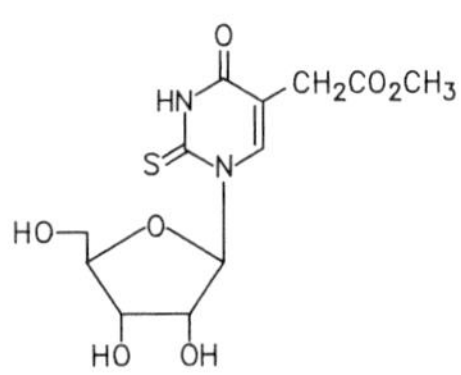 mcm^{5}s^{2}U	5-Methoxycarbonylmethyl-2-thiouridine	mcm^{5}s^{2}U	3
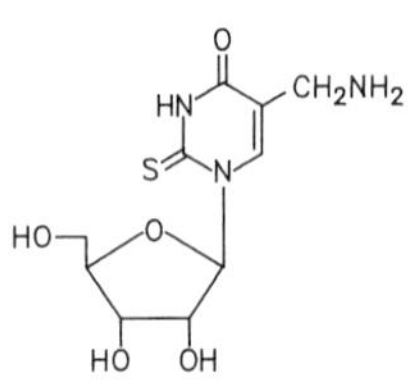 nm^{5}s^{2}U	5-Aminomethyl-2-thiouridine	nm^{5}s^{2}U	ù
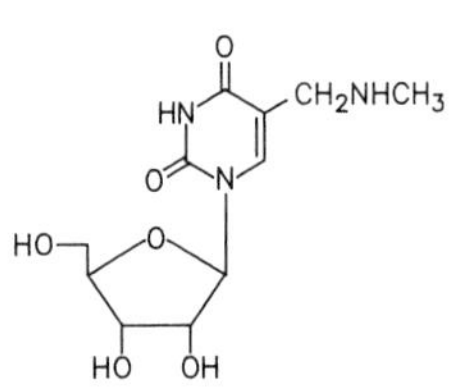 mnm^{5}U	5-Methylaminomethyluridine	mnm^{5}U	{
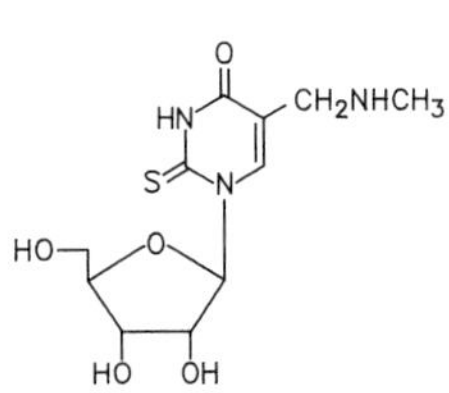 mnm^{5}s^{2}U	5-Methylaminomethyl-2-thiouridine	mnm^{5}s^{2}U	S

Structure	*Common name*	*Symbol*	*One-character abbreviation*
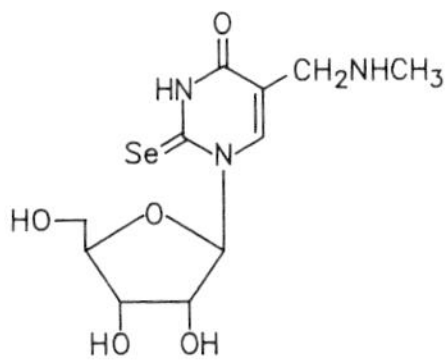 mnm^5se^2U	5-Methylaminomethyl-2-selenouridine	mnm^5se^2U	Φ
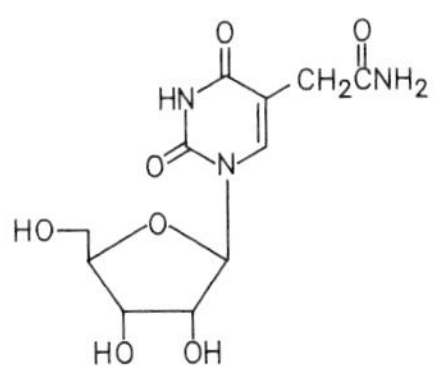 ncm^5U	5-Carbamoylmethyluridine	ncm^5U	&
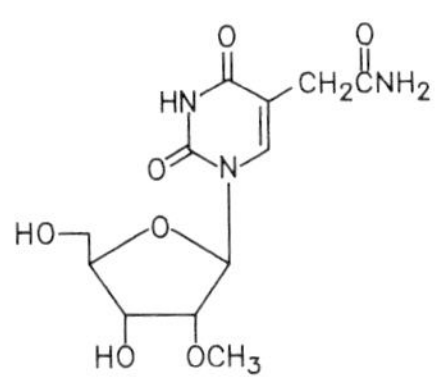 ncm^5Um	5-Carbamoylmethyl-2′-*O*-methyluridine	ncm^5Um	~
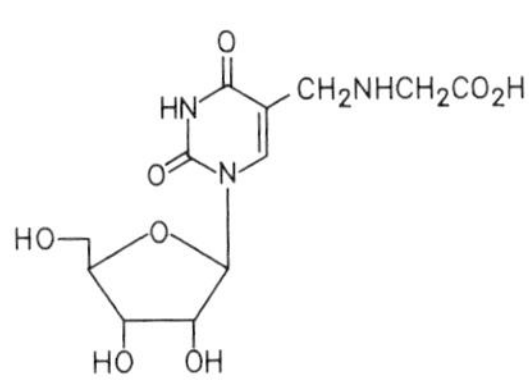 $cmnm^5U$	5-Carboxymethylaminomethyluridine	$cmnm^5U$	!
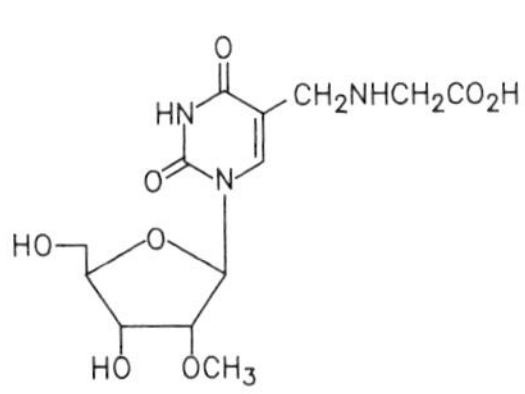 $cmnm^5Um$	5-Carboxymethylaminomethyl-2′-*O*-methyluridine	$cmnm^5Um$	)
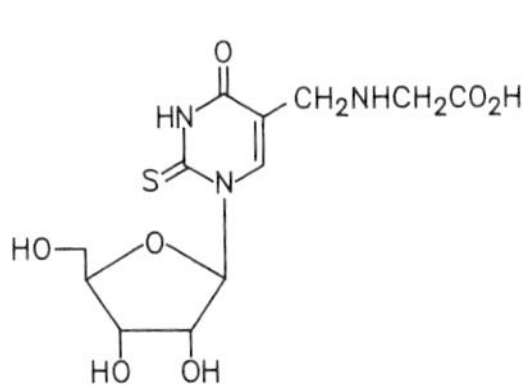 $cmnm^5s^2U$	5-Carboxymethylaminomethyl-2-thiouridine	$cmnm^5s^2U$	$

ACKNOWLEDGMENTS

Structural drawings from S.C. Pomerantz. The RNA Modification Database is maintained in part through support from The National Institute of General Medical Sciences, and the efforts of P.F. Crain and J. Rozenski.

REFERENCES

1. P.A. Limbach, P.F. Crain and J.A. McCloskey, Summary: the modified nucleosides of RNA. *Nucleic Acids Res.*, 1994, **22**, 2183–2196.
2. M. Sprinzl, C. Horn, M. Brown, A. Ioudovitch and S. Steinberg, Compilation of tRNA sequences and sequences of tRNA genes. *Nucleic Acids Res.*, 1998, **26**, 148–153.
3. H. Grosjean and R. Benne, eds. "Modification and Editing of RNA", ASM Press, Washington, DC, 1998.

Subject Index

PHILIP and LESLEY ASLETT
Marlborough, Wiltshire, UK

Every effort has been made to index as comprehensively as possible, and to standardize the terms used in the index in line with the IUPAC Recommendations. In view of the diverse nature of the terminology employed by the different authors, the reader is advised to search for related entries under the appropriate headings.

The index entries are presented in letter-by-letter alphabetical sequence. Compounds are normally indexed under the parent compound name, with the substituent component separated by a comma of inversion. An entry with a prefix/locant is filed after the same entry without any attachments, and in alphanumerical sequence. For example, 'diazepines', '1,4-diazepines', and '2,3-dihydro-1,4-diazepines' will be filed as:

diazepines
1,4-diazepines
1,4-diazepines, 2,3-dihydro-

The index is arranged in set-out style, with a maximum of three levels of heading. Location references refer page number (separated by a comma); major coverage of a subject is indicated by bold, elided page numbers, for example,

triterpene cyclases, **299-320**
  amino acids, 315

See cross-references direct the user to the preferred term, for example,

olefins *see* alkenes

See also cross-references provide the user with guideposts to terms of related interest, from the broader term to the narrower term, and appear at the end of the main heading to which they refer, for example,

thiones
  see also thioketones